Integrated Broadband Networks

TCP/IP, ATM, SDH/SONET, and WDM/Optics

For a listing of recent titles in the *Artech House Telecommunications Library*,
turn to the back of this book.

Integrated Broadband Networks

TCP/IP, ATM, SDH/SONET, and WDM/Optics

Byeong Gi Lee
Woojune Kim

Artech House
Boston • London
www.artechhouse.com

Library of Congress Cataloging-in-Publication Data
Lee, Byeong Gi.
Integrated broadband networks: TCP/IP, ATM, SDH/SONET, and WDM/Optics/
Byeong Gi Lee, Woojune Kim.
p. cm. — (Artech House telecommunications library)
Includes bibliographical references and index.
ISBN 1-58053-163-6 (alk. paper)
1. Broadband communication systems. I. Kim, Woojune. II. Title. III. Series.
TK5103.4 .L44 2002
621.382—dc21

2002023674

British Library Cataloguing in Publication Data
Lee, Byeong Gi
Integrated broadband networks: TCP/IP, ATM, SDH/SONET, and WDM/Optics. —
(Artech House telecommunications library)
1. Broadband communication systems
I. Title II. Kim, Woojune
621.3'821

ISBN 1-58053-163-6

Cover design by Igor Valdman

685 Canton Street
Norwood, MA 02062

International Standard Book Number: 1-58053-163-6
Library of Congress Catalog Card Number: 2002023674

10 9 8 7 6 5 4 3 2 1

Contents

Preface

In recent years, data traffic has been growing exponentially over all communication networks. Use of the Internet, which initially started with e-mail and simple file transfer, has been expanding to encompass *World Wide Web* (WWW)–based information services and new multimedia streaming services that deliver video and audio traffic in addition to data. This increase in services pushed the growth of network bandwidth, which, in turn, pushed the growth of services by enabling new applications that had never been possible previously.

While numerous technologies have been involved in this network evolution, *transmission control protocol* (TCP)/*Internet protocol* (IP), *asynchronous transfer mode* (ATM), and new optical technologies such as *synchronous digital hierarchy* (SDH)/*synchronous optical network* (SONET) and *wavelength division multiplexing* (WDM) are the fundamental technologies on which the new broadband networks have been constructed. TCP/IP was a key technology in the birth of the Internet, and the flexibility of TCP/IP to work over different types of physical links and devices was a key driving force in its growth. ATM is another key technology that evolved from the telephone networks and that supports both *constant-bit-rate* (CBR) and *variable-bit-rate* (VBR) services in a reliable and flexible manner. The optical technologies SDH/SONET and WDM are important as underlying transmission technologies for current and future networks with their ability to scale with bandwidth demands and provide survivability and longevity.

Particularly considering the progress of new WDM components and systems, optical technologies have become the fundamental driver behind the network bandwidth growth.

The aforementioned basic technologies are frequently used in combined forms. In fact, IP over ATM, IP over SDH/SONET, and IP over WDM are all currently hot topics. Operators use each technology to gain advantages that the other technologies cannot offer by themselves. For example, optical WDM networking offers very large bandwidth, while IP offers well-defined user *application program interfaces* (APIs) for ease of application development. SDH/SONET offers protection switching and reliability, while ATM offers *quality of service* (QoS) and flexibility.

Understanding such diverse sets of broadband technologies as TCP/IP, ATM, SDH/SONET, and WDM/optics and their interaction/integration is not a simple matter. There is much demand for a book that presents a systematic framework for understanding the fundamental differences in their approaches and compares their relevant advantages and disadvantages. This book aims to satisfy this demand by considering all those major streams of changes in an integrated manner.

Accordingly, *Integrated Broadband Networks* presents the key network technologies in a technology-functionality matrix approach. We deal with each technology in a separate chapter, analyzing it from different functional aspects—that is, considering how each technology solves the fundamental problems that a network technology must solve. This book describes the protocol layering and basic architecture; how to handle the problems of routing, switching, and multiplexing; the available methods for network control; and traffic management techniques. In addition, whenever possible, the book shows how QoS is offered for each particular technology. However, in the case of the optical network technologies, we look at relevant transmission issues such as add/drop multiplexing, cross-connect, and network reliability instead. We believe that a functionality-based investigation of the network technologies will help readers gain a comprehensive understanding of the broadband networks. In addition, readers should realize that network engineering is simple in that the problems to investigate are basically the same for all the network technologies but complex in that the answers to the problems differ depending on the available technologies or user requirements.

This book will be among the first to deal with the key architectures and technologies of broadband networks in a comprehensive and organized manner based on the technology-functionality matrix, which deals with the integration of the constituent technologies to realize IP traffic over optical networks. We hope that this book will serve as a valuable guide to this fast-evolving broadband network field, helping students and practicing engineers to stay abreast of the developments.

Acknowledgments

Our initial goal was to write the third edition of the book *Broadband Telecommunication Technology* (Artech House) whose first and second editions were published in 1993 and 1996, respectively. However, changes in the communication world during the past several years have been so extensive and fundamental that the plan was soon switched to authoring a new book that would present a comprehensive and focused description of the most fundamental facets of today's communication networks. This statement itself implies, to our disappointment, that the contents of this book—while they may describe today's state-of-the-art broadband network technologies—will merely represent a snapshot in time once again. We thank the coauthors of the previous book, Minho Kang and Jonghee Lee, for their extended support and encouragement.

We are also grateful to the organizations for which we worked during the authoring period, namely George Washington University and Seoul National University for Byeong Gi Lee and Bell Laboratories at Lucent Technologies and Samsung Electronics for Woojune Kim, for supporting the authoring work directly or indirectly. In fact, the publication of this book has been much delayed due to the job transfers we each experienced in the middle of the process. We thank Artech House for kindly accepting this delay with patience.

This publication has been made possible thanks to the help of several graduate students at the Telecommunications and Signal Processing (TSP)

Laboratory at Seoul National University. In particular, we are indebted to Hyun-Yong Choi, who supported the writing process by connecting the two authors in separate geographical locations, drawing the figures, and doing various miscellaneous book-related tasks. Without his help, this publication might have been delayed even more. In addition, we thank In-Soo Yoon and Won-Ick Lee for their assistance in the proofreading process.

We are grateful to Dr. Tsong-Ho Wu at Transtech Networks and Professor Suresh Subramanian at George Washington University for reviewing and helping to improve Chapters 5 and 6. We are also thankful to Professor Saewoong Bahk at Seoul National University for reviewing the whole text of the book. We thank In-Soo Yoon and Won-Ik Lee at Seoul National University for jointly reviewing and cross-checking the contents of the book.

Most of all we would like to thank our wives, Hyeon Soon Kang and Heejung Kim, whose love and support at home encouraged and enabled us to carry out this multiyear authoring project while conducting the related research and our regular jobs.

1

Introduction

Today we are in the midst of the information revolution, a revolution that is propelled by two old inventions: the first is the telephone, an invention of the nineteenth century, and the second is the computer, an invention of the twentieth century. The convergence of these two inventions in the last decade of the twentieth century has resulted in the third revolution in human history, the so-called information revolution.

The information revolution manifests itself in various ways in the everyday life of everyone today. One of the most obvious is the spread of the WWW, commonly known as the Web, into the lives of ordinary people everywhere. The Web was originally developed at the European Organization for Nuclear Research (CERN) as a way of enabling physicists to easily share information. With the development of the graphical interface, Mosaic, and the Netscape browser that followed, the Web has become part of the lives of ordinary people. We are now advancing to the stage where Web access is ubiquitous—Web access anytime and anywhere. The popularity of Internet cafes is an example at the most consumer-friendly end of the spectrum, and wireless Web access such as the new *wireless application protocol* (WAP) phones and NTT DoCoMo's I-Mode phones is a more advanced example. Further down the road are intelligent appliances that also function as Web servers. In fact, it is expected that in the future every appliance will be Web-accessible. All these lead to another major development in the form of e-commerce. Note that the recent ".com" revolution is essentially based on the belief that businesses in the future will all be Internet-related.

The developments discussed above are all based on data communications, which is firmly rooted in computers. Data is a single-medium expression of information that is orthogonal to the more traditional, universal form of single-medium expression, voice. Video is another single-medium expression that complements voice. There have been strenuous efforts to combine these three media, ever since computers and communications merged, thereby giving birth to multimedia communications. In particular, as a consequence of the efforts for the voice-data convergence, the *voice-over-IP* (VoIP) technology finally started to gain popularity in the late 1990s. VoIP—even though it appears to be a single-medium communication—actually unites real-time voice streams with data transmission by transmitting the former over the latter. VoIP initially gained popularity as a cheap way of making long-distance phone calls, but it is expected to lead to a feature-rich phone technology—much more powerful and flexible than ever before.

Both the Web and multimedia communications will expand the demand for bandwidth in the basic communication networks. Additionally, they will drive developments in other areas of network technology that go hand-in-hand with the increased bandwidth. Those areas encompass the fundamental facets of communications, including architecture, routing, network management, and signaling.

1.1 Technological Driving Forces

What are the fundamental technological forces behind those revolutions? It is commonly believed that there are three main driving forces: IP technology, ATM technology, and optical technology.

The first driving force is the widespread WWW/Internet and its underlying technology, the *transmission control protocol/Internet protocol* (TCP/IP) network protocol suite. Even before the WWW became popular the Internet was growing exponentially. Originally, e-mail was the "killer application" for the TCP/IP suite, the basis of the Internet. It was invented during the early days of the Internet and has emerged as one of the first applications to really become popular to all users. However, the real explosion in Internet growth occurred with the appearance of the WWW. Along with the global spread of the WWW, the base protocol suite, TCP/IP, has been globally deployed, from desktops to network equipment to mobile phones. This has led to the expectation that TCP/IP may eventually become the single unified communication protocol used in all networks.

The second factor is the development of *asynchronous transfer mode* (ATM) technology and its consequent deployment. ATM was originally developed and deployed by telephone service providers and equipment vendors as a means for integrating all services in *variable bit rates* (VBRs) and *constant bit rates* (CBRs). Now it has become a mature, stable technology that retains many advanced features. For example, the traffic engineering capability of ATM is very beneficial to network operators. However, the most frequently discussed advantage of ATM is its built-in support for *quality of service* (QoS), which is notably lacking in the traditional TCP/IP suite.[1] While ATM has not been widely accepted in the major data markets, its widespread acceptance and deployment by telephone companies and other communications providers make it a firm basis for the integration of traditional voice-related services and new data services.

The third factor is the optical revolution made possible by *synchronous digital hierarchy* (SDH)/*synchronous optical network* (SONET) and *wavelength division multiplexing* (WDM)/optics. The information revolution is built on bits, and optics has provided the means for the transport of the bits. Traditional transmission technology based on copper wire was fundamentally limited in bandwidth and transmission distance. The only viable solution that has been found to carry the increasing number of data bits so far is the use of optical technology. Today, optical technology is based on lasers and optical fibers and rather focused on point-to-point links. Basically it is regarded more as a transport technology than as a networking technology. The first stage of an optical technology–based solution was SDH/SONET. Originally SDH/SONET emerged as a means for integrating the existing transmission signals, accommodating all voice-derived digital hierarchical signals in North America and Europe. The integrated digital signals were then converted into broadband optical signals in the gigabits-per-second range. This broadband integration capability of the SDH/SONET has been strengthened by the introduction of optical WDM technology. While introduced as an extension of SDH/SONET optical links, the WDM systems are now evolving into the next-stage all-optical networks.

1. Unfortunately, ATM's ability to support QoS has not really been used very often in real networks, though it has been an important catalyst for developing many important QoS ideas.

1.2 History and Recent Developments

To gain a thorough understanding of current and future technological trends it is important to understand how we have arrived at the current technological status. Sections 1.2.1 through 1.2.5 provide a quick overview of the history of, and recent developments in, the four major network technologies: TCP/IP, ATM, SDH/SONET, and WDM/optics.

1.2.1 The Internet and TCP/IP

In 1961, Paul Baran at the RAND Corporation studied the problem of designing a communication network that would be robust enough to withstand nuclear attacks. The invention that resulted at that time was the concept of store-and-forward packet switching. Packet switching had three main advantages. First, it could efficiently utilize available transmission facilities. Second, it offered reliability in the use of retransmission mechanisms. Third, if implemented in a mesh-topology network it could ensure the maintenance of communications even if any single link or node was destroyed by an attack.

Baran's idea lay dormant for several years until the U.S. Defense Advanced Research Projects Agency (DARPA) decided to use it for organizing a network composed of expensive computer facilities from various vendors and researchers located around the country. Packet switching was chosen at that time as the preferred method for interconnecting the computers and allowing scattered researchers to share computing resources. The network was proposed in 1966 and was built in 1969 among a small group of universities and companies. The group developed the basic ideas and applications of the Internet, including telnet, file transfer protocol (ftp), flow control, and decentralized routing. The network originally connected four universities in the United States and was called the Advanced Research Projects Agency (ARPANet).

Initially ARPANet was only a single network, but its concept was soon applied to other networks and the concept of the gateway was developed to support connections among different networks. This is the basis of the construction of an internet, a network of networks. At this stage some more key decisions were made with regard to the design of the network. First, the network would be administered in a distributed manner. Second, the network would be able to scale. Third, the basic IP packet protocol should be usable over almost any type of lower layer technology. These fundamental decisions still form the groundwork of the Internet today and in fact are the very features that have enabled the tremendous growth of the Internet over the past decade [1].

In the late 1970s, the basic TCP/IP protocols were also designed and deployed. In 1978, TCP/IP was officially adopted by the U.S. government as the "preferred" networking protocol. By the 1980s there were about 10 ARPANet-based networks all over the United States. Around that time the National Science Foundation (NSF) agreed to fund the establishment of a national backbone network based on TCP/IP that would interconnect universities and other research centers around the country. By 1995, the Internet had grown to such an extent that federal funding for the network could be withdrawn, and the operation of the backbone was soon handed over to private companies. Of course, by that time, the WWW had been introduced to the world, and the use of the Internet had exploded.

While TCP/IP was being developed, various other packet network solutions were also being developed by a number of other companies. Leaders among them were IBM's *system network architecture* (SNA) and DEC's DECnet. Both were successful as proprietary network technologies built on large customer bases. However, as the WWW spread the popularity of the Internet around the world, customers gradually switched to the open standards of TCP/IP. Nevertheless, many important networking concepts and technologies developed by those two companies—namely, flow control, congestion avoidance, and link state routing protocols, for example—still play seminal roles in networking today.

On the smaller network side, many *local area network* (LAN) technologies were developed during the 1970s and 1980s. Among the technologies that attained wide popularity are the token ring, token bus, and *fiber-distributed data interface* (FDDI) LAN technologies, but the most distinguished one was Ethernet developed by Metcalfe at Xerox PARC in 1970s. During the same period the personal workstation, the *graphical user interface* (GUI), and the mouse were developed. Originally, Ethernet was conceived as a cheap and easy way of connecting the personal workstations that had been developed at PARC. It was based on the *carrier-sense multiple access/collision detection* (CSMA/CD) technology, which was an advanced version of the ALOHA protocol. Offering a speed of 10 Mbps, Ethernet has become ubiquitous and has been extended to speeds as high as 100 Mbps and 1 Gbps in recent years. As switching hubs (or layer 2 switches) were developed, Ethernet technology has partially overcome the inherent scaling problems caused by its shared-medium nature.

Conceptually the WWW had its origins in the hypertext machine conceived by Vannevar Bush during the 1950s. This machine could link documents together by the use of links in sentences. The WWW was first designed and built by Tim Berners-Lee at CERN as a way for physicists to share information. WWW technology comprises three important components—the *universal resource locator* (URL), a way to universally identify items on the Internet; the *hypertext markup language* (HTML), a subset of the *standard generalized markup language* (SGML) for describing documents; and the *hypertext transfer protocol* (HTTP), a way of transporting HTML documents across the Internet. The WWW spread rapidly, especially after the development of the Mosaic browser by some students of University of Illinois, including Marc Anderson, who later set out and started Netscape, the first real Internet success story.

1.2.2 ATM Technology

Asynchronous time division multiplexing (ATDM) was originally proposed by W. W. Chu at Bell Laboratories in 1968 as a way of realizing faster multiplexers [2]. His solution was to make use of headers added to the time-slots. A. G. Fraser realized that this could be used to build networks of faster switches if the headers could also be swapped. Based on this concept Fraser developed Spider (1972), which first introduced the concepts of virtual circuits and small fixed-size packets. He further developed Datakit based on the ideas of Spider with several refinements. He realized that if the endpoints of the network were made to be service-specific, but with the internal switch backplane-fixed, then voice and data could all be carried in one format over that network. Technologically, Spider and Datakit are equivalent to ATM in various aspects, including the use of virtual circuits, small fixed-packet sizes, and statistical multiplexing [3].

During the late 1970s, engineers at telephone companies around the world foresaw that, in the future, communication users would demand not only voice communications but also video and data communications. In answer to this forecast, the International Telegraph and Telephone Consultative Committee (CCITT) (later reorganized into the ITU-T in 1993) began to develop techniques to integrate voice, image, and data signals in one format, and came up with the standards that comprise the *integrated services digital network* (ISDN). The fundamental products of this standardization effort were the basic access at the 144-Kbps bit rate and the primary access at the DS-1 rate (more specifically, at 1.536 Mbps for the North American digital hierarchy and at 1.920 Mbps for the European digital hierarchy) announced in 1984. The basic access carried two 64-Kbps channels and one 16-Kbps channel, the so-called "2B+D" channel, within which voice, image, and data signals could be mapped into. The ISDN standards soared as the most popular key word in the 1980s in the communications sector but the boom soon died out even before seeing its practical deployment in the field when *broadband ISDN* (B-ISDN) standardization began to take place. In the late 1990s, there was a brief resurgence in its use when the popularity of the Internet spurred the demand for high-speed local digital links. This movement, however, did not last long due to the spread of the *digital subscriber line* (DSL) technologies and cable modems.

The B-ISDN standardization efforts, which began in the mid 1980s, originally intended to expand the ISDN standards to accommodate broadband services such as videophones and teleconferences. Later, however, the standardization of the B-ISDN, an effort also undertaken by the CCITT, was strongly influenced by the standardization of the SDH/SONET. In addition, the ATDM technique also influenced the B-ISDN standardization significantly, thereby yielding the new communication mechanism, ATM. For example, the transmission rate of 155.52 Mbps of ATM communication was taken from the transmission rate of the STM-1/OC-3 signal of the SDH/SONET, whereas the fixed-size packet and statistical multiplexing–based conveyance of the integrated information was taken from the ATDM technique. In addition, the *dynamic time division multiplexing* (DTDM) techniques proposed by Sanghoon Lee at Bellcore in 1988 also contributed to establishing the ATM standards [4]. The cell-level broadband transport architecture of the DTDM solved the slot synchronization problem of the ATDM technique.

While ATM standardization, as well as the subsequent product development, was successful, its practical use was much delayed due to the delay in the deployment of optical fibers and the installation of ATM equipment. In the mean time, engineers in computer companies realized that ATM could be very useful as a data networking technology, for a number of reasons. First of all, it was basically a packet-switching technology, a technology with which they were familiar because of Ethernet. It offered scalable large bandwidth, in contrast to Ethernet, which at that time offered only 10-Mbps speeds. At the same time the bandwidth granularity was unlimited compared to the traditional circuit switches, which were fixed at 64-Kbps. Also, ATM had the built-in ability to support voice, video, and data, as it was designed from the beginning for such support. As a result, the computer vendors and telecommunication service providers organized the ATM Forum in 1991, an industry-based consortium, to accelerate ATM product development and ATM service diffusion. Later, the consortium grew to become the real driving force behind many ATM standards and generated many ATM-related specifications under close collaboration with other formal standards organizations. Among the various specifications defined by the ATM Forum are the standards on signaling, physical layers, and different types of interworking. In addition, the Internal Engineering Task Force (IETF) has generated several IP suite–related ATM standards.

1.2.3 SDH/SONET

In 1982, Jan D. Spalink and a forward-looking research group at AT&T Bell Laboratories began to develop Metrobus, an optical transmission system that operated in internally synchronous mode. The prime objective of Metrobus was to develop an optimal communication system that takes full advantage of optical communications and the evolution in communication networks, device technology, and service growth. The name Metrobus signifies the fact that it was originally developed to accommodate all communication services in the metropolitan area. In the process of developing the Metrobus system, a number of novel communications concepts and techniques were introduced. Among the most representative ones were the point-to-multipoint optical network, internally synchronous operation, visibility of *digital signal level zero* (DS-0) at the exterior frame level, one-step multiplexing, simultaneous accommodation of tributaries by controlling the number of containers, establishment of 150 Mbps as the internal signal standard, and maximum utilization of overhead. The Metrobus system was fully developed and put into its first office application in 1987 but its commercial market penetration did not follow due to a failure in standardization.

The idea of point-to-multipoint optical networking was considered revolutionary when it was first introduced because at that time all existing optical systems employed point-to-point communications. In addition, one-step multiplexing was another revolutionary concept that enabled the direct multiplexing of DS-1 signals at 1.544 Mbps to the 150-Mbps internal standard signal without passing the intermediate DS-2 and DS-3 signals at the time. These two concepts contributed to efficiently establishing the add/drop and cross-connect capabilities of optical networks.

Choosing 150 Mbps (146.432 Mbps, to be precise) as the internal signal standard of the communication network was another key notion of Metrobus. It accommodated all existing North American and European hierarchical tributaries on one side and allowed users to take full advantage of the supporting technologies on the other. In fact, 150 Mbps was the signal rate that could accommodate all signals from DS-1 at 1.544 Mbps to DS-4E at 139.264 Mbps. From the service application standpoint, the 150-Mbps rate accommodated all voice, data, and video signals (including the compressed HDTV signals), and it was expected that *complemetary metal-oxide-semiconductor*

(CMOS) technology could be easily used up to this rate. From the subscriber loop viewpoint, at this data rate, *light-emitting diodes* (LEDs) and pin diodes could be used instead of laser diodes and *avalanche photodiodes* (APDs), and optical fiber coupling efficiency could be increased by employing graded-index multimode fibers instead of single-mode fibers.

The introduction of internally synchronous operation signaled the beginning of the synchronous communication network. The internally standardized communication network was designed to serve a specific metropolitan area and was therefore adequately equipped to establish communication with plesiochronous neighbors within the area. The clock signal could be chosen from among the *basic synchronization reference frequency* (BSRF), the local oscillator frequency, and the frequency derived from the received signal.

The visibility of the DS-0 at the exterior frame level can be said to be a direct consequence of constructing a 125-μs-based frame structure. In other words, if the construction of the frame structure and the mapping of the tributaries into respective containers are all performed on a 125-μs-unit basis, the DS-0 signals obtained from the 8-Kbps sampling rate are transparent or can be directly accessed from higher-order signals. Hence, a DS-0 signal at 64 Kbps can be efficiently extracted from the 150-Mbps internal standard signal.

In late 1984, Rodney Boehm and Yau Chau Ching at Bellcore made a proposal for a synchronous optical transmission system called SONET to the T1 Committee of North America. Initially, SONET had a bit rate of 50.688 Mbps and targeted a goal of "midspan meet." At that time, there was skepticism as to whether this goal was achievable, so little progress was made in its standardization during the early stage. However, the announcement of Metrobus in September 1985 rejuvenated the standardization efforts, and some innovative improvements were made to the original SONET proposal in this process. Some of the most distinctive contributions were the concepts of layered system structure and pointer-based synchronization proposed by J. Ellson at Bell Northern Research. The SONET proposal was then modified in frame structure and bit rate, from 50.688 Mbps to 49.92 Mbps, and competed with the Metrobus-based proposal at the 146.432-Mbps rate that Bell Laboratories proposed later.

Heated debates followed in the T1 Committee between the two alternate signal standard candidates; then SONET defeated Metrobus in the standardization race in 1986. At that time, CCITT was engaged in the standardization of broadband optical channel standards, and T1 Committee proposed a SONET-derived 149.976-Mbps (= 3 × 49.92 Mbps) signal as the North American standard. Afterward, the SONET standardization process within North America sailed on smoothly, even standardizing the specifics by early 1987. Subsequently, however, a full-scale mediation with CCITT ensued regarding the B-ISDN interface standard, and a large-scale modification took place before the SONET standards got finalized at the rate of 51.84 Mbps (a third of the rate 155.52 Mbps) in 1988.

In July 1986, CCITT began the process of standardizing SDH to be used for the *network node interface* (NNI) independently of the *user network interface* (UNI). This was the first step toward a full-scale SDH standardization, and for that purpose the T1 Committee and CCITT maintained a close partnership. For developing the STM-1 signal standards, which is the essential core of the B-ISDN's NNI standardization, North America introduced a 149.976-Mbps bit rate and a frame structure of 13 bytes (B) × 180, while the European Conference of Postal and Telecommunications Administrations (CEPT) responded by proposing a 9 B × 270 frame structure at the 155.520-Mbps rate. Following some heated discussions, the 9B × 270 structure was finally chosen as the STM-1 standard in 1988, and thereby the rate of 155.520 Mbps came into existence. This is fundamental to all SDH, SONET, and ATM. This decision kicked back the SONET standardization with the restructuring of the STS-1 signal to the rate of 51.84 Mbps.

The contributions of Metrobus to SONET, and of SONET to the standardization of SDH, cannot be overemphasized. Among the unique features of SDH, the concepts of point-to-multipoint optical networking, internally synchronous operation, visibility of DS-0 through a 125-μs time unit, one-step multiplexing, accommodation of multirate signals by controlling the number of containers, the establishment of 150 Mbps as the internal signal standard, and the enhancement of network adaptability and reliability through versatile use of overheads are all derived from Metrobus. In addition, the layered system structure, systematic overhead organization, synchronization via pointers, and establishment of global networking are the by-products of SONET standardization. On this foundation, SDH has completed the years-long standardization process by finding a harmonization point to the North American and European digital hierarchies, consequently enabling a global communications network.

1.2.4 WDM/Optics

When the laser was first developed in the 1950s, researchers first considered the possibility of using them for high-speed optical transmission over free space, a plan that later turned out to be unsuccessful. A better possibility for optical transmission was later found in guided media with the invention of low-loss fibers and semiconductor laser diodes. This possibility turned into reality in the early 1970s when the attenuation of optical fibers dropped below 20 dB/km.

The first commercial optical system was deployed in the early 1980s on the east coast of the United States, and the 1980s saw competitive growth in the optical transmission systems market among a multitude of manufacturers.

In the 1980s, the main thrust of optical communication was developing technologies to transmit higher bit rates over longer distances. In support of this, the copper-based transmission medium was replaced with optical fiber, but the processing for switching and transmission all remained in the electronic domain. There were no standards for optical transmission, and all the optical transmission systems were proprietary, point-to-point systems. Such primitive nonstandard systems existed until the standard systems based on SDH/SONET were introduced in the 1990s. In the meantime the Metrobus optical transmission systems developed in the mid-1980s experienced a turning point with the development of point-to-multipoint optical systems, opening a new era for network-oriented synchronous optical systems and leading to SDH/SONET systems.

The CCITT standardization works of the 1980s gave birth to the SONET and SDH optical systems, which are closely compatible. These standard optical systems were built based on the point-to-multipoint network concept, but the transmission and switching processes were still done in electrical domain. Such SDH/SONET-based *first-generation optical networks* have been built throughout the 1990s, and since then they have formed the core of the telecommunications infrastructure in Europe, Asia, and North America. The first-generation optical network distinguishes itself from the primitive optical network in that it is network-oriented, wherein the add/drop and cross-connect functions are realized efficiently and reliably.

As demand for bandwidth increased, solutions have been sought in two different directions. The first was to increase the transmission bit rate, and the second was to increase the number of optical wavelengths into which user signals were multiplexed. In the first solution, multiple lower-rate data streams are multiplexed into a higher-speed stream by means of electronic *time-division multiplexing* (TDM). In the second solution, WDM, multiple wavelengths of optical carriers are multiplexed together, with each wavelength carrying independent bit streams. As WDM is orthogonal to TDM, it is commonplace to combine the two multiplexing technologies to maximize the transmission capacity.

WDM is the process of combining multiple optical signals in different wavelengths into one optical signal. WDM is essentially the same as *frequency division multiplexing* (FDM), which multiplexes different data streams at different carrier frequencies simultaneously, but the term wavelength is used in place of frequency. As optical signals in different wavelengths normally interfere with each other if the wavelengths are closely spaced, the wavelengths for WDM should be arranged sufficiently far apart. Thus, in earlier WDM technology, the number of wavelengths was limited to about 10, but now *dense WDM* (DWDM) technology is available with a larger number of wavelengths, usually 40 or more, by combining wavelengths more densely.

As the optical WDM networks evolve, wavelength routing functionality can be added in the form of *optical add/drop multiplexing* (ADM) (OADM) and the *optical cross-connect* (OXC) system. In such second-generation optical networks, *optical time-division multiplexing* (OTDM) or WDM can be employed to increase transmission capacity, and the wavelength converter systems can be employed to realize efficient add/drop and cross-connect functions. As the bandwidth grows even higher, the relevant switching and routing processes can be arranged to take place in the optical domain. It is advantageous to adopt optical switching and routing means over the electronic counterparts in that repeated optical-to-electrical and electrical-to-optical conversions are not needed. In such a *third-generation optical network*, optical switching may be further expanded to optical packet or burst switching. The so-called all-optical networks belong to this third-generation optical network.

While SDH/SONET systems were being developed, optical networks also gained the interest of the local network sector in the mid 1980s. Unlike the case of SDH/SONET, this approach tried to extend the optical communications paradigm from point-to-point links to mesh networks. The first such network was basically the broadcast-and-select based network, whose typical example was the LambdaNet. The networks used tunable lasers and passive optical couplers to realize a logical mesh network. Later, the second type of network was developed based on the multihop and wavelength-routing concepts.

Today OADMs and OXCs are available, even if in limited capability, but fully optical switches, logic devices, and storage devices are still under active development. The result is that advanced optical transmission products are widely available, with the expectation that advanced optical networking products will also become available in the near future.

1.2.5 Recent Developments

In the past, the evolutionary trend in communication networks has been in the direction of the digitalization of transport networks. More recently the trend has been toward the increased support of data services. This is manifested in the evolution of the voice-oriented telecommunication networks to accommodate ever-increasing amounts of data traffic and the growth in the bandwidths and the coverage of the data networks themselves. This has reached the stage where the telephone and data networks are now being integrated with each other and thereby accommodating various types of broadband multimedia Internet services.

These recent trends may be represented by three keywords: *bandwidth*, *QoS*, and *wireless*. First of all, the major accelerator of the bandwidth demand has been the growth of the Internet for commercial, business, and recreational purposes. Today many businesses use the Internet as the basis of communication, resulting in a continuous flow of e-mails, documents, and other forms of business information. However, as the WWW and the Internet have become popular with general ordinary users, we have seen the introduction of new applications, many of which are rich in audio, graphics, and video. This has led to an increased deployment of broadband multimedia services such as streaming video and radio. These new broadband services demand substantially larger amounts of bandwidth than conventional e-mail or other data services did. In response to such increased bandwidth demand, optical transmission technologies such as SDH/SONET and WDM have been widely deployed.

In addition to increased bandwidth, many of the new multimedia applications mentioned above demand QoS, or guarantees in delay and/or bandwidth, since the applications will not function correctly or satisfactorily without such guarantees. Typical examples of this would be the VoIP and videoconferencing services, for which reasonable level of delay and bandwidth guarantees are a prerequisite for making the services meaningful. In the case of time-critical applications, in particular, users may be willing to pay for such preferential services. A typical example of this would be data transactions for stock market day traders. While the application may work without guarantees, the user in many cases will be willing to pay a premium to get a guarantee in service quality, as even a delay of a couple of seconds could be absolutely critical. To *Internet service providers* (ISPs) and network operators, such time-critical applications may bring about a new way of generating revenues that is not possible in the existing flat-rate fee structure.

Finally, another important trend has been the wireless revolution. While we do not directly deal with this technology in this book, it has been adding a new dimension to communication services by enabling communications without restriction in time and place. During the last decade, mobile communications have been developing very quickly in technology and in the number of users—even faster than the Internet. Ubiquitous connections, mobility, and personalization, all contributed to the success of the cellular phone and its ilk. More recently, data services have been becoming more and more important, when compared with the traditional voice services. In particular, wireless connection to the Internet has been helping to launch such services as real-time stock quotes, location-specific Web surfing, and voice/e-mail.

Our main interest in the wireless revolution is how it will affect the networking infrastructure. The current wireless infrastructure is based on leased lines and expensive ubiquitous networks (as the operator needs base stations everywhere). The networks are optimized to an extraordinary degree to carry voice over unpredictable wireless environments. Basically these networks are based on the circuit paradigm and therefore are not built to carry data in efficient ways. However, it is obvious that the main growth of traffic will be in data. Therefore it is necessary to introduce a more packet data–oriented restructuring of the wireless network and to introduce IP technology into the fundamental core of the wireless network. By doing so, the networks can carry IP traffic more easily and can achieve cost-effectiveness by converting the control mechanism from a proprietary control network to a more open IP-based control paradigm. In addition, this increase in the need for data at the edge of the network results in increased demand for bandwidth from the core network. As such, the wireless revolution is also pushing the first two trends of extending the reach of IP technology and the demand for bandwidth.

1.3 Organization of the Book

This book is composed of three sections. The first section deals with a framework of network functionality, the second deals with the technological components for broadband communications, and the third deals with the integration of those technological components.

In Chapter 2 of the first section, we discuss a framework of network functionality for analyzing and comparing basic network technologies. The framework consists of six basic network functions—layering, multiplexing and switching, routing, network control, traffic management, and QoS. Layering is concerned with how the network functions are divided among various constituent layers. Multiplexing and switching define the methods by which the network resources are shared and switched. Network control discusses how the network's operations are controlled either by the operator or by the user. Traffic management is concerned with how to ensure that the resources are not overused. QoS discusses how to enable service providers to provide customers with a certain level of service.

Subsequently, in in the second section, Chapters 3 through 6 introduce and analyze four basic network technologies—TCP/IP suite, ATM technology, SDH/SONET transmission, and WDM/optics technology—based on the framework established in Chapter 2. These four technologies, the driving forces behind the information revolution discussed in this chapter, are now commonly aiming to provide *broadband* services. Each of these technologies offers different solutions to the basic problems in networking. Each particular technology brings in its own specific solutions to the different facets of the six principles discussed in the framework. Frequently there is an overlap in the solutions as each technology was developed independently of the others and thus approaches each problem from its own perspective. In the case of SDH/SONET and WDM/optics, however, the overlap is less than that in the TCP/IP and ATM cases, as they are more transmission-oriented than network-oriented and more physical/optical layer–dependent. In particular, we add the topic of synchronization in the case of SDH/SONET discussions and emerging optical component technologies in the case of WDM/optics discussion, considering the distinguished features of each technology, while we do not discuss traffic management or QoS issues.

The third section, which consists of Chapters 7 and 8, discusses how to *integrate* the four technologies in various different ways and discusses the distinctive characteristics of each integration type. In many cases, networks are hybrids, with a mixture of TCP/IP, ATM, SDH/SONET, and WDM/optics. For example, TCP/IP traffic frequently runs over ATM networks, while ATM traffic normally runs over SDH/SONET networks, and SDH/SONET traffic runs over WDM/optics. Basically, this section examines how the four technologies interact with one another and the different types of integration that are possible. First, Chapter 7 deals with the integration of IP and ATM, which is in use today in diverse forms. Then, Chapter 8 discusses the integration of IP and WDM/optics, which is currently in the research stage. In each discussion of the integration topics, we return to the framework of Chapter 2 and examine how to solve the various problems of networking in each integration method, investigating the advantages and disadvantages of each case. One of the most important points here is to identify how the solutions for solving similar problems in each technology interact, with a typical example being how the traffic management mechanisms of ATM and IP interact.

References

[1] Lynch, D. C., and M. T. Rose, *Internet System Handbook*, Reading, MA: Addison-Wesley, 1993.

[2] Chu, W. W., "A Study of the Technique of Asynchronous Time Division Multiplexing for Time-Shared Computer Communications," Bell Laboratories Memorandum, September 1968.

[3] Fraser, A. G., "Early Experiments with Asynchronous Time Division Networks," *IEEE Network Magazine*, Vol. 7, No. 1, January 1993, pp. 12–26.

[4] Lee, S. H., "Dynamic TDM—An Integrated Transport Technique for Circuit- and Packet-Switched Traffic," *Proc. IEEE INFOCOM'88*, March 1988, pp. 110–118.

2

The Framework

The basic problem of network design is how to facilitate communication or the transfer of information between two or more entities. Though the problem statement is simple, it is a problem with many ramifications, as there can be many variables in the main subjects and multiple solution spaces. The subjects are diverse as they may be humans, computers, or simple machines. The type of information being transferred may also be diverse and include speech, video, or data. The solution space is complex, because one may use any property of the physical world, including electronics, optics, acoustics, and electromagnetism. In addition, abstract mathematical tools such as the field of algorithms may be brought to bear on the subject to solve various problems.

The complexity and broadness of this subject area raise the basic problem of how to understand this field. Is it desirable to understand the bits and parts of the network first? Does a network designer need to understand how lasers work, or how network switches are built out of *application-specific integrated circuits* (ASICs), *field programmable gate arrays* (FPGAs), and *microprocessor units* (MPUs)? Or, does the designer need to understand the complicated mathematical control algorithms that work underneath the network structure?

We believe that the key is in understanding that the field of network design is in fact a subfield of system design. The main issue of system design is how to group and connect a number of objects to organize a system in such a way that it can achieve the goals of the system designer. For example, the government may be viewed as a system for governing a state, and a car may be viewed as a system for transportation. Likewise, a network may be viewed as a system for communication.

A system designer, especially when the target system is large and complex like a network, therefore, must look at the problem as a whole and try to devise a solution that is composed of multiple small solutions, each solving a different aspect of the problem. In this process the system designer often realizes that many problems or facets of the system design are common to many real-life systems. A commonly used approach to a system design is first to build up an abstract model of the problem in question and then to devise specific solutions to cover different aspects of the modeled problem. Identifying the problems and then devising specific solutions are therefore the major jobs of a system designer.

This book approaches the issues of *integrated broadband networks* from a system designer's point of view. We categorize the broadband networks from the aspects of the governing technological components such as TCP/IP, ATM, SDH/SONET, and WDM/optics. Based on this categorization we analyze each technological component from different aspects of network functionality, such as layering, routing, multiplexing/switching, network control, traffic management, and QoS. When considering the interaction and integration among the technological components in later chapters, we compare them from the aspects of such network functionality again.

The six functional components—layering, routing, multiplexing/switching, network control, traffic management, and QoS—can be viewed as individual facets of the total network architecture, with each solving a specific aspect of the whole network problem. They are fundamental to any network but the degree of importance varies depending on the targeted role of the networks. For example, the channel-oriented functions like routing, switching, network control, traffic management, and QoS are more significant to IP and ATM networks than to SDH/SONET and WDM/optics networks. In the case of the latter transmission-oriented networks, multiplexing, add/drop and multiplex, and network survivability are of more significant interest, instead.

In analyzing or integrating different types of networks it is important to identify how each network arranges those six functional components, to compare how the solutions differ, and consider how much they have in common. For example, the problem of multiplexing is solved differently in WDM optical networks than it is in IP networks. The former relies on the physical characteristics of optics to achieve multiplexing of multiple wavelengths, while the latter uses complex electronics to multiplex IP packets in a time-division fashion. On the other hand, SDH/SONET networks also use electronics and TDM but are in fact more similar to WDM networks than IP networks from an abstract point of view. This is because WDM and SDH/SONET networks are both transmission-oriented optical networks that rely on circuit-switched network architecture.

As such, given an arbitrary network, we can grasp a clear view of the network by identifying how the six functional components (and their derivatives) are arranged in the network. In addition, by comparing how the six facets are arranged in different networks, we can highlight the key differences among the networks and hence their advantages and disadvantages. Therefore, this chapter discusses the principles of these six functional components in detail, as they will establish a groundwork for the material in the subsequent chapters in the second and third sections.

2.1 Layering

Network architecture layering is a fundamental concept in system design, and it is also used in many other system designs, such as computer operating systems. In a layered system, each upper layer builds its services upon the services provided by the lower layers. Each lower layer offers to its upper layer a certain range of services. The upper layer only needs to understand how to use the services, without worrying about their implementation.

More specifically, in communication networks, layering is a way of dividing up the tasks of communications into a reasonable number of parts by way of using an abstract service model. This arrangement helps to hide the complexity of certain functions in a layer from the other functions in different layers that do not need to know about the functions. As such, layering is an ideal method for incrementally building and independently operating a communication system.

2.1.1 Advantages of Layering

The postal system, whose primitive goal is the delivery of letters, illustrates the layering concept very clearly. A person who writes a letter puts it in an envelope and then puts it into a mail receptacle, which is a "black box" as far as the sender is concerned because the sender does not know what happens beyond that point. A postal worker then collects the letters from the postbox and brings them to the post office, but he or she does not know how the individual letters in their bags will be handled in the post office. In the post office, the letters are sorted for proper distribution to their destinations in a process that corresponds to the switching function in communications. The sorted letters are then carried to the destination post office in a process that corresponds to the transmission function in communications. The remainder of the operations until the letter is delivered to the receiver is a reverse process of the above.

The layering concept in the postal system closely applies to communication networks. There is usually an application or a high layer at which the original message is generated. The message is then enclosed in an envelope (i.e., a packet) that becomes an independent unit of handling in the network. The series of steps that follow ensure secure delivery of the packet between each pair of adjacent nodes as well as between the source and the destination nodes. Finally, the processed packets are passed to the physical layer for transmission. Of course, in a real communication network, the previously mentioned functions are divided into a number of different layers. Functions such as networking and routing that arrange the actual delivery of the message to the destination point are performed at the network layer in the middle of the layer stack.

As can be seen in the example of the postal system, the basic reason for employing the layering concept is that it allows a *divide-and-conquer* approach to system design. An important advantage of the divide-and-conquer approach is that each layer can be tackled by the experts of that particular layer. Given the many different technologies that today's communication systems need, it is a significant advantage that each layer can be handled without having the knowledge pertaining to other layers. For example, in normal communication networks, we may expect that the physical and data link layers can be well handled by digital communication engineers, wireless RF engineers, and/or optical engineers. In contrast, the network and transport layers require different types of engineers, such as network protocol engineers and routing protocol experts. Further, the layers at the top may best be handled by application engineers and certain computer scientists.

Another advantage of the divide-and-conquer approach is that each layer can be improved independently, without adversely affecting the other layers. Examples of such cases abound in the communications industry. We have seen continuous improvements in the area of the physical layers, from the copper-based T1 and T3 systems to SDH/SONET systems to WDM optical systems, without affecting the upper layer functions. Modems have also improved from 9.6-Kbps systems to 33.4-Kbps and 56-Kbps systems, and currently to xDSL technologies. On the other side, there have also been applications improvements, including *e-mail*, *telnet*, and now the WWW.

2.1.2 Examples of Layering

There are many examples of layering in communications, including those in communications protocols: For example, *signaling system no. 7* (SS7) is the layered signaling protocol stack used in modern telephone networks. The *open system interconnection* (OSI) seven-layer protocol model is the layering model defined by the *International Standard Organization* (ISO) for data networks. The TCP/IP protocol suite defines the model for the Internet. In ATM, the three-layer protocol model defined by the ITU-T is used.

Sections 2.1.2.1 through 2.1.2.5 examine five examples of the layering model, starting from the very formally generated OSI seven-layer model, to the less formal but practically designed TCP/IP protocol model, to the recently developed ATM and SDH/SONET layering models, and finally to the still-evolving WDM optical layering model.

2.1.2.1 OSI Seven-Layer Model

The OSI seven-layer model establishes a conceptual foundation for today's data communications. Even though the related OSI protocols themselves are not widely used in practice, this reference model is a popular way of understanding and viewing communication protocols. The seven layers in the model are the physical layer, data link layer, network layer, transport layer, session layer, presentation layer, and application layer. Figure 2.1 illustrates this layer stack.

In the low end, the *physical layer* is concerned with the transmission of an unstructured bit stream over a physical medium.[1] The *data link layer* provides the reliable transfer of information across the physical link and sends blocks (or frames) with the necessary synchronization, error control, and flow control. It transmits frames from one point to another and recovers frames from the bits transported by the physical layer. The data link layer usually needs a local addressing scheme. The *network layer* provides the upper layers with independence from the data transmission and switching technologies used to connect systems. It transmits frames from one point to another over many other points. The network layer needs a network-addressing scheme.

The middle layer, or *transport layer*, provides reliable, transparent transfer of data between end points. It also provides end-to-end error recovery and flow control and basically ensures the end-to-end transmission of frames.

1. Note that the OSI seven-layer model is for data networks, so it is assumed that the basic unit of communication is a bit. That is, digital rather than analog communication is assumed in the physical layer.

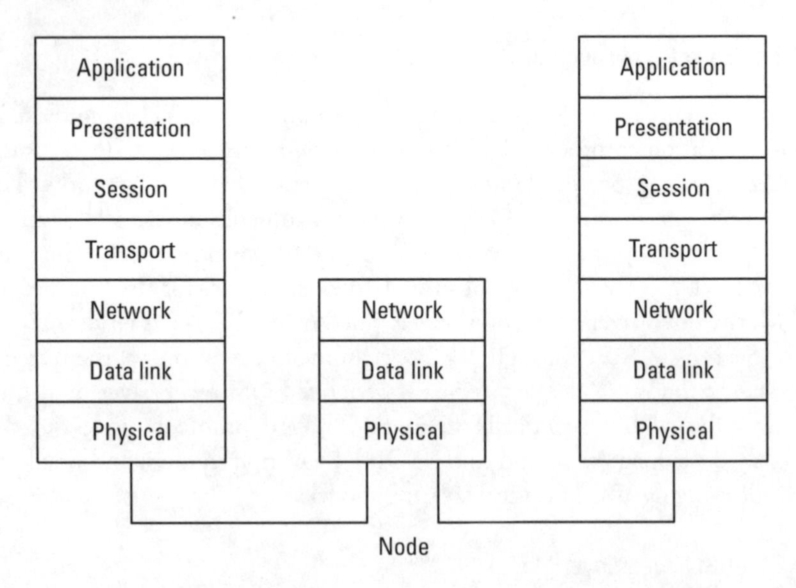

Figure 2.1 OSI seven-layer model.

In the case of the upper layers, the *session layer* provides the control structure for communication between applications and establishes, manages, and terminates connections (i.e., sessions) between cooperating applications. The *presentation layer* provides independence to the application processes from differences in data representation (i.e., syntax). The *application layer* provides access to the OSI environment for users and provides distributed information services.

While the OSI seven-layer model is useful as an abstract model for analyzing and understanding communications protocols in general, very few real-life protocol suites or applications follow this model. The main reason may be because the model was defined by a committee without taking the practical environment into account. For example, one of the main questions asked about the OSI model is whether it is necessary to separate the session and presentation layers from the application layer. Some applications may need them but many applications do not. Even when needed they are frequently built in as part of the application layer. It is because in many ways these functions are application-specific so each application defines its own.[2]

2. For example, *coder-decoder* (codec) is a representative of the presentation layer. Codecs exist for many different types of applications, so it would be unreasonable to try to limit the types. There may be other reasons for implementing the three top layers together. A

2.1.2.2 TCP/IP Model

The TCP/IP model contains only four layers. These four layers roughly correspond to the "seven-layer" OSI model discussed above. They are the link layer (which includes the physical layer and the data link layer), the IP layer, the transport layer [for TCP or user datagram protocol (UDP)], and the application layer. This is shown in Figure 2.2. (Refer to Chapter 3 for more details.)

Actually this protocol model was not strictly defined for the TCP/IP protocols. Instead of strict layering, rather fluid and functional definitions were used. The end result is a tool-based rather than a model-based model. It reflects in many ways the computer science background of many of the original TCP/IP protocol developers.

In contrast to traditional communications networks such as telephones, TCP/IP was not defined by a standards development organization first and then implemented, but rather implemented first and then defined (to put it

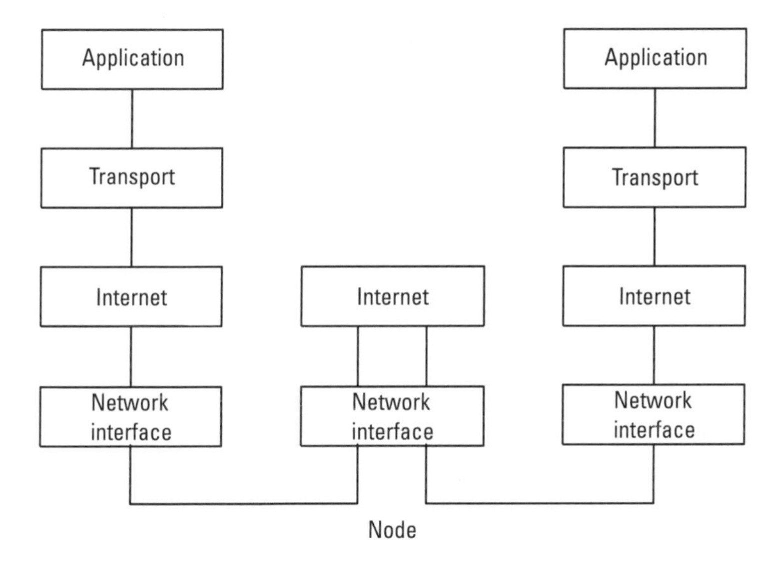

Figure 2.2 TCP/IP layer model.

good case in point is the use of *real-time transport protocol* (RTP) and the application framing principle. The application framing principle states that by intricately binding the application's coding (which is a presentation layer function) with the network layer packet actually transmitted, it is possible in many cases to get better performance from a given network.

extremely). More exactly, the implementation and definition were done in a more organic manner, with a continuous feedback path between preliminary implementations and standards definitions.

2.1.2.3 ATM Model

The ATM protocol reference model basically contains only three layers, which, however, may be expanded to four layers. The three layers are the *ATM adaptation layer* (AAL), the ATM layer, and the physical layer. The ATM layer can be subdivided into the *virtual channel* (VC) layer and the *virtual path* (VP) layer, thereby yielding a four-layer model. This ATM protocol model was defined by the ITU-T. Figure 2.3 shows the protocol stack of the ATM model. (Refer to Chapter 4 for more details.)

Depending on the level of processing in the ATM layer, the ATM switching is divided into VP switching and VC switching. VP switching is done in the VP layer and VC switching is done in the VC layer. This is illustrated in Figure 2.3.

The three layers of the ATM model may be said to be equivalent to the four layers of the TCP/IP model in the sense that both ATM and TCP/IP can carry all different types of application services. However, the three ATM layers may also be put underneath the TCP/IP layers since ATM can provide transport means to TCP/IP.

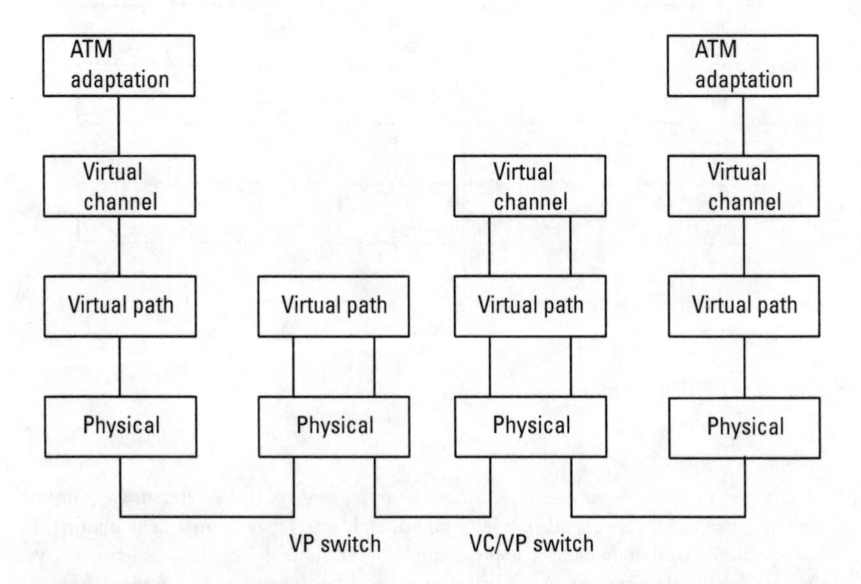

Figure 2.3 ATM layer model.

2.1.2.4 SDH/SONET Model

The SDH/SONET layer model contains four layers: the path layer, the multiplexer section layer, the regenerator section layer, and the physical layer. The path layer may be subdivided into the lower-order path layer and the higher-order path layer. Figure 2.4 depicts the SDH/SONET layer model. (Refer to Chapter 5 for more details.)

The four layers in the SDH/SONET layer model can provide a physical layer for both the TCP/IP and ATM models. In the case of stacking ATM layers on top of the SDH/SONET layers, the physical layer path of the ATM is serviced by the path layer of the SDH/SONET. Likewise, the SDH/SONET itself may be transmitted over the WDM/optics layers.

2.1.2.5 WDM/Optics Model

Another case to consider is the problem of layering in WDM/optics networks. As the field of optical communications is still evolving there may be further changes in the optical layer models. Nevertheless, the layered architecture defined by the ITU-U for the *optical transport network* (OTN) is well-established. It consists of three layers, namely, the *optical channel* (OCh) layer, the *optical multiplex section* (OMS) layer, and the *optical transmission*

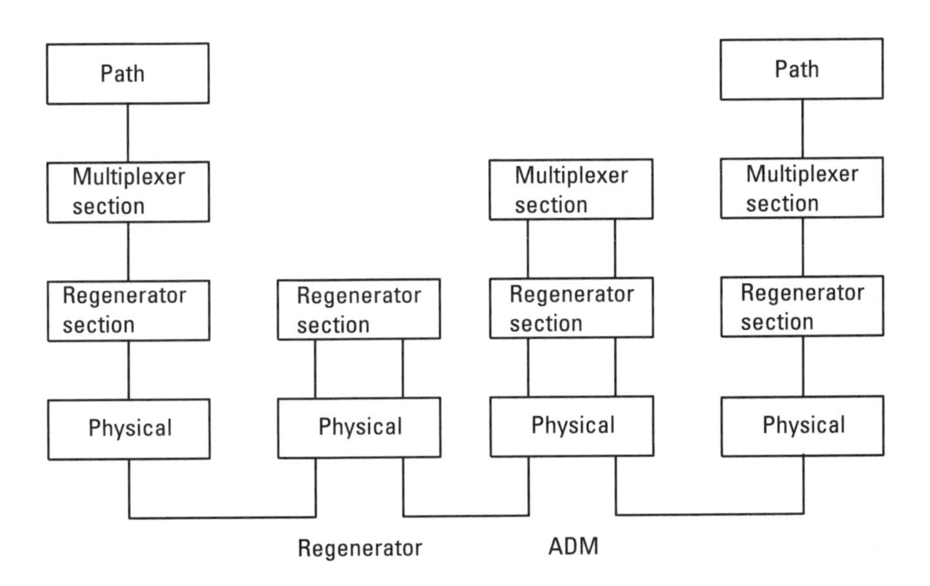

Figure 2.4 SDH/SONET layer model.

section (OTS) layer. Figure 2.5 depicts the layer stack of this WDM/optics model. (Refer to Chapter 6 for more details.)

Among the three layers in the WDM/optics model, the OCh layer provides lightpaths for data transport, the OMS layer provides wavelength multiplexing function, and the OTS layer provides amplification and transmission functions on the optical signals. Thus, a WDM ADM (or OADM) device contains the OMS and OTS layers.

2.1.3 Problems of Layering

As discussed above, layering has many advantages that are closely coupled with the divide-and-conquer approach. However, there are also some disadvantages. The main problem is that network performance may be sacrificed to some extent while protocols are being implemented according to a complete layering model. The main reason for this is that layering basically hides information/complexity in other layers. Ironically, this was the source of the advantages of layering to some extent.

Network performance can be improved if the lower layers understood more about the data being sent to them by the upper layers. A typical example of this is the use of the *end-of-frame* (EOF) bits in ATM AAL-5, which indicate the end of the frame carried by the ATM. By observing the EOF bits

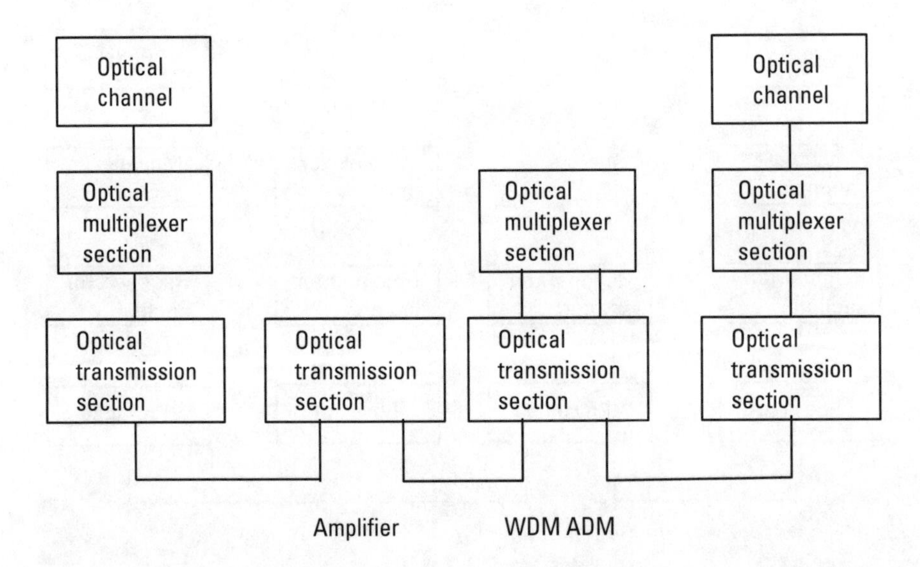

Figure 2.5 WDM/optics layer model.

in ATM cells, ATM switches can improve the data throughput by dropping cells in units of AAL-5 frames. This is because a single cell loss in a frame renders the rest of that frame useless, so once a cell is dropped, it makes sense for the switch to drop all the remaining cells belonging to the same frame (refer to Chapters 4 and 7 for further discussion).

Another example is the use of the *application-layer framing* concept in many new protocols such as the RTP. In the RTP the frame size for delivering each chunk of data is defined by the application layer. This results in saving communication resources in cases where network congestion causes some packets to be dropped. As the packet size is the same as the dropped frame size, this ensures that the effects of packet dropping on decreasing congestion will be maximized, while also ensuring that the usefulness of any packets that arrive at the destination is maximized.

2.2 Routing

Routing is the process that decides how to control the route for delivering packets or setting up circuits from the source to the destination. Routing is a fundamental networking problem that must be solved in all networks. It exists in telephone networks where the initial signaling message generated by the user must be sent from the source to the destination. It also exists in packet networks where the packet switches must decide to which links to send the arriving packets.

As far as routing is concerned, the main difference between telephone networks and packet networks is when the routing decision is made. For telephone networks that operate in circuit mode, the decision is made during the signaling phase—that is, before the delivery of the user information—so once the call is set up the route cannot be changed. For packet networks that operate in the datagram mode, however, the decision is made when each packet arrives at each packet switch. The decision is made dynamically by looking up a routing table at every packet arrival.

2.2.1 Routing Methods

As in most cases, the simplest method of routing is *centralized routing*. In this case, a central control entity builds up a complete map of the whole network and generates routing tables for each node. This information is then provided to each node and used for setting up connections. More specifically, the central control point gathers all the information on the connected nodes in the network, and based on this information it generates a graph connecting all the nodes. Out of this graph a routing database is made and downloaded into every node in the network. Beyond this, each node makes its routing decisions based on this database.

Each node, upon receiving an incoming message, looks up the database to determine how to send the message to its destination node and then sends it out along the path indicated in the database. In this arrangement, the central node may become complicated, but the intelligence needed at other nodes is very little. Of course, this method has the downside of necessitating a very large workload at the central control node, as well as the subsequent problems related to the distribution of the routing information. Consequently, this centralized method has obvious problems in scalability and configuration.

An improved method of routing that resolves the problems of centralized routing can be found in *distributed routing*. Distributed routing takes advantage of the results from the graph theory and its derivative algorithms. Two main algorithms that are commonly used for distributed routing are Bellman-Ford's algorithm and Dijkstra's algorithm.

In distributed routing the routing database is not calculated at a central point but, instead, each node calculates its own routing database based on the information received from its neighbors. The main results are the distance vector routing protocol based on Bellman-Ford's algorithm and the link state routing protocol based on Dijkstra's algorithm.

The decision as to whether distributed or centralized routing should be used will be a major design point for any network. It may also depend on the sort of solutions that the network designer is seeking during the design of the network. For the case of telephone networks that are centrally administered, a centralized routing methodology is a logical fit in many ways. In fact this is the way it is actually carried out in today's telephone networks. For packet networks like the Internet, which is distributed in nature and not completely centrally administered, a distributed routing methodology is a logical fit.

2.2.2 Issues in Routing

In addition to the centralized and distributed routing issue discussed above, there are various other issues to consider when deciding routing schemes. For example, it is important to decide whether to use source routing or a hop-by-hop routing. It is also important to consider whether to prepare a single path or multiple paths for the routing and whether to make the routing state-dependent or state-independent.

2.2.2.1 Source Routing or Hop-by-Hop Routing

In the case of the *source routing*, the source of the message decides what path to take to send the message through the network. This method requires each source node to have complete knowledge of the network so as to decide on the path before sending out packets. When source routing is used, it is easy to introduce new services or policies to the network. Depending on the type of services or policies that the sender adopts, the path taken by the messages can vary. For example, a personal telephone call can be routed through one network provider's network, while a business telephone call is routed through another network provider's network.

In the case of the *hop-by-hop routing*, the routing decision is made by each node along the path to the destination. This method requires each node to have knowledge regarding which link to forward a message destined for a particular destination. The main advantage of hop-by-hop routing is that the end nodes are free from the burden of knowing the whole network. Consequently, the edge devices can be kept simple. However, this very fact can also be considered a disadvantage since it would require that the network switches be more intelligent. The cost of having more intelligence in the network switches could be more than the savings earned at the end nodes.

2.2.2.2 Single Path or Multiple Path

Another decision point is whether to arrange the routing protocol to select and maintain a single routing path or multiple routing paths. Obviously, offering only a single path would be simpler and require less processing power, memory, and time. However, what will happen if the single path breaks down? It will cause either a complete loss of communication or, possibly, a long wait for the connection to be resumed with an alternate path. If multiple paths are available, this problem would be avoided, as an alternate path is readily available. Furthermore, multiple paths can help to balance the traffic load among the different paths. It would be also possible to arrange different paths to have different levels of QoS support and thereby provide different classes of services.

2.2.2.3 State-Dependent or State-Independent

A third point to consider is whether the routing will be done in a *state-dependent* or *state-independent* manner—that is, whether to take into account the current state of the network when deciding a route. This is a problem that is attracting more attention from packet-switching research groups today due to the current interest in QoS support. It is reasonable to set aside certain paths from consideration of routing path selection if they are congested. However, the state-dependent routing requires more processing and memory than state-independent routing. In addition, it requires the use of source routing, as otherwise the edge node would not know whether or not a path is suitable. Note that this problem was already considered and implemented in circuit-switching networks as is shown in the telephone-network routing example in Section 2.2.3.

2.2.3 Examples of Routing

Telephone networks exemplify traditional centralized routing schemes. Telephone networking is based on a three-tier architecture. In the first tier there are edge devices, telephones, and modems that are connected to the network. In the second tier there are central offices (or local exchanges) that usually employ class 5 switches. In the third tier there are toll switches. Toll switches are connected in mesh topology and carry the long-distance traffic, while the central office switches exchange calls within the local area and connects long-distance calls to the toll switches.

For a central office the routing decision is trivial: When a call comes in to the central office, it is connected to another user internally if it is a local call, but the decision is passed to a toll switch if it is a long-distance call. For the toll switch, however, the routing is slightly more complicated. Due to the mesh interconnection among toll switches, both one-hop and two-hop paths are possible to the destination, so the toll switch must decide which way to send the traffic. There are many possible examples of the routing algorithms used by the toll switches, among which are *dynamic nonhierarchical routing* (DNHR), *trunk status map routing* (TSMR), and *real-time network routing* (RTNR).

Examples of distributed routing are abundant in data networks. The most typical among them are the distance vector routing protocol based on Bellman-Ford's algorithm [e.g., routing information protocol (RIP)] and the link state routing protocol based on Dijkstra's algorithm [e.g., open shortest path first (OSPF)], which are explained in detail in Section 3.2.2.

2.3 Multiplexing and Switching

Two of the most basic communications functions are transmission and switching. These fundamental functions have existed since the beginning of multiple-user communications and remain as the minimum core even in today's complicated and diversified communication networks. Multiplexing is a means of resource sharing in transmission, and switching is a means for providing connections among the multiple users in the networks. Thus, multiplexing may be needed only when transmission-channel sharing is pursued, but switching is required as a fundamental function in any type of network.

Conceptually, multiplexing and switching carry out different tasks, but they are closely related and operate in a harmonized manner. The two are related in such a way that the unit of data for multiplexing is usually the same as the unit of data for switching. If multiplexing is done in a time-division fashion, then it is most effective to carry out the switching function in a time-division fashion. Likewise, if multiplexing is done in packet mode, it is most appropriate to use packet-mode switching. On the other hand, as the degree of transmission-channel sharing (or the degree of multiplexing) increases, the degree of complexity in switching increases as well, in general. The type of switching can vary depending on the network topology and operation mode.

2.3.1 Multiplexing and Switching Functions

In a primitive, full-mesh type of network, there is no multiplexing function involved because there is no sharing of transmission channels. In addition, there is no switching function in the network because each user node has a dedicated line to every other user. In this extreme case, multiplexing is not employed, as there is no resource sharing in transmission.

In reality, however, switching is involved somewhere in the overall network since connection is necessary in every form of multiple-user communications. Indeed, switching takes place at the premises of each sender's end node. Among the multiple lines that originate from the end node, the sender has to select one particular line that connects it to the desired destination, and this very selection is the switching operation.

Another extreme case on the opposite side is represented by the shared-medium packet networks, such as CSMA/CD bus and token ring networks. In this case the transmission channels are shared among all end users. Each user accesses the common shared channel through contention or token passing, and this effort in packet-mode networks is comparable to the signaling operation in circuit-mode networks (i.e., orthogonal to transmission or switching). The bus or the ring, which renders a medium for transmission-channel sharing, may be viewed as proving multiplexing in a sort of TDM manner. Switching then takes place at the receiver station where the end station selects the packets that are destined to itself. This may be viewed as distributed switching.

In typical public telephone networks or wide area networks (WANs), multiplexing is always involved in the long-haul or backbone lines, and switching also takes place in all central offices or backbone nodes. In the case of circuit-mode operation, multiplexing may take the form of FDM, TDM, or WDM in dedicated-line accesses and the form of FDMA, TDMA or CDMA in the case of multiple accesses.

If we consider switching in relation to multiplexing, each input line to a switch usually carries the multiplexed data of end-user signals. Likewise, an individual output line of a switch carries a multiplexed traffic out of some selected input lines. From each individual output line's point of view, therefore, a switch is like a multiplexer, and consequently a switch may be viewed as a group of multiplexers.

This points to the fact that the unit and method chosen for multiplexing essentially defines how switching must be carried out in the network. This further affects how switches can be built and how resources can be shared in the network. As such, the choice of the multiplexing technology can be viewed as being tantamount to defining the central aspect of a communications network. This is well demonstrated in the example of ATM networks, where the choice of cells (or fixed-sized packets) as the basic unit of multiplexing resulted in a whole new network architecture and switches that are completely different from the traditional circuit-based telephone switches.

As such, when designing a communication system, the method of multiplexing and switching is a fundamental choice to the designer. In many ways this may be dictated by the service model that the network designer considers. For example, in the case of the *public switched telephone networks* (PSTN), network designers would consider the importance of voice calls for which physical circuits are most adequate, and therefore would come up with circuit mode–based multiplexing and switching schemes. In contrast, in the case of the Internet, the designer would consider the importance of the efficient sharing of links by bursty services, for which dedicated user channels do not provide the desired efficiency, and would therefore come up with packet mode–based switching mechanisms.

2.3.2 Examples of Multiplexing and Switching

In communication networks, examples of multiplexing abound—including TDM, which is commonplace in circuit-mode communications. In TDM switches, the time slots on an outgoing line are divided up among users, resulting in the same effect as the multiplexing of multiple user signals onto a single output line. TDM can be divided into two types: one is synchronous, and the other is asynchronous. Synchronous multiplexing periodically allocates time slots to the same user, while asynchronous multiplexing does not allocate time slots in a periodic manner.

FDM is another method of multiplexing that is similar to synchronous TDM. However, the multiple-user channels are differentiated by frequencies. Optical WDM is basically the same as FDM, as a wavelength has a one-to-one correspondence with frequency. In optical networking, the term *wavelength* is opted over the term *frequency*.

In all the above examples, the basic resource that is shared in the network is the link bandwidth. Accordingly, the units of bandwidth are determined by the units of multiplexing. In addition, the above models are mostly based on the circuit-based multiplexing model in which multiplexing is carried out in units of dedicated channels, or "circuits." The user is given a periodic time slot, a frequency in the radio band, or a wavelength in the optical network, which are all logically equivalent in that the user is given part of the shared resource indefinitely.

In contrast, the multiplexing for packet services uses packet mode. This is the abstract model that is enabled by the asynchronous TDM method. This method does not allocate any fixed part of the shared resources to a certain user. Instead, the resources are allocated in a nonperiodic manner. Over time, the user may be logically allocated a certain amount of resources, but there is no easy one-to-one mapping as in the synchronous TDM case.

As for the case of routing, there are two different types of switching models—centralized switching and distributed switching. Centralized switching sends data only to the intended output lines, so it is efficient. Most circuit switches as well as ATM switches belong to this category. Distributed switching usually relies on the broadcast-and-select type of operation, sending data to every node in the network. The obvious problem with the distributed switches is the lack of scalability, as the size of the network is limited.

2.4 Network Control

Network control can be viewed from two perspectives: one is from the network operator's point of view, based on the need for controlling the network operation. The other is the user's point of view, based on the need for demanding and receiving a certain level of services from the network.

Obviously the network operator needs control of the network to be able to provide proper communication services and keep the network healthy. This functionality is frequently called *network management.* For an operator, network management is in many ways the most important functionality. Even if certain network equipment is state-of-the-art and implements the latest technology, it is unlikely to be successful unless it supports adequate management and control facilities. The demand for the ability to manage and control the network is the basis behind the development of *network management systems* (NMSs).

However, there is another facet of network control that is involved with the establishment, maintenance, and release of connections. In this context, "control" means that the user requests a certain service from the network and that the network takes actions to offer that service. An example would be the steps involved in the setup of a telephone call. The caller notifies the network that he or she would like to call another party. On receiving such requests, the network allocates appropriate resources and sets up the requested connections. Once these resources are set up, callers are able to communicate with the parties they have contacted. In communications networks, this connection-related network control is usually called *signaling.*

2.4.1 Network Management Functions

Network management is composed of a number of components. Though different terminology may be used in different networks, there are five main components to network management in general, namely fault management, accounting management, configuration and name management, performance management, and security management. These functions, as a whole, enable the network operator to configure the network, understand the state of the network, get feedback on problems in the network, and take corrective actions as needed. In the following we explain the functions in more detail.[3]

Fault management is the function to detect, isolate, and correct any abnormal operation of network elements. Fault management facilities determine exactly where the fault is located, and then isolate the rest of the network from the failure so that the network can continue to function without interference. They can reconfigure or modify the network in such a way that the impact of operation is minimized without the failed components and repair or replace the failed components to restore the network to its initial state, with human involvement as needed.

3. This follows the definition of the OSI management functional areas.

Accounting management is the function that establishes charges for using the network resources and identifies the costs of using those managed objects. Besides such basic functionality, it also offers a number of advantages such as pointing to end users who may be abusing access privileges and burdening the network at the expense of other end users. The information on the end users acquired through accounting management helps network managers to plan better for network growth.

Configuration management is the function that controls the network resources for the purpose of providing for the continuous operation of services. It also involves identifying, collecting data from, and providing data to the network components. Configuration management facilities are used to initialize networks and gracefully shut down parts of the network. Also, they are used to maintain, add, and update the relationships among the network components and the status of the components themselves.

Performance management is the function that evaluates the behavior of the network components and the effectiveness of communication activities. In support of this, a monitoring function tracks the activities on the network, and a controlling function enables making adjustments to improve network performance. Performance management and fault management are the key elements of the so-called *operation and management* (OAM) function.

Security management addresses the security aspects essential to operating network management correctly and protecting the network components. It is concerned with managing information protection and access-control facilities and with generating, distributing, and storing the encryption keys. In addition, it is involved with maintaining and distributing passwords and other authorization or access-control information.

2.4.2 Signaling

As mentioned above, signaling refers to the functions needed to establish, maintain, and release connections, which may be extended to other advanced and intelligent functions related to connection setup. From the user's perspective, signaling can be viewed as a way to exert control over the network to obtain the desired service. A typical example of this can be found in the telephone network where a signaling message is automatically generated at the moment the user picks up the handset and the signaling message triggers all the follow-up network control actions within the telephone network needed to establish the requested connection.

In contrast, in the case of packet networks, such as the "classical" best-effort Internet, signaling was not used or needed. The network was not designed to receive users' inputs as to the level of services the users need or to set up paths inside the network for the delivery of user services. The user never got guarantees as to the delivery of the packets, delay of the packets, or the loss rate in packet delivery. The network simply exerted its best efforts to deliver the packets to the desired destination without fail, yet without paying any attention to timely delivery. This resulted in simple networks, but the applications will become complicated if this factor is taken into account.

In recent years, as the methods for introducing QoS into the Internet have been explored, it has been recognized that a method for signaling user requirements would be helpful in providing QoS. This has been the essential driving force for the development of the *resource reservation protocol* (RSVP). Signaling becomes important in a network when the network begins to introduce the concept of QoS.

If we take a slightly more generic viewpoint, we realize that signaling is needed whenever the network needs a method for allowing dynamic resource reservation or allocation by network users. In general, resource allocation is not done for the benefit of any single user but carried out at the judgment of the network operator to ensure smooth network operation under changing network conditions. For example, when a certain path gets congested with too much traffic, another path may be set up to offload the excess traffic. This may be done by using new signaling protocols such as the newly extended *label distribution protocol* (LDP) defined for *generalized multiprotocol label switching* (G-MPLS) (refer to Chapters 3 and 8). This is an example of *traffic engineering* (TE).

An important item to consider when discussing signaling is how the signaling messages are to be conveyed between the user and the network node, and between the network nodes themselves. The essential question is whether, in relationship to the data channel, the signaling messages will be sent *in-band* or *out-of-band*. In-band methods result in the signaling messages using the same traffic channel as the user data, while out-of-band methods result in the signaling messages using a separate traffic channel from the user data.

2.4.3 Examples of Network Control

We first consider network management and the signaling aspects of circuit-switched networks, and then we consider the network control of packet-switched networks.

2.4.3.1 Network Management

In the case of telephone networks, *telecommunication management network* (TMN) supports the network management functions. One of the most important objectives of network management through the TMN is the survivability of network services, or uninterrupted service provision. For such service survivability, a real-time management capability is necessary to be able to react to network environment changes actively. For example, it is necessary to be able to identify the degraded network elements and set up detouring paths. Also, it is necessary to be capable of evaluating the performance quality of a network element by measuring the failure occurrences and then taking appropriate reactive actions.

In the case of the SDH/SONET network, *SDH/SONET management network* (SMN) supports network management as an extended part of the TMN. The SMN is constructed among synchronous network elements over the *data communication channels* (DCCs) in the section overhead of the SDH/SONET frame. The SMN is furnished with a management application function, a network element function, and a message communication function that contribute to the generation, termination, collection, and transfer of the TMN messages.

In the existing plesiochronous transmission environment, the network management system has to be newly defined whenever a new system is developed and the OAM information gathered from different transmission equipment has to be transformed to the data formats adequate for integrated transmission network management. The SMN improves such inefficiency and equipment dependency of network management by employing an integrated management information system in the synchronous transmission environment.

2.4.3.2 Signaling

If we consider the signaling aspect, digital *channel-associated signaling* (CAS) had been dominant until *common channel signaling* (CCS) was introduced in the late 1970s. CAS used the channels formed by the "robbed" bits inside the DS-1 frame or the out-of-band bits in the DS-1E frame. In this case the signaling information is carried over the signaling channel that travels with the user information. Thus, it renders relatively simple surveillance and addressing functions for call setup, but it also has the drawbacks of slow call setup and insecure signaling information transport. Moreover, the CAS does not adequately support the flexible operation of communication networks or intelligent provision of communication services. In contrast, CCS deploys separate signaling channels independently of the traffic channels. As a result, it can help increase the call setup speed and the signaling channel security. Furthermore, it helps to improve network flexibility and intelligence, thereby enabling the development of intelligent networks.

2.4.3.3 Network Control in IP Networks

In TCP/IP networks, network control is quite different. The original Internet design did not consider network control as one of its prime factors. In fact, the whole design principle was for a noncontrolled or completely distributed network that could survive any single point of failure. From the beginning, all controls of communications were based on the end terminals, with no control being exercised in the core of the networks or by any of the edge components. As such, the TCP/IP network had no signaling component in the network, relying completely on the distributed, connectionless routing architecture and the intelligence in the end host applications.

However, control mechanisms have been gradually developed in recent years. First, the *simple network management protocol* (SNMP) was developed as a method of checking and controlling IP network nodes. It consists of a simple set of commands for retrieving, setting variables, and setting up traps. Second, RSVP was developed as a signaling protocol for IP networks. As mentioned earlier, RSVP is basically a new type of signaling protocol specifically designed for IP networks. The basic design philosophy differs from the signaling in the telephone networks in many aspects, but the end results are similar in that both methods result in state setup in the network. This consists of basically reserving resources along the path to provide the service quality users want. (Refer to Chapter 3 for more details.)

2.5 Traffic Management

Traffic management refers to a set of policies and mechanisms that enable a network to meet a diverse range of service requests in efficient ways. It also tries to offer tools to achieve the basic objectives of satisfying diverse user requirements and maintaining efficiency in satisfying them. These two aspects of traffic management—satisfying users and maintaining efficiency—often act in tension due to their different goals.

Traffic management relies on a number of components, which can be divided into three groups. The first group is related to defining the traffic of the user, the transport level offered by the network, and the traffic contract between the user and the network. The second group consists of specific mechanisms inside the network to ensure that the user receives the level of services agreed upon. The third group contains the mechanisms to help the network to recover from the congestion in the network.

ATM is a good model of a complete traffic management framework. Basically, ATM traffic management begins with a traffic contract that defines the traffic characteristics between the user and the network. The user agrees to transmit and receive traffic only in accordance with the traffic contract, while the network uses *usage parameter control* (UPC) and/or *network parameter control* (NPC) to check that the user traffic meets the contract. On the other hand, the network takes various actions, such as priority queuing, prior reservation, and allocation of network buffers and bandwidth, to ensure that the user receives the agreed-upon services.

In the case of many conventional networks, traffic contracts are either nonexistent or done implicitly. For example, in the case of the best-effort service offered by the Internet there is no explicit traffic contract that binds the user to sending packets at less than a certain rate or size. On the other side, the network does not guarantee anything on the packet delivery. The traffic contract in this case is an implicit one that each side will offer their best efforts whenever possible. As another example, in the *plain old telephone service* (POTS) service of the telephone network, the network handles only one type of service, so the traffic contract does not have to be negotiated for each call.

2.5.1 Traffic Models and Contracts

As mentioned above, traffic contracts define the type of services that a network will provide to a user upon the request of the user. It can be implicitly defined so that there is no need for any definition or exchange between the user and the network. However, even in such cases the implicit contract must follow.

To define a traffic contract, we first need a method for accurately describing the traffic. This usually takes the form of a traffic model and its constituent parameters. Accordingly, a traffic contract basically defines the characteristics of the traffic to send and receive in terms of the traffic parameters. In support of the traffic contract, we need a method for monitoring the traffic to ensure that the actual traffic conforms to the agreed upon traffic contract.

The traffic model in the telephone network was very simple as only a single 64-Kbps line was needed for the time of the call. A method for monitoring was not explicitly needed as it was a dedicated-line service. For traffic control on the network level, traditional telephone networks used traffic models to model the call attempts that come into a switch, thereby taking a traffic-engineering or capacity-planning approach (see Section 2.5.2).

The traffic model is much more complicated in the case of a packet network. The traffic model is defined based on traffic shaping methods such as the token bucket or leaky bucket algorithms (refer to Section 4.5.2). These algorithms determine how a packet stream will be controlled before being sent by the user into the network. They include various parameters such as peak rate, average rate, and packet size. These algorithms also have the advantage of simultaneously serving as a traffic-monitoring algorithm to check whether the user traffic is conforming to the traffic contract.

2.5.2 Traffic Control Mechanisms

Traffic control mechanisms are used primarily to ensure that user traffic conforms to the level agreed upon at connection setup, thereby ensuring that traffic congestion does not occur.

In circuit-switched networks traffic control is based on signaling, admission control, and traffic-capacity planning. In packet-switched networks, traffic control relies on priority buffer control, flow control, rate control, traffic shaping, admission control, scheduling, and occasionally signaling. Note that depending on the specific protocol suite or family, the same function may have different names. For example, the functions of UPC/NPC in ATM networks are comparable to the token bucket/leaky bucket functions in IP networks.

Traffic control mechanisms start operation at the moment a connection is first requested and contention for network resources begins. As most of the specific traffic control mechanisms will be covered in subsequent chapters, this section simply offers an overview of where and how they are used.

Admission control is carried out at the beginning of connection setup to figure out whether there are enough resources in the network to enable the connection setup. Signaling mechanisms are used in this stage to first find out if enough resources are available and, if so, to reserve resources along the path. Telephone networks basically use only this traffic control mechanism. Signaling can also be used to indicate flow-control or congestion-control information.

Flow control or *rate control* is used to control the flow of packets being sent into a network by the source. The initiator of the flow/rate control may be the destination host or other network nodes. The former would be the case where the end node does not have enough buffers to buffer the packets it is receiving. As such, it is more of an end node traffic control or congestion-control problem. Network nodes can also cause the flow/rate control to slow down the source generator. This would be done if the network node detects that the user is violating the traffic contract or if the node is starting to get congested.

From the individual packet level, many of these actions are related to the *buffering* (or *queuing*) of packets. If the throughput of the nodes were fast enough there would be no need for buffering, as the packets would be transmitted before being queued. Otherwise, there would be a transient burst of traffic, necessitating that the packets be buffered. There are many different ways of buffering the packets. First, it is necessary to decide which packets can be buffered depending on various criteria such as the current buffer backlog, the packet's priority levels, or the number of packets that the call has already buffered. There are many other possibilities, as buffering is one of the most actively researched areas of traffic control.

Whereas buffering methods decide which packets will go into a buffer, scheduling decides which packets will be removed from the buffer and transmitted. As such, it is critical to determine the delay that a packet will see, as it consequently affects the final QoS that an end user will experience. Again, this area has a very wide solution space and has recently been an actively researched area.

2.5.3 Congestion Control

Congestion occurs when too many users compete for a single resource in the network, which is caused by too many calls or packets coming into the network simultaneously. Even if traffic management was readily conducted in the connection setup stage to prevent congestion, congestion can still occur due to the transient characteristics of user traffic. Consequences of traffic congestion are the loss of data, delay increases, and failures in meeting traffic requirements.

As mentioned previously, congestion control essentially tries to decongest a readily congested network node. To carry out this task, it is necessary to first clear up the original congested node and at the same time notify the source node to stop sending packets. For the former step, various congestion control actions are taken to alleviate the congestion by dropping or marking some low-priority packets. For the latter step, some sort of signaling mechanisms are activated to deliver the congestion notification messages to the source node.

There are two different ways of signaling to the source node: one is an explicit method, and the other is an implicit method. In *explicit* congestion signaling, congestion notification to the source is done in an explicit manner, such as the transmission of a notification packet back to the source node. Similarly, it is possible to mark the packets passing through the congested node so that the source node can recognize the congestion status upon receiving the packets. In *implicit* congestion signaling, congestion notification is done in an implicit manner such as by dropping packets. When the source or the destination node determines that packets were dropped midway it interprets this as indicating that congestion has occurred. In fact, this is the method used in TCP/IP networks.

2.5.4 Capacity Planning and Traffic Engineering

The mechanisms of traffic management that we have considered above try to ensure that network resources are not overbooked or abused. However, all those mechanisms make sense only when the network resources are adequately sized with respect to the number of users. If the number of users is exceedingly large, then none of the above traffic management mechanisms would be able to solve the problem. Many of the users would inevitably get either unsatisfactory service or no service. This is frequently the case in the networks with limited telecommunications facilities. The only solution in this case is to increase the size of the network, adding more resources to the network.

One of the problems faced in such a case is how to plan or size the network. This is the area covered by *capacity planning;* the network operator first analyzes the network usage patterns from which a traffic matrix is derived. This traffic model also depends on the traffic source models and the traffic parameters discussed above. The operator may also carry out real-world traffic data measurements to use in the matrix. Based on this traffic matrix the operator then considers various network topologies while experimenting with various values or functions for the network links. When a satisfactory topology with an appropriate link bandwidth and functionality is determined, the network operator can finally implement the scheme.

TE can be viewed as another facet of capacity planning. It tries to use the current available network resources in the most efficient manner. It carries out many of the steps of capacity planning again and, based on the result, changes the network topology or link values. To use TE effectively the network must support certain functions such as the ability to set up, reroute, and release the transport paths. Traditionally these functions relied on manual operations through a network management station, but techniques for automating this function are being studied in the newly evolving networks (refer to Chapter 8).

A good example of capacity planning and traffic engineering is the design of switches in the telephone networks. Much of the theory and practical uses of capacity planning were developed in the telephone networks, whose primary example is the Erlang formulas. The formulas were developed by Erlang in answer to the question of how to size switch capacity to ensure that the call blocking probability drops below a certain level.

In general, capacity planning is a problem that is difficult to solve completely or accurately, mainly because of the many variables and possible solutions it is necessary to consider. Consequently, many operators use simple heuristics or hit-and-miss methods instead. For example, if the capacity of a certain link is insufficient, fiber links are added to increase the link capacity or rerouting mechanisms are activated to detour some traffic to other links.

2.6 QoS

The definition of QoS is a complex problem as the level of the desired quality will differ from user to user and from service to service. In analog telephone networks, noise was the biggest concern; this problem, however, was completely resolved in the digital telephone networks. Instead, the bit error problem is the major issue in the digital networks, because digital network impairments such as loss of synchronization, inaccuracy of clocks, bit slips, jitter, and wander all led to bit errors. In the case of packet networks, packet error and packet loss problems are the important problems as all different forms of bit errors eventually lead to packet errors and packet losses. If multimedia services are to be provided over the packet networks, packet delay and delay variation, caused by packetized processing of continuous-time signals, emerge as the foremost problems to solve.

As such, there are several performance factors that are perceived as the key elements of QoS, including those exemplified above—bit error rate (BER), packet loss rate, delay of services, and delay variation. This raises a number of issues in defining QoS items, determining the performance factors that dictate the QoS, and measuring what QoS is actually offered. If the QoS items can be clearly defined, it will be possible to reflect them in the traffic contract between the network and the user. The traffic monitoring and policing functions must also be expanded to encompass those items.

The next question is how to actually implement QoS—that is, how to operate the network to meet the required QoS conditions. This is a question that as yet does not have clear-cut answers. Nevertheless, the QoS implementation issue is closely related to the performance parameters that govern the QoS items. The implementation methods themselves should differ for each network technology, as will be discussed in later chapters covering different network technologies.

If viewed from a high-level, abstract point, there are two options for solving the QoS problem. We may either overprovision the network or use some sort of intelligent traffic management functionality in the network. The first, overprovisioning solution allocates more resources to each connection than is minimally needed, while installing more capacity in the network than is minimally needed. The second intelligent network management solution, on the other hand, devises intelligent management mechanisms in buffer management, scheduling, and other functions to enable each user to obtain the desired QoS level.

Ultimately the decision depends on whether to invest in increasing the raw capacity of the network or to invest in building intelligent traffic management schemes. This is in many ways the essential question to answer when investigating the QoS issue in the IP network or seeking its integrated solution in the IP, ATM, SDH/SONET, and WDM/optics networks. For example, many IP network researchers suggest that QoS may be achieved by increasing the bandwidth while keeping the intelligence inside the network simple. On the other hand, others have sought solutions that devise intelligent schemes to manage the multiple-component integrated network in QoS-maximizing ways.

Selected Bibliography

Ahmadi, H., and W. E. Denzel, "A Survey of Modern High-Performance Switching Techniques," *IEEE Journal on Selected Areas in Communications*, Vol. 7, No. 9, September 1989, pp. 1091–1103.

Aidarous, S., and T. Plevyak, *Telecommunications Network Management*, New York: IEEE Press, 1998.

Ash, G., *Dynamic Routing in Telecommunications Networks*, New York: McGraw-Hill, 1998.

Bellamy, J., *Digital Telephony*, Third edition, New York: John Wiley and Sons, 2000.

Bellcore, *Telecommunications Transmission Engineering*, Vols. 1–3, Bellcore Technical Publications, 1990.

Bux, W., et al., "Technologies and Building Blocks for Fast Packet Forwarding," *IEEE Communications Magazine*, Vol. 39, No. 1, January 2001, pp. 70–77.

Caric, A., and K. Toivo, "New Generation Network Architecture and Software Design," *IEEE Communications Magazine*, Vol. 38, No. 2, February 2000, pp. 108–114.

Chen, T. M., "Evolution to the Programmable Internet," *IEEE Communications Magazine*, Vol. 38, No. 3, March 2000, pp. 124–138.

Clark, D. D., "The Design Philosophy of the DARPA Internet Protocols," *Proceedings of ACM SIGCOMM '88*, August 1988, pp. 106–114.

Clark, D. D., "Modularity and Efficiency in Protocol Implementation," *IETF RFC817*, July 1982.

Clark, D. D., S. Shenker, and L. Zhang, "Supporting Real-Time Applications in an Integrated Services Packet Network: Architecture and Mechanism," *Proceedings of ACM SIGCOMM '92*, August 1992.

Clos, C., "A Study of Nonblocking Switching Networks," *Bell System Technical Journal*, Vol. 32, No. 3, March 1953, pp. 406–424.

Day, J., and H. Zimmerman, "The OSI Reference Model," *Proceedings of the IEEE*, Vol. 71, No. 12, December 1983.

Demers, A., S. Keshav, and S. Shenker, "Design and Analysis of a Fair Queuing Algorithm," *Proceedings of ACM SIGCOMM '89*, September 1989.

Doverspike, R. D., S. Philips, and J. R. Westbrook, "Future Transport Network Architectures," *IEEE Communications Magazine*, Vol. 37, No. 8, August, 1999 pp. 96–101.

Fraser, A. G., "Early Experiments with Asynchronous Time Division Networks," *IEEE Networks Magazine*, Vol. 7, No. 1, January 1993, pp. 12–26.

Gavrilovich, C. D., Jr., "Broadband Communication on the Highways of Tomorrow," *IEEE Communication Magazine*, Vol. 39, No. 4, April 2001, pp. 146–154.

Girard, A., *Routing and Dimensioning in Circuit-Switched Networks*, Reading, MA: Addison-Wesley, 1990.

Hasegawa, T., Y. Tezuka, and Y. Kasahra, "Digital Data Dynamic Transmission Systems," *IEEE Transactions on Communications Technology*, 1964.

Hassan, M., A. Nayandoro, and M. Atiquzzaman, "Internet Telephony: Services, Technical Challenges, and Products," *IEEE Communications Magazine*, Vol. 38, No. 4, April 2000, pp. 96–103.

Hegering, H., S. Abeck, and B. Neumair, *Integrated Management of Networked Systems*, San Francisco, CA: Morgan Kaufmann, 1998.

Huitema, C., *Routing in the Internet*, Englewood Cliffs, NJ: Prentice Hall, 1995.

Iida, K., et al., "Performance Evaluation of the Architecture for End-to-End Quality-of-Service Provisioning," *IEEE Communications Magazine*, Vol. 38, No. 4, April 2000, pp. 76–81.

ITU Recommendation E.164 / I.331, *Numbering Plan for the ISDN Era.*

Keshav, S., *An Engineering Approach to Computer Networking*, Reading, MA: Addison-Wesley, 1997.

Konrad, J., "Visual Communications of Tomorrow: Natural, Efficient, and Flexible," *IEEE Communications Magazine*, Vol. 39, No. 1, November 2000, pp. 126–133.

Kumar, V. P., T. V. Lakshman, and D. Stiliadis, "Beyond Best Effort: Router Architectures for the Differentiated Services of Tomorrow's Internet," *IEEE Communications Magazine*, Vol. 36, No. 5, May 1998, pp. 152–64.

Lee, B. G., M. Kang, and J. Lee, *Broadband Telecommunications Technology*, Norwood, MA: Artech House, 1996.

Lynch, D. C., and M. T. Rose, *Internet System Handbook*, Reading, MA: Addison-Wesley, 1993.

Mathy, L., C. Edwards, and D. Hutchison, "The Internet: A Global Telecommunications Solution?" *IEEE Network*, Vol. 14, No. 4, July/August 2000, pp. 46–57.

Maxemchuk, N., and M. E. Zarki, "Routing and Flow Control in High-Speed Wide Area Networks," *Proceedings of the IEEE*, Vol. 78, No. 1, January 1990.

Mockpetris, P., and K. J. Dunlap, "Development of the Domain Name System," *Proceedings of ACM SIGCOMM '88*, August 1988.

Modarressi, A., and R. A. Skoog, "Signaling System No. 7: A Tutorial," *IEEE Communications Magazine*, Vol. 28, No. 7, July 1990, pp. 19–34.

Moral, A. R., P. Bonenfant, and M. Krishnaswamy, "The Optical Internet: Architectures and Protocols for the Global Infrastructure of Tomorrow," *IEEE Communications Magazine*, Vol. 39, No. 7, July 2001, pp. 152–159.

Moridera, A., K. Murano, and Y. Mochida, "The Network Paradigm of the 21st Century and Its Key Technologies," *IEEE Communications Magazine*, Vol. 38, No. 1, November 2000, pp. 94–98.

Nilsson, T., "Towards Third Generation Wireless Communication," *Ericsson Rev.*, 1998.

Perlman, R., *Interconnections: Bridges and Routers*, Second edition, Reading, MA: Addison-Wesley, 1999.

Psounis, K., "Active Networks: Applications, Security, Safety, and Architectures," *IEEE Communications Surveys*, Vol. 2, No. 1, 1st quarter, 1999.

Roberts, J. W., "Traffic Theory and the Internet," *IEEE Communications Magazine*, Vol. 39, No. 1, January 2001, pp. 94–99.

Rey, R. F., *Engineering and Operations in the Bell System*, Second Edition, Murray Hill, NJ: AT&T Bell Laboratories, 1983.

Saltzer, J. H., D. P. Reed, and D. D. Clark, "End-to-End Arguments in System Design," *ACM Transactions on Computer Systems*, Vol. 2, No. 4, November 1984, pp. 277–288.

Toga, J., and J. Ott, "ITU-T Standardization Activities for Interactive Multimedia Communications on Packet-Based Networks: H.323 and Related Recommendations," *Computer Networks*, Vol. 31, No. 3, 1999, pp. 205–23.

Wetherall, D., et al., "Introducing New Internet Services: Why and How," *IEEE Network*, Vol. 12, No. 3 May/June 1998, pp. 12–19.

Zhang, H., "Service Disciplines for Guaranteed Performance Service in Packet Switching Networks," *Proceedings of the IEEE*, October 1995.

Zimmerman, H., "OSI Reference—The ISO Model of Architecture for Open Systems Interconnection," *IEEE Transactions on Communications*, Vol. 28, No. 4, April 1980, pp. 425–432.

3

TCP/IP

TCP/IP is the basic protocol suite on which the Internet and the WWW are built. Originally started as a small government funded research network, TCP/IP has now grown to be the foundation of the information age.

TCP/IP has its origin in the U.S. DARPA's basic research on computer communications and networking. As a part of the research, computer networks associated with several universities and research institutes were linked with DARPA's computer network, and this internetwork was called ARPANet.

The original aim of the ARPANet was to study the possibility of communications between computers. By the mid 1970s, this experiment had evolved into a system for interconnecting various networks of computers and into an architecture with different manufacturers' proprietary networks, data transmission protocols, hardware, and operating system software. In response to this demand, the TCP/IP suite was developed, and later officially adopted by the U.S. government as the "preferred" networking protocol.

The TCP/IP suite includes both network-oriented protocols and application support protocols. TCP/IP is broadly used in the existing Internet, and a considerable portion of the TCP/IP suite was used as the basis for OSI standards.[1] Furthermore, since TCP/IP does not require a license fee, all

1. By the time the ISO began to develop an open data-networking standard, the OSI suite of protocols, the TCP/IP protocol suites, were already available as open standards and were widely deployed. However, at that time, TCP/IP needed enhancements and were dominated by U.S. manufacturers. Nevertheless, the TP4 protocol was basically based on TCP, and the CLNP on IP.

the associated protocol specifications are in the public domain. While both TCP/IP and the OSI suite were to be used to create open system networking environments, the TCP/IP suite is much more widely deployed than the OSI protocol suite today.

This chapter examines the TCP/IP suite. It first discusses the protocols under the networking principles detailed in Chapter 2, considering the key concepts, layering model, and architecture of the TCP/IP suite. Then the discussion expands to cover routing concepts, switching and multiplexing, network control functions, and QoS-related functions.

3.1 Concepts, Layering, and Architecture

Before going into detailed discussions, we examine the concepts and protocol structures that form the basis of the Internet[2] architecture. We discuss the features and design philosophy of the TCP/IP protocols and then the overall protocol structure and the network/transport layer protocols.

3.1.1 Features and Design Principles

The Internet has grown exponentially due to several attractive features of the TCP/IP suite. The distinctive features may be summarized by the following three points [1]: First, members of the TCP/IP suite are open protocol standards; they are not under the ownership of any one company or group; they are freely available to anyone interested; and anyone can contribute to the standardization process. Second, members of the TCP/IP suite are independent of actual physical networks and independent of computer hardware and software. They can run over Ethernet, token ring, X.25, modems, UNIX workstations, PCs, and others. Third, TCP/IP uses a common global address scheme that enables worldwide connectivity. Since the address is globally unique, anyone can connect to anyone else on the Internet worldwide.

The first and second features received full support from both academic and industrial sectors. This made TCP/IP an ideal basis for interconnecting heterogeneous computer systems. It also gave birth to many popular applications of TCP/IP protocols, such as *ftp, telnet, e-mail, usenet,* and the WWW. The third feature is important in that it allows for the growth of the Internet's connectivity in line with the growth of network size. This has increased

2. Note that an *internet* differs from the *Internet.* An internet is any group of networks interconnected by a common protocol, whereas the Internet (with a capital I) refers to the worldwide interconnection of networks based on the TCP/IP suite.

the usefulness of the Internet for all users connected, resulting in a positive feedback effect.

The TCP/IP suite was designed based on two basic principles—namely, the "IP over everything" principle and the "end-to-end" principle. The former enables a single network layer protocol IP to connect any communication entity. It bases the TCP/IP suite on an internetworking protocol layer that would overlay all networks to be interconnected.[3] Owing to this principle it has become extremely easy to adopt new network technologies. Moreover, a unique global addressing method has been defined so that all hosts can be easily reached.

The latter principle states that the network should offer only the basic transport functions, and any final decisions regarding extra communication functions should always be made by the end users themselves. That is, the networking functions should be delegated as much as possible to the outside of the network, not inside. For example, for error checking and control functions, which can be executed either hop-by-hop or end-to-end, it is natural to use only end-to-end error control in TCP.[4]

3.1.2 TCP/IP Structure

The TCP/IP suite does not have an exact layered model as the OSI seven layer reference model does, but it can be divided into four layers, namely the *link layer, network layer, transport layer,* and *applications layer,* as shown in Figure 3.1. While a direct comparison of the TCP/IP suite with the OSI seven layer model is not easy, the internal layers of the two are roughly matched in the following way: The TCP/IP link and network layers correspond to the physical, data link, and network layers of the OSI reference

3. There are basically two methods for interconnecting networks: *translation* and *overlay.* Interconnecting by translation means that gateways are used to map the data and control information of similar services from one network to another. For example, mail gateways between TCP/IP networks and OSI networks translate TCP/IP mail into OSI mail, and vice versa. Obviously, translation has limitations as are observable in human language translations. Consequently, the Internet architects have chosen the overlay method, which uses a single unifying network, on which all higher layer protocols are based, to interconnect all networks.
4. Note that this end-to-end principle does not result in the same answer to all network problems as the application would depend on the network technologies used. For example, error correction may be performed inside the network if it would result in a critical improvement in performance. So hop-by-hop error correction makes sense over extremely lossy wireless links. However, end-to-end error control is desirable for low BER fiber links.

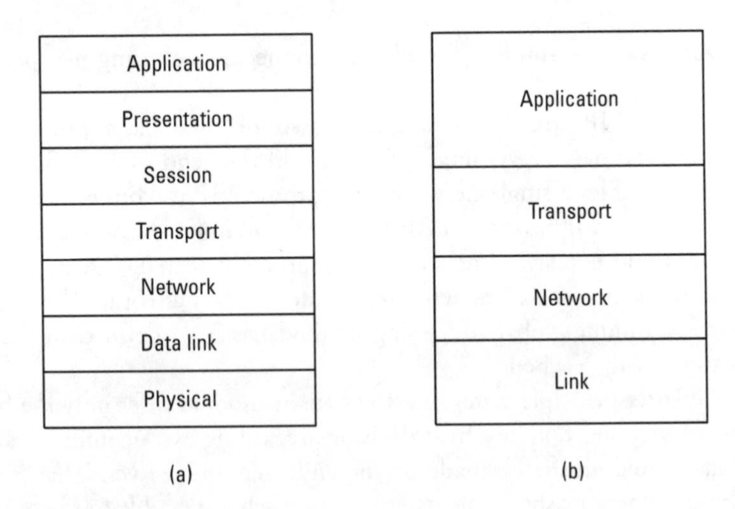

Figure 3.1 Comparison of (a) OSI, and (b) TCP/IP layers.

model; the TCP/IP transport layer corresponds to the OSI transport and session layers; and the TCP/IP applications layer corresponds to the OSI session, presentation, and application layers. Figure 3.2 shows the main protocols related to these four layers.

3.1.2.1 Link Layer

The link layer is the lowest layer of the TCP/IP suite. It provides the means for the transmission and reception of data over a physically connected transmission media. As such, encapsulating IP datagrams into the actual transmission frames of the link layer network and mapping the global IP address to the link layer network address are two main functions of the link layer.

Due to its strong dependence on the actual physical network, the link layer protocols must be developed in accordance with the corresponding physical networks. For example, the standards for encapsulating IP datagrams in Ethernet frames are defined in RFC894 and RFC1042, while the *address resolution protocol* (ARP) and *reverse ARP* (RARP) protocols for converting between IP addresses and Ethernet addresses are defined in RFC826 and RFC903, respectively.

3.1.2.2 Network Layer

Whereas the link layer is used to transfer data between IP nodes, the network layer transfers data between IP end nodes by using those link layer functions. There are three protocols defined in the TCP/IP network layer: IP, the *Internet control message protocol* (ICMP), and the *Internet group management*

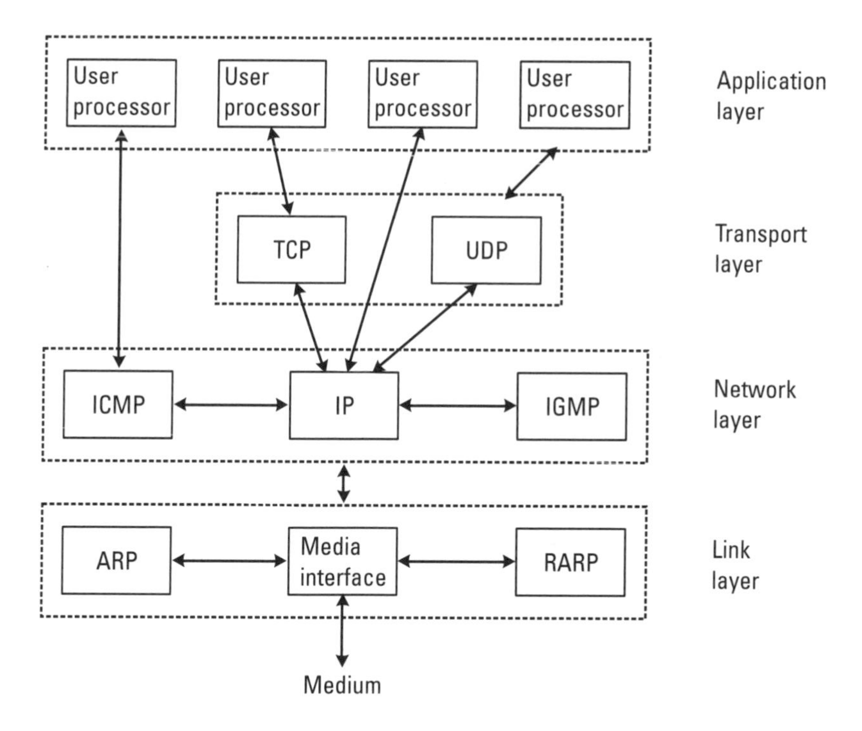

Figure 3.2 Layers and protocols related to the TCP/IP suite.

protocol (IGMP). Among the three protocols, IP is the most important protocol in the TCP/IP network layer. IP routes (i.e., transports) the IP datagrams over various networks from the transport layer of the sender system to the transport layer of the receiver system.[5] ICMP and IGMP protocols are used to provide other functions in support of this basic transport function.

3.1.2.3 Transport Layer

The transport layer services are the actual end-to-end data transfer services seen by the end user. Depending on the type of user services, two different types of protocols—TCP and the UDP—are used. UDP is used for a simple datagram transfer service, while TCP is used for a reliable connection-oriented service. Because the service offered by UDP is not complicated, it is realized by adding a minimal amount of control information to the basic IP. In contrast, the

5. Note that the end user of the datagram is an application process above the transport layer, not the transport layer itself.

service offered by TCP is more complicated than that of UDP, so more functions such as error detection and control must be provided.

3.1.2.4 Applications Layer

The application layer, which is located above the transport layer, is composed of application programs. Applications can be divided into two types—user applications and infrastructure applications, which are used by other applications to carry out their functions. The *domain name service* (DNS) and the SNMP service are examples of infrastructure applications, whereas ftp, telnet, rlogin, e-mail, and the WWW are the examples of user applications.

3.1.3 Network Layer Protocols

Among the three protocols defined in the network layer (i.e., IP, ICMP, and IGMP), the IP protocol is of primary importance, as it is the core protocol that all TCP/IP nodes must implement to ensure end-to-end connectivity.

3.1.3.1 IP Network Layer Characteristics

The IP network layer is connectionless and unreliable. The fact that IP is connectionless means that there is no need to exchange any signaling or control information before the source sends a datagram to the destination. IP end user just sends the packets with the destination's address attached. If a connection-oriented mechanism is needed, a higher layer protocol such as TCP must be added. The fact that IP is unreliable means that there is no error correction, retransmission of lost datagrams, or guarantee of in-order delivery. That is, the IP layer tries its best to deliver datagrams correctly, but it does not guarantee anything for it (i.e., best-effort service).[6]

3.1.3.2 IP Datagram Format

The IP datagram consists of a variable-length header and a data field as shown in Figure 3.3. The fields of the header are aligned to 32-bit words for ease of handling in a computer. The version field contains the version number of the protocol.[7] The *header length* (HL) field contains the variable length of the header. The *type of service* (TOS) field was originally defined for allocating network resources according to the needs of the user but in fact is almost never

---

6. These characteristics reflect the "end-to-end" design principle. That is, in both cases the end user may or may not need the functions related to these characteristics, so only the minimal requirements are specified in the IP layer, with more specific functions left to higher layer protocols [2].
7. The current version is IPv4. The next-generation IP version is termed IPv6.

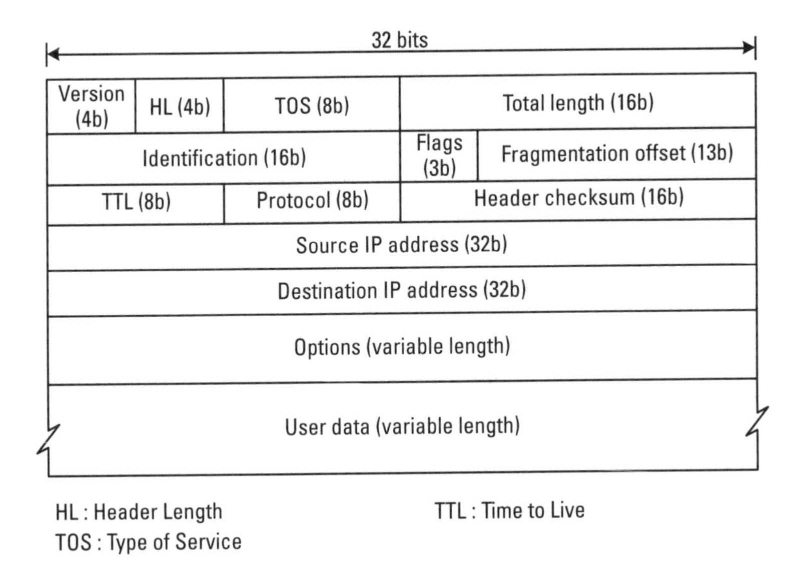

Figure 3.3 IP datagram format.

used. The *total length* field indicates the whole length of the IP datagram in bytes. The *identification, flags,* and *fragmentation offset* fields are used in the fragmentation of IP datagrams (see Section 3.1.3.4). The *time to live* (TTL) field is used to ensure that no packet loops forever in the network. Basically it is a hop count and the packet is dropped when it becomes zero. The *protocol* field is used to differentiate end-user transport layers. The *header checksum* is a 16-bit one's (1's) complement sum of the IP header.[8] The *source IP address* and the *destination IP address* are 32-bit IP addresses. The *options* field is of variable length and can contain various options.

3.1.3.3 Routing of IP Datagrams

A TCP/IP network is composed of two types of nodes: *hosts* and *routers.*[9] A router is a node with two or more network interfaces that are capable of

8. A "16-bit one's complement sum" means that the data is divided into 16-bit unit slices, then the one's complement is calculated for each 16-bit slice, and finally all the one's complement numbers are summed together in a modulo-2 operation.
9. Originally, TCP/IP networks were defined to be composed of hosts and gateways, with gateways being the routers defined above. Today, the term "gateway" is usually reserved for the devices that translate data between similar services in different protocol families, such as a mail gateway between TCP/IP and OSI networks.

forwarding a datagram from one network to another, and a host is a node that is not a router.

A simple example of a TCP/IP internetwork is shown in Figure 3.4. In this example there are three networks, namely, the InmacNet, the SnuNet, and the EngNet. The InmacNet is a 10-Mbps Ethernet network, the SnuNet

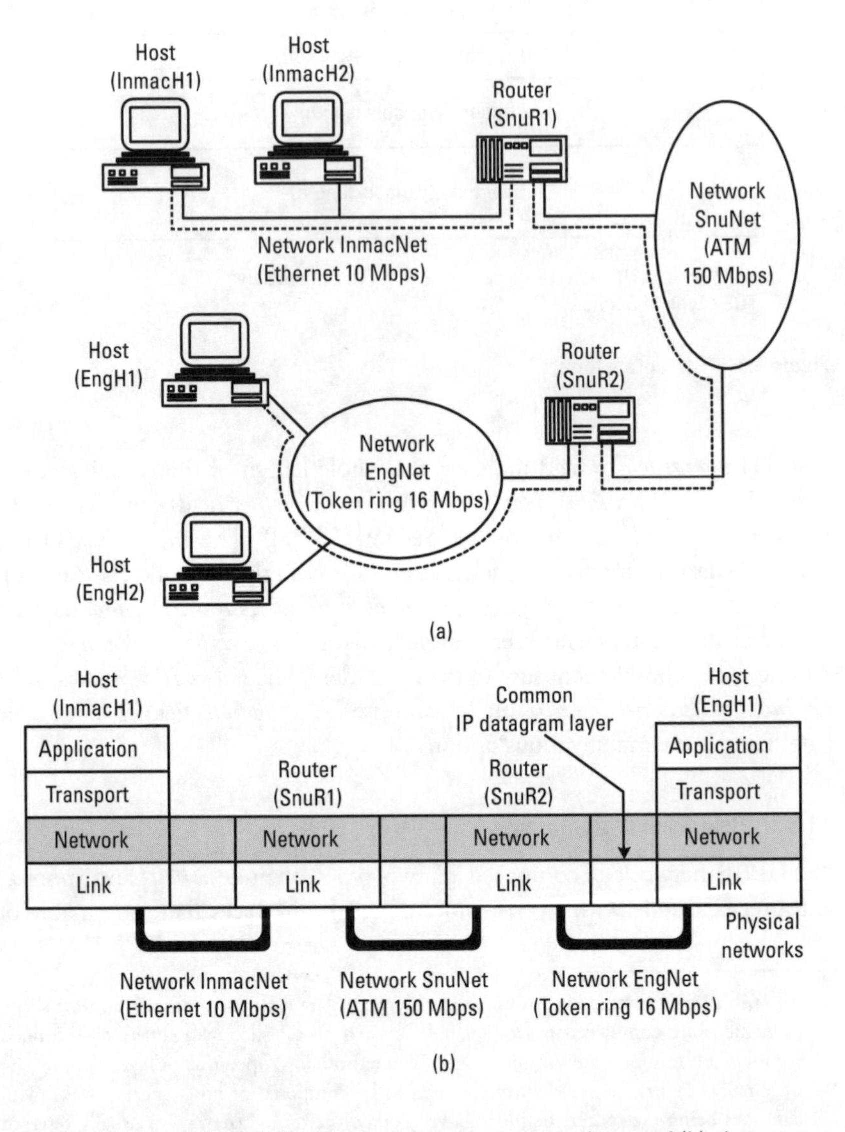

Figure 3.4 Example of a TCP/IP network: (a) physical connections, and (b) abstract protocol connections.

is a 150-Mbps ATM backbone network, and the EngNet is a 16-Mbps token ring network. The hosts on each network are named InmacH1, SnuH1, and EngH1. The routers connecting InmacNet and EngNet to the backbone SnuNet are called SnuR1 and SnuR2, respectively. In this example, if the host InmacH1 wants to send a datagram to the host EngH1 on EngNet, it must send the IP datagram to the router SnuR1 first. SnuR1 then sends the datagram to the router SnuR2, which then forwards the datagram to EngH1. For EngH1 to send data to any host on SnuNet, it would send the datagram to SnuR2, which would then send it to the appropriate host.

In all cases the routing of the datagram is based on the destination IP address in the IP header and the routing information in the routing tables that are maintained by each host and router. The hosts do not know anything about networks beyond their own. If the destination is on their network, it is forwarded to the specific host. Otherwise, based on the information in its routing table, the host or router just sends the datagram to the next router on the route to the destination. For cases in which the specific destination is not in the routing table, a default router is defined in each routing table. In Figure 3.4, SnuR1 and SnuR2 are the default routers for the networks InmacNet and SnuNet, respectively. They are the routers to which hosts will send the datagrams that they do not know where to forward (refer to Section 3.2 for discussions on routing).

Once the data arrive at the destination host, they must be reassembled into the application data units and passed on to the appropriate users. This means that the datagrams must be demultiplexed and sent to the appropriate software or protocol modules. Multiplexing and demultiplexing occur in three places in the TCP/IP suite—at the link layer-network layer boundary, the network layer-transport layer boundary, and the transport layer-application layer boundaries. This is indicated in Figure 3.5.

At each boundary an identification field in the lower layer's protocol header is used to differentiate between the higher level users. If Ethernet is used as the link layer, the *frame type* field in the Ethernet header is used to differentiate IP datagrams, ARP packets, and RARP packets. The *protocol* field value in the IP header is used to identify the type of protocol used by its payload among TCP, UDP, ICMP, and IGMP. The *destination port* number in the TCP or UDP header is used to differentiate various user processes.

3.1.3.4 Fragmentation of IP Datagrams

In the course of IP datagram routing, IP datagram fragmentation can also occur. This is because the various networks over which a TCP/IP operates may use different-size transmission frames.

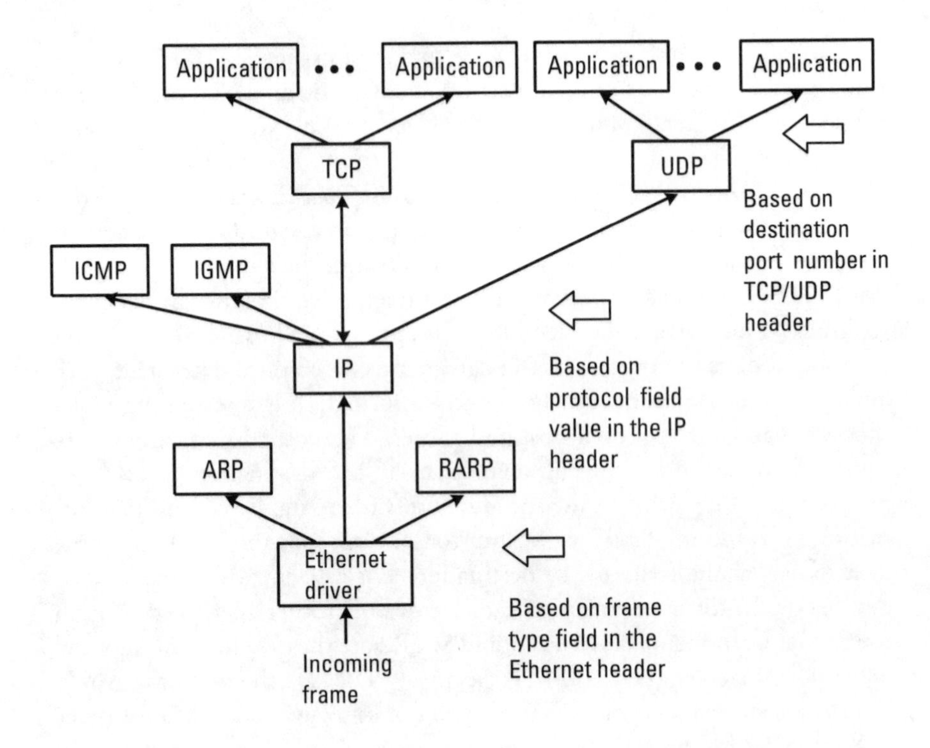

Figure 3.5 Demultiplexing procedure in the TCP/IP stack (with an Ethernet link layer).

Each link layer network has a maximum transmission frame size. For example, Ethernet has a maximum frame size of 1,518 bytes including the header and trailer, and the maximum-size data unit that can be transmitted over it is 1,500 bytes [3, 4]. This is the *maximum transmission unit* (MTU) of Ethernet. The MTU differs among link layer networks. For example, the ATM MTU is 4,352 bytes, the X.25 MTU is 576 bytes, and the SLIP line MTU is 256 bytes.

Fragmentation occurs when an IP datagram passes over a network whose MTU size is smaller than the size of the datagram. The IP datagram must then be broken up into smaller pieces (or fragmented), to be transported over this network, and then reassembled at the receiver system. This will result in the original IP datagram being delivered to the destination host in fragments. The identification, flag, and offset fields in the IP header are used for this IP fragmentation mechanism.

As fragmentation usually results in the degradation of network performance, its use is not recommended in the future. Recently a *path MTU*

(PMTU) discovery method has been developed for finding the MTU of the whole path to the destination. The source can then send datagrams as large as this PMTU without fragmentation on the way to the destination.

3.1.3.5 ICMP and IGMP

Besides IP, two more protocols are defined in the network layer, ICMP and IGMP. Both protocols use the basic IP datagram with special protocol field numbers in the header. ICMP datagrams are used primarily for diagnosis or control. For example, the echo message is used to check whether or not the destination node is operating, while the destination unreachable message signals that the destination node is unreachable. The IGMP datagram also uses the IP datagram. The IGMP is used to form and maintain multicast groups.

3.1.4 Transport Layer Protocols

The TCP/IP suite has two transport layer protocols, UDP and TCP, which are run directly over the IP layer (see Figure 3.2). The main differences of UDP and TCP are in the type of services they offer. UDP offers a simple datagram-like service while TCP offers a more complicated, reliable byte stream service. For simple applications or for cases where the application wants maximum flexibility, UDP is the preferred transport layer protocol. For applications requiring reliable delivery, in contrast, TCP offers a simple-to-use universal transport layer.

3.1.4.1 UDP

UDP is the IP datagram service that has the minimal control information needed for an end-to-end transport protocol. Because no special functions are added other than those offered by the IP layer, UDP is an unreliable, connectionless datagram service.[10]

The structure of the UDP datagram format is shown in Figure 3.6. The *source port* and *destination port* numbers of 16 bits each are used to identify the user application at the source and the destination, respectively. The length field contains the *length* of the whole UDP packet. The *checksum* is a

10. The term "unreliable" must be understood correctly. UDP is deemed unreliable only because it does not have any built-in features to ensure reliable transmission, as TCP does. But UDP by itself actually offers a minimal basis on which more complicated protocols satisfying different criteria may be more easily built.

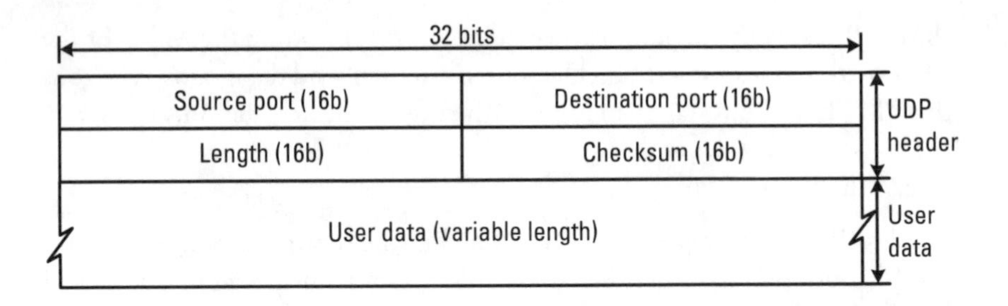

Figure 3.6 UDP datagram format.

16-bit one's (1's) complement sum of the UDP header, the pseudo-UDP header,[11] and the UDP data.

There are several applications that frequently use UDP. The first type is the application that uses UDP because the application itself is based on a very simple model of exchange such as query and reply, where a connection setup and release procedure would be burdensome. A typical example of this would be the *ping* service. The second type is the application that uses UDP because very low protocol overhead is needed—that is, the amount of resources consumed by the protocol must be minimized. An example of this is diskless workstation booting. The third type of application is for its own type of transmission control such as SNMP. The fourth type is the streaming-media application, for which delivery time is more important than reliable delivery. The fifth type is the reliable multicast application that, due to its multicast nature, is not suitable for TCP connections. The last two applications would be especially good examples in which UDP enables the user to build more complicated application-specific protocols.

3.1.4.2 TCP

Applications that require reliable delivery of data use the TCP instead of the UDP. It is a reliable, connection-oriented, byte-stream transport protocol. The structure of the TCP segment is shown in Figure 3.7. The *checksum*, *source port*, and *destination port* fields are the same as those in the UDP datagram. The *sequence* and *acknowledgment* fields are used for reliable data transmission, and the *window* field is used to indicate the receiver window size

11. The pseudo UDP header consists of the source and destination IP addresses of the IP datagram header and the actual UDP header. It is used only for calculating the UDP checksum.

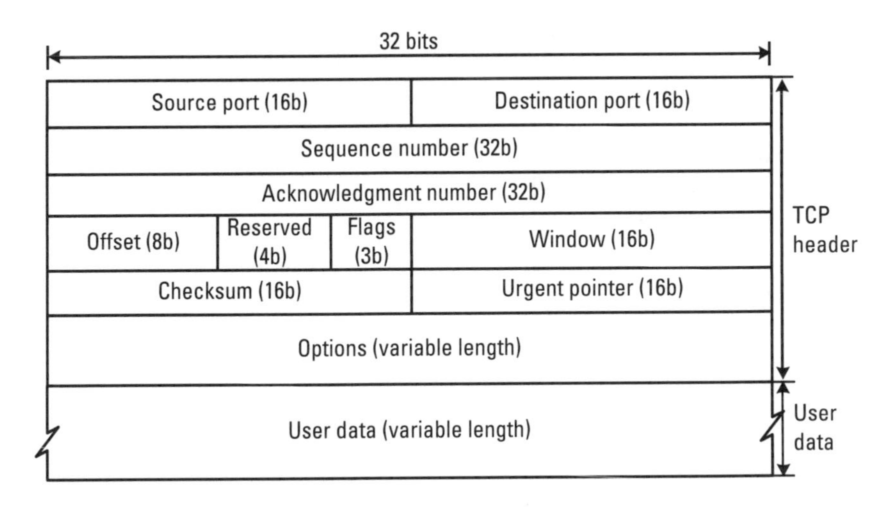

Figure 3.7 TCP segment format.

in the sliding window algorithm. The *urgent pointer* field is used to indicate the position of data, which must be processed immediately at the receiving station.

TCP is connection-oriented because a logical end-to-end connection must be set up between the two end hosts before data transmission. This means that a three-way handshake must take place between the two end points to synchronize them before data transfer can occur. This is illustrated in Figure 3.8. In the figure, SYN is the sequence number to be synchronized, and ACK is the acknowledgment number. End points are uniquely identified by their *IP address* and *port number* pairs.[12]

TCP is reliable because it relies on positive acknowledgments with retransmission. This means that every data segment sent by the sender must be checked and acknowledged by the receiver. If the receiver does not acknowledge the segment within a timeout, the transmitter assumes that the segment has been lost and retransmits the segment.

TCP is a byte-stream protocol, because, unlike UDP, it regards the data it sends as a continuous stream of bytes. A TCP host assigns a number

12. To be more accurate, the *protocol field* value in the IP header is also needed, but this is needed only for differentiating between the users of TCP and UDP protocols. Accordingly, a TCP service user would actually need only the IP address and port number to uniquely identify its corresponding host, since they both must be using TCP.

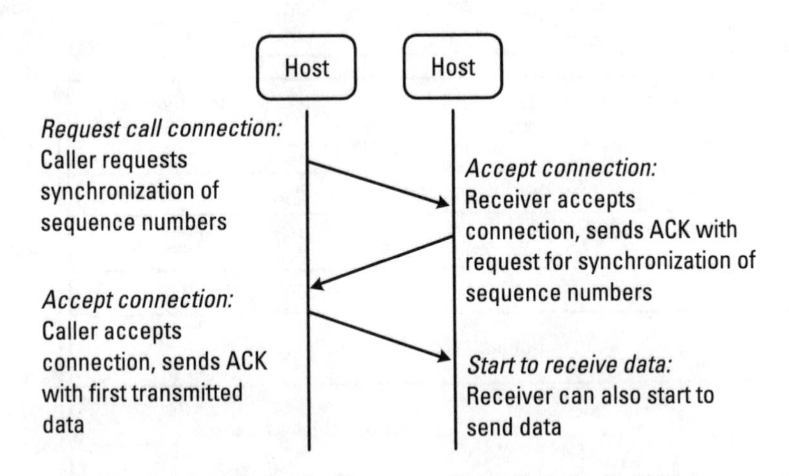

Figure 3.8 TCP connection setup by three-way handshake.

to each byte it sends, and the positive acknowledgment and retransmission mechanisms are executed based on these numbers.

TCP also contains a flow-control mechanism modeled on the sliding window method. It is realized by using the window and acknowledgment fields in the TCP header. The amount of data that a host can send without receiving an acknowledgment is set by the last window advertised by the receiver. An example is shown in Figure 3.9. In the figure, the transmitter sends data bytes 1–6, but receives acknowledgment (ACKs) for only bytes 1–3. Since the advertised window size is six bytes, the sender may send only three more bytes, 7–9, waiting for a new acknowledgment.

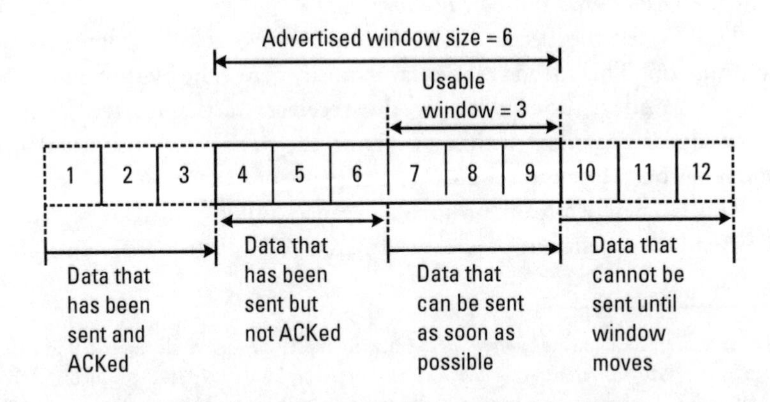

Figure 3.9 Example of the flow-control mechanism used in TCP.

TCP also contains sophisticated congestion control algorithms based on implicit probing of the network congestion status (see Section 3.5 for details). Recently, methods of extending TCP to gigabits-per-second rates have also been developed.

3.2 Routing

One of the most important parts of the IP architecture is the mechanism of IP packet routing. In this section we discuss how routing is carried out in TCP/IP networks. We first describe in detail the steps involved in routing a single IP packet from its source to destination. The description includes the addressing structure and the actual methods used to route a packet based on the routing tables that are built and maintained in the hosts and routers. Then we examine the routing algorithms that are practically in use to build the routing tables in the hosts and routers.

3.2.1 Routing IP Packets

To examine the routing mechanism in the TCP/IP network, we first consider an example for the routing operation and then discuss IP addresses, routing tables, address resolution protocols, and routing protocols.

3.2.1.1 Example of IP Packet Routing

Figure 3.10 depicts an example that illustrates how routing is conducted in IP-interconnected networks. This example is an extension of the example shown in Figure 3.4. If a user wants to send data from the host InmacH1 on the network InmacNet to the host EngH1 on the network EngNet, then the following processes will take place for the relevant routing operation:

1. The host InmacH1 finds out the IP address EH1[13] of EngH1.
2. InmacH1 makes an IP packet with the source IP address IH1 and the destination IP address EH1.
3. InmacH1 searches for the host or router on InmacNet to which it must send the packet to route it to EngH1. It then finds that the

13. The IP address is a 32-bit number, but for ease of explanation, the names EH1, IH1, SR1, and SR2 are used for the addresses. A similar principal is also applied to the Ethernet, token ring, and ATM addresses in the example.

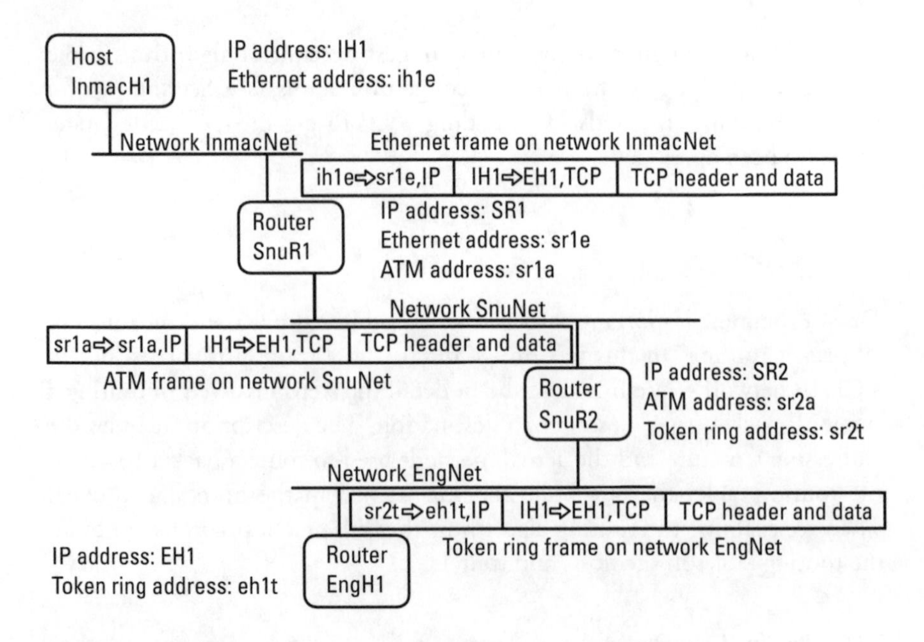

Figure 3.10 Example of routing operation in a TCP/IP network.

packet must be sent to the router SnuR1, which has an IP address SR1.

4. To actually send the data to SnuR1, InmacH1 then finds out the Ethernet address, sr1e, of SnuR1.
5. InmacH1 encapsulates the IP packet in an Ethernet packet addressed to sr1e, and transmits it on the InmacNet network.
6. The router SnuR1 receives the Ethernet packet, and checks the destination IP address. SnuR1 must also find out (just as InmacH1 did) to which host or router on SnuNet it has to send the packet to route it to EH1. It finds out that it has to send it to router SnuR2.
7. SnuR1 finds out the ATM address of SnuR2, which is sr2a, and sends the packet to the router SnuR2 encapsulated in an ATM frame.
8. SnuR2 receives the ATM frame, and checks the destination IP address. SnuR2 determines which host or router on SnuNet to which it has to send the packet to route it to EH1. It then finds that SnuR2 must send the packet directly to the host EngH1, because both are directly connected to the same network, EngNet.

9. SnuR2 finds the token ring address of EngH1, which is eh1t, and encapsulates the packet in a token ring packet transmitted to EngH1.
10. The host EngH1 receives the packet and passes the data to the appropriate user application.

In the preceding steps, various problems arise. For example, in step 1, how does the host find the IP address EH1 of EngH1 from its name EngH1? In step 3, how does the host find out that it must send to the router SnuR1 to send the datagram to EngH1? In step 4, how does the host find the Ethernet address sr1e of SnuR1? Variations of these same problems occur in the other steps as well. We will consider how these problems are solved in TCP/IP, after reviewing the IP address structure.

3.2.1.2 IP Address

All hosts on a TCP/IP network receive unique 32-bit IP addresses. This address is used in all communications. If the host computer moves to another network, this host IP address also changes.

The 32-bit IP address uniquely identifies a host or router on the Internet. The IP address is made up of *network identification number* (netid) and *host identification number* (hostid). This hierarchical structure is used because it simplifies routing.

Various differently sized networks may exist depending on the usage and objective of the network. So, as shown in Figure 3.11, five types of IP addresses, classes A, B, C, D, and E, are defined. The first few bits of the class A, B, and C addresses are fixed as 0, 10, and 110, respectively, with the rest of the address bits being divided into netid and hostid. The first 4 bits of the class D and E addresses are fixed as 1110 and 1111, respectively. The class D addresses use the rest of the address bits as a multicast groupid (i.e., group number), while the usage of the remaining bits is not defined for class E addresses. In Figure 3.11, the group of numbers to the right represents the range of possible 32-bit IP addresses in that class. Note that the normal notation for IP addresses is four decimal numbers separated by dots, with each decimal number representing 8 bits of the IP address.[14]

---

14. Organizations are given class A, B, or C addresses according to their needs. A class C network can have 254 hosts, a class B network can have roughly 64,000 hosts, while a class A network can have about 16,000,000 hosts. Very few organizations are given class A addresses, and most are given class B or class C network addresses.

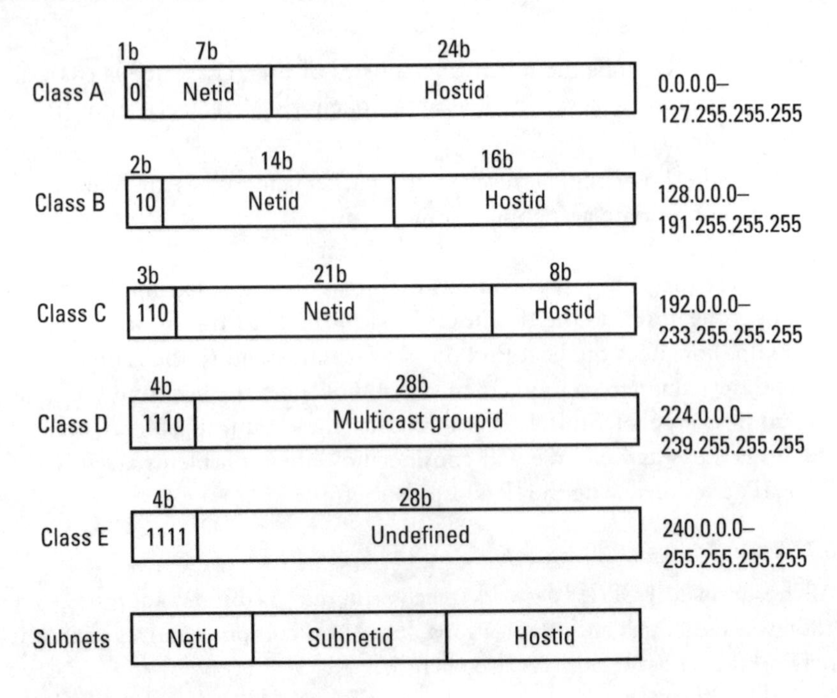

Figure 3.11 Internet address classes.

Due to the spread of LANs and private company networks during the 1980s, it turned out that a more structured address space was needed. Thus, as shown at the bottom of Figure 3.11, *subnet* fields were added to the address structure to ease the management and routing problems. This is accomplished by using 32-bit subnet masks to divide the IP address hostid field into a number of subnets. The bits that are 1's in the 32-bit subnet mask mark the bits that can be used as the netid or subnetid. Examples of subnets and subnet masks are shown in Table 3.1. In the first example, the subnet mask is 255.255.0.0, while the address is of class A since the first bit is 0. Therefore the netid is 18, the subnetid is 18.20, and the last two bytes 16.91 represent the hostid on that subnet. In the third example, the subnet mask is 255.255.255.192, so 26 bits out of the 32 bits are used as the netid or subnetid, while only the last 6 bits are used as the hostid. Also as the first 3 bits are 110, the address is a class C type, and the first 3 bytes 192.90.88 are the netid, 192.90.88.128 is the subnetid, and the hostid is 4.

The subnet concept essentially is a way of enabling the efficient use of a class B type address. The class B address could be divided up into 256 class C type networks, each capable of supporting 256 hosts. But, as the Internet

Table 3.1
Examples of Subnets

IP Address	Subnet Mask	Interpretation
18.20.16.91	255.255.0.0	Host 16.92 on subnet 18.20.0.0
147.46.66.19	255.255.255.0	Host 19 on subnet 147.46.66.0
192.90.88.132	255.255.255.192	Host 4 on subnet 192.90.88.128

spread, there was a shortage of class B type addresses.[15] Initially this problem was handled by handing out class C addresses to organizations, but it led to an explosion in the size of routing tables that the core routers had to maintain. The solution to this problem was found in *classless Internet domain routing* (CIDR). Essentially it extended the subnet mask solution so that a variable length mask can be defined for the network field of all IP addresses. This means there would no longer be any meaning in the terms of class A, B, or C addresses. In other words, the concept of class is now removed from the IP address structure, and routing is done based on network addresses expressed in terms of IP addresses with prefix lengths. For example, in the case of the second entry of Table 3.1, we may use 147.46.66.0/24 to indicate the network on which 147.46.66.19 is located. The number after the slash indicates the number of bits that are part of the network address.

3.2.1.3 Routing Tables

The routing table contains information on where the IP packet must be sent next to get to the desired destination. When an IP packet is received, the destination IP address is used to look up the relevant routing table entry and the packet is sent to the next hop destination (host or router) in that entry.

The destination host may be on a network directly connected to the host or the router that has just received the packet.[16] In this case, the datagram can be routed to the destination host by just using the physical network address of the final destination. This is called *direct routing*. In case the

15. Most organizations found class C networks (256 hosts) too small for their needs, while class A networks (more than a million hosts) were too large. Class B networks (64,000 hosts) were usually thought to be about the right size.
16. This is easily found out by examining the routing table (explained further in this section).

destination host is on a different network, it is necessary to determine the next hop router to send the packet first. The routing table normally contains the IP address of this next hop router. The packet is then sent to this next hop router, which will finally forward the packet to the appropriate network or host according to its own routing table. This is called *indirect routing.*

We now describe the routing function in detail through the example of the routing table in Table 3.2. The first column of Table 3.2 lists the address of the host or network to which the datagram is destined. The second column lists the address of the next hop router or host to which the datagram has to actually be sent. The third column contains information on columns 1 and 2: H means that the destination address type in column 1 is a single host, otherwise it is a network; U means that the router or host in the second column is in operation; and G means that the next hop destination in column 2 is a router, otherwise it is a host. The fourth column indicates the actual network interface to which the datagram has to be sent. This is necessary because routers usually have more than one network interface.

The first entry of Table 3.2 is for local loopback. It is used by the host to send datagrams back to itself for such purposes as diagnostics and debugging. The second entry is the default entry. If no match for the destination address is found in the routing table, the datagram is forwarded to this router. The third entry is for hosts on the subnet 147.46.66.0.[17] Because the

Table 3.2
Example of Routing Table

Destination	**Gateway**	**Flags**	**Interface**
127.0.0.1	127.0.0.1	UH	lo0
Default	147.46.66.1	UG	emd0
147.46.66.0	147.46.66.19	U	emd0
147.14.148.0	147.46.66.1	UGH	emd0

17. The fact that this is a subnet is easily recognized as follows. Since the first byte is 147, this is a class B address, which uses the first two bytes for netid, and the last two bytes for hostid. But because there is no H flag in the third column of the routing table, the corresponding entry in the first column must be a network address. This means that three bytes are used to identify this network, so the third byte must be a subnetid that is used to define 256 subnets.

host is directly connected to this network, the destination gateway entry is itself. This means that any packets destined for hosts on this network should be sent by just using the link layer network. The fourth entry is for hosts on subnet 147.46.148.0. Since these hosts are on a different network, the destination is a router (with the IP address 147.46.66.1) connected to the host's own network 147.46.66.0.

Figure 3.12 shows how to perform table-based routing in relation to Figure 3.4. Figure 3.12 shows routing tables for each host and router on the path from InmacH1 to EngH1. Figure 3.4 used IH1 and EH1 to signify their host addresses, but Figure 3.12 uses their actual numerical IP address values, which are 147.46.66.19 and 147.46.148.12 respectively.

As in the original example, a packet must be sent from InmacH1 to EngH1. The IP address of EngH1 is 147.46.148.12 (how InmacH1 knows this is discussed later in this section). InmacH1 examines its routing table and finds that to send the packet to hosts on network 147.46.148.0, it must forward the packet to the router with the address 147.46.66.1, which is the router SnuR1. After SnuR1 receives the datagram, it examines its routing table, and finds that to forward the datagrams destined for network 147.46.148.0, it must send the datagrams to the router 147.46.80.99 (i.e., the router SnuR2). Router SnuR2 finds that it is directly connected to the network 147.46.148.0, so it just needs to send the datagram to its interface 147.46.148.1. The interface then transmits the datagram on the EngNet to reach its final destination EngH1.

3.2.1.4 ARP

In the preceding example, InmacH1 knows that it must forward the IP datagram to the router SnuR1 on the InmacNet network. InmacH1 also knows the destination IP address of SnuR1 to be 147.46.66.1, as it was in the routing table. However, for InmacH1 to send the datagram to SnuR1, it must first know the physical network (i.e., Ethernet) address of SnuR1. This is done by using the ARP.

ARP is a protocol enabling a host to find the physical address of another host on the same physical network by only using the IP address of the other host. This enables any IP address to be allocated to any host, and thereby hides the physical network protocols.

In the preceding example, the protocol for the case of an Ethernet network operates as follows: First, InmacH1 broadcasts a packet containing the IP address 147.46.66.1 on the InmacNet network queueing for the Ethernet address of the host or router. Every host on the Ethernet will hear the

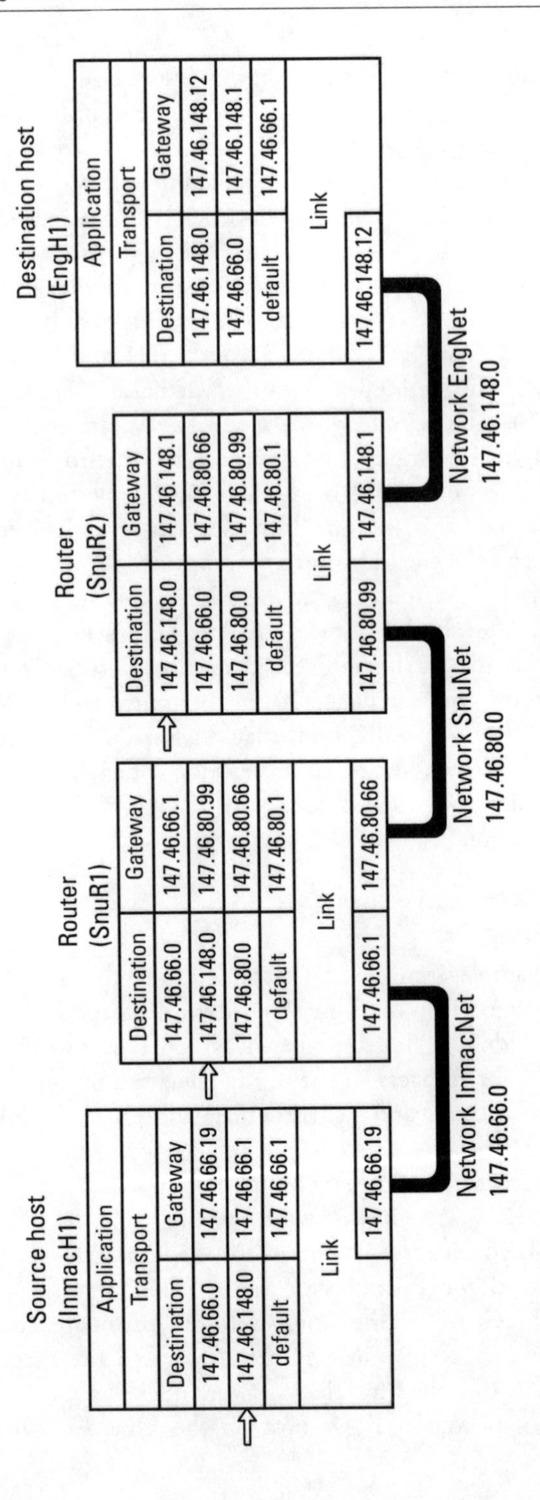

Figure 3.12 Example of table-based routing.

broadcast and the correct one—that is, SnuR1—will answer with an ARP reply message containing its own Ethernet address. Usually the results of an ARP request and reply sequence are kept in an ARP cache, so in most cases the ARP protocol is used only a few times and most link layer addresses are obtained from the ARP cache maintained by the host.

3.2.1.5 DNS

All Internet interfaces have a numerical 32-bit IP address. This is the actual numerical value used inside the Internet to route packets. But IP addresses are hard to remember and to use. So usually a nonnumeric host name is given to each host interface. For example, a host may have an IP address, 147.46.66.19, and a host name, tsp.snu.ac.kr. In this case when a user uses telnet, instead of typing "telnet 147.46.66.19" the user types "telnet tsp.snu.ac.kr." The TCP/IP software will then automatically translate tsp.snu.ac.kr into 147.46.66.19. When the user uses e-mail, in the same manner the user types "mail blee@tsp.snu.ac.kr" and the software translates this into "mail the message to user blee at the host 147.46.66.19."

However, two problems arise with the use of names. First, how will the system translate the nonnumeric name into the numerical IP address? Second, how can host names be globally known? Basically there are two ways to solve these problems: one is by using host tables, and the other is to use the DNS. The host table method is the intuitive but naive solution. Each host maintains a host table that contains a simple mapping between host names and host IP addresses. Whenever a name is used for communication, the system looks up the name in the host table and uses the IP address in there. There are problems with this method, with scalability being the most obvious one. Also this does not define a global host name to IP address translation method, so we can only connect to hosts defined in the host table. However, if DNS is used, both problems can be solved, and consequently DNS is the standard Internet name service.

Basically, DNS is a name service that uses a very large distributed database containing all Internet host names and their corresponding IP addresses. DNS is based on a hierarchy. There are "top-level domains" based on countries such as .kr, .uk, .jp, and some based on classes such as .com, .gov, .org, and .edu. Each top-level domain can then assign subdomains such as ac.kr. This is repeated at each layer in the hierarchy. Due to this hierarchical structure, a unique domain name may be obtained for everyone, and IP addresses can also be easily found by searching the corresponding DNS name servers.

All DNS servers are organized in a hierarchy based on the structure examined above. When a DNS server is requested to resolve a domain name, if it knows the name, it will send back the IP address. Otherwise, it will ask another server higher up—that is, a DNS server of a domain that is higher up in the hierarchy.

3.2.2 Routing Protocols Used in TCP/IP Networks

In some networks, routes are calculated centrally and distributed to all nodes, either automatically or manually. However, in the Internet most routes are calculated in a distributed fashion. More specifically, dynamic routing protocols are used by routers to exchange routing information and construct routing tables. Static routing, where the routing tables are configured manually by the operator/administrator, are also used in many small networks.

In small scale TCP/IP networks, RIP and OSPF protocols are commonly used for routing.[18] In a single routing domain all the constituent routers must use the same routing protocol, whether it is RIP or OSPF. Each router exchanges routing data and calculates the routes to all destinations in the domain based on the data it receives.

As a single routing domain becomes larger, routing protocols such as RIP and OSPF become unsuitable to use due to the scaling problem. So the concepts of *autonomous systems* (ASs) and *exterior gateway protocols* (EGPs) were developed to handle the routing problems in the Internet; an AS refers to a single or group of routing domains operating under the control of a single organization.[19] Multiple ASs are interconnected through exterior gateways. Exterior gateways must exchange routing data among themselves and calculate routes so that packets can be routed to destination hosts in different ASs. Exterior gateways use EGPs to exchange this data. The most popular EGP is the *border gateway protocol* (BGP), whose current version is BGP-4 [7]. In contrast to EGPs, RIP and OSPF are called *interior gateway protocols* (IGPs). In the following we consider the two IGPs (i.e., RIP and OSPF) and one EGP (i.e., BGP-4).

---

18. There are two other protocols that are also commonly used. One is the *Enhanced Interior Gateway Routing Protocol* (EIGRP) used by Cisco routers, which is based on RIP but contains many proprietary enhancements. The other is *ISO 10589: Information Processing Systems - Data Communications - Intermediate System to Intermediate System* (*IS-IS*) *Intra-Domain Routing Protocol,* which was developed by the OSI. It is a link state protocol, similar to OSPF and is used by many ISPs in place of OSPF [5, 6].
19. Note that a single routing domain uses a single routing algorithm.

3.2.2.1 RIP

RIP is one of the most widely used IGPs.[20] It is based on the distance-vector algorithm and is extremely simple, but inadequate for large networks because of its long convergence time and the possibility of loops. A router running RIP in active mode broadcasts a message once every 30 seconds. The message contains information taken from the router's current routing database. Each message consists of pairs, with each pair consisting of an IP network address and the hop count to that network. Both active and passive RIP participants listen to all broadcast messages and update their tables according to the distance-vector algorithm.

In contrast to OSPF, RIP only uses routing metrics based on hop counts.[21] To prevent routes from oscillating between two equal-cost paths, RIP specifies that existing routes should be retained until a new route has a strictly lower cost. RIP specifies that all listeners must timeout routes they learn via RIP after 180 seconds.

RIP must handle all possible errors that can be caused by the underlying distance vector algorithm. The most significant of these is the slow convergence or the count-to-infinity problem. This problem is solved by choosing a small number, 16, to indicate infinity.

Figure 3.13 shows the RIP packet format. It contains a command field (usually request or reply), version number (1 or 2), address family (which is always 2 for IP addresses), and IP routes. IP routes consist of IP addresses with a metric that is a hop count. An RIP message essentially transfers the current routing table from one router to another. When used in IP networks, RIP uses UDP.

3.2.2.2 OSPF

OSPF is another widely used IGP. It includes support for TOS routing, load balancing, and authenticated routing. OSPF was designed by the IETF with the goal of making a routing protocol that would converge more quickly

---

20. It is also known by the name of the program that implements it "routed." This program was included in the 4BSD UNIX system, which made RIP popular. This also caused RIP to be widely adopted before a formal standard was written in 1988 as an RFC standard.

21. Hop count is a crude measure of network response or capacity that does not produce optimal routes. Further, it makes routing relatively static because routes cannot change in response to network load. When these factors are taken into account, actual delay may render a more desirable metric than the hop count does.

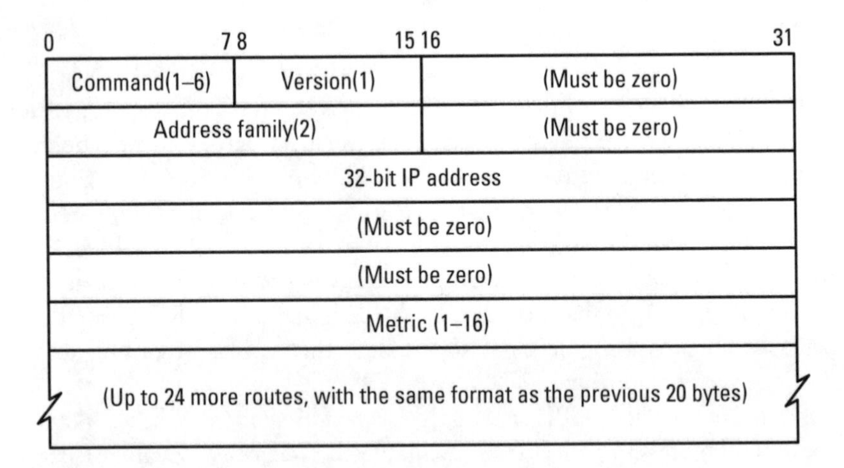

Figure 3.13 RIP packet format.

than the distance-vector algorithm–based RIP protocol. Consequently, the linkstate algorithm is taken for the design of OSPF.

All link state protocols share the same basic principles. Most importantly, they are based on the construction of a network tree at every single node. There are three phases in constructing a network tree. First, each node must find its neighbors. Second, the node must share the information with all the other routers on the network. Third, the node must combine the information about individual neighborhoods to construct consistent routes (usually by way of Dijkstra's algorithm).

Neighbor greeting is carried out by the use of *hello* packets that each node sends to all neighbors. Synchronization of the database is performed by using *link state advertisement* (LSA) packets along with a flooding mechanism. Once each node gets a complete neighborhood map, the *shortest path first* (SPF) route is found by way of Dijkstra's algorithm.[22]

Among the key ideas of OSPF is the fact that it allows networks to be partitioned into areas. By partitioning the network into areas OSPF limits the topology map required in each router and thereby increases the scalability of the network. Each area is assigned a number, as illustrated in Figure 3.14. Every OSPF network must have at least one area with the 0.0.0.0

22. The name OSPF comes from the fact that most implementations use the SPF algorithm to calculate routes. The "open" concept comes from the fact that its specification is available in the public domain for free.

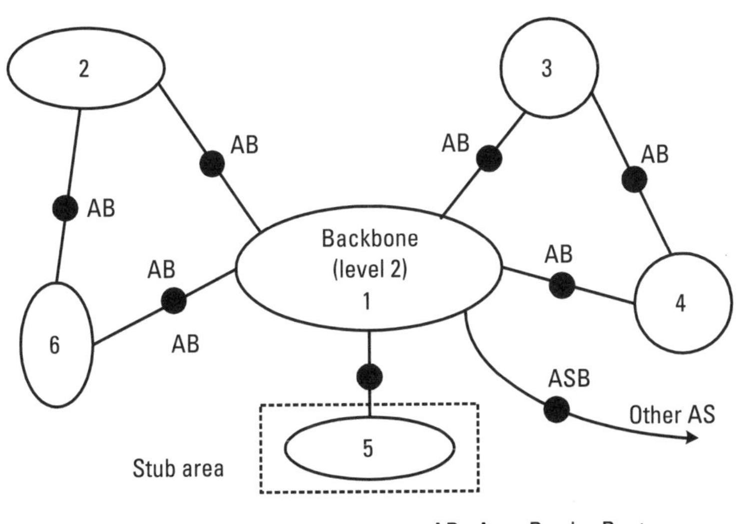

Figure 3.14 Example of OSPF hierarchy.

address, denoting it as the *backbone area.* It is essentially the hub of the AS that contains the link state information for the entire network. All other areas in the AS must connect to this backbone area.

The above structure leads to three special types of routers: First, *internal area* (IA) router is responsible for routing inside one area. Each IA knows the topology of its area. Second, an *area border* (AB) router is a router that connects two areas. An AB router maintains the topology of all areas to which it is connected. Additionally, AB routers flood summary LSAs so that internal routers can learn the interarea routes. Third, an *autonomous system boundary* (ASB) router connects two autonomous systems. An ASB router uses an external LSA when flooding information about the external networks connected to the OSPF network. The ASB routers import non-OSPF routing information from such protocols like RIP, hello, and EGP, and then redistribute them as OSPF LSAs to the OSPF network.

A *stub area* is a small OSPF area that has only one area border router. A stub area functions like any other OSPF area except for the propagation of external routes—external routes are not propagated inside a stub area. A stub area database contains only a default route to all the external routes.

By limitng the number of external routes maintained by routers in the stub area, the memory and processing requirements for internal routers can be lessened.

OSPF has other various features that differentiate itself from other link state routing algorithms. It supports TOS routing based on various parameters, including low delay and high throughput. This feature has never been used in commercial networks, and its use is now criticized. In contrast, the load-balancing feature of OSPF is used quite frequently. Routing messages are an important network control message and are of concern to many security experts. OSPF addresses security issues by having all exchanges between routers authenticated. Support for CIDR addressing and routing is also built into OSPF. This is usually termed as support for *variable-length subnet masks* (VLSMs) as the CIDR routes are expressed as the network IP addresses with the subnet mask length indicators (refer to Table 3.1).[23]

Five types of messages are used to realize the functionality of the OSPF, namely, *hello*, *database description*, *link status request*, *link status update*, and *link status acknowledgment* packets. The link status packets contain up to five types of link state messages—the router link LSA, the network LSA, the summary LSA-IP network, the summary LSA-ASB router, and the AS external LSA.

Figure 3.15 shows the OSPF packet format. The *type* field in the format identifies one of the five packet types, the *router ID* field the source of the packet, and the *area IP* field the area to which the packet belongs. In addition, the *authentication type* field contains the authentication type information and the *authentication* field authentication information itself.

3.2.2.3 BGP-4

BGP-4 is responsible for the exchange of routing information between ASs. It is used to exchange network reachability information with other BGP systems, which is sufficient to construct a graph of AS connectivity. It supports classless interdomain routing based on the distance-vector algorithm. A significant characteristic of BGP-4 that distinguishes it from other distance-

23. OSPF was originally designed for point-to-point networks. But due to the ubiquitous nature of Ethernet and other shared media multiaccess networks, OSPF was enhanced to support such networks. This was done by making every multiaccess network have a designated router that sends and receives all link status messages on behalf of all other routers on the multiaccess network. OSPF also allows routers to exchange routing information learned from other (external) sites. This means that even if a neighboring route domain uses a different routing protocol OSPF is able to import routing information from that protocol.

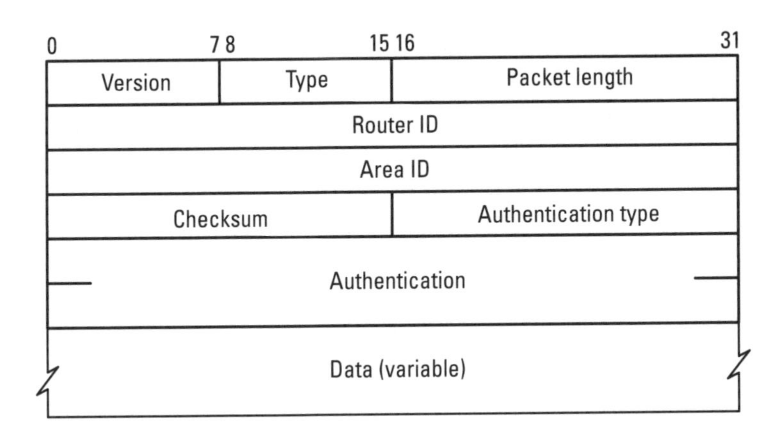

Figure 3.15 OSPF packet format.

vector algorithms is that BGP-4 implements what is called *path vector*—that is, it advertises the full path information (i.e., a list of all the ASs used on the path from the source to the destination) [7]. By using the path vector mechanism it can ensure that no routing loops are formed and problems such as the count-to-infinity problem are avoided.

BGP-4 uses four main types of messages, namely, *open*, *update*, *notification* and *keepalive*. All BGP-4-enabled external routers must open TCP connections with their neighbors to exchange the above messages. By using TCP connections BGP-4 is able to ensure the reliable and lossless delivery of these messages. Once a connection is opened the above messages are exchanged to setup a BGP-4 session between the nodes so that routing information may be exchanged between the nodes.

Each routing update sent by BGP contains the path vector. Duplicate AS entries in the list would imply a loop in the path. Each BGP checks to see whether its name appears in each routing update. If it does, it means that the routing update information for that path is looped, so it must be silently discarded. Once the routers have exchanged their routing tables at least once, the routers will only send updates to each other, not the whole table.

3.3 Switching and Multiplexing

In TCP/IP networks, the fundamental unit of multiplexing is the IP packet, as all TCP/IP traffic is transported as IP packets. However, IP packets are themselves transported over many different types of data link layer packets.

Thus, depending on the protocols used in each layer, switching and multiplexing occur at the network and/or data link layer.

LANs provide a means for switching and multiplexing packets at the data link level, so LANs can be considered as the basis for layer 2 networking. Initially, the broadcast and multiple access characteristics limited the use of LANs, but later, bridges and switches were introduced as means for increasing the range and applicability of LANs. In contrast, routers can be viewed as layer 3 switches. Today, *multiprotocol label switching* (MPLS) switches are being developed; these switches can be viewed as a mix of layer 2 and layer 3 technologies.

3.3.1 LAN

A LAN is the basic data communication network formed among computers and peripheral equipment in a local area. In a LAN, computers and equipment are connected using coaxial cables or twisted-pair copper wires as the transmission medium. The transmission rate is about 10 to 100 Mbps and the transmission coverage is usually limited to a radius of about 2 km.

Numerous types of LANs exist but, in general, LANs share several characteristics. For example, multiple systems are connected to a shared medium, total bandwidth is shared among multiple stations, and *media access control* (MAC)/multiple access is employed.[24] In addition, LANs have low delay, low error rate, broadcast capability, limited geographical coverage, and limited numbers of stations.

The communications characteristics and the performance of a LAN depend on the network topology and the transmission medium. The topology is usually categorized into ring, star, tree, and bus, with the most frequently used one being the bus and ring topologies.

Standards for various LAN protocols are well organized by the IEEE 802 committee. In particular, the basic standards IEEE 802.3 through 802.5 prescribe the protocols for CSMA/CD bus, token bus, and token ring, respectively.

3.3.1.1 MAC Protocol and LLC

The data link layer consists of two sublayers, namely, the MAC and *logical link control* (LLC) layers. The MAC layer is concerned with the medium access

24. MAC addresses are based on the 48-bit IEEE format.

functions such as frame structure and error checking, which vary depending on the network topology.[25]

In contrast, the LLC layer controls the transmission and reception of frames, which is independent of the network topology. LLC allows multiple higher-layer protocols to share a single data link. The two main protocols are LLC type 1, which provides a simple datagram service, and LLC type 2, which provides a reliable connection-oriented service. LLC type 1 is most widely used today.

In general, there are two methods available for controlling medium access—centralized control and distributed control. A LAN employs distributed control, which is subdivided into a contention scheme and a token control scheme. In the *contention scheme* each station can attempt to access the medium at any time on contention basis, but in the *token control scheme* a station can access the medium at a preregulated time and order. The most typical contention scheme is the CSMA/CD scheme, and the token control scheme includes the *token bus* and *token ring* schemes. The CSMA/CD and the token bus schemes are used in bus topology networks, and the token ring scheme is used in ring topology networks. In both the token bus and token ring schemes, medium access is controlled by a special frame called a *token*; the station that catches the token owns the exclusive privilege to transmit frames for a preregulated duration of time. If the station finishes transmitting data, or if the allowed time is consumed, the station passes the token to the next station.

3.3.1.2 CSMA/CD Scheme

CSMA/CD is currently the most widely used scheme, and its most representative product is the *Ethernet* [3]. In this CSMA/CD scheme, a station that wants to transmit can transmit a data frame if the medium is not being used by another station. The transmitted frame has the format shown in Figure 3.16.

The media access operation of the CSMA/CD scheme is done in the following manner: If the medium is "idle," then it commences transmission. Otherwise, it monitors the medium until an idle state is sensed and then commences transmission. If collision is detected during transmission, then it sends a jamming signal that informs all the stations of frame collision

---

25. Stations connected to a LAN share the transmission medium and its transmission capacity. To share transmission capacity among stations, it is necessary to control the access of the stations to the common medium. The protocol designed for this purpose is the MAC protocol.

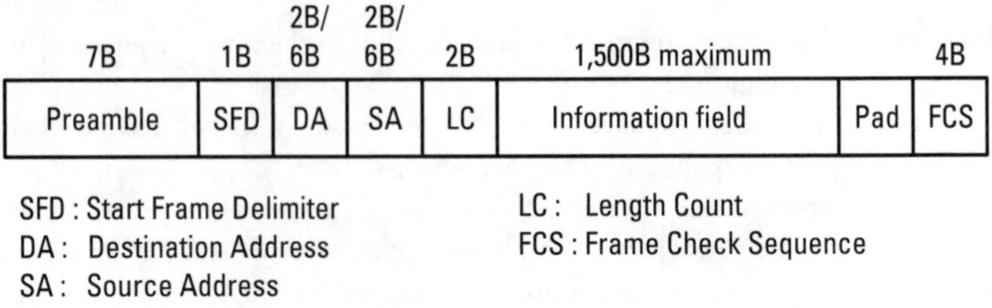

7B	1B	2B/6B	2B/6B	2B	1,500B maximum		4B
Preamble	SFD	DA	SA	LC	Information field	Pad	FCS

SFD : Start Frame Delimiter
DA : Destination Address
SA : Source Address
LC : Length Count
FCS : Frame Check Sequence

Figure 3.16 IEEE 802.3 CSMA/CD MAC frame format.

and discontinues transmission. It then reattempts transmission sometime after the jamming signal has been transmitted. After a given frame has collided with another frame, the source station that has transmitted the colliding frame continues the transmission until the frame returns to discern whether the frame has been in collision.

The collision detection time becomes maximum when a collision occurs between the two frames emitted by the two stations located on the opposite ends of the network. In this case if the frame emitted by one station collides with the frame emitted by the other station in front of the other station, then the collision detection time reaches the maximum, which is twice the transmission delay time between the two farthest stations in the network. This implies that a transmitting station has to continue transmitting data at least for the maximum collision detection time before it can detect whether or not the transmitted data collided. Consequently, the minimum frame size of the MAC protocol should be determined such that its equivalent time length is twice the maximum propagation delay within the network.

These constraints result in limiting the maximum possible size of a CSMA/CD-based network. Ethernet networks consequently have packets with a minimum size of 64 bytes and can cover a maximum distance of up to 2 km depending on the type of cable and the transmission rate used.

The concepts of a MAC access domain and MAC broadcast domain are also important for this discussion. The *MAC access domain* is the domain that encompasses all the nodes that join contention for network access, whereas the *MAC broadcast domain* refers to the area over which the packets are broadcast. In initial Ethernet networks, these two domains were the same. But, by using bridges or routers, it is possible to segment those domains and thereby realize larger LAN networks. This is discussed in more detail in the next section.

3.3.2 Interconnecting LANs and Layer 2 Switching

While LAN technology is an efficient and effective communication means, it also has a number of problems. Only a limited number of stations are allowed on a single LAN, the physical size of a single LAN is limited, and the number of simultaneous users allowed on a LAN is also limited. Basically, LANs are limited in the area and the number of nodes that can be connected to a single node. Because of these problems, there appeared strong demands for lifting the limitation on the size of the LAN network. This triggered the development of a number of methods for extending and interconnecting LANs.

The first type of device for extending a LAN is the *repeater hub.* This is a layer 1 interconnection device that simply regenerates electrical signals for LAN interconnection. It allows easy configuration and maintenance of network connections by following a hub-and-spoke model. Each repeater hub would be located at an easy-to-reach central position. A LAN that is extended through the use of repeater hubs is a *flat LAN,* which refers to a LAN built of one large single MAC access and broadcast domain. Consequently it is easy to reach all the other devices on the domain. However, this also results in some major problems. The larger the MAC access domain is, the harder it becomes to get reasonable performance with the contention-based access methods. In addition, the larger the MAC broadcast domain is, the more resources are consumed by the MAC broadcast message. Unless the network is planned carefully, it is likely to degrade the performance of the network. In other words, repeater hub-connected LANs enable the construction of large flat LANs with large MAC access and broadcast domains at the cost of degraded performance.

LAN segmentation aims to solve this problem by limiting the MAC access domain. Basically, it manages this by segmenting the MAC access domain, while leaving the MAC broadcast domain alone. Figure 3.17 shows how this becomes possible by the use of a bridge. A *bridge* is a layer 2 interconnection device that looks at the source and destination addresses of each packet to decide whether or not to pass it to the other LAN segment. The LAN bridge installed between two LAN segments can increase the efficiency because it passes only the traffic that is destined for other LAN segments. The main type of bridge in use today is the transparent bridge. Other types include translating and speed-buffering bridges. A *transparent bridge* learns the MAC addresses of the nodes that are connected to each port and forwards packets to the links for which the destination node is known to exist. The addresses of the nodes connected to each port are learned by examining the packets received on each port. The use of bridges

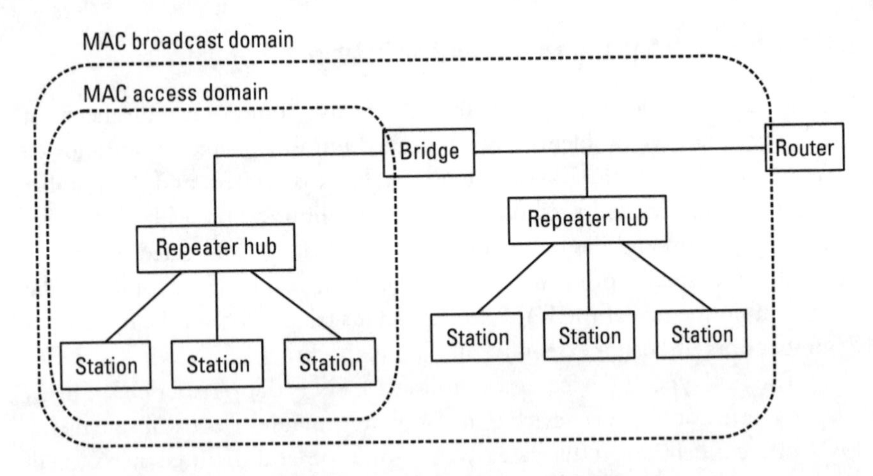

Figure 3.17 MAC access domain versus MAC broadcast domain.

allows the expansion of LANs over larger physical areas or numbers of stations. In other words, it helps to improve the performance of the LANs for any given physical size and number of nodes.

Even though the main aim of bridges was originally in LAN extension (i.e., the physical extension of the coverage area of a LAN), LAN segmentation has another important goal of high-speed operation. This goal can be accomplished mainly by connecting each port on a switch to a single node. Each link between the switch and a node then becomes a single MAC access domain. As a result, there is no access contention between end stations, though there still exists contention between the node and the switch port. This enables the possibility of full-duplex operation, removing contention or collision completely. The per-port dedicated bandwidth for end systems leads to an increased aggregate bandwidth/capacity that is independent of the data rates among ports. Further, this change extends the distance limitations and the number of nodes that can be connected.

The bridges, in their original form, just divided up a large MAC access domain into multiple smaller segments and MAC access domains, but each port was still connected to a single MAC access domain with multiple user nodes. Conceptually this can be viewed as "segment switching." With the improvement discussed above, such bridges evolve into "port switching," with each port connected to only a single station. This brings to us an interesting aspect of the difference between a switch and a bridge—a *layer 2 switch* (e.g., the fast Ethernet switch) and the bridges in existence today may be viewed the same in functionality but different in performance and implementation. A layer 2 switch may be viewed as a high-performance implementation of a bridge, built out of ASICs with a high level of integration. In many ways the distinction of the two could be purely a matter of marketing.

3.3.3 Routers and LAN Segmentation at Layer 3

The LAN segmentation methods described in Section 3.3.2 only segment the MAC access domain, not the MAC broadcast domain. Consequently, once the domain grows beyond a certain size, the performance of the network will drop abruptly due to the enlarged broadcast area. In addition, as LANs do not use an address strategy that allows routing information to be grouped, the MAC address tables that a bridge maintains grow in proportion to the size of the LAN's broadcast area. Furthermore, this happens even after the MAC access domain has been segmented by using bridges. This problem can be resolved in the next logical step by segmenting the MAC broadcast domain. This task is done by employing routers.

A *router* is a layer 3 interconnection device that, based on its routing database and the information in the packet headers, decides where to send the packets to deliver them to their final destination. A router can be programmed to be able to transfer packets according to the desired objectives, as is usually the case in complex internetworking.

3.3.3.1 Basic IP Router Functions

In a traditional router, the IP header was verified first; then using the destination address as the key, the routing table was searched to find the appropriate forwarding path. Finally, the packet was sent to an output queue. Modern routers carry out much more complicated packet-processing steps consisting of IP packet verification, classification, route lookup, packet buffering, and transmission scheduling, as shown in Figure 3.18.

First, IP packets are verified and classified. Verification is usually carried out at the data link frame level by methods specific to the data link used. One item that is checked by the network layer is the TTL field. Further, the packet is classified according to various parameters, including IP address, port number, TOS, and protocol type.

Based on the above classification parameters a routing table lookup is carried out. Usually it is a high-speed lookup in a routing table that may contain up to 256K entries. As with all Internet routing lookups, it must be a longest-prefix-match lookup. Traditionally, Radix trees and Patricia tree

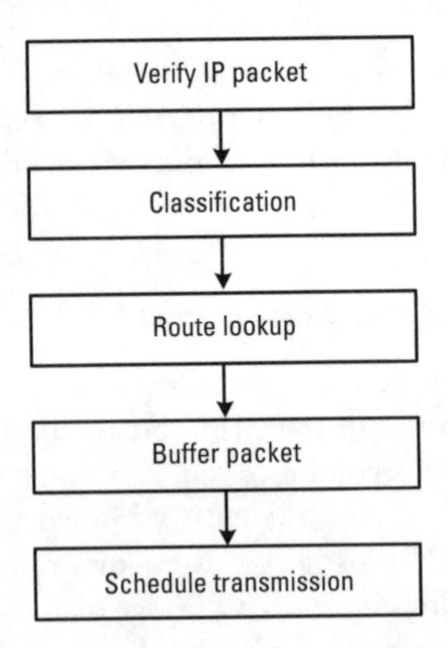

Figure 3.18 Packet processing structure of router.

algorithms have been the basis for routing table lookup schemes.[26] Based on this result, a forwarding path is selected.

Then the packets are queued in the output ports. During this stage several constraints must be considered to support QoS. These result in the need for implementing complex queuing and scheduling algorithms, which may include per-flow queuing, per-class queuing, priority queuing, and weighted fair queuing (refer to Section 3.5.3). Additionally buffer congestion control mechanisms such as *random early detection* (RED) may be used.

3.3.3.2 Router Structure

Figure 3.19 shows the basic structure of a router.[27] A router consists of an input processor that carries out packet classification and route lookup for the incoming packet. Once the packet is classified and its output link is identified, it is passed through a switching fabric to the appropriate destination. There it is put into an appropriate output buffer depending on how it was classified. A scheduler decides the time at which the packet is to be transmitted onto the output link. In all these decisions, high-level router policy rules are applied.

---

26. Recently, various fast routing table lookup methods have been investigated. Among the methods developed are variants of these two schemes as well as more hardware-intensive schemes.
27. This structure, while generic, is most similar to the third-generation router structures to be discussed in this section.

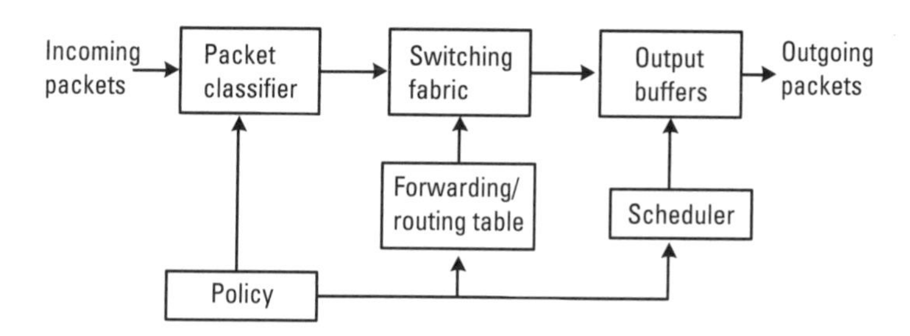

Figure 3.19 Basic router structure.

The structure of routers in general can be divided into three different generations as shown in Figure 3.20 [8]. The first-generation routers were based on a centralized structure, basically being single CPU-based systems. They were built so that a single central CPU stores and routes packets from line card to line card. The CPU carries out the packet-processing steps discussed above. In this sort of router the CPU is the obvious bottleneck of the system.

The second-generation routers tried to remove this bottleneck by employing a decentralized structure. A CPU was put on each *line interface* (LI) card to handle the packet processing, and a shared bus or ring is put in place to transfer packets between the LI cards. A central CPU is maintained as the route protocol and management agent. In this structure, the shared bus or ring is the bottleneck in the router performance.

The third-generation routers employ switch fabrics to switch packets between LI cards simultaneously and thereby remove the bottleneck of the shared-bus/ring structures of the second generation. It is an extremely hardware-intensive method but results in the highest performance. In general, the fastest routers of today are of this third-generation structure, and are in many cases called *layer 3 switches.*

3.3.3.3 Router Versus Layer 3 Switch

As with the bridge versus layer 2 switch case, one can question how a router is different from a layer 3 switch. Basically, the two are functionally the same but the layer 3 switch may be viewed as being leaner and faster—but with less intelligence. The main reason is that a layer 3 switch is usually a hardware-based integrated solution. This implies that a layer 3 switch sacrifices flexibility for the higher performance.

Layer 3 switches have come into existence due to the following three reasons. First, silicon integration has advanced to the degree that makes it feasible to build very large chips. Second, the IP protocol has become the dominant

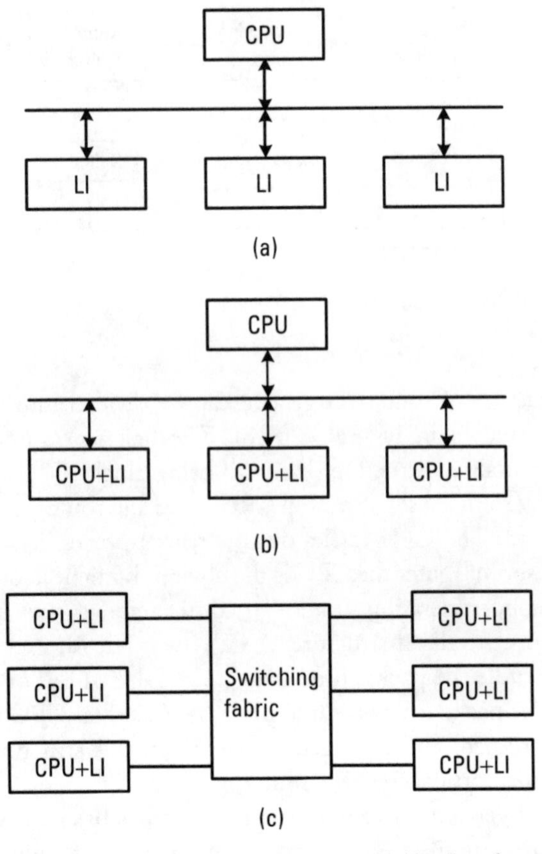

Figure 3.20 Router architecture: (a) first-generation, (b) second-generation, and (c) third-generation.

protocol in the networking world, so that multiprotocol operation is not so important. Third, the IP protocol has matured and its functionality has become stable.[28] These reasons form the basis for constructing a hardware-based solution to building a layer 3 switch.

As a specific example, consider the modules that must be on the *fast path* for packet processing. The fast path refers to the sequence of operations

28. By stability we mean not only that there are no bugs in the protocols but also that the features and options that are used by implementations and applications are the same in almost all cases.

that a packet with no abnormal processing would take. As this is the sequence of operations that the vast majority of packets will take, optimizing the processing carried out in this step would increase the speed significantly, hence leading to the name "fast path." The modules related to the following functionality do not need to be on this fast path—fragmentation and reassembly of IP packets, source routing, route recording, timestamp options, and the handling of various ICMP messages. Also routing protocols (i.e., RIP, OSPF, BGP-4, etc.), network management (e.g., SNMP) and configuration (e.g., DHCP) packets do not need to be put on the fast path. This is because these functions are needed in a minority of the packets received or do not need to be processed quickly. These choices usually lead to a layer 3 switch, which usually handles only IP and Internetwork packet exchange (IPX) network layer protocols, fast Ethernet physical interfaces, RIP, OSPF, and DVMRP routing protocols, and only simplified QoS. In contrast, a full-fledged traditional router will usually handle multiple types of network protocols, multiple types of media interfaces, a full set of routing protocols including RIP, OSPF, BGP-4, PIM, and so on.

3.3.3.4 Comparison of Bridges and Routers

Bridges have the ability to look inside the data frames of the data link layer and identify the packets inside the frame. This means that bridges are essentially looking at layer 2 information only. Bridges (or layer 2 switches) usually come with high performance for a given cost. In addition, due to the simple LAN protocols, plug-and-play installation is guaranteed in most cases. As the Ethernet packet supports multiple protocols, installing a bridge ensures multiprotocol support. Also bridges employing the spanning tree algorithm have automatic configuration ability.

Routers can forward the network layer packets, as opposed to the data link layer frames. This means that routers see layer 3 information. Based on this difference, routers usually have additional functionalities beyond those of bridges. A typical example among them is the ability to support fragmentation and the processing of TTL counters to avoid routing loops. Routers also efficiently support administrative isolation and full mesh topologies. Limited multicast propagation can also be easily established. The ability to support complex routing protocols means that routers can have multiple paths to a destination and can discriminate among them.

In many cases, the choice of a bridge or a router will depend on the level of the organization of desktop/workgroups in an office. However, routers are gaining advantageous positions for several reasons. First, the cost advantages that bridges used to have over routers are disappearing. Second, in many cases the increased network complexity demands the use of routers. Third, the increased security concerns need certain functions that only routers can support.

3.3.4 Layer 4–7 Switching

Layer 2 and layer 3 switchings basically rely on hardware to increase performance. In contrast, layer 4–7 switching is intended to utilize the processing information found in transport-level and application-level protocol headers as part of packet classification. For example, a layer 4 switch may give priority to UDP packets that are carrying RTP streams due to the real-time characteristics of RTP applications. Originally layer 4 switches first appeared, followed recently by layer 5–7 switches.

While the terms are all different slightly, the key concept of layer 4–7 switches is that the realization of routing need not be limited just to the layer 3 information. For example, different paths could be chosen based on TCP/UDP port numbers. The network layer router implements routing policies based on higher-level information such as the type of application flow. Specific examples include giving SNMP a high priority, video a medium priority, and ftp a low priority. When the network is congested, delay-sensitive applications may be moved to the head of queue. Another example may be that security policies are enforced so that telnet sessions from outside are not allowed into the network and external Web site accesses are also prohibited. This is basically a firewall. Other possible uses of layer 4 switches include the ability to provide QoS considerations by routing individual flows based on the URL of the destination and the ability to provide server load balancing. In the latter case, the layer 4 switch would be put at the front end of server clusters, where it would then direct requests to the most appropriate server based on the current load of each server.

Whether a switch is termed layer 4 or layer 7 depends on how much of the packet header information is analyzed and used in the switching/routing decision. Additionally, as in the example of the firewall above, layer 4–7 switches also frequently include methods for defining policies for handling the application traffic. In many cases a layer 4–7 switch may be viewed as an *application-level gateway* (ALG). Nevertheless, it is called a switch because it is faster and is constructed based on silicon integration.

3.3.5 IP Switching, Tag Switching, and MPLS

During the last several years, intense efforts have been exerted to develop innovative solutions that would enable the IP architecture to solve various interrelated problems. First, there was the need to evolve the routing architecture of IP networks. It has been extremely hard in the past to develop and deploy new routing mechanisms in the Internet because the components for packet forwarding and the component for packet routing process control are interrelated. That is, there must be a separate forwarding component and control component for unicast routing and multicast routing.[29] Second, there was the need for greater performance in routers, basically getting better performance-to-price ratios. Third, the problem of integrating IP and ATM networks was becoming very complex. Numerous solutions were developed but mostly involved the use of many servers and complex protocols for the synchronization of those servers. Fourth, the scalability of many of the overlay network models that were being examined by the IP over ATM solutions was also a major problem, as the overlay network routers must maintain links to all the routers on the same ATM cloud, which leads to problems in scaling the routing algorithms.

Various methods and solutions have been put forth. Among the factors considered were the scalability, cost, simplicity, and extensibility of the proposed solution. The main solutions that gained popularity are shown in Figure 3.21. Among the integrated models that were examined were IP switching, tag switching, and MPLS [9].[30] Among the overlay models were LAN emulation (LANE), multiprotocol over ATM (MPOA), classical IP over ATM, and next-hop resolution protocol (NHRP).[31] In the following, we shall concentrate on the integrated models.

3.3.5.1 IP Switching Protocol

One of the first solutions that was developed was the IP Switch by Ipsilon. It used a separate IP Switch controller along with two new protocols, *general switch management protocol* (GSMP) and the *Ipsilon flow management protocol* (IFMP) to set up VCs for long-lived IP packet flows [10, 11].

The GSMP is used for the communication between an external switch controller and the ATM switch—for example, connection setup/release and connection management on the ATM switch. Any ATM switch may operate as an IP Switch if the switch can understand the GSMP. The IFMP is the

29. Forwarding and control components will be explained in more detail in the next section.

30. The development of high-performance IP routers (the so-called *gigabit* and *terabit routers*) was also considered a solution to some of the above problems. While they did not solve the problems related to running IP over ATM, it could be argued that they ensured that these problems need not be addressed by making pure IP networks run faster and more efficiently than ATM networks.

31. The overlay models and other details pertaining to the IP over ATM models will be examined in detail in Chapter 7.

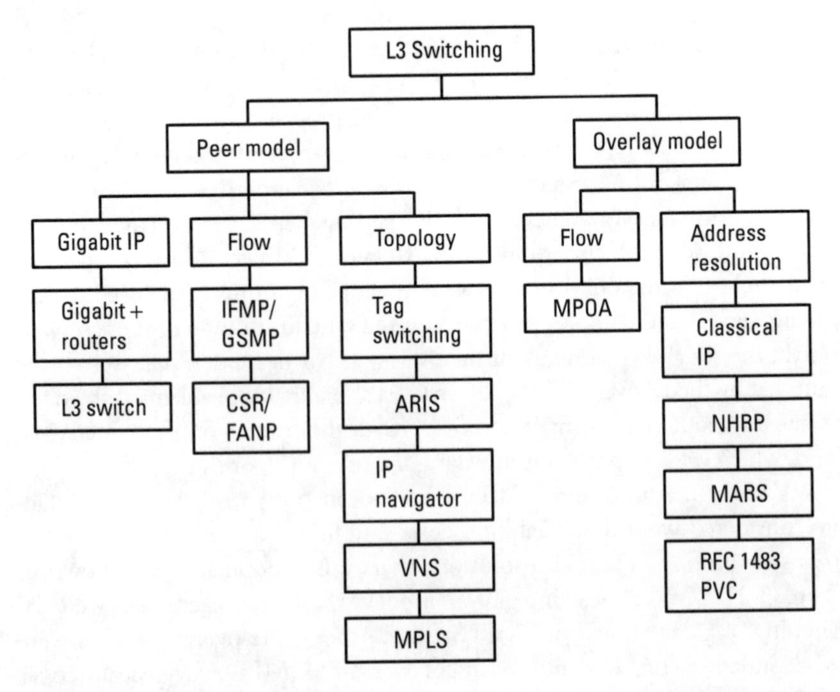

Figure 3.21 A taxonomy of L3 switching solutions.

protocol used for communication between the hosts and IP switches to transport flow redirection message. It is basically a simple way of setting up and releasing VCI links between ATM switches for IP packet flows. Figure 3.22 shows the elements and the basic operation of the IP switching.

IP Switch operation. An IP Switch basically operates in four steps as shown in Figure 3.23. Initially, the ATM cells carrying the IP packets are all forwarded to the IP Switch controller. All ATM cells are carried over one default VC channel. The controller reassembles these packets, analyzes the destination address, and then routes the packets to the appropriate VC so that the ATM switch can send it to the correct next hop ATM switch. This operation is the same as in a normal router [see Figure 3.23(a)].

Once packets from the same source going to the same destination are received several times, the switch controller decides that the packets are part

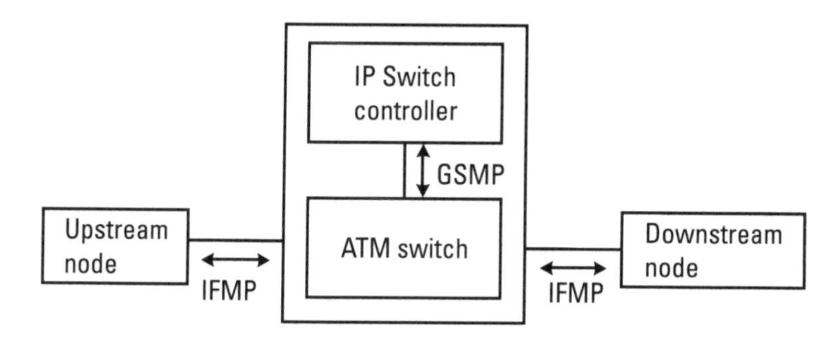

Figure 3.22 IP switching protocol.

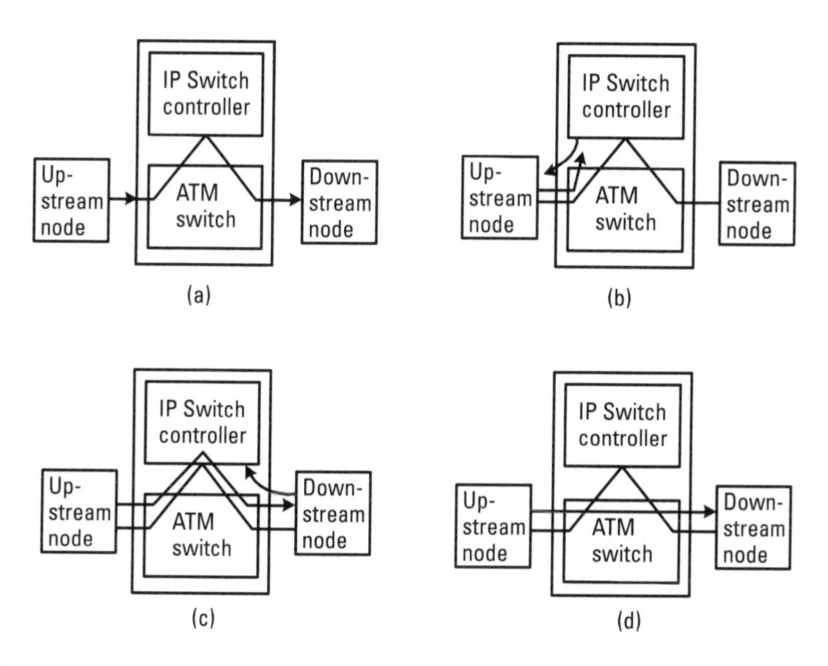

Figure 3.23 IP Switch operation: (a) default operation, (b) flows upstream labeled, (c) flows downstream labeled, and (d) cut-through switching.

of a flow that should be IP-switched. It then assigns an appropriate VCI for this flow and notifies the upstream switch that all packets for this flow in the future should be sent with this VCI. Then the ATM switch forwards all cells that come over this VC to the IP switch controller [see Figure 3.23(b)].

The key to IP Switch operation occurs in the next step, when the downstream ATM switch and its IP Switch controller also classify this packet flow as being a long-lived flow and indicate the VCI to be used. This means that the ATM switch may now just convert the VC label on all cells received from the upstream ATM switch to the VC label received from the downstream switch [see Figure 3.23(c)].

The end result is as shown in Figure 3.23(d), where the cells carrying the particular IP packets are just switched by the ATM switch without going through the IP Switch controller. This cut-through switching operation continues until the end of the IP packet flow.

Flow classification. IP Switch relies on the accurate classification of flows. There is no explicitly defined algorithm, but the aim is to identify the long-lived flows that would benefit from IP switching. Examples of long-lived flows include the ftp data, telnet data, http data, Web image downloads, and multimedia audio/video. It would be obviously advantageous to classify these flows as being suitable for IP switching. In contrast, short-lived flows that should not be IP-switched include DNS queries, SMTP data, *network timing protocol* (NTP) data, and SNMP queries.

The IP Switch had various problems, the main one being that it uses too many virtual circuits. For example, many different virtual circuits could be made along one path, and even for the same route different virtual circuits may be made for each traffic flow.

3.3.5.2 Tag Switching

Tag switching was developed by Cisco as another solution to the problems detailed above. In many ways it can be viewed as being ATM reinvented to fit with TCP/IP networks.

Tag switching uses tags in between the layer 2 header and the layer 3 header to enable a simpler and, presumably, faster method of packet switching. The concept of a tag is similar to the concept of virtual path identifier (VPI)/virtual channel identifier (VCI) identifiers in ATM cells. In fact, when tag switching is applied to ATM hardware, tags are carried in the ATM cell's VPI/VCI fields. By using short fixed-sized tags, tag switching is able to carry over many of the advantages of ATM to TCP/IP networks.

Figure 3.24 illustrates the operation of tag switching. Each packet contains a tag, and when a tag switch receives a packet, it first looks up the tag in its routing table. If a match is found, there will be an indication as to which

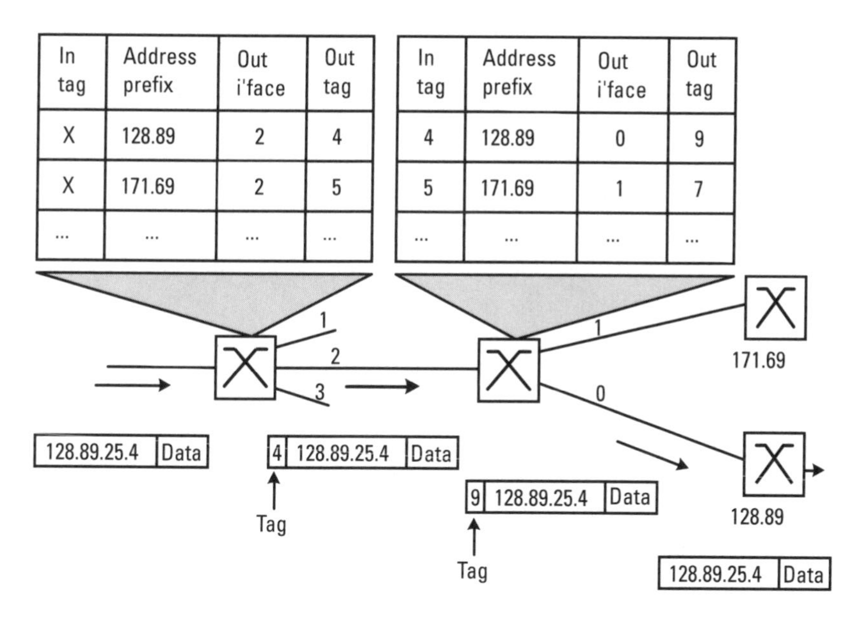

Figure 3.24 Tag switching example.

output port to use and which output tag to use. Obviously, it would be easy to extend this model to include QoS. For example, the edge *tag switch router* (TSR), or the first tag switch that a user's packet hits in the network, uses different tags for different precedences.[32]

3.3.5.3 MPLS

MPLS is the label-swapping routing technology that was developed by the IETF in response to the intense interest in IP switching, tag switching, and other related protocols [12]. The goals of MPLS were to extend traditional IP in a number of ways. First, it aimed for a simplified forwarding model based on labels instead of longest prefix-match. Second, it aimed to support efficient explicit routing by making it possible for the route to be specified once by the source at the path setup time. Third, it aimed to efficiently support traffic engineering by enabling the splitting of traffic load over multiple parallel or alternate routes. Fourth, it aimed to efficiently support QoS routing by allowing the

32. As much of the MPLS work detailed next is based on that of tag switching, we omit further discussions on tag switching.

selection of routes based on QoS requirements. By these extensions MPLS was able to solve many of the original problems that were discussed earlier.

One of the questions that is often asked is the specific relationship between MPLS and ATM. MPLS is being designed to explicitly allow *label switch routers* (LSRs) to be based on both traditional router platforms and ATM switch platforms. In addition, MPLS and MPLS devices incorporate many of the basic building blocks developed by ATM for simplified forwarding, explicit routing, and traffic engineering.[33]

MPLS network architecture. Figure 3.25 shows how MPLS-based packet forwarding differs from that of traditional routers. In a traditional router, packet forwarding is carried out based on the destination address. A longest prefix match lookup is carried out against the entries of the routing table to find the output link onto which the packet should be forwarded. In contrast, packet forwarding in an MPLS switch occurs in three steps. First, at the ingress node

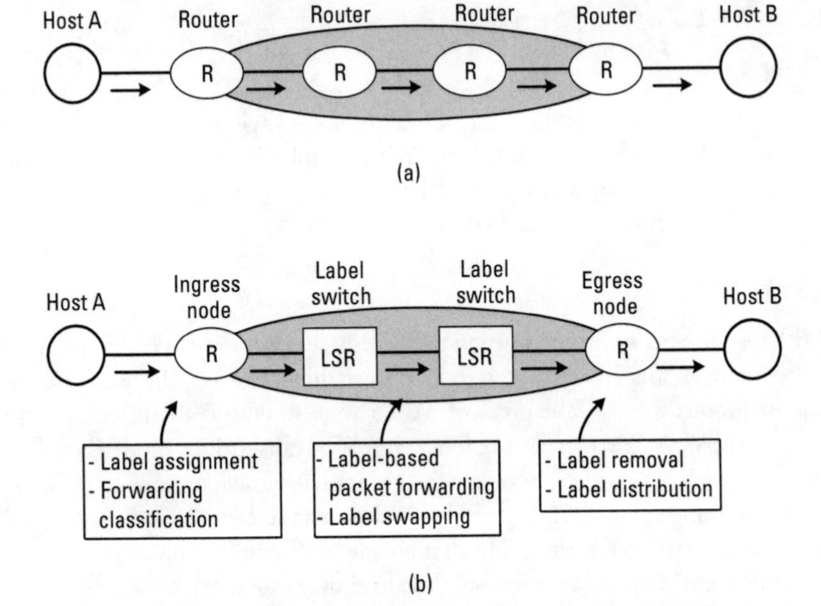

Figure 3.25 MPLS-based routing in comparison with conventional routing: (a) packet forwarding by router, and (b) packet forwarding by label switch.

33. Refer to Chapter 7 for more detailed discussions on how MPLS works with ATM.

the incoming packets are classified according to the *forwarding information base* (FIB) entries. Based on this information, the packets are assigned the appropriate labels. Second, in the MPLS switches inside the network, packets are forwarded based on the labels contained in their headers. At each MPLS switch, label swapping occurs, with the labels in the headers replaced with the labels from the FIB entries. Third, in the egress node, the labels are removed and the packets are forwarded as normal IP packets.

An MPLS domain consists of two types of MPLS switches, *label edge routers* (LERs) and LSRs. An LER is basically a full-function layer 3 router. It is located at the edges of the MPLS domain and, based on the FIB entries, it binds labels to the packets received. When the packets are leaving the MPLS domain, it removes the label bindings. This indicates that the ingress and the egress nodes must be LERs. In contrast, all switches inside the MPLS networks are LSRs. LSRs carry out switching on labels, concentrating on label switching. The core switches are in many cases built with an MPLS control component and an ATM switch fabric for data forwarding.

The problem of label encoding is solved in MPLS depending on whether the data link layer supports label switching natively. For cases such as the Ethernet, where no field exists in the data link frame to support labels, the MPLS architecture defines "shim" fields that are to be included between the layer 2 and layer 3 headers. This will be possible only in MPLS-specific hardware or software routers. The generic MPLS encapsulation "shim" consists of a 20-bit label, a 3-bit experimental field, a 1-bit indication of the bottom of the stack, and an 8-bit TTL.

When the data link supports label switching, methods have been defined for mapping the labels to the appropriate fields in the data link layer. For example, in ATM networks the labels will be mapped to the VPI/VCI fields (refer to Chapter 7).

Label assignment in the MPLS architectures can be done in several different ways. The first method is the topology-driven label assignment model, where labels are added to data exchanged through the routing protocol control traffic such as RIP and BGP. As the labels are assigned to the destinations or routing table entries in this case, this method is not adequate for link state protocols that distribute link states only. Note that only distance-vector routing protocols such as RIP or BGP explicitly exchange routing table entries. In the case when the network uses link state protocols, a second method that employs the LDP may be used to facilitate the exchange of MPLS labels. A third method is the request-driven label assignment that employs a request-based control of traffic, such as RSVP. In this case, LSRs in the MPLS networks process the control traffic only.

3.4 Network Control

For the Internet to be able to support applications that demand guaranteed level of services, it is necessary to devise a means for reserving resources along the path between the source and the destination. RSVP is the Internet signaling protocol devised for such resource reservation, which supports unidirectional resource reservation. We consider the RSVP as the means for Internet network control in this section [13].

3.4.1 RSVP Design Features and Basic Operation

RSVP has a number of design features that distinguish it from other previous signaling and reservation protocols. Accordingly, the operation of RSVP takes a unique form that reflects the design features.

3.4.1.1 RSVP Design Features

RSVP was designed, from the beginning, to support multicast. RSVP is a *receiver-initiated* resource reservation protocol in which the receiver is responsible for initiating the resource reservation. This reservation scheme enables the accommodation of heterogeneous receivers within a single multicast group. Specifically, each receiver can reserve the amount of resources that best fits to its own environment and can select and change the information streams that it needs.

RSVP provides various different reservation types and allows for each application process to specify how to combine reservation resources for the multicast group aggregated at the intermediate switches. This feature helps to efficiently use network resources.

RSVP employs the *soft-state* concept that enables easy updating of the switch state. This feature enables dynamic changes in the multicast group membership and to the automatic adaptation of this routing change. It also enables RSVP to deal effectively with dynamic reconfiguration of the multicast groups. Note that RSVP signaling is basically used to set up the reservation state for a session, where a session means a simplex data flow sent to a unicast or a multicast address.

RSVP employs two main messages that are exchanged over the path. One is the *PATH message* that a sender sends periodically via the data delivery path. This message is to set up the path state of each router, including the address of the previous hop. The other is the *RESV message* that each receiver sends via the reverse path. This message is to specify the reservation style, the desired QoS, and other factors, and it also sets up the reservation state at each router.

The above two messages and their operations bring up a couple of items to note. The first is that the RSVP is basically a receiver-initiated reservation and the second is that the routing is decoupled from the actual reservation. In fact, RSVP is designed to work with a variety of routing protocols by requiring only a minimal routing service; RSVP only requires information on how to route a PATH message.

3.4.1.2 Basic RSVP Operation

We examine the basic operation of RSVP by considering the simple example illustrated by Figure 3.26(a) [14]. In the figure, H1 and H2 are data sources, H3, H4, and H5 are receivers, and S1, S2, S3, and S4 are switches where resources are reserved. The set of links formed by the solid arrows, or that

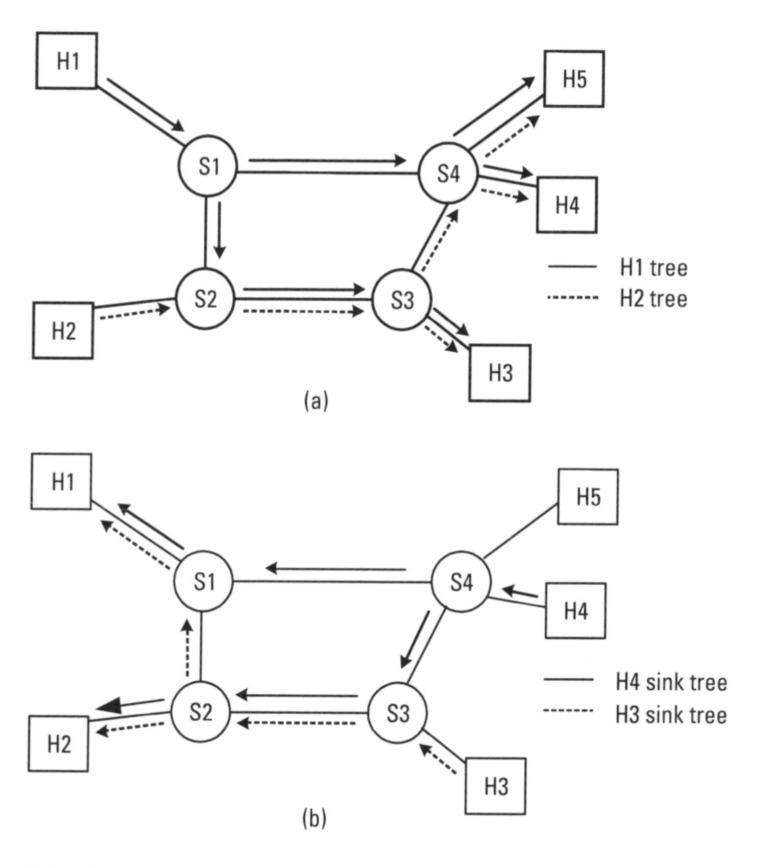

Figure 3.26 Illustration of RSVP setup: (a) path message flow, and (b) setup message flow [14].

formed by dashed arrows, indicates the multicast routing tree for the relevant multicast communication.

Before or when a data source starts transmitting, it sends a PATH message that contains the flow specification of the data source to the target multicast receiver group along the routing tree. The switch that receives this routing message checks if the routing information of the multicast destinations readily exists and creates one if it is nonexistent. The routing information includes the basic information such as the incoming and outgoing links for the sender and the receiver. In Figure 3.26(a), the senders H1 and H2 send PATH messages to the receivers H3, H4, and H5 along the multicast routing trees depicted as solid and dotted lines, respectively.

When a receiver receives a routing message from a receiver that wants to reserve resources, the receiver issues a resource RESV message based on the flow specification in the routing message. This reservation message is transmitted along the *sink tree*, which is the reverse route of the PATH message, to reach the data source. In Figure 3.26(b), the dotted lines depict the sink tree of H3 and the solid lines the sink tree of H4. Any switch on the route, if it cannot meet the resource reservation request, sends an RSVP reject message back to the data receiver and discards the reservation message.

The switches on the sink trees merge the reservation requests for the same multicast group by pruning those that carry a request for reserving resources smaller than, or equal to, the amount of resources of some previous request. For example, in case H1 is a video source—if H4 has reserved enough bandwidth to receive the full video data stream while H5 wants to receive only low-resolution video data—then the requested reservation of H5 is discarded as a result of these readily reserved sufficient resources. Once the reservation is established, the receiver periodically sends a reservation refresh message, but otherwise the reserved resources will be released.

3.4.2 RSVP Messages and Performance Guarantees

In support of the RSVP operation discussed above, the PATH and RESV messages play important roles. They contain several different objects and parameters such as TSPEC and RSPEC.[34]

34. Refer to [13, 14] for details of RSVP messages and specifications.

3.4.2.1 PATH Messages

The basic PATH message contains a *sender template* and a *SENDER_TSPEC.* The PATH message always follows the same path as that taken by the data packets. The PATH message leaves the path state in the form of the previous hop information at each router in the downstream route. This information is used by the RESV message sent back by the receiver so that it may follow the same path as the data packets but in the opposite direction. The sender template is a filter spec that identifies the sender. It is the basis on which the routers along the path decide whether to forward a RESV message to this sender. The SENDER_TSPEC describes the data traffic generated by the sender. This must never be modified in the network.

The most important parameter in the PATH message is the SENDER_TSPEC. It is delivered to both the intermediate network elements and the receiving applications. Each intermediate network element uses the SENDER_TSPEC and information arriving in the RESV messages to make an appropriate resource reservation from the desired service. Figure 3.27 shows the format of the SENDER_TSPEC, in which Tb TSpec stands for the token bucket TSpec.

One more object that may be optionally included in the PATH message is the *ADSPEC* object. It describes the services and parameters of the path and may be updated by the network elements. It is used in the *one pass with advertisement* (OPWA) method of reservation. Basically this method uses the ADSPEC to keep a record of the resources that are available

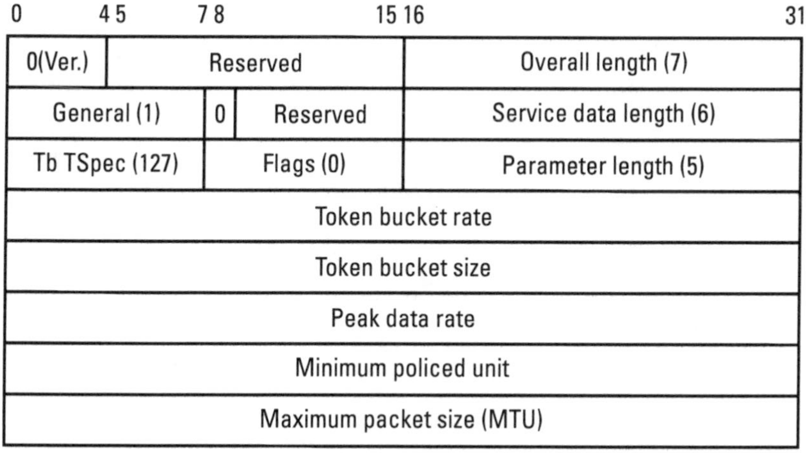

<table>
<tr><th>0   4</th><th>5   7</th><th>8</th><th>15</th><th>16   31</th></tr>
<tr><td>0(Ver.)</td><td colspan="3">Reserved</td><td>Overall length (7)</td></tr>
<tr><td colspan="2">General (1)</td><td>0</td><td>Reserved</td><td>Service data length (6)</td></tr>
<tr><td colspan="2">Tb TSpec (127)</td><td colspan="2">Flags (0)</td><td>Parameter length (5)</td></tr>
<tr><td colspan="5">Token bucket rate</td></tr>
<tr><td colspan="5">Token bucket size</td></tr>
<tr><td colspan="5">Peak data rate</td></tr>
<tr><td colspan="5">Minimum policed unit</td></tr>
<tr><td colspan="5">Maximum packet size (MTU)</td></tr>
</table>

Figure 3.27 SENDER_TSPEC format.

along the path from the source to the destination so that the receiver may know, upon receiving the PATH message, the amount of resources available on the path. Based on the receiver's requirements and the information in the ADSPEC option, the receiver is able to request an optimal amount of resources in its RESV message.

3.4.2.2 RESV Messages

The RESV message is transmitted by the receiver after it receives a PATH message. This is basically a reservation request. It contains both flow spec and filter spec. In handling these RESV messages the intermediate routers must be aware of the various reservation styles and the merging rules that RSVP permits.

A RESV message contains the FLOWSPEC object. The FLOWSPEC object contains the TSPEC and RSPEC. The TSPEC describes the traffic flow desired, while the RSPEC describes the parameters required to invoke the desired service. These two parameters are the basis on which the routers along the path back to the source router are able to decide how much resource to reserve. The RESV message is transmitted upstream to the intermediate network nodes and the senders. It may be merged at the intermediate network nodes.

As previously mentioned, the RESV message can specify the reservation. A reservation request or flow descriptor has two parts, the FLOWSPEC and FILTERSPEC. The FLOWSPEC specifies the desired QoS parameters used by the admission control module during the setup time and by the traffic control module or the packet scheduler. The FILTERSPEC specifies the set of data packets that can use the reservation. It consists of a list of <IP source address, port number>, with the empty list implying a *wildcard* (meaning that any source can use the resource).

As mentioned above, the FLOWSPEC is formed by a combination of the TSPEC and the RSPEC. The TSPEC, the traffic specification, consists of peak rate of flow, bucket depth of maximum burst size, token bucket rate or long term average rate, minimum policed unit, and maximum datagram size. The RSPEC, the resource specification, consists of the required service rate and a slack term.

When the upstream routers receive the FLOWSPEC the router may merge multiple FLOWSPECs and make appropriate service reservation. When the FLOWSPECs arrive at the sender, the sender takes new minimum MTU value for sending data. Figure 3.28 shows the format of the FLOWSPEC.

0 — 4	5 — 7	8	9 — 15	16 — 31
Ver.(0)	Reserved			Overall length (7)
Serviced id (5)		0	Reserved	Service data length (6)
Tb TSpec (127)		Flags (0)		Parameter length (5)
Token bucket rate				
Token bucket rate				
Peak data rate				
Minimum policed unit				
Maximum packet size (MTU)				
Rspec for invoking QoS control (parameter header, parameter) pair				

Figure 3.28 FLOWSPEC format.

3.4.2.3 Performance Guarantees

Based on the mechanisms and messages described above, RSVP can offer performance guarantees. The performance guarantees are based on the traffic characteristics of the source (carried in the TSPEC) coupled with the minimum amount of resource reservation needed by the receiver to get reasonable services.

The delay has two components. The first component, or the fixed component, is the sum of propagation and switching delays along the path, so it is a function of the path. The second component, or the variable component, is the sum of queuing delays along the path, so it is a function of the service and current state of the network. The guaranteed service provides a bound on the variable component, the queuing delay. It is a function of the TSPEC and RSPEC of the connection and the parameters of all the nodes along the path. It is calculated based on the fluid model.

3.4.3 Characteristics of RSVP

As stated earlier, RSVP has a number of interesting characteristics that distinguish it from other previous signaling or control protocols. They include the use of soft-state mechanisms, multiple reservation styles, receiver-initiation, and reservation merger.

3.4.3.1 Soft State

A major characteristic of RSVP is that it uses *soft state.* Each per-session state, such as path state and reservation state, has a timer associated with it and the state is lost when the timer expires. The sender/receiver periodically refreshes the state, resends PATH/RESV messages, and resets the timer. The states can also be explicitly deleted by a teardown message. There are several advantages of using soft state: it is not necessary to clean up dangling states after failure, it enables tolerating lost signaling packets, leading to the characteristic that the signaling message need not be reliably transmitted, and it helps to easily adapt to route changes.

3.4.3.2 Reservation Styles

Another major characteristic of RSVP is that it offers multiple reservation styles. The key motivation for this is to achieve more efficient resource utilization in many-to-many multicast applications. It stems from the observation that in videoconferencing only a few senders can be active simultaneously and, consequently, multiple senders can share the same reservation. The various reservation styles specify different rules for sharing among senders. Different reservation styles use different filter specifications to specify which senders can use the reservation.

There are three reservation styles. The first is the *fixed filter* (FF), which permits no sharing among senders and under which senders are explicitly identified for the reservation. It is indicated by the use of filters of the form <IP source address, port number>. The second reservation style is the *wildcard filter* (WF), which permits sharing among senders and explicitly identifies senders for the reservation. The filters are of the form of an empty list or *. The third reservation style is the *shared explicit* (SE) filter, which permits resource sharing among senders that are explicitly specified. Figure 3.29 shows examples of the FF and WF reservation styles.

3.4.3.3 Merging of Reservations

Another characteristic of RSVP is that the merging of a number of reservations into a single reservation is allowed. This follows from the multicast and receiver-initiated model of the reservation. In reservation, merging can occur when multiple receivers of the same session may request reservation for the same sender. An intermediate router merges the requests and passes the merged message to the upper link. The RESV message is forwarded back along the original downstream path according to the previous hop path state that was set up by the PATH message.

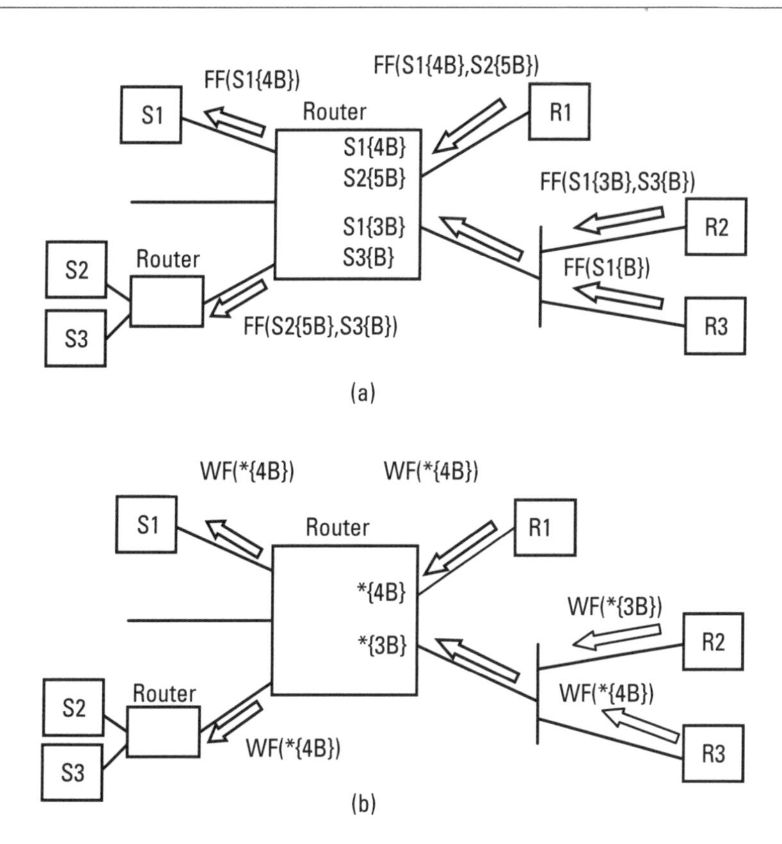

Figure 3.29 Examples of reservation styles: (a) FF, and (b) WF [13].

One problem may occur when different receivers request different QoS. A number of rules define the actions to take in this case. The most important rule among them is that conflicting reservation styles may not be merged. Basically, merging reservations means that the maximum value of the reservations requested is calculated and is used for a new merged reservation request. Consequently, this can lead to cases where larger requests may have to be rejected because of insufficient resources.

3.4.3.4 OPWA

In the basic RSVP model, the receiver does not know what reservations are supportable by the network; even though it can send back an RESV message with a requested FLOWSPEC, there is no guarantee that resources will be available. OPWA aims to improve this basic model by collecting the information along the path such that receivers can make a better reservation request. This is made possible by including the ADSPEC object in the PATH message sent by the sender. The ADSPEC object contains a default general parameter fragment. Service-specific fragments such as guaranteed service or controlled-load service fragments are also allowed. At each network element each service updates its section of ADSPEC. On receiving a PATH message containing the ADSPEC option the receiver is able to find out whether RSVP is fully supported along the downstream path and what services and resources are available. The receiver then uses SENDER_TSPEC and ADSPEC to pick the right service and the amount of resources needed.

3.5 Traffic Management

The major traffic management tools or functions in TCP/IP networks are the TCP congestion control mechanisms. The basic philosophy underlying TCP congestion control was developed under the assumption that all the users of the network are not greedy and are willing to share network resources; when there is no congestion each user is supposed to use the maximum bandwidth available, but, when congestion occurs, all users involved are supposed to decrease their transmission rates accordingly.

The TCP congestion control methods applied at the end stations are all indirect methods for detecting congestion in that the end user studies the pattern of acknowledgments or packet losses and then extrapolates the state of the network from them. Gateways actually know whether or not congestion is occurring, and as such are able to detect the onset of congestion much better, so feedback mechanisms should be more effective in congestion management.

There are various mechanisms available for gateway congestion control. These can be grouped into three types—notification schemes, active queue (buffer) management schemes, and bandwidth management or scheduling schemes. *Notification schemes* mean that gateways use some sort of congestion notification schemes. *Active queue management schemes* decide how to queue packets and how to allocate the buffer space for these queues. *Bandwidth management schemes* control the traffic by applying scheduling schemes that determine the order of packets to transmit.

3.5.1 TCP Congestion Control Mechanisms

The seminal work on TCP congestion control mechanisms is the one invented by Jacobson and Karol [15], which introduced the basic adaptive window adjustment mechanisms used in today's implementations.[35] Though various other algorithms have been proposed, none of them have been universally adopted.

3.5.1.1 Basic Mechanism

In the TCP congestion control mechanism, there are two stages in the window adjustment as shown in Figure 3.30, *slow start* and *congestion avoidance.* The window size used in sending packets is the minimum of the window size

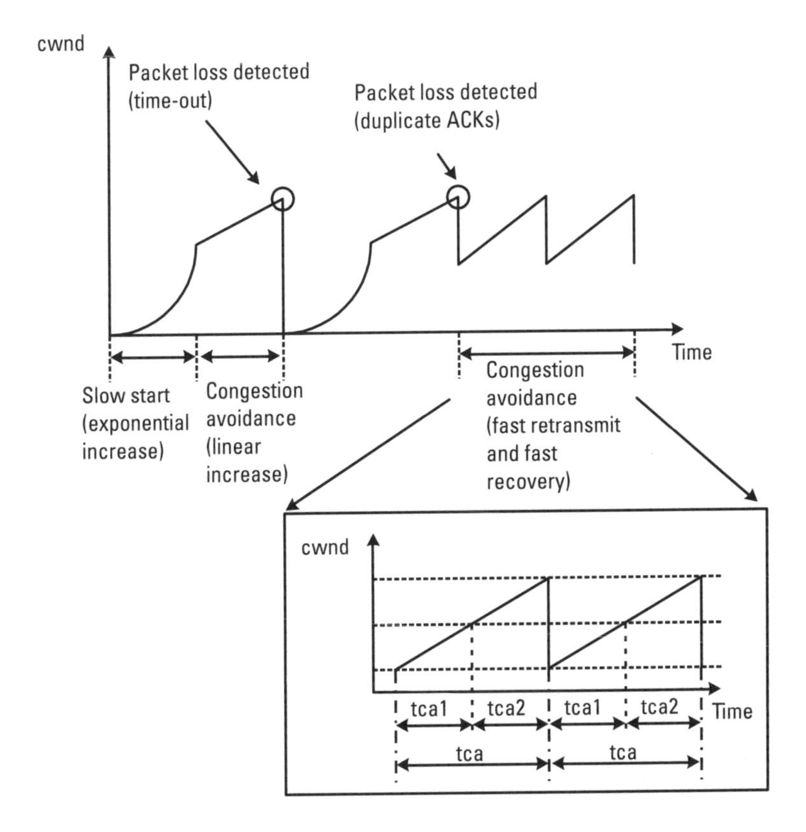

Figure 3.30 TCP cwnd adjustment (TCP Reno).

---

35. TCP implementations were distributed in the releases of 4.3 BSD Unix. The release was generally known by the names Tahoe and Reno, rather than their official names of BSD Network Release 1.0 and 2.0. Following this convention we use the terms TCP Tahoe and TCP Reno in this book.

advertised by the receiver and the *congestion window* (cwnd), which is adjusted by the window adjustment algorithm.

The slow start stage is entered whenever the connection first starts to send packets, and after a retransmission timeout. The cwnd is first set to 1, and increased by 1 for every new acknowledgment packet received. This results in an exponential increase in the window size after every round trip time—that is, the successive window sizes are 1, 2, 4, 8, etc. This stage continues until a timeout occurs, indicating a packet loss, or until the cwnd equals the slow-start threshold (ssthresh) value. If a timeout occurs the slow start stage is entered again with the cwnd set to 1, and half of the former cwnd value is saved in ssthresh. If the cwnd equals ssthresh the congestion avoidance stage is entered.

In the congestion avoidance stage, the window is increased by 1/cwnd for every new acknowledgment, resulting in a linear increase of 1, approximately, in the window size in every round trip time.

Basically, the slow start stage aims to quickly fill up the empty pipe with data by exponentially increasing its window size every round trip time. In contrast, the congestion avoidance stage aims to probe the limits of the pipe size and increase the window size if possible.

3.5.1.2 TCP Tahoe and TCP Reno

The basic mechanism was further refined after it was observed that the loss of a single packet would result in the pipe emptying and consequently bandwidth waste. The loss of a packet can be detected in two ways—through a *timeout*, or the reception of a *duplicate acknowledgment*.[36]

While the use of a retransmission timeout is a basic and fail-safe method of detecting packet loss, it wastes time in that a time much longer than the estimated RTT must pass before timeout occurs. Packet loss can be detected much more quickly by using duplicate acknowledgments. For example, if only the *n*th packet were lost, the reception of the $n+1$th and $n+2$th packets would trigger duplicate acknowledgments, all acknowledging the reception of the *n*th packet. On receiving a certain number of duplicate acknowledgments, the transmitter may assume that the *n*th packet was indeed lost. In this case the transmitter can quickly retransmit the packet without waiting for a timeout. This is called the *fast retransmit mechanism.*

A further improvement may be realized by using the fact that usually duplicate acknowledgments signify that only a single packet was lost, so

36. Duplicate acknowledgment indicates the case where the sender receives the same acknowledgment message as was previously acknowledged.

entering the slow start stage would needlessly empty the pipe. Instead, after the fast retransmit, half of the current cwnd value is saved in ssthresh, and the cwnd is set to ssthresh plus 3. Every time another duplicate acknowledgment is received the cwnd is increased by 1, and a new packet is transmitted if allowed by the new value of the cwnd. When an acknowledgment that acknowledges new data is received, the cwnd is set to ssthresh and the congestion avoidance stage is entered. In this way the cwnd is effectively halved after a packet loss, and the slow start stage is avoided. This is called the *fast recovery mechanism.*

The actual operation of the TCP transmitter depends on what type of implementation it is. In TCP Tahoe, the transmitter only does fast retransmit and immediately enters the slow start stage after saving half of the cwnd value in ssthresh. In TCP Reno, the fast recovery scheme is also used.

3.5.1.3 TCP Vegas

There have been various other proposed improvements to the basic mechanisms. They can be grouped into the changes in the end host operation and the improvements depend on new TCP options. The following gives short summaries of both of these approaches.

Many are based on adding preventive control like mechanisms to the TCP window adjustment policy by checking whether the window increase has any adverse effects. The Tri-S scheme measures the throughput after every window increase and sets the window size back to the previous value if it detects a drop in throughput. The dual scheme combines the DECbit algorithm with the TCP/IP scheme. Basically it measures the delay after a window increase, and decreases the window size if the delay increases. The TCP Vegas scheme that we explain next is also similar to these schemes.

TCP Vegas aims to control the amount of extra data that the connection has in transit, where by extra data we mean the data the source would not have transmitted if it had been trying to exactly match the available bandwidth of the network. The goal of TCP Vegas is to maintain the right amount of extra data in the network [16]. If the source is sending too much extra data, it will cause long delays and will possibly lead to congestion. Also if a connection is sending too little extra data, it cannot respond rapidly enough to transient increases in the available network bandwidth. TCP Vegas first defines an expected rate based on the minimum of all measured roundtrip times. Second, it calculates the current sending rate once per roundtrip time by recording the sending time of the current packet and the time its acknowledgment is received. Third, TCP Vegas compares the actual rate with the expected rate and then linearly increases or decreases the cwnd accordingly. The overall effect is to keep the actual rate from straying too far from the available network bandwidth.

As mentioned in Section 3.5.1.2, it is important to detect lost packets and retransmit them quickly to avoid emptying the network pipe. The fast retransmit and recovery method aims to achieve this just by adjusting the end-user implementation, while at the same time not introducing any new options into the actual TCP protocol specification. However, in a number of studies, it has been shown that when multiple losses occur, the TCP Reno algorithm may have adverse effects. A general solution for these cases could be found in incorporating some sort of selective acknowledgment procedure, where all packets successfully received are acknowledged. This is studied for the case of TCP in *selective ACK* (SACK) [17]. By using SACK, the end source may retransmit only the packets that were lost, instead of following the go-back-*N* method of the basic algorithm.

3.5.2 Active Queue Management

The basic function of queue management is to add a packet to a queue, drop a packet if the queue is full, and remove a packet from the queue (if ordered to do so by the scheduler). Optionally, the queue management scheme may try some proactive functions such as monitoring the queue occupancy, maintaining the history of congestion, dropping packets before the queue is full (according to some criteria), and marking packets (according to some criteria).

The most intuitive and simplistic queue management scheme is the *drop-tail* scheme. The router drops all packets that arrive after a queue is full. Due to its simplicity, many routers originally implemented this scheme.

However, the drop tail scheme has many problems. The main problem is that the drop-tail detects congestion after the queue is full. Consequently, in many cases it maintains a high average queue length, resulting in a large average delay. Also once congestion occurs, the drop-tail mechanism frequently results in global synchronization of all the affected TCP connections simultaneously going into slow start. This is because the drop-tail scheme will drop all packets as they arrive at the queue and thereby affect connec- tions indiscriminately. The drop-tail mechanism also has a bias against bursty traffic, causing bursty traffic sources to loose more packets than smooth traffic sources. In addition, as the queue management scheme cannot control which packets to drop, the operator experiences difficulty in offering differentiated services to users.

The Internet Research Task Force (IRTF) has studied this problem and has recommended that a certain type of queue management be implemented, with the RED algorithm being used as a specific example [18]. RED is an example of a queue management algorithm that involves statistical, randomly distributed feedback signals whose strength (or probability of occurrence) is an increasing function of the average queue occupancy. Various other queue management schemes based on RED have also been developed, fixing problems or deficiencies in the operation of RED.

3.5.2.1 RED

The RED algorithm aims to avoid congestion by detecting congestion before it occurs and to proactively stop congestion from happening. Usually this is achieved by dropping packets before congestion actually occurs. This achieves the desired effect because the RED algorithm assumes that most of the traffic coming into the network implements feedback-based transport protocols such as TCP.

The design goals of RED were to achieve congestion avoidance (rather than congestion control which is an a priori mechanism), to avoid global synchronization, to avoid a bias against bursty traffic, and to bound on average queue length (which would lead to lower average latency).

The RED algorithm maintains the average queue length and, based on this queue length, calculates a drop probability for each incoming packet when the average length is larger than a certain value *min_th*. When the average length is larger than another threshold *max_th*, the packet is always dropped. When the average length is between *min_th* and *max_th*, the probability of dropping the packet increases. Specifically, for each packet arrival, the following algorithm is applied.

```
if avg_len < min_th
     queue the packet
else if min_th < avg_len < max_th
     calculate probability P
     drop the arriving packet with probability P
else if avg_len ≥ max_th
     drop the arriving packet
```

This simple algorithm results in high throughput, while maintaining a low average delay for all TCP-like traffic.

Two methods for calculating the packet drop probability P were suggested in the original proposal. The first method is to take it as a linear function of the average queue size Q_{avg}. Specifically, $P_b = max\{0, (Q_{avg} - min\_th)/(max\_th - min\_th)\}$. This gives geometric (or roughly exponential) distribution of packet intermarking intervals (or the number of packets between marked/dropped packets). The second method is to replace P_b with P_a, where $P_a = P_b / (1 - count \times P_b)$. In this case the packet-marking probability increases with the value of *count*, which is the number of packets transmitted since the last marked packet. This gives packet intermarking intervals with a uniform distribution over $[1, 1/P_b]$.

The average queue length can be calculated by an *exponentially weighted moving average* (EWMA) of the instantaneous queue occupancy, or $Q_{avg} = (1 - W_q) \times Q_{avg} + Q_{inst} \times W_q$, where Q_{inst} is the instantaneous queue length and W_q is the exponential weight. The choice of W_q is critical to the performance of RED, as it will determine how quickly the algorithm will react to fluctuations in the queue length.

3.5.2.2 WRED

Weighted random early detection (WRED) is a modified version of RED deployed by a major router vendor. It aims to drop packets selectively based on IP precedence. It provides different thresholds and weights for different IP precedences.

The main advantage of WRED is that it can make the queue management algorithm become unfair in a controlled manner. This means that it can treat some packets with preference with respect to others. This type of "controllable unfairness" is an important stepping stone toward implementing QoS in the network.

3.5.2.3 RIO

RED with IN/OUT (RIO) bit is another method of implementing "controlled unfairness" in the queue management routines of a router. RIO maintains twin RED algorithms for calculating the drop probabilities of incoming packets. All incoming packets are assumed to be marked with a bit, indicating whether it is an IN packet or an OUT packet. The bit is set by upstream routers, usually the ingress router or edge device. These upstream routers are assumed to mark the packets according to whether it is "in" profile or "out" of the profile. The aim is to ensure that the "in" profile users get better service than the "out" of the profile users. When there is no congestion the two types of packets receive the same service, but when network congestion occurs the packets are treated differently.

The IN packets drop probability is calculated based on the average queue length of only the IN packets, whereas the OUT packets drop probability is calculated based on the average queue length of both the IN and OUT packets. More specifically,

```
For each packet arrival
if it is an IN packet
     calculate the average IN queue size avg_in;
     calculate the average queue size avg_total;
     if min_in < avg_in < max_in
          calculate probability P_in;
          with probability P_in, drop this packet;
     else if max_in < avg_in
          drop this packet.

if it is an OUT packet
     calculate the average queue size avg_total;
     if min_out < avg_total < max_out
          calculate probability P_out;
          with probability P_out, drop this packet;
     else if max_out < avg_out
          drop this packet.
```

3.5.3 Schedulers

There are several desirable attributes to any scheduling algorithm [2]. First, a scheduling algorithm must be easy to implement; the complexity and state information of any practical algorithm should be kept to a minimum to make implementation easy. Second, it should ensure fairness and protection between flows; all flows over a constrained network resource should share the flows fairly, and additionally no flow should suffer from the misbehavior of other flows. Third, it should offer performance bounds—deterministic or stochastic performance bounds—for constrained user traffic. This condition is absolutely necessary if the network is to support QoS. In addition, it should be scalable and support admission control and network operation efficiency.

The basic vanilla type of scheduler is the *first-come-first-served* (FCFS) scheduler. In this case the scheduler just receives and transmits packets in the order of arrival. FCFS is basically a work-conserving scheduling algorithm. An FCFS scheduler has the advantage of being simple to implement and consequently has been used in most routers in the past. If the network can be engineered to have sufficient bandwidth and not operate in a state of congestion, FCFS schedulers will be more than adequate.

The problem with the FCFS schedulers appears when the network becomes congested and sustained nonempty queues are maintained in the network nodes. In this case, the offered service levels will degrade. However, this degradation can be expected to affect users in an unpredictable manner, with some users getting near-normal service, while others suffer severe degradation in service levels.

Dynamic priority type schedulers have been developed to fix these problems with the FCFS queues. While various types exist, here we examine two main types, the frame-based schedulers and the *generalized processor sharing* (GPS)–based sorted-priority schedulers [19].

3.5.3.1 Frame-Based Schedulers

A frame-based scheduler operates by dividing time into frames. Each user is allocated a fixed number of transmission slots per frame. Note that the actual length of the frames may vary but the fixed number of transmission slots allocated per user does not. A simple example of a frame-based scheduler is the round-robin scheduler.

Round robin. In a round-robin scheduler each user is allowed to send one packet per round. Each user's queue will be checked at the start of each *round-robin* frame. The scheduler will go in sequence through the queues that have a packet to send and transmit one packet from each frame. The service interval, or the interval between the times when successive packets in the same queue are serviced, depends on the number of flows. This is a simple scheme that is easy to implement and ensures that all flows get a fair share of the bandwidth, as each flow will be ensured at least $1/N$ of the allowed bandwidth when there are N flows.

Round-robin schedulers have two problems. First, it is hard to guarantee latency bounds. It can happen that a packet is transmitted almost immediately after arriving at a queue but it can also happen that the packet has to wait a full maximum service interval. As a result this fluctuation can cause a large jitter. Second, the round-robin scheduling is fair only at time scales larger than the service interval.

Weighted round robin. One of the problems of the simple round-robin scheduler is that it gives each flow the same share of the bandwidth. *Weighted round robin* (WRR) improves upon this by allowing each flow to be serviced *round robin* in proportion to a weight assigned for each flow or queue. For example, if there are three users, with the weights of 0.2, 0.3, and 0.15, then the weights can be met by assigning 4, 6, and 3 timeslots per each round respectively.

If the users have variable-length packets, then the above weights must be adjusted to account for this. One way is to first adjust the weights by the average length of the packets per each user and then normalize the weights to determine the number of transmission slots per user. However, since some users will not know the average length of their packets it is not easy to use the WRR algorithm in this manner in variable-packet-length environment. Another algorithm, called the *deficit round robin*, waives the need for knowing the average length of the packets by introducing the concepts of deficit counter and service quantum.

If the packet length is small and fixed, the WRR algorithm yields a simple and efficient scheduler. ATM networks fit this criteria ideally. The small, fixed-size ATM cells along with the high bandwidth ensure that bandwidth can be shared efficiently with minimal jitter in ATM networks.

3.5.3.2 GPS-Based Sorted-Priority Schedulers

In a GPS-based scheduler, the outgoing link schedule is defined by a global variable that is used to prioritize the packets to transmit.

GPS. GPS is the basis of many sorted-priority schedulers. It uses a fluid-flow model of multiplexing user data. Figure 3.31 shows the model of an ideal scheduler implementing GPS. The scheduler assumes that traffic can be cut up into infinitely small bits and served in a bit by bit round robin fashion.

The basic service principles are that each session receives service according to the minimum rate allocated, and any leftover bandwidth is divided up among the sessions. As with other cases, it is easy to introduce weighting into this model. In this case, the user traffic is served in a bit-by-bit WRR

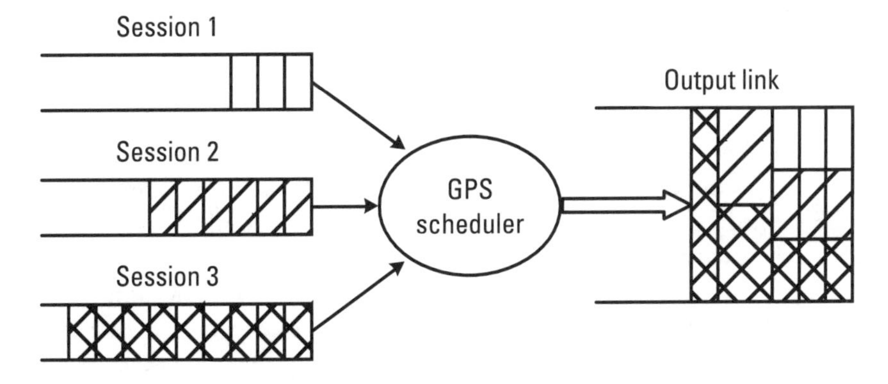

Figure 3.31 Illustration of the GPS.

method. Any leftover bandwidth is shared among the sessions in proportion to each session's allocated bandwidth.

The main problem with GPS is that it cannot be implemented in real networks. Servers cannot service multiple sessions at the same time in a bit-by-bit manner, as service units are always packets. This means that the packet of a session must be completely served (i.e., transmitted), before the service of the next session may begin.

Weighted fair queuing. Due to the above problems of implementing GPS in real networks, several approximations to GPS have been developed. Included among them are *weighted fair queuing* (WFQ), *self-clocked fair queuing* (SCFQ), *virtual clock*, *frame-based fair queuing* (FFQ), *starting potential-based fair queuing* (SPFQ), *worst-case fair weighted fair queuing* (WF^2Q), and WF^2Q+.

WFQ can be applied to real packet-based traffic, wherein it is also known by another name, *packet-by-packet generalized processor sharing* (PGPS). The main principle of WFQ is that it tries to ensure that packets are transmitted in the order they would have been served if an ideal GPS server had serviced them.

Specifically, when the scheduler chooses a packet to transmit it chooses the session with the smallest *finish tag* $F(i,k,t)$. A finish tag approximates the time when the packet would have been transmitted, had an ideal GPS server serviced the packet. A WFQ scheduler keeps track of the *virtual time* $V(t)$ of the corresponding GPS scheduler for this queue. This virtual time is a measure of time in a GPS scheduler that is used as the basis for calculation of the finish tag when a packet arrives at an empty queue. In terms of equation, the finish tag takes the expression

$$F(i,k,t) = \max\{F(i,k-1,t), V(t)\} + P(i,k,t)/r(i) \tag{3.1}$$

where $P(i,k,t)$ denotes the size of a packet that arrives at a queue, and $r(i)$ the rate user i is to be serviced. So the value $P(i,k,t)/r(i)$ is the amount of time a GPS scheduler using bit-by-bit round-robin service would have taken to service the packet. When the queue is nonempty, the packet must be queued behind the packet that is already in the queue. As this packet had a finish tag of $F(i,k\text{-}1,t)$, the finish tag of the new packet becomes $F(i,k\text{-}1,t) + P(i,k,t)/r(i)$. When the packet arrives at an empty queue, the packet would finish at $P(i,k,t)/r(i)$ after the current time of the GPS scheduler, $V(t)$. As a result, the finish time for the packet becomes $V(t) + P(i,k,t)/r(i)$.

WFQ has several advantages. It can protect connections from other misbehaving connections. In addition, it is possible to use WFQ to guarantee both bandwidth and delay bounds for a connection when the user's traffic is bounded. However, WFQ also has various disadvantages. It requires keeping states per connection, making scalability a problem. Calculation of the virtual time $V(t)$ is another big problem. Since the virtual time depends on the number of active connections in the scheduler, it must be updated continuously as the number of active connections changes continuously. The problem is that an update of $V(t)$ may result in more queues being in an active state, causing another cycle of calculation. This is called the *iterative deletion problem* and is a major disadvantage of the WFQ algorithm.

3.6 QoS

The existing Internet model is based on a best-effort service paradigm. It assumes that most applications will gracefully share whatever resources are available and that congestion may occur from time to time, during which every application will get a similar amount of degraded service.

However, the incredible growth and commercialization of the Internet have strained this model. Many users are now using both data and real-time applications. Additionally, due to the commercialization aspect, the traditional problems of the Internet have become more glaring. These include the lack of support for support QoS, the nonavailability of variable grades of services, and the lack of support for charging. More specifically, we can describe the limitations of the current IP networks as follows. First, they only provide best-effort service. Second, they do not participate in resource management. Third, they cannot provide service guarantees on a per-flow basis. Fourth, they cannot provide service differentiation among traffic aggregates.

These problems are currently being studied in the IETF and research community. Their efforts have resulted in two different QoS architecture initiatives—the *integrated services* (IntServ) QoS architecture and the *differentiated services* (DiffServ) QoS architecture.

3.6.1 IntServ

The IntServ Internet architecture was designed to enhance the basic IP service model. The old Internet model was based on a single best-effort service class with basically no resource management equipped at the IP level, but this new model replaces this with multiple service classes, including the best-effort and multiple QoS classes. In addition, explicit resource management techniques have been developed at the IP level to support the new service models.

The key architecture difference between the old model and this new model is that the old model was stateless, but the new model maintains per-flow state at the routers. Maintaining of state in the routers drives two critical aspects of the new model. First, as per-flow state is available, complicated admission control and traffic scheduling mechanisms are feasible. Second, to maintain and set up this state, a new signaling protocol is needed. The IETF developed the RSVP protocol in response to these needs.

3.6.1.1 IntServ Architecture

Figure 3.32 depicts the IntServ architecture. It consists of a classifier and route selector, a packet scheduler, a traffic control database, a reservation protocol, and an admission control. Management agents and routing protocols may be regarded as part of the architecture, as both affect the provisioning and availability of QoS services.

The IntServ model assumes that there are various components in the routers and nodes of the network. First, it is assumed that every node implements RSVP (refer to Section 3.4). Second, all routers are assumed to run an admission control mechanism. This mechanism determines whether the

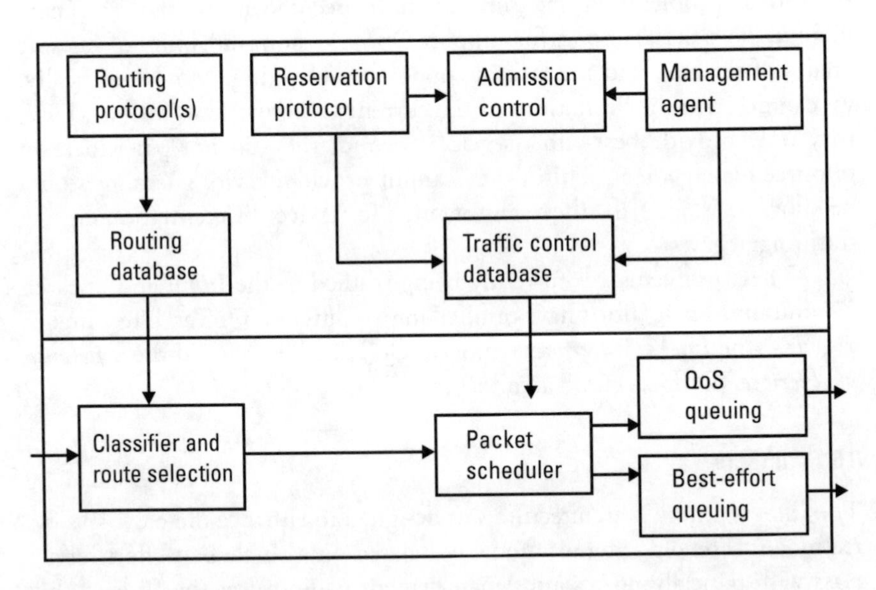

Figure 3.32 IntServ architecture.

node has sufficient available resources for admitting a call and, if so, the user's traffic may receive appropriate service. Along with the admission control, some sort of policy control is usually implemented to determine whether the user has an administrative permission to make the reservation. Third, it is assumed that each router has a packet classifier. The packet classifier analyzes all incoming packets to determine the route and the QoS class for each packet. Based on this it maps each packet to an appropriate class. Fourth, packet schedulers are assumed to make forward decisions on every packet to achieve the promised QoS on the particular link-layer medium. The packet schedulers may also perform a policing function to determine whether a packet may be allowed to enter the network.

3.6.1.2 IntServ Service Classes

Two service classes are defined in the IntServ architecture, besides the default best-effort model. They are the guaranteed service and the controlled-load service.

The *guaranteed service* is defined to support hard real-time applications. These applications would need a guaranteed deterministic or probabilistic upper bound on the delay for each packet in a session provided that the session does not send more traffic than is specified during the initial connection setup. The admission control is based on a worst-case analysis, as the guarantee must be kept under all conditions. To implement this service, per-flow queuing mechanisms and signaling support must be implemented at all routers in the network.

The *controlled-load service* is defined to support soft real-time applications and customers who wanted a "better" best-effort service. This service does not offer any hard guarantees but tries to ensure that each user gets service closely approximating the service that the same flow would receive from an unloaded network. Note that "unloaded network" service must be available even when the network itself is under congestion. The controlled-load service model is able to use more relaxed mechanisms than guaranteed service does, including measurement-based admission control and aggregate scheduling mechanisms in the routers.

3.6.1.3 Issues of IntServ

Overall, the IntServ model has not really succeeded in being widely deployed. There are a number of reasons for this. First, it relies on the use of RSVP and widespread state information. IntServ relies on all the nodes in the end-to-end path understanding and implementing the IntServ architecture. Given the size and heterogeneous state of the Internet, this could be a major stumbling block in the widespread deployment of IntServ. Second, it has many parameters, which make it complex to configure and operate. IntServ requires many components including signaling, packet classification, and scheduling. It is not clear how the various parameters used in these components should be set. Third, the amount of states that an IntServ router would have to maintain is very large. The more state that a router has to maintain, the more problems the routers will have in supporting more connections. However, all these problems are not necessarily unsolvable, so the IntServ architecture may become popular as more operators and users gain experience with it. Also, there have been various attempts to use IntServ and the following DiffServ mechanisms to solve these problems.

3.6.2 DiffServ

The DiffServ model for QoS services has been developed by the IETF in response to the observation that the complexity and end-to-end nature of the IntServ model would make it very hard to deploy. Service providers required that any architecture be scalable so that it could be easy and cheap to implement in the high-speed core routers that are in the backbones of the Internet. Additionally, it should be simple to manage and possible to differentially price the services so that users can receive differentiated services. On the other hand, users required something better than best-effort service. This may include not only some sort of bandwidth guarantee but also low delay, jitter, and loss. The key property from the user's perspective is that the user gets predictable services.

3.6.2.1 Scalable Service Differentiation

The DiffServ solution to the above problems was to put *differentiated services code points* (DSCPs) in every IP header. The values set in the DSCP would indicate the type or the level of service the packet is to receive. A second key part of the solution was to use an edge-core router model, where the edge routers would do the complicated analysis of the packets to set the DSCP, while the core routers would only classify and forward packets according to the DSCP field.

The use of DSCP as the basis for deciding the service level results in a number of advantages. The amount of state that must be kept in the core routers reduces to a minimum, which is equal to the number of DSCP code points. This also means that the amount of traffic classification state that must be kept in the core routers is also significantly smaller. The use of the

DSCP code points is a method of traffic aggregation that ensures that no microflow state needs to be kept in the core routers. Instead, the edge routers may have to keep a large number of states and probably have to keep track of microflows. This, however, is not perceived as a big problem as the edge routers are at the edge of the network where the traffic speed is comparatively slow. This basic innovation of using DSCP value in the headers results in a model that enables scalable service differentiation.

3.6.2.2 DiffServ Architecture

The DiffServ architecture is made up of a number of components such as DiffServ domain, DiffServ edge node, and DiffServ core node.

A *DiffServ domain* is a single domain that uses the DiffServ mechanism to provide edge-to-edge QoS. It consists of a set of continuous DiffServ nodes that are under the administrative control of a single management entity. Each DiffServ domain is normally composed of a number of DiffServ edge nodes and DiffServ interior nodes.

A *DiffServ edge node* is a node on the edge of a DiffServ domain. Depending on the direction of the traffic flowing into or out of the DiffServ domain, the edge node may either be a *DiffServ ingress node* or a *DiffServ egress node*. The DiffServ edge node enforces the service agreements between the DiffServ domains and carries out traffic conditioning on any packets leaving or entering the network as needed.

A *DiffServ core node* is a node inside a DiffServ domain. Depending on the *per-hop-behavior* (PHB) groups supported in the DiffServ domain, it is able to transfer packets according to the DSCP. An interior node is able to perform limited traffic conditioning on traffic flowing through it.

3.6.2.3 Key Elements of the DiffServ Architecture

The DiffServ architecture uses a number of elements such as DSCP, SLA, TCA, and PHB.

DSCP. The DSCP field is the basis of the DiffServ architecture. It uses the TOS field in the IPv4 header and the 8-bit traffic class field in the IPv6 header. Both fields are 8-bits-long, but the DSCP field uses only 6 bits, with 2 bits being reserved as *currently unused* (CU). Hence the name "field" is used instead of DSCP byte.

SLA and TCA. Service agreements between network entities may involve quantitative service and qualitative service agreements. Either type of agreement must be made between adjunct DS domains, as they will most likely have to transmit and receive traffic to/from one another. This entails the use of a *service level agreement* (SLA) between the two domains along with a *traffic conditioning agreement* (TCA). The TCA determines the traffic conditioning mechanisms, such as the classification, policing/marking/dropping of packets, that are to be applied at a DS edge node to particular flows. The SLA may be dynamic as well as static.

PHB. One of the key architectural definitions in DiffServ is the concept of PHB. It defines how a packet will be processed when it arrives at a node. The PHB is selected based on the DSCP value in the arriving packet.

PHBs are the building blocks of an end-to-end model of service for the end user. Depending on the type of service wanted, there are basically three different types of PHBs—*default* PHB, *assured forwarding* (AF) PHB, and *expedited forwarding* (EF) PHB. Based on these PHBs the service provider should be able to define a number of edge-to-edge services.

For the default PHB, only minimal amounts of resources need to be allocated. It is basically the best-effort traffic class that exists in the current Internet. It is used for non-DiffServ traffic and has the DSCP value 000000.

The key idea of the AF PHB is to divide traffic into a number of classes depending on the traffic characteristics. By using differentiated drop levels, it should be possible to guarantee a rough minimal rate even during congestion. Currently, the AF PHB is divided into four classes, each having three drop precedences. Within each class the drop levels increase between the different drop precedence levels. So when congestion occurs the packets with the highest drop preferences should be dropped more frequently than the lower drop precedence packets. Note that DiffServ does not make an absolute guarantee with respect to the drop precedences, but the relative frequency will reflect the precedence levels. This necessitates that each class has a different queue. The QoS that a packet gets will depend on the amount of resources allocated to that class as well as the number of active connections in that class.

The key idea of the EF PHB is to implement a very low loss, low delay, and low jitter traffic class such that the user can get service quality that is similar to that of a leased-line service. Hence it is also known by the names of *premium service* or *virtual leased-line service.* The delay or jitter experienced by the traffic should be minimal. EF PHB can be implemented by arranging the appropriate queuing and scheduling methods at each node. EF PHB packets are treated as top-priority traffic. Additionally, at each node it must be ensured that the maximum arrival rate into an EF PHB queue is less than the minimum departure rate. To ensure this behavior, the user's bandwidth usage must be heavily regulated. Any excess traffic exceeding the user's quota must be dropped at the ingress filter so as not to affect any of the conforming flows.

3.6.2.4 DiffServ Edge Node

The DiffServ edge node is where most of the complex functions of the DiffServ architecture reside. It is at the edge node that the ingress packets are analyzed and given the DSCP values. It is also at this point where nonconformant packets are marked or dropped according to the TCA. Figure 3.33(a) shows the structure of a DiffServ edge node, which is composed of a classifier, meter, marker, and policer (or shaper/dropper).

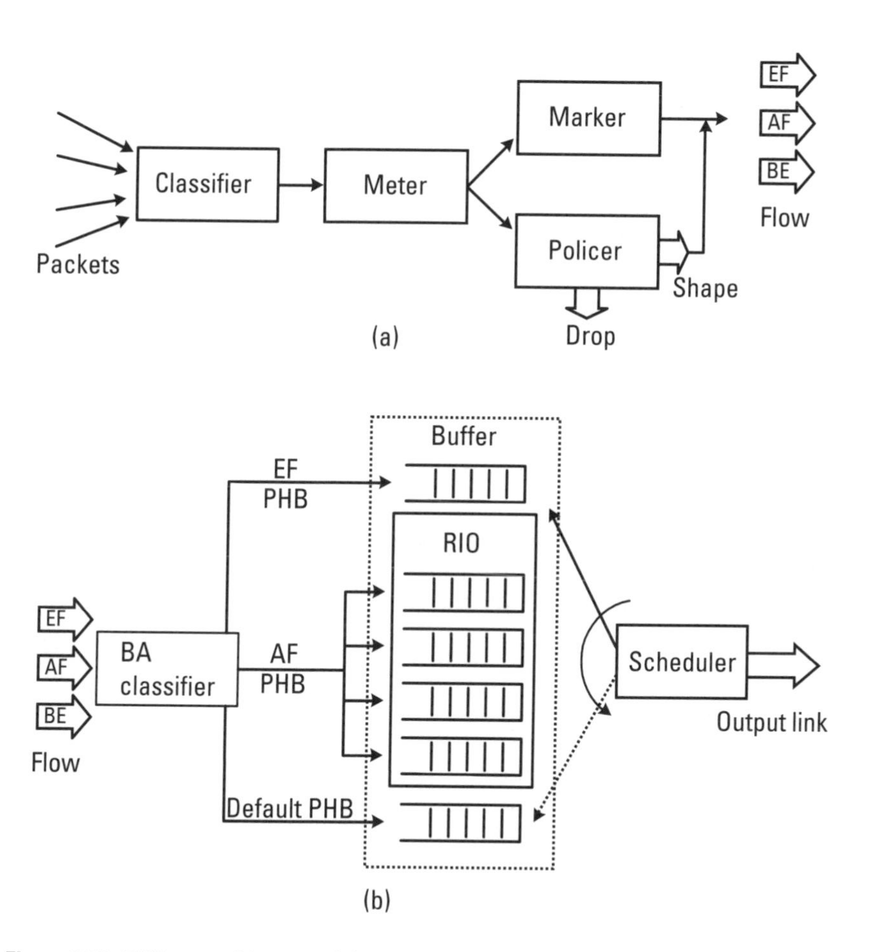

Figure 3.33 DiffServ architecture: (a) DiffServ edge node, and (b) DiffServ core node.

Classifier. When a packet arrives at a node, the packet header is first analyzed by the classifier to determine what PHB should be allocated to this packet. There are two types of classifiers possible, the *multifield* (MF) classifier and the *behavior aggregate* (BA) classifier. The MF classifier analyzes various fields in the packet to determine what PHB to allocate to the packet. Among the fields that may be analyzed at this stage are the source and destination addresses and the port numbers. The BA classifier classifies packets by looking at the DSCP value in the packet, so it is appropriate for core routers.

Packet classification tries to find the object that contains a point in multidimensional space. Traditionally, MF classifiers have been applied to firewall and security functions, but in QoS-enabled networks, it is being applied to the identification and classification of packets so as to provide the appropriate level of services to meet customer-specific differentiated service requirements. Among the attributes that a classifier should have are the ability to analyze multiple fields and ranges, and the ability to manage a large number of classification rules.

Meter. Based on the TCA profile selected by the classifier for a packet, the meter tries to decide whether the packet is an in-profile packet or an out-of-profile packet. This may involve measuring the rate of the traffic as well as its burstiness. Methods for measurement and determination are numerous including the *time sliding window* (TSW)-based method and the leaky/token bucket–based method.

Marker. The marker is the entity that decides what codepoint value to put into the packet header. Currently there are two recommended codepoint values: 000000 for the best-effort service and 101000 for the EF PHB service. For the AF PHB, currently four classes and three drop precedences are defined with a recommended DSCP codepoint assigned to each of the resulting 12 different types. These DSCPs can also be defined by each operator as needed, but the mapping of the PHBs to the DSCPs must be kept consistent at the edge routers.

Shaper. A shaper delays and buffers the packets of a traffic stream to make sure that they fit to the traffic profile that was agreed on for the traffic flow. It tries to reduce the burstiness of traffic flowing into the network by smoothing it out.

Dropper. A dropper can be easily implemented by setting the buffer size to zero or a very small value. Then any packets discarded by the policer will be dropped.

3.6.2.5 DiffServ Core Node

As shown in Figure 3.33(b), a DiffServ core node is usually composed of a BA classifier, various buffers for each service class (EF, AF, BE), and a scheduler. The BA classifier analyzes all incoming packets and allocates them to the appropriate queue based on the DSCP values. Depending on which PHB is used, different schemes are used for buffer management or scheduling.

For the AF type, an active queue management scheme that can have up to three different drop precedences is needed. Among the possible methods are RIO, WRED, and FRED. For EF types, it is important to make sure that the packets are only minimally delayed. Therefore, a simple fixed-priority queue or WRR scheduler, with the EF types being allocated the highest priority, should be sufficient.

References

[1] Hunt, C., *TCP/IP Network Administration,* Sebastopal, CA: O'Reilly & Associates, 1992.

[2] Keshav, S., *An Engineering Approach to Computer Networking,* Reading, MA: Addison-Wesley, 1997.

[3] IEEE P802.3: Part 3, "IEEE Standard: Carrier Sense Multiple Access with Collision Detection (CSMA/CD) Access Method and Physical Layer Specifications," *IEEE Std. 803.3,* 1998 edition.

[4] IEEE P802.6, "IEEE Standard: DQDB Subnetwork of a Metropolitan Area Network," 1991.

[5] Malkin, G., "RIP Version 2," *RFC 2453,* November 1998.

[6] Moy, J., *OSPF: Anatomy of an Internet Routing Protocol,* Reading, MA: Addison-Wesley, 1998.

[7] Rekhter, Y., and T. Li, "A Border Gateways Protocol 4 (BGP-4)," *RFC 1771,* May 1995.

[8] Perlman, R., *Interconnections: Bridges and Routers,* Reading, MA: Addison-Wesley, 1992.

[9] Davie, B., P. Doolan, and Y. Rekhter, *Switching in IP Networks—IP Switching, Tag Switching, and Related Technologies,* San Francisco, CA: Morgan Kaufmann, 1998.

[10] Newman, P., et al., "Ipsilon Flow Management Protocol Specification for IPv4 Version 1.0," *RFC 1953,* May 1996.

[11] Newman, P., et al., "Ipsilon's General Switch Management Protocol Specification Version 1.1," *RFC 1987,* August 1996.

[12] Davie, B., and Y. Rekther, *MPLS Technology and Applications,* San Francisco, CA: Morgan Kaufmann, 2000.

[13] Braden, R., et al., "Resource Reservation Protocol (RSVP)—Version 1 Functional Specification," *RFC 2205,* September 1997.

[14] Zhang, L., et al., "RSVP: A New Resource Reservation Protocol," *IEEE Network Magazine,* Vol. 7, No.5, September 1993.

[15] Jacobson, V., and M. Karol, "Congestion Avoidance and Control," *Proceedings of SIGCOMM'88,* August 1988.

[16] Brakmo, L. S., S. O'Malley, and L. L. Peterson, "TCP Vegas: New Techniques for Congestion Detection and Avoidance," *Proceedings of SIGCOMM'94,* August 1994.

[17] Fall, K., and S. Floyd, "Comparisons of Tahoe, Reno, and SACK TCP," *Technical Report LBL,* 1995.

[18] Floyd, S., and K. Fall, "Promoting the Use of End-to-End Congestion Control in the Internet," *IEEE/ACM Transactions on Networking,* Vol. 7, No. 4, August 1999, pp. 458–472

[19] Demers, A., S. Keshav, and S. Shenker, "Analysis and Simulation of a Fair Queuing Algorithm," *Internetworking: Research and Experience,* Vol. 1, No. 1, John Wiley & Sons, September 1990, pp. 3–26.

Selected Bibliography

Ahn, J. S., et al., "Experience with TCP TCP Vegas: Emulation and Experiments," *Proceedings of SIGCOMM'95,* August 1995.

Armitage, G., "MPLS: The Magic Behind the Myths," *IEEE Communications Magazine,* Vol. 38, No. 1, January 2000, pp. 124–131.

Awduche, D., "MPLS and Traffic Engineering in IP Networks," *IEEE Communications Magazine,* Vol. 37, No. 12, December 1999.

Balakrishnan, H., and V. N. Padmanabhan, "How Network Asymmetry Affects TCP," *IEEE Communications Magazine,* Vol. 39, No. 4, April 2001, pp. 60–67.

Banerjee, A., et al., "Generalized Multiprotocol Label Switching: An Overview of Routing and Management Enhancements," *IEEE Communications Magazine,* Vol. 39, No.1, January 2001, pp. 144–150.

Barakat, C., E. Altman, and W. Dabbous, "On TCP Performance in a Heterogeneous Network: A Survey," *IEEE Communications Magazine,* Vol. 38, No. 1, January 2000, pp. 40–46.

Bates, B., and D. Gregory, *Voice and Data Communications Handbook,* New York: McGraw-Hill, 1996.

Bernet, Y., "The Complementary Roles of RSVP and Differentiated Services in the Full-Service QoS Network," *IEEE Communications Magazine,* Vol. 38, No. 2, February 2000.

Beyda, W. J., *Data Communications,* Second edition, Englewood Cliffs, NJ: Prentice-Hall, 1996.

Braden, R., "Requirements for Internet Hosts—Communication Layers," *RFC 1122,* October 1989.

Braden, R., "TCP Extentions for High Performance: An Update," *Internet draft,* June 1993.

Braun, T., M. Guenter, and I. Khalil, "Management of Quality of Service–Enabled VPNs," *IEEE Communications Magazine,* Vol. 39, No. 5, May 2001, pp. 90–98.

CCITT Rec. I.370, "Congestion Management for the ISDN Frame Relaying Bearer Service," 1991.

Chase, J. S., A. J. Gallatin, and K. G. Yocum, "End System Optimizations for High-Speed TCP," *IEEE Communications Magazine,* Vol. 39, No. 4, April 2001, pp. 68–74.

Cheung, N. K., "The Infrastructure for Gigabit Computer Networks," *IEEE Communications Magazine,* Vol. 30, No. 4, April 1992, pp. 60–68.

Cisco white paper, "Enterprise IP Packet Telephony Solutions Guide," 1998.

Clark, D., and J. Wroclawski, "An Approach to Service Allocation in the Internet," *Internet draft,* draft-clark-diff-svc-alloc-00.txt, July 1997.

Comer, D. E., *Internetworking with TCP/IP, Volume 1,* Third edition, Englewood Cliffs, NJ: Prentice Hall, 1995.

Davidson, R. P., and N. J. Muller, *Interworking LANs: Operation, Design, and Management,* Norwood, MA: Artech House, 1992.

Feldmann, A., et al., "NetScope: Traffic Engineering for IP Networks," *IEEE Network,* Vol. 14, No. 1, March/April 2000, pp. 11–19.

Ferguson, P., "Simple Differentiated Services: IP TOS and Precedence, Delay Indication, and Drop Preference," *Internet draft,* draft-ferguson-delay-drop-00.txt, November 1997.

Ferguson, P., and G. Huston, *Quality of Service: Delivering QoS on the Internet and in Corporate Networks,* New York: John Wiley & Sons, 1998.

Fink, R. L., and R. E. Ross, "Following the Fiber Distributed Data Interface," *IEEE Network Magazine,* Vol. 6, No. 2, March 1992, pp. 50–55.

Floyd, S., "A Report on Recent Developments in TCP Congestion Control," *IEEE Communications Magazine,* Vol. 39, No. 4, April 2001, pp. 84–90.

Floyd, S., "TCP and Explicit Congestion Notification," *Computer Communications Review,* Vol. 24, Oct. 1994, pp. 8–23.

Floyd, S., "TCP and Successive Fast Retransmits," *Technical Report LBL,* May 1995.

Floyd, S., and V. Jacobson, "Random Early Detection Gateways for Congestion Avoidance," *IEEE/ACM Transactions on Networking,* Vol. 1, No. 3, August 1993, pp. 397–413.

Forouzan, B., *Introduction to Data Communication and Networking,* New York: McGraw-Hill, 1998.

Halabi, B., *Internet Routing Architectures,* Indianapolis, IN: New Riders Publishing, 1997.

Halsall, F., *Data Communications, Computer Networks and Open Systems,* Fourth edition, Reading, MA: Addison-Wesley, 1995.

Hardy, K. J., *Inside Networks*, Englewood Cliffs, NJ: Prentice-Hall, 1995.

Hassan, M., A. Nayandoro, and M. Atiquzzaman, "Internet Telephony: Service, Technical Challenges, and Products," *IEEE Communications Magazine,* Vol. 38, No. 4, April 2000.

Hedrick, C., "Routing Information Protocol," *RFC 1058,* June 1998.

Hegering, H., S. Abeck, and B. Neumair, *Integrated Management of Networked Systems,* San Francisco, CA: Morgan Kaufmann, 1998.

Held, G., *Voice Over Data Networks,* New York: McGraw-Hill, 1998.

Hioke, W., *Telecommunications*, Second edition, Englewood Cliffs, NJ: Prentice-Hall, 1995.

Huitema, C., *Routing in the Internet*, Englewood Cliffs, NJ: Prentice-Hall, 1995.

Iida, K., et al., "Performance Evaluation of the Architecture for End-to-End Quality-of-Service Provisioning," *IEEE Communications Magazine,* Vol. 38, No. 4, April 2000.

Jacobson, V., R. Braden, and D. Borman, "TCP Extension for High Performance," *RFC 1323,* May 1992.

Johnson, W. H., *Fast Ethernet,* Englewood Cliffs, NJ: Prentice-Hall, 1996.

Katz, D., "IP Router Alert Option," *RFC 2113,* February 1997.

Kawamura, R., and R. Stadler, "Active Distributed Management for IP Networks," *IEEE Communications Magazine,* Vol. 38, No. 4, April 2000, pp. 114–170.

Keshav, S., "On Efficient Implementation of Fair Queueing," *Internetworking: Research and Experience,* Vol. 2, No. 2, New York: John Wiley & Sons, 1991, pp. 157–174.

Kumar, V. P., T. V. Lakshman, and D. Stiliadis, "Beyond Best Effort: Router Architectures for the Differentiated Services of Tomorrow's Internet," *IEEE Communications Magazine,* May 1998.

Kung, H. T., "Gigabit Local Area Networks: A System Perspective," *IEEE Communication Magazine,* Vol. 30, No. 4, April 1992, pp. 134–142.

Lawrence, J., "Designing Mulitiprotocol Label Switching Networks," *IEEE Communications Magazine,* Vol. 39, No. 7, July 2001, pp. 134–142.

Lefelhocsz, C., et. al., "Congestion Control for Best Effort Service: Why We Need a New Paradigm," *IEEE Network,* Vol. 12, No. 1, January/February 1996.

Li, B., et al., "QoS-Enabled Voice Support in the Next-Generation Internet: Issues, Existing Approaches, and Challenges," *IEEE Communications Magazine,* Vol. 38, No. 4, April 2000.

Lin, D., and R. Morris, "Dynamics of Random Early Detection," *ACM SIGCOMM'97,* September 1997.

Mankin, A., and K. Ramakrishnan, "Gateway Congestion Control Survey," *RFC 1254,* August 1991.

Mathis, M., et al., "TCP Selective Acknowledgment Options," *RFC 2018,* October 1996.

Metz, C., "AAA Protocols: Authentication, Authorization and Accounting for the Internet," *IEEE Internet Comp.*, Vol. 3, No. 6, November/December 1999, pp. 75–79.

Miller, P., *TCP/IP Explained,* Woburn, MA: Digital Press, 1997.

Mir, N. F., "A Survey of Data Multicast Techniques, Architectures, and Algorithms," *IEEE Communications Magazine,* Vol. 39, No. 9, September 2001, pp. 164–170.

Morley, J., and S. Gelber, *The Emerging Digital Future,* Boston, MA: Boyd & Fraser, 1996.

Moy, J., "OSPF Version 2," *RFC 2328,* April 1998.

Naugle, M. G., *Network Protocol Handbook,* New York: McGraw-Hill, 1994.

Nichols, K., V. Jacobson, and L. Zhang, "A Two-Bit Differentiated Services Architecture for the Internet," *Internet draft,* draft-nichols-diff-svc-arch-00.txt, November 1997.

Padhye, J., et al., "Modeling TCP Throughput: A Simple Model and Its Empirical Validation," *Proceedings of ACM SIGCOMM'98,* September 1998, pp. 303–314.

Parekh, A. K., and R. G. Gallager, "A Generalized Processor Sharing Approach to Flow Control in Integrated Services Networks—The Single Node Case," *IEEE/ACM Transactions on Networking,* Vol. 1, No. 3, June 1993, pp. 344–357.

Parekh, A. K., and R. G. Gallager, "A Generalized Processor Sharing Approach to Flow Control in Integrated Services Networks—The Multiple Node Case," *IEEE/ACM Transactions on Networking,* Vol. 2, No. 2, April 1994, pp. 137–150.

Park, J. –T., J. –W. Baek, and J. W. Hong, "Management of Service-Level Agreements for Multimedia Internet Service Using a Utility Model," *IEEE Communications Magazine,* Vol. 39, No. 5, May 2001, pp. 100–106.

Partridge, C., *Gigabit Networking,* Reading, MA: Addison-Wesley, 1994.

Partridge, C., et al., "A 50-Gb/s IP Router," *IEEE/ACM Transactions on Networking,* Vol. 6, No. 3, June 1998.

Pearson, J. E., *Basic Communication Theory,* Englewood Cliffs, NJ: Prentice-Hall, 1992.

Perkins, C., "IP Mobility Support," *RFC 2002,* October 1996.

Perloff, M., and K. Reiss, "Improvements of TCP Performance," *Communications of the ACM,* Vol. 38, No. 2, February 1995, pp. 90–100.

Pras, A., et al., "Internet Accounting," *IEEE Communications Magazine,* Vol. 39, No. 5, May 2001, pp. 108–113.

Rajan, R., et al., "A Policy Framework for Integrated and Differentiated Services in the Internet," *IEEE Network,* Vol. 13, No. 5, September/October 1999, pp. 36–41.

Ramakrishnan, K. K., and S. Floyd, "A Proposal To Add Explicit Congestion Notification (ECN) to Ipv6 and TCP," *Internet draft,* draft-kksjf-ecn-00.txt, November 1997.

Rannsom, M. N., and D. S. Spears, "Applications of Public Gigabit Networks," *IEEE Network Magazine,* Vol. 6, No. 2, March 1992, pp. 30–41.

Rayes, A., and K. Sage, "Integrated Management Architecture for IP-Based Networks," *IEEE Communications Magazine,* Vol. 38, No. 4, April 2000, pp. 48–53.

Schonwalder, J., "Emerging Internet Management Technologies," *IEEE IM '99 (tutorial),* October 1999.

Siyan, K. S., *Inside TCP/IP,* Third edition, Indianapolis, IN: New Riders, 1997.

Smith, P., *Frame Relay,* Reading, MA: Addison-Wesley, 1993.

Stallings, W., *Data and Computer Communications,* Fifth edition, Englewood Cliffs, NJ: Prentice-Hall, 1997.

Stallings, W., "IPv6: The New Internet Protocol," *IEEE Communications,* Vol. 34, No. 7, July 1996, pp. 96–108.

Stevens, W., *TCP/IP Illustrated, Vol. 1,* Reading, MA: Addison-Wesley, 1994.

Swallow, G., "MPLS Advantages for Traffic Engineering," *IEEE Communications Magazine,* Vol. 37, No. 12, December 1999.

Takashima, S., "Network," *NTT R&D,* Vol. 38, No. 4, 1989, pp. 441–458.

Tanenbaum, A. W., *Computer Networks,* Third edition, Englewood Cliffs, NJ: Prentice-Hall, 1996.

Thomas, S. A., *Ipng and the TCP/IP Protocols,* New York: Wiley, 1996.

Trimintzios, P., et al., "A Management and Control Architecture for Providing IP Differentiated Services in MPLS-Based Networks," *IEEE Communications Magazine,* Vol. 39, No. 5, May 2001, pp. 80–88.

Vetter, R. J., D. H. C. Du, and A. E. Kleitz, "Networking Supercomputing: High Performance Parallel Interface (HIPI)," *IEEE Network Magazine,* Vol. 6, No. 3, May 1992, pp. 38–44.

Wang, Z., "User-Share Differentiation (USD): Scalable Bandwidth Allocation for Differentiated Services," *Internet draft,* draft-wang-usd-00.txt, November 1997.

Washburn, K., and J. Evans, *TCP/IP: Running a Successful Network,* Second edition, Reading, MA: Addison-Wesley, 1996.

Xiao, X., and L. Ni, "Internet QoS: A Big Picture," *IEEE Network,* Vol. 13, No. 2, March/April 1999, pp. 8–18.

Zhang, H., "Service Disciplines for Guaranteed Performance Service in Packet Switching Networks," *Proceedings of IEEE,* Vol. 83, No. 10, October 1995, pp. 1374–1396.

Zhang, L., "Virtual Clock: A New Traffic Control Algorithm for Packet Switching Networks," *Proceedings of the SIGCOMM'90,* September 1990.

4

ATM

One of the predominant networking concepts of the 1990s was the term *ATM networking*. This was a networking technology defined by the ITU-T as a basis for supporting the broadband communications networks of the future. It was an immensely complex endeavor leading to many advances in many fields of telecommunications engineering. In many ways it proved to be the catalyst for finally bringing voice and data networks together into a single network.

ATM networking was a hybrid of both packet-based and circuit-based communication technologies. As such, it used the concepts and techniques taken from both fields to solve various networking problems. For example, the addressing formats and the basic signaling mechanisms were built on the foundations deployed previously in circuit-oriented networks. However, the mixing of the two fields also introduced new problems to study anew, with a typical example being the support of real-time traffic in packet-based networks.

This chapter surveys the multiple technical areas that ATM covers. We will consider the issues of layering, routing, multiplexing and switching, network control and management, and QoS, as we did for TCP/IP in Chapter 3. We will discuss the most important points of ATM networking technology in this chapter, referring the reader to other references and standards for more details.

4.1 Concepts, Architecture, and Layering

ATM was originally presented as a method of supporting all types of services, including voice, video, data, and multimedia traffic in the B-ISDN. In the ATM networks, ATM cells, or fixed-size packets, are the basis of switching, multiplexing, and transporting traffic. As ATM is based on a sort of packet technology, it yields an efficient use of network resources through dynamic allocation of resources and statistical multiplexing. In addition, the concept of virtual circuits results in a clear and simple basis for QoS guarantees.

ATM technology is an integration of the existing circuit-mode digital communications technique with the packet-mode communications technique. In the sense that ATM uses ATM cells as its basic unit of transmission, it has a close connection with packet-mode communication. However, there is a significant difference in that ATM is designed from the beginning to manage real-time CBR/VBR traffic, whereas traditional packet networks were designed primarily for non-real-time data traffic. Also, as ATM was originally designed for use in public networks, the requirements on the architecture and specifics of address assignment, access and flow control, switching, and transmission are all different from those for the packet technology. On the other hand, ATM is fundamentally different from the circuit mode in which information signals are transmitted over separate dedicated channels, because no separate channel is specifically allocated; only virtual streams exist in the case of ATM. This results in more efficient use of network resources and flexibility in the granularity and the type of services that the network can offer to users.

4.1.1 Characteristics of ATM

ATM technology has four main characteristics that differentiate it from other network technologies. First, ATM networks transport traffic in the form of cells; second, ATM networks form virtual circuits using these cells; third, ATM technology uses a multiplexing technique called *asynchronous time division multiplexing* (ATDM); and fourth, it was designed to support the concept of QoS. We examine these characteristics more closely in the following sections.

The *first* characteristic of ATM technology is that cells are the basis for multiplexing, switching, and transport. Cells are small fixed-sized packets, so ATM can be regarded as a sort of packet-based network technology. Fixed-sized packets were chosen, as they would enable simple and cheap equipment design, including large and fast switches.

The reason for choosing a small cell size of only 53 bytes was to accommodate voice traffic. If the packet size were large, due to the small size of compressed voice samples, unacceptably large packetization delays would occur while filling its payload with voice samples.

The *second* main characteristic is the use of virtual circuits. ATM is a connection-oriented method that transfers service information through the establishment of VCs. A connection identifier is assigned whenever a VC is established and is removed when the connection is released. This identifier, located at the header of each cell, contains information on the virtual connection and is utilized for the multiplexing and switching of the cell. Signaling information for call setup is delivered via dedicated ATM cells. One additional characteristic is that a two-tier virtual circuit architecture is used, with the lower layer comprised of VCs, and the upper layer being comprised of virtual paths. This two-tier architecture was seen as a way of aggregation. By aggregating many VCs into a VP connection, the switches in the core network could be built in a simpler manner, yet enable faster operation.

The main reason for adopting virtual circuits was that it is the easiest way to enable easy implementation of QoS guarantees. Since the virtual circuit architecture would set up connections along the path before data transport, it would be easy to negotiate and allocate resources along the network path. This contrasts to the connectionless packet network architecture that has no obvious means of guaranteeing such resource allocation. A disadvantage of this approach is that it requires complicated signaling for the setup/release of connections along with the resource allocation/negotiation methods.

The *third* characteristic of ATM is that it makes use of ATDM. ATDM is a TDM technique that is based on fixed-sized time slots, but the time slots are allocated to users not on a periodical basis but on an occasional basis. The ATM cells of various users are chosen, one at a time, to be transported over the time slots, but no user is guaranteed a fixed time slot on a regular basis.[1]

1. ATDM is a multiplexing technique that stores each of the incoming low-speed signals inside a buffer, then retrieves and inserts the stored signals one by one into a multiplexing slot according to a priority-scheduling principle. Therefore, the low-speed input signals do not occupy locations inside the ATDM signal in a well-regulated manner, and thus behave *asynchronously* compared to their TDM equivalent.
ATDM is superior to TDM in that it has a higher channel-utilization factor. TDM assigns an exclusive channel to each of the incoming service signals; thus, even when a given channel is in a vacant state containing no effective information, it is not possible to pass other service information through it. However, since there is no exclusive channel allocation in ATDM, a blank channel can be taken by any incoming signal, resulting in a higher channel-utilization factor.

The main advantage of ATDM is that it enables an efficient use of transport and switching resources by allowing intelligent multiplexing of user traffic.

A *fourth* characteristic is that ATM has been designed to support QoS. This is basically enabled by a combination of the afore-mentioned characteristics; virtual circuits result in fixed routes through the network, allowing easy network resource allocation. The use of fixed-size cells enables easy network resource allocation. Additionally, designing hardware with sophisticated resource management schemes is simpler due to the fixed size. The use of ATDM enables flexible allocation of network resources, allowing the support of a much more varied and flexible QoS infrastructure.

These characteristics of ATM allow the integration of various B-ISDN services possessing many different characteristics. That is, broadband and narrowband services can coexist within the same communications network in the form of ATM cells. Also, the delay problem associated with real-time services is solved through the use of VCs, and hence it becomes possible to provide real-time services in the ATM networks.

4.1.2 ATM Architecture

ATM technology has brought in a three-layer protocol architecture, which comprises an AAL, an ATM layer, and a physical layer. The AAL layer provides an adaptation function that adapts higher layer packets to the payload of ATM cells; the ATM layer does the "network layer" functions of ATM networks; and the physical layer performs ordinary "physical layer" functions tailored for ATM networks. This ATM architecture is incorporated in ATM cells in such a way that AAL and ATM layer functions are embodied, respectively, in the payload space and the header space of ATM cells. Sections 4.1.2.1 and 4.1.2.2 examine the ATM cell format and the ATM protocol architecture based on the ATM protocol reference model, respectively.

4.1.2.1 ATM Cell Format

The ATM cell acts as the basic unit of information transfer in the ATM communication. As shown in Figure 4.1(a), the ATM cell is composed of 53 bytes. The first 5 bytes are for the cell-header field, and the remaining 48 bytes form the user information field.[2] The cell-header field is divided into

2. The 53-byte ATM cell size including 48 bytes of payload field is an awkward number. It is the result of a compromise between two proposals, the United States' proposal of 64 bytes and the European proposal of 32 bytes, made at the ITU-T standardization meetings in the early 1990s.

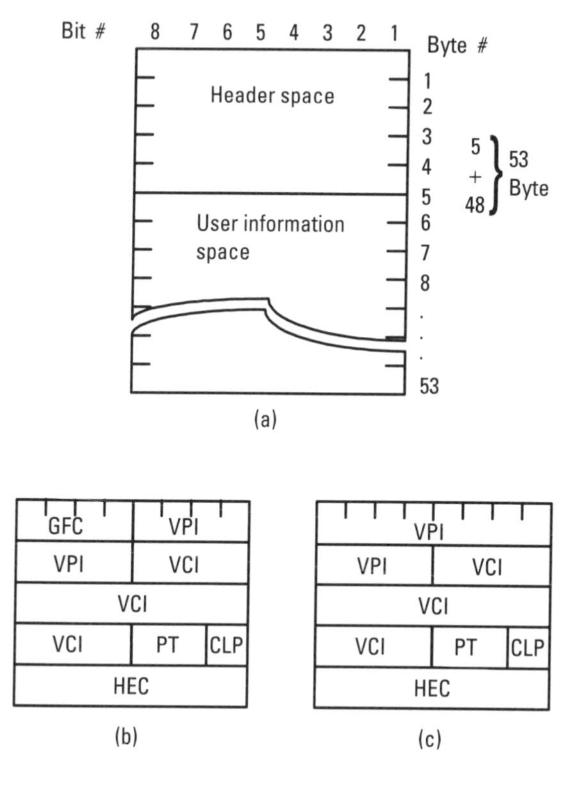

Figure 4.1 ATM cell structure: (a) cell structure, (b) header structure at UNI, and (c) header structure at NNI.

generic flow control (GFC), VPI, VCI, *payload type* (PT), *cell loss priority* (CLP), and *header error control* (HEC) fields. The cell format differs slightly at the NNI and the UNI. As shown in Figure 4.1(b, c), the NNI interface cell has a longer VPI field but no GFC field.

The GFC field is defined for use in physical access control. It is used only at the UNI interface cell format and the corresponding ATM switch overwrites this field when transmitting at the NNI. It is intended to control the transmission of the ATM terminal/endpoint. It can also be used to reduce cell jitter for CBR services, to fairly allocate capacity for VBR services, and to control traffic for VBR flows. Currently it is not used in any major systems.

The role of the VPI/VCI field is to indicate VC or VP identification numbers to distinguish cells belonging to the same connection. The field size is either 24 bits (at UNI) or 28 bits (at NNI). The values have only local significance—that is, they are only meaningful between the relevant nodes.

Some VPI/VCI values are preassigned for special purposes such as indicating unassigned cells, physical layer OAM cells, metasignaling channels, and generic broadcast signaling channels.[3]

The PT field is used for indicating the presence of user information and for indicating whether the given ATM cell suffered from traffic congestion. The CLP bit is used for indicating whether the corresponding cell may be discarded during the time of network congestion. HEC is a CRC byte for the cell-header field and is used for detecting and correcting cell errors and delineating the cell header.

ATM cells can be classified according to the associated layers and functions. For instance, the *ATM layer cell* is a cell that is formed at the ATM layer, and the *physical layer cell* is similarly formed at the physical layer. The ATM layer cells are divided into assigned cells and unassigned cells, and physical layer cells are divided into idle cells and physical layer OAM cells. *Assigned cells* refer to those cells that are allocated to ATM layer services and the *unassigned cells* refer to the remaining types. *Idle cells* are created to fill the vacant space that results when there are no cells to be transmitted and *physical layer OAM cells* are used for the transfer of OAM information of the physical layer.

4.1.2.2 ATM Protocol Reference Model

ATM *protocol reference model* (PRM)[4] was originally adopted to perform the various functions required by the B-ISDN in an organized manner.[5] The PRM involves separating the overall communication functions into AAL, ATM, and physical layers, vertically, and defining the appropriate set of functions and protocols at each layer.

The PRM for the B-ISDN consists of three planes to the horizontal direction, namely, the *user*, *control*, and *management* planes, as shown in Figure 4.2 [2]. The management plane can be further divided into *layer management*

---

3. Refer to Table 4.3 for details.
4. Though the overall B-ISDN architectural model has become less important, the ATM UNI remains critical to understanding ATM networks. The UNI basically means the interface between user and network equipment. There are two types of UNI, *private UNI* defined by the ATM Forum and *public UNI* defined by the ITU-T. The private UNI is defined between user equipment and private ATM switches, whereas the public UNI is defined between user equipment and public ATM switches.
5. We omit discussions on the functional architecture and interface configuration of the B-ISDN. Refer to Section 4.2 of [1] for those topics and the related reference points (e.g., S_B, T_B).

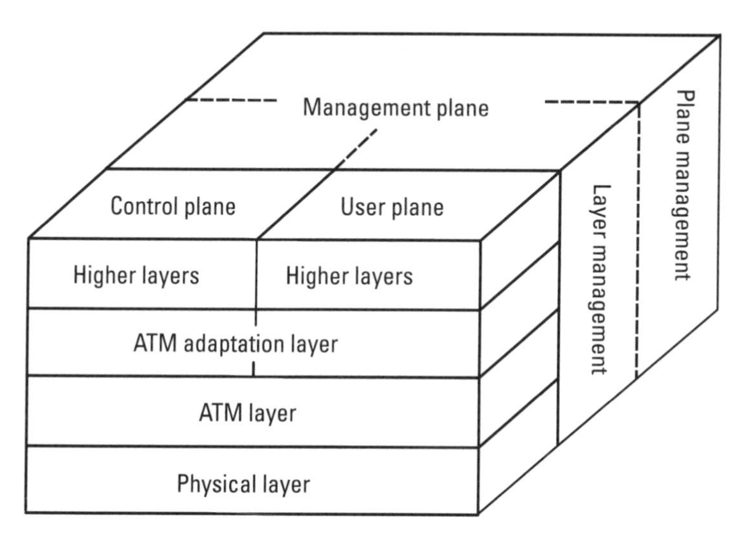

Figure 4.2 B-ISDN PRM.

and *plane management*. The user plane provides user information-related functions; the control plane performs various control functions for service provision; and the management plane provides network management functions.

The *user plane* provides functions for transferring the flow of user information, as well as the associated control functions such as flow control and error correction. *User information* here includes voice, image, data, text, and graphics. User information can be passed transparently through the B-ISDN, or it can be transferred after appropriate processing procedures.

The *control plane* provides call connection and connection control functions. That is, the control plane provides the functions associated with call establishment, call monitoring, and call release. It can also provide a control function for changing the characteristics of readily established services. This is accomplished mainly by the use of signaling functions that are defined for both the UNI and NNI interfaces. This functionality is most important for *switched VCs* (SVCs), as all SVCs can only be controlled by using these control plane functions. Additionally, functions related to routing, such as PNNI functionality, are also a part of the control plane functions.

The *management plane* provides the communications network monitoring function associated with the transfer of user information and control information. The management plane is divided into *plane management* and *layer management*. The plane management function performs management of the entire network through its role as an interplane arbitrator. Basically it can be seen as the management of the whole system from a particular aspect. Layer management function refers to the management of network resources such as metasignaling channels and protocol entity parameters. The management of each layer is done by layer-specific OAM flows.

4.1.3 ATM Layering

The user plane and the control plane each consist of the *physical layer*, the *ATM layer*, the AAL, and the upper layers. The physical layer provides the physical medium and transmission functions, the ATM layer provides the data transfer function for all of the B-ISDN services, and the AAL provides service-related adaptation functions for its upper layers, as discussed above. The upper layer of the user plane provides service information management functions, and the higher layer of the control plane provides functions associated with call control and connection control. Table 4.1 lists the functions of each layer [3].

4.1.3.1 Physical Layer

The main function of the physical layer is to transmit and receive ATM cell data over the physical media–connecting ATM equipment. It is basically a two-stage problem. The first stage concentrates on recovering the bit stream from the physical media and the second stage on retrieving cells from the bit

Table 4.1
Functions of Each Layer of the B-ISDN PRM

Layer	Sublayer	Functions
AAL	Convergence sublayer (CS)	Convergence functions
	Segmentation and reassembly (SAR)	Segmentation and reassembly functions
ATM layer		Generic flow control Cell-header generation/extraction Cell VPI/VCI translation Cell multiplex and demultiplex
Physical layer	Transmission convergence (TC)	Cell rate decoupling HEC header sequence generation/verification Cell delineation Transmission frame adaptation Transmission frame generation/recovery
	Physical medium–dependent (PMD)	Bit timing Physical medium

stream. For the second stage, there are two methods available—the frame-based approach, which relies on the frame format of some transmission system such as SDH/SONET (refer to Section 5.2.3); and the HEC-based approach, which relies on using a HEC-based cell delineation algorithm (see below).

The above two-stage approach leads to the following two sublayers. The *physical medium–dependent* (PMD) sublayer and the *transmission convergence* (TC) sublayer. The PMD sublayer has responsibility for the transmission and reception of bits over specific physical media (bit synchronization, clock recovery). ANSI, CCITT/ITU-T, and the ATM Forum have defined various sublayers such as SDH/SONET, FDDI, E1, and UTP. The TC sublayer maps cells to a transmission frame such as SDH/SONET for transmission and recovers cell boundaries on reception of the bit stream. It also carries out the HEC byte calculation and the corresponding error detection/correction based on the HEC field. In addition, decoupling takes place in this layer between the physical transmission rate and the actual transmitted cell rate.

The physical layer provides several functions such as bit timing, transmission frame generation and extraction, cell scrambling and descrambling, transmission frame adaptation, HEC signal generation and confirmation, cell delineation, and cell rate decoupling.

The bit timing information function involves the conversion of data bit flow into a waveform adapted to a particular physical medium or the reverse conversion process, the insertion or the extraction of timing information, and line coding or decoding. As a result, the information transferred from the PMD sublayer to the transmission convergence sublayer consists mainly of data bit/symbol stream and the corresponding timing information.

Transmission frame generation and extraction function involves the generation and extraction of the transmission frame. This function does not apply in cell-based transmission, since a separate transmission frame is not needed in this case. However, STM-*n* frames are required in SDH-based transmission, and DS-3 signal frames are required in G.702-based transmission.

The transmission frame adaptation function involves the mapping of ATM cell flow into the payload of the transmission frame, or, conversely, the extraction of ATM cell flow from the transmission frame. This function is required in an SDH-based network or in a G.702-based network.[6]

The cell scrambling and descrambling function is applied to ATM cells for the purpose of randomizing of the contained bit arrangement. In the case of the SDH-based transmission, *self-synchronous scrambling* (SSS) is

6. Refer to Section 4.5 of [1] for more discussions on the SDH-based and cell-based transmissions.

conducted on the cell header, with the user information scrambled in the STM-*n* frame level after frame generation. In the case of the cell-based transmission, *distributed sample scrambling* (DSS) is conducted on both the header and the user information.[7]

The HEC signal generation and confirmation function handles the HEC signal of the ATM cell header. In the transmitting direction, it generates the CRC for the first four bytes of the ATM cell header, which is called the HEC signal, and inserts it into the fifth byte of the header. Conversely, it applies an identical procedure to the received signal to inspect whether the CRC is correct, and in the event an uncorrectable error is detected, the cell in question is discarded.

The cell delineation function identifies ATM cell boundaries in the ATM cell flow in the following manner. It first applies the HEC operation for an arbitrarily chosen starting point and shifts the starting point by one bit until it reaches a correct CRC matching. Then it jumps by 53 bytes and repeats the HEC operation. If it gets correct CRC matching again it jumps by 53 bytes and applies the HEC operation again; otherwise it resumes the initial search operation above. It repeats the 53-byte based CRC matching several times (for example, eight times) and finally declares completion of the cell delineation.[8]

The cell rate decoupling function augments the ATM cells that are carrying valid information using idle cells to match the overall cell rate to that of the corresponding payload capacity, or, conversely, extracts cells with effective information by removing idle cells.

4.1.3.2 ATM Layer

ATM layer corresponds to the "network layer" of the ATM network. It is at the ATM layer in which ATM cell switching, multiplexing, and demultiplexing takes place based on VPI/VCI. In addition, cell rate decoupling takes place by inserting unassigned cells, and traffic shaping/policing is also performed in the ATM layer. Basically all functions related to the ATM cell

---

7. Basically, the SSS processing is done by a scrambler with the characteristic polynomial $x^{43} + 1$. The scrambler runs continuously without being reset per frame. The follow-up scrambling in the STM-*n* frame level is based on *frame synchronous scrambling* (FSS). In the case of the DSS, the scrambler takes the characteristic polynomial $x^{31} + x^{28} + 1$. In this case, the state information of the scrambler in the transmitter is conveyed to that in the receiver in a distributed manner to aid the synchronization. Refer to Section 4.5 of [1] for more details.
8. Refer to Section 4.5 of [1] for more discussions on the cell-delineating mechanism. A similar type of delineation mechanism applies to the *simple data link* (SDL) protocol. Refer to Figure 8.13 and the related explanations in Section 8.3.4.

header except for the HEC byte are performed in the ATM layer. In this process, ATM cells are classified according to the cell (refer to Section 4.4.2) header fields and dropped according to the CLP bit field.

ATM layer functions. The ATM layer functions can be grouped into the cell multiplexing and demultiplexing function, the cell VPI/VCI translation function, the cell-header generation and extraction function, the generic flow control function, the payload-type indication function, and the cell loss priority function. We briefly discuss these functions one by one below.

The cell multiplexing and demultiplexing function provides the capability of multiplexing ATM cells from different VPs and VCs to form a composite cell flow, or the reverse demultiplexing capability. Here, the multiplexed cell flow does not have to be continuous.

The cell VPI/VCI translation function is required at the ATM switch or the ATM cross-connect node. Its role is to map the values stored in the VPI/VCI field of each ATM cell header into a new set of values.

The cell-header generation and extraction function applies at the ATM layer's terminations, and involves the generation or extraction of the first four bytes in the ATM cell header. For the generation of the cell header, the associated information received from the upper layer is mapped into the corresponding field, and the opposite is executed for the cell-header extraction process. This function also encompasses the translation of *service access point identifier* (SAPI) into VPI and VCI.

The GFC function controls access and information flow at the UNI. Here, the information is transferred via assigned cells or unassigned cells (see Section 4.1.3).

The payload-type indication function handles the PT field. This field indicates whether the contents of payload consist of user information or network information and provides indications of network congestion experience and ATM layer user-to-ATM layer user indication. Among the three bits in the PT field, the first bit is used to indicate whether it is user information (0) or network information (1). In the case of user information, the second bit is an *explicit forward congestion indication* (EFCI) bit, which indicates whether the relevant cell experienced congestion (1) or not (0) while traversing the network along the specific connection; and the third bit is an AAL-indicate bit that indicates whether the information belongs to the tail part of an AAL-5 PDU (1) or not (0). In the case of network information, the PT field indicates whether the relevant cell is an OAM flow F5 cell (see Section 4.4.2) or *resource management* (RM) cell. A summary of this is listed in Table 4.2.

Table 4.2
Usage of the PT Field

PT Field	Indication
000	User cell, EFCI = 0, AAL indicate = 0
001	User cell, EFCI = 0, AAL indicate = 1
010	User cell, EFCI = 1, AAL indicate = 0
011	User cell, EFCI = 1, AAL indicate = 1
100	OAM F5 segment associated cell
101	OAM F5 end-to-end associated cell
110	Resource management cell
111	Reserved for future use

Another function of the ATM layer is the CLP function. That is, the priority level to be used for cell loss (or cell discard) is recorded in the CLP field of each ATM cell employed for VBR services, and when congestion occurs, the cells with lower priority are discarded first. If the CLP bit indicates 1, then it represents a cell with a lower priority that can be abandoned. The CLP function must be provided in conjunction with the QoS determined at the time of establishing VPC/VCC. That is, it must be possible to provide the minimum guaranteed bit rate even after the cell loss processing, and the prescribed service quality must be maintained. Consequently, the network must determine the bit rate of the cells with the higher priority at the time of establishing the connection, and the rate must be negotiable even after the connection is completed. The network must constantly monitor via usage parameter control whether the number of cells corresponding to a given connection exceeds the prearranged value. When the cell traffic exceeds the negotiated level, even the cells that have been preassigned with higher priority can be discarded.

Preassigned cell headers. Cells reserved for physical layer have preassigned values in the whole header. Here, the ATM cells used by the physical layer include *idle cells* and *physical layer OAM cells.*

ATM cells in the ATM layer with preassigned cell headers include unassigned cells, signaling cells, OAM F4/F5 cells, and RM cells. Signaling cells consist of metasignaling cells, general broadcast signaling cells, and point-to-point signaling cells. Cell headers of these preassigned ATM layer cells are as listed in Table 4.3.

Table 4.3
ATM Layer Cells with Preassigned Cell Headers

ATM Layer Cells	VPI	VCI	PT	CLP
Unassigned cell	00000000	00000000 00000000	XXX	0
Metasignaling	X	00000000 00000001	0A0	C
General broadcast signaling cell	X	00000000 00000010	0AA	C
Point-to-point signaling cell	X	00000000 00000101	0AA	C
OAM F4 flow cell: Segment	Y	00000000 00000011	0A0	A
End-to-end	Y	00000000 00000100	0A0	A
OAM F5 flow cell: Segment	Y	Z	100	A
End-to-end	Y	Z	101	A
Fast RMcell	Y	Z	110	A

(Note): A: a bit to be filled in the ATM layer; C: a bit to be filled by the signaling entity; X: "Don't care"; Y: any VPI value; Z: any nonzero VCI.

VCs and VPs. VC refers to a logical unidirectional connection between two endpoints for the transfer of ATM cells, and VP implies a logical combination of VCs. Each VC is assigned a VCI and each VP is assigned a VPI. Figure 4.3 illustrates VP switching and VC/VP switching as well as the related VCI/VPI interchange. In VP switching the switching decision is based only on the VPI field, while in VC/VP switching both fields are used in the decision process. VP switching, by analogy, can be viewed as offering a function similar to a *digital cross-connect system* (DCS).

There are two kinds of ATM connections—*VC connection* (VCC) and *VP connection* (VPC). VCC refers to a concatenation of VC links for achieving connection between ATM service access points. Here, the term *VC link* implies the unidirectional virtual connection for enabling the transport of ATM cells between the point where VCI is assigned and the point at which the VCI gets translated or removed. Likewise, VPC refers to a concatenation of VP links for connecting the points at which a VPI is assigned with those at which a VPI is translated or removed. Here, the term *VP link* implies the VC link groups that join a VPI assignment point with the corresponding setup/removal point. Inside a VPC, VC links that are different from one another can exist, and each is differentiated through the use of VCI.

A VCC can be established by switching equipment on a permanent or semipermanent basis. Integrity of cell sequence is ensured within the same VCC. The network provides a VCC user with a set of QoS parameters such

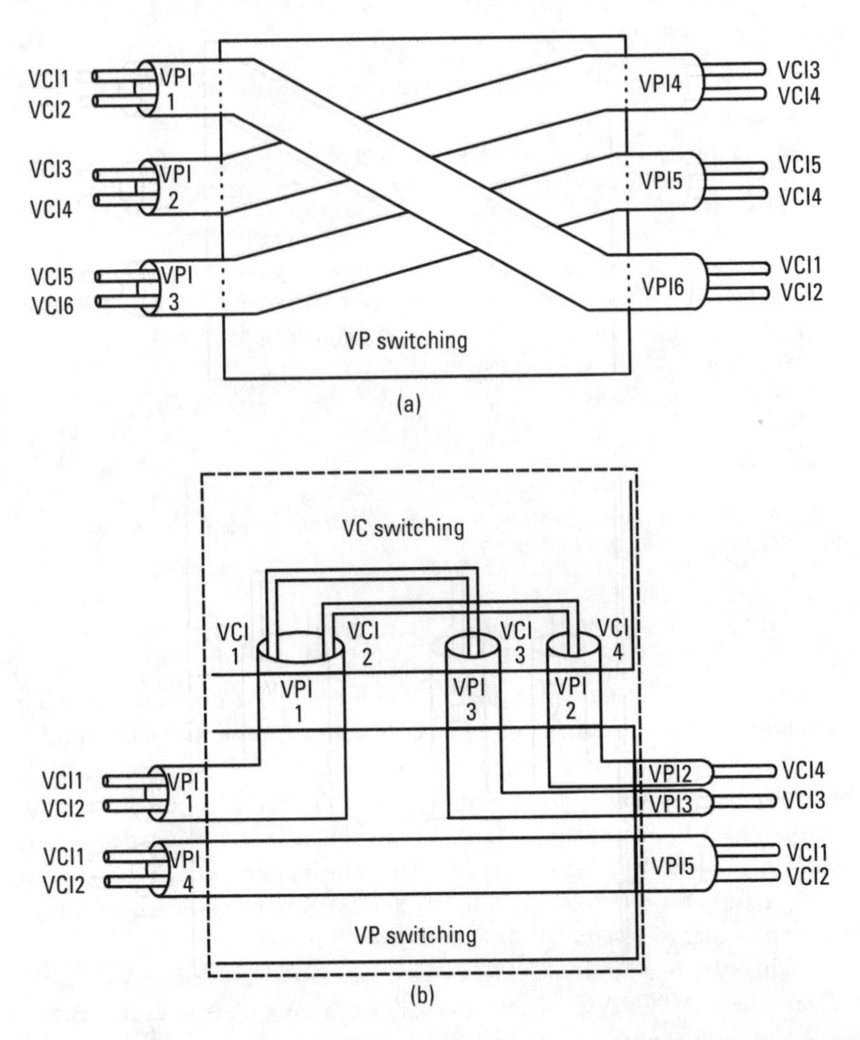

Figure 4.3 Illustration of VP and VC switching: (a) VP switching, and (b) VC/VP switching.

as cell loss and cell delay. Such user traffic parameters are prescribed at the time of VCC setup through negotiation between the user and the network, and the network monitors the observance of these parameters.

There are four different methods for establishing or removing VCC at the time of VCC setup. First, the signaling procedure can be bypassed if connection setup or release is achieved through reservation. This method applies to both permanent and semipermanent connections. The second is the use of a metasignaling procedure—that is, a signaling VC is established or removed through the use of a metasignaling VC. Third is the use of a user-to-network signaling procedure, which involves the use of a signaling VCC to establish or release a VCC for end-to-end communication. The fourth is the use of a user-to-user signaling procedure, which employs a signaling VCC to establish or release a VCC internal to a VPC that is pre-established between two UNIs. Accordingly, there are four possible methods of assigning VCI values at the UNI—assignment by the network, assignment by the user, assignment through a network-user negotiation, and assignment using a standardized method.[9]

4.1.3.3 AAL

The AAL of the B-ISDN is a layer positioned between the ATM layer and the higher user service layer. Its main function is to resolve any disparity between the service provided by the ATM layer and the service demanded by the user. For that purpose, the AAL adapts user service information with the ATM cell format and performs handling of transmission errors, misinserted or lost cells, and the cells with errors. It also provides a flow control function to meet the QoS demanded by the user and a timing control function to restore the user signal.

The user information field of the ATM cell is fixed at 48 bytes, while the user service information to be adapted to this space is extremely diverse in character. Under such a constraint, the AAL performs various functions, such as ATM cell adaptation, transmission error processing, lost cell and inserted cell processing, flow control, and timing information control.

The AAL is divided into the CS and SAR sublayer. At the CS, the function for converting the user service information coming from the upper layer into a protocol data unit, or the reverse process, is performed. At the SAR sublayer, the function for segmenting the PDU to form the user information field of the ATM cell, or the reverse process, is performed. A generic model of these operations is shown in Figure 4.4. The actual functions performed in the CS and SAR layers of the AAL vary depending on the type of upper layer services. As a result there are four different types of AAL layer: AAL-1, AAL-2, AAL-3/4, and AAL-5 [4].

AAL-1 [5]. The delivery of constant-rate *user services data units* (U-SDU), along with the associated timing information using a common bit rate, and the indication of uncorrectable errors are some of the services that AAL-1

9. We omit further discussion of VPC as it is similar to that of the VCC.

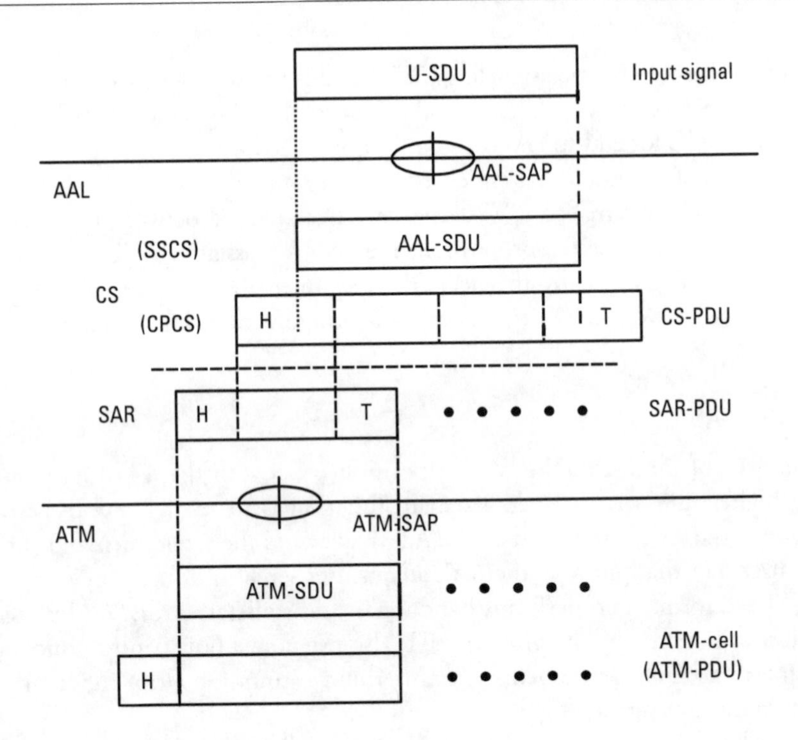

Figure 4.4 Processing of data at AAL and ATM layer.

provides to the upper layers. AAL-1 provides a function for partitioning and reassembling user information. It also provides a function for handling cell delay variations and lost and inserted cells, and enables the receiver to extract timing information from the information source.

When transferring *constant bit rate* (CBR) data, the timing information is delivered by the *synchronous residual time stamps* (SRTSs). This arrangement enables the provision of circuit-emulation services for the DS-1 or DS-3 signals. For the octet-structured CBR signals such as $n \times$ DS-0 ($n \leq$ 92), circuit-emulation service becomes possible by employing the *structured data transfer* (SDT) scheme. For error detection and correction, the *Reed-Solomon* (RS) code is used.

The function of AAL-1's SAR sublayer is to segment the CS-PDUs and then add a header to form the SAR-PDUs and send them to the ATM layer. Also, through a reverse process it reassembles the SAR-PDUs to recover the CS-PDU. Figure 4.5 shows the SAR-PDU formed at the SAR sublayer. The number of bits assigned to the *sequence number* (SN) and *sequence number protection* (SNP) fields is four each; consequently, the size of the SAR-PDU

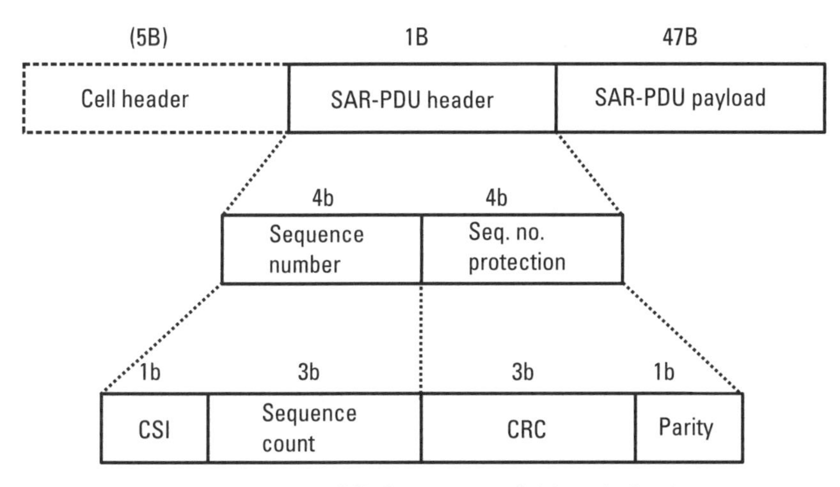

Figure 4.5 SAR-PDU format for AAL-1.

payload space becomes 47 bytes. SN is used for inspecting whether a cell loss or cell insertion has occurred, and the SNP is used for error correction to protect SN from errors.

More specifically, the *convergence sublayer indication* (CSI) bit is used for special purposes such as indicating the presence of the CS functions; the *sequence count* (SC) bits indicate a serial SAR-PDU count in modulo-8; the CRC bits represent the CRC-3 code for the four bits in the SN field; and the *parity* (P) bit denotes the parity check for the preceding seven bits in the SAR-PDU header. The CRC and P bits determine whether or not the SN bits are valid.

AAL-1 CS reconstructs the original CBR data stream by eliminating cell jitter through buffering, and by properly handling the lost and misinserted cells through a sequence number checking process. When using SRTS, it inserts and recovers timing information for source clock recovery, and when using SDT, it transfers information on the user data structure. By employing the RS code-based *forward error correction* (FEC), it can also monitor and improve the error status of the end-to-end VC.

Through the SN checking process, AAL-1 CS can determine whether a SAR-PDU is in normal state, or lost, or misinserted. For the lost SAR-PDUs, it can determine the position and the number.

For timing clock recovery, AAL-1 CS can employ the SRTS scheme. If the jitter and wander requirement is not stringent, it can also adopt the adaptive clock recovery method.

AAL-2 [6]. The delivery of connection-oriented variable-rate SDUs along with a means for identifying and multiplexing multiple users over a single ATM connection is the basic service that AAL-2 provides to the upper layers. AAL-2 has a number of advantages compared to AAL-1 or AAL-5 in transporting real-time voice or video traffic. It makes more efficient use of bandwidth by using silence detection and suppression; also it supports idle voice channel deletion. It is expected that these advantages will spur its wide use in voice trunking applications.

One of the distinctive characteristics of AAL-2 is that it does not have a SAR sublayer or CS sublayer. It has only what is called a *common part sublayer* (CPS). Figure 4.6 shows the basic protocol structure of AAL-2. As shown, there is only CPS sublayer between the *service-specific convergence sublayer* (SSCS) and the ATM cell.[10]

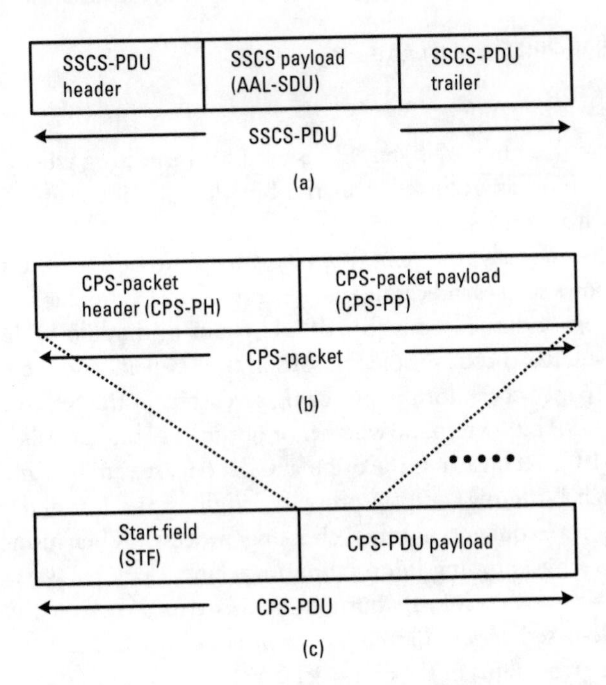

Figure 4.6 Protocol structure of AAL-2: (a) SSCS PDU, (b) CPS packet, and (c) CPS PDU.

10. As yet there are no SSCSs defined but one specifically targeted for narrowband ISDN traffic over ATM applications is being developed by the ITU-T. In addition, to effectively carry variable-rate video over AAL-2, a corresponding SSCS must be defined for the transfer of timing information.

The CPS layer basically provides a means to transport multiplexed user traffic over a single AAL-2 ATM connection. It supports the SAR of variable-length payloads. Each AAL-2 user generates CPS packets, which are multiplexed into CPS-PDUs and finally into a single ATM cell. Essentially the CPS packets are another layer of ATM-like cells inside a single ATM cell.

The CPS packet format is shown in Figure 4.7(a). The size of a CPS packet may be 64 bytes at maximum. Each individual user in a CPS is identified by the *channel identifier* (CID) field. The *length indicator* (LI) field contains the length of the variable length packet, the *user-to-user indication* (UUI) field provides a means for identifying SSCS layers.

The CPS sublayer multiplexes multiple CPS packets into a single CPS-PDU. Figure 4.7(b) shows the format of a CPS-PDU. The PDU is divided into *start field* (STF) and *CPS-PDU payload*. The STF is composed of the *offset field* (OSF), which indicates where in the CPS-PDU payload the first new CPS packet starts; the SN and the P fields provide error detection and recovery. By combining the OSF and the LI field values in the CPS packets, it is easy to multiplex and demultiplex multiple user traffic streams.

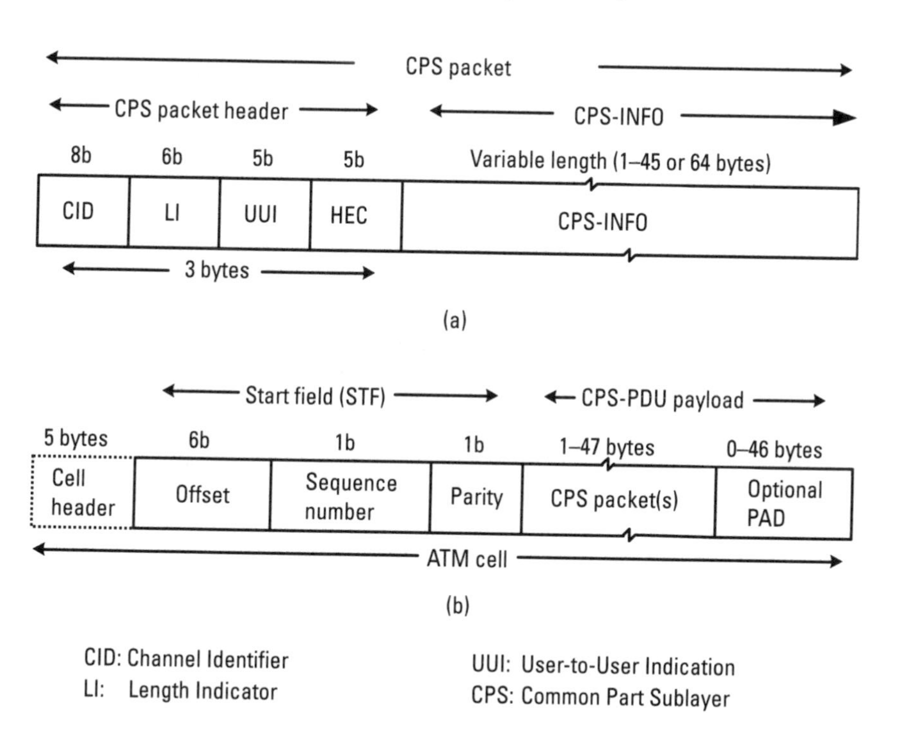

Figure 4.7 AAL-2 CPS packet formats: (a) CPS packet, and (b) CPS PDU.

A *padding* (PAD) field exists at the end of the CPS-PDU so that an empty field may be filled when the protocol times out while waiting for real-time data.

Figure 4.8 illustrates the AAL-2 multiplexing process for the case where four user traffic streams are multiplexed. The first CPS packet from user A starts at the beginning of the first CPS-PDU payload. The first three CPS packets do not completely fit into the first ATM cell, so part of the third CPS packet (from user C) is divided up between the first and the second ATM

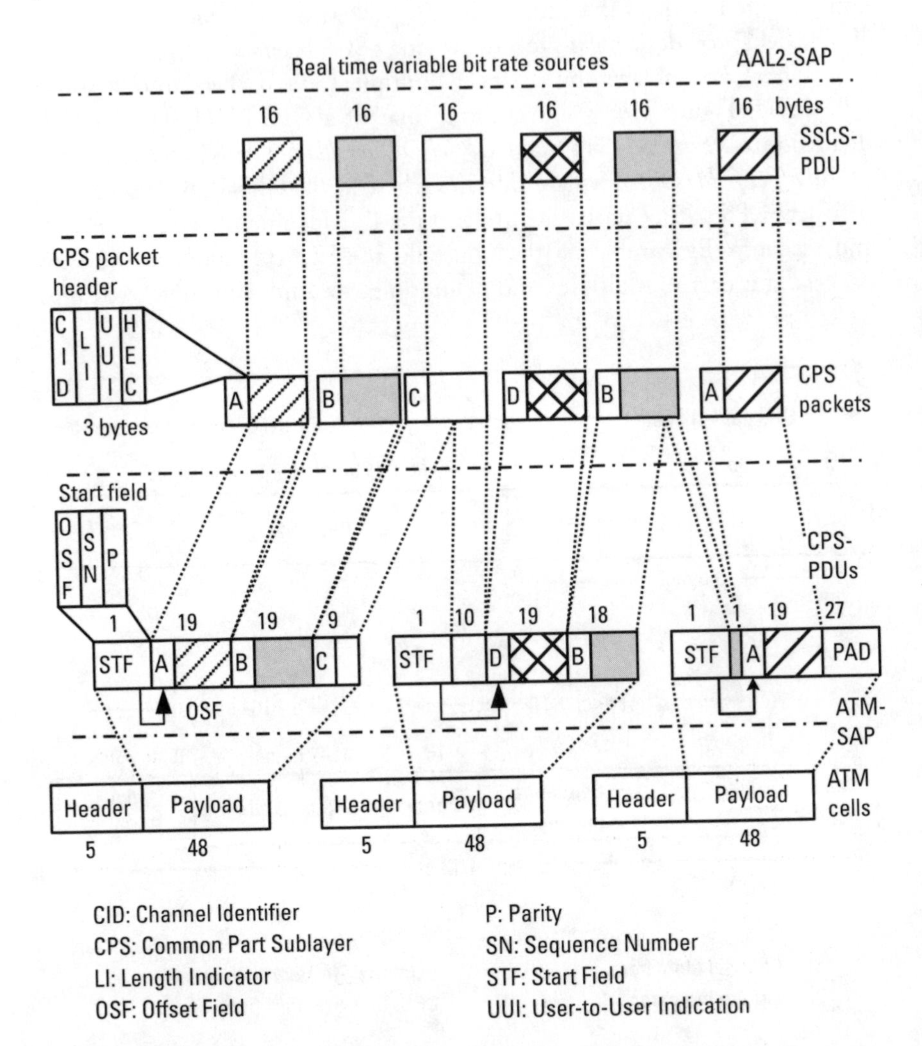

Figure 4.8 Example of AAL-2 multiplexing.

cells. The OSF field in the second CPS-PDU header points to the starting position of the next CPS packet. A similar action occurs for the fifth CPS packet that is divided between the second and third ATM cells.

AAL-3/4 [7]. AAL-3/4's function is to establish an adaptation layer connection prior to the transmission, then transport class C and class D service data with VBR characteristics.[11] The services provided at AAL-3/4 can be divided into *message-mode* services and *stream-mode* services. In the message mode, an AAL-SDU passes across the AAL interface in exactly one AAL *interface data unit* (IDU), whereas, in the streaming mode, it does so in one or more AAL-IDUs. Here, an internal pipelining function can be applied, and an AAL entity can initiate data transfer to the receiving AAL entity before it has the complete AAL-SDU available.[12]

The AAL-3/4 SAR sublayer receives variable-length CS-PDUs from the CS and then segments and appends a header and trailer to form the SAR-PDUs, which are then sent to the ATM layer. It can also reassemble SAR-PDUs through a reverse process and recover the CS-SDUs.

Figure 4.9(a) shows the structure of the SAR-PDU of AAL-3/4. In Figure 4.9(a) the *segment type* (ST) field indicates whether the corresponding payload is the *beginning of message* (BOM), a *continuation of message* (COM), the *end of message* (EOM), or a *single segment message* (SSM), and an SN field indicates the serial number of each message. The *multiplexing identification* (MID) field is used when multiple *common part convergence sublayer* (CPCS) (to be discussed below) connections are multiplexed through one ATM layer connection; the *length indication* field indicates the length of the SAR-PDU payload in octets; and the CRC field contains the CRC code for the entire SAR-PDU including the header.

The AAL-3/4 CS provides various functions for AAL-3/4 service users, including transparent delivery of AAL-SDUs, mapping between AAL-SAP

---

11. In the early stage of the B-ISDN standardization, broadband services were classified into four classes, namely, class A, B, C, and D, depending on the timing relation, bit-rate variability, and connection mode. Among them, class C and class D referred to variable connection-oriented services and variable connectionless services, respectively.

12. Originally planned operations also include the assured and nonassured modes. The assured mode includes retransmission and flow control functionality. These functions, however, were not defined in the actual standards because such functionality can be better dealt with in the upper layers. The single example of assured mode protocol that has been developed is the *service-specific connection-oriented protocol for the signaling AAL* protocol stack.

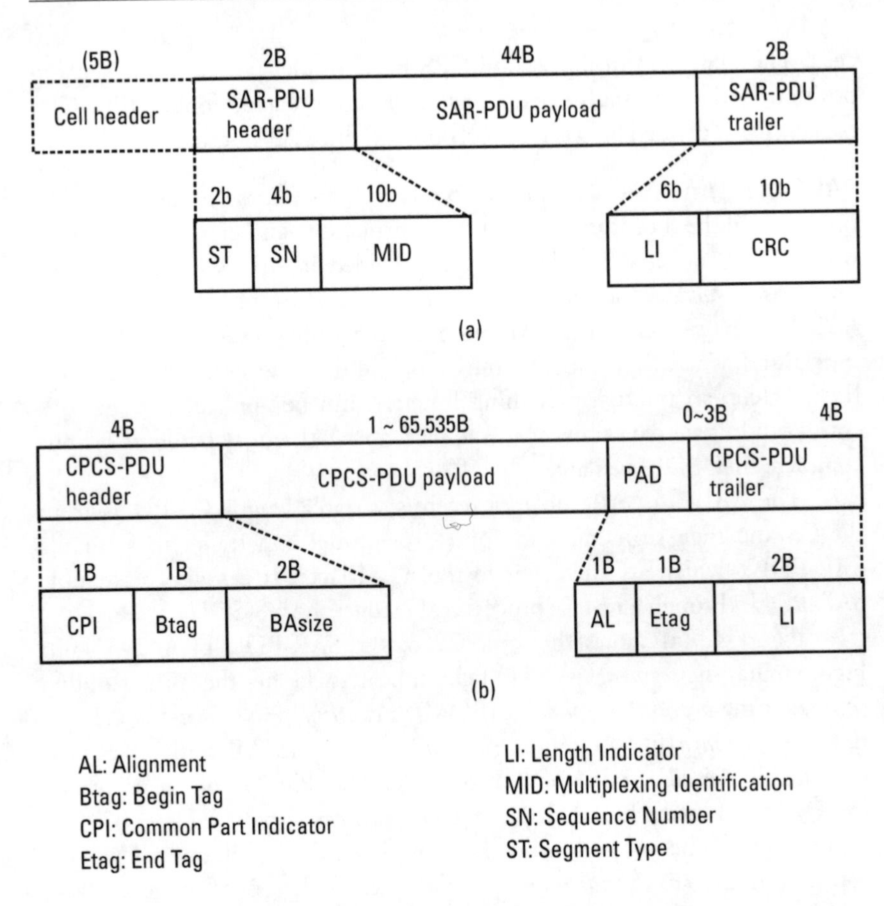

Figure 4.9 Frame formats of AAL-3/4: (a) SAR-PDU, and (b) CPCS-PDU.

and ATM layer connections, error detection and treatment, message SAR, information identification, and buffer allocation. It also provides special functions specific to class C and class D AAL-3/4 services. The CS functions of AAL-3/4 can be rearranged into the CPCS, which is common to all class C and class D services, and the SSCSs services, which differ depending on each specific service.

The structure of the CPCS-PDU is as shown in Figure 4.9(b). In Figure 4.9(b), the CPI field indicates whether the corresponding PDU belongs to the common part. The *begin tag* (Btag) and *end tag* (Etag) are identical tags attached to the header and the trailer of a CPCS-PDU so that the start/end of a CPCS-PDU payload can be matched. The *buffer allocation size* (BAsize) field indicates the size of buffer to be allocated in the receiver and PAD is a padding to make a CPCS-PDU payload sized in multiples of four bytes. Finally, *length indication* (LI) field indicates the length of the CPCS-PDU payload, and *alignment* (AL) is a filler to make the CPCS-PDU trailer size 32 bits.

AAL-5 [8]. AAL-3/4, standardized for connection-oriented and connectionless services, is not adequate for high-speed data communication because of its heavy protocol overhead. AAL-5 is a derivative of AAL-3/4, which is endowed with a simplified protocol and is thus more suitable for high-speed processing.

AAL-5 protocol has a simple header structure, enabling simple protocol processing and, consequently, fast and efficient data communication. AAL-5, like AAL-3/4, supports connection-oriented and connectionless VBR data services, and consists of SAR, CPCS, and SSCS sublayers. In reality, however, the SAR sublayer of AAL-5 is a blank layer that has no particular header functions. As a consequence, the SAR-PDU multiplexing capability associated with the MID field is no longer available for AAL-5, and the SNP and CRC functions are not supported. Instead, a single CRC-32 function is provided at the CPCS sublayer.

The AAL-5 SAR sublayer segments the CPCS-PDU in units of 48 bytes to produce SAR-PDUs, and, conversely, reassembles the original CPCS-PDU out of SAR-PDUs. Figure 4.10(a) shows the structure of SAR-PDU. The SAR sublayer has no protocol overhead, so the protocol processing becomes very simple. Instead, an indication is put on the ATM header to indicate whether the particular ATM cell carries the rear-end portion of a CPCS-PDU. More specifically, the *AAL-indicate* bit in the PT field of the ATM cell header is set to 1 if the corresponding ATM cell contains the rear-end portion, and to 0 otherwise.

The AAL-5 CS consists of AAL-5 CPCS and AAL-5 SSCS. AAL-5 CPCS provides the mapping function between CPCS-SDU and CPCS-PDU, the error detection and correction function, and so on. AAL-5 SSCS is used in the form of null SSCS in most cases.[13]

Figure 4.10(b) shows the structure of AAL-5 CPCS-PDU. It has no header but has a trailer of 8 bytes. PAD is the padding field of 0 to 47 null bytes to make the length of CPCS-PDU a multiple of 48 bytes.

The CPCS-UU (user-to-user) in the CPCS trailer conveys user-to-user information transparently. CPI indicates whether or not the corresponding

13. In the ITU-T standard, SSCS is defined only for the frame relay service. FR-SSCS performs the multiplexing and demultiplexing functions, FR-SSCS PDU length indication, congestion control, and so on.

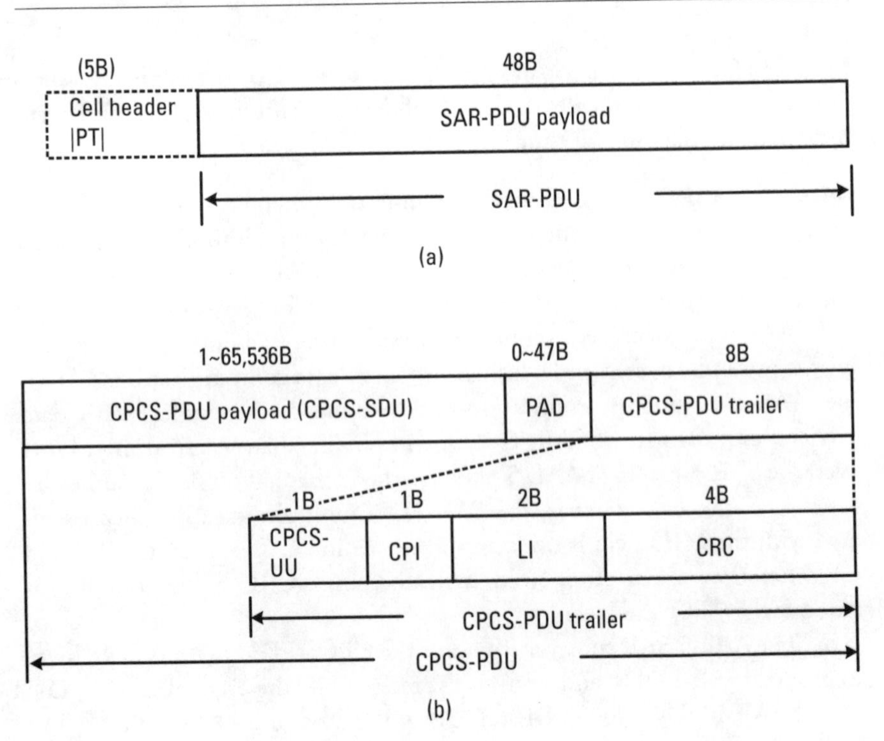

Figure 4.10 AAL-5 PDU structures: (a) SAR-PDU, and (b) CPCS-PDU.

PDU is a common part and can also indicate the unit of the LI count. LI indicates the length of the CPCS-PDU payload, whereas CRC carries the CRC-32 code for the CPCS-PDU (including the payload, PAD, and the first 4 bytes of the trailer).

Figure 4.11 illustrates the AAL-5 multiplexing process for two VBR data packets of 78 bytes and 121 bytes, respectively. The 78-byte data packet is appended by a 10-byte padding and an 8-byte trailer, whereas the 121-byte data packet is appended by a 15-byte padding and an 8-byte trailer. The resulting lengths of the CS-PDUs are 96 bytes and 144 bytes, respectively. These are segmented in units of 48 bytes to produce SAR-PDUs, which become the ATM cell payloads themselves without further processing. The AAL-indicate bit in the PT field of ATM header is set to 1 only for the ATM cell which contains the rear-end portion of a CS-PDU, as is illustrated in Figure 4.11.

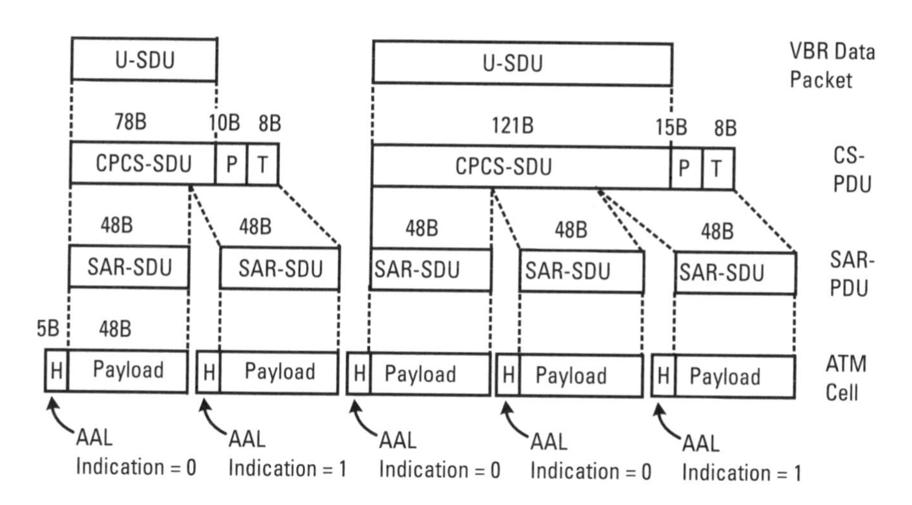

Figure 4.11 Illustration of AAL-5 multiplexing.

4.2 Routing

Though ATM networks were designed to represent a completely new network technology, they still needed to solve the same problem of routing that had been faced in other previous network technologies. The problem of routing is also closely related to the problem of addressing. How to get from point 1 to point 2 is often defined by how to define the addresses of sources and destinations. In addition, as the ATM network would interact with legacy telephone networks, the addressing architecture would have to be compatible with existing telephone networks as well.

Once the address formats are clearly defined, there are a number of methods for finding routes from any source to any destination. Preferably the method would be nearly automatic, flexible, QoS-supportive, and scalable to large networks. In ATM networks addressing and routing architecture are all designed within these constraints.

In this section we first discuss how ATM addresses are defined and then describe the *private network-network interface* (PNNI) signaling and routing protocol defined by the ATM Forum.

4.2.1 Addressing

An ATM address identifies the location of an interface, most often a UNI but possibly including NNI also. In ATM networks, connections do not necessarily terminate at that interface, so addresses should have significance within the network topology.

In contrast to the IP networks, which have only one type of address format, a number of formats are possible in ATM networks. There are two different types of addresses defined by the ATM Forum: one is the *ATM end system address* (AESA) and the other is the E.164 format address. The AESA format is based on ISO *network service access point* (NSAP) address format. There are three variations of the AESA format currently defined, namely, *data country code* (DCC), *international code designator* (ICD), and E.164 (i.e., E.164 address contained in AESA). The E.164 format is the same type of address used by telephony that is often called "native E.164" to distinguish itself from the above E.164 AESA variation.[14]

4.2.1.1 Native E.164 Address

The ATM Forum specifications state that only international E.164 numbers (including country code) are to be supported. Also, all E.164 numbers should only be used as addresses and never be used as service identifiers. The format for an E.164 number is the same as normal international telephone numbers (e.g., +82-2-880-7276).

4.2.1.2 AESA Format

The AESA format is shown in Figure 4.12. It is composed of two components—the *initial domain part* (IDP) and the *domain-specific part* (DSP). The IDP is composed of the *authority and format identifier* (AFI) and the *initial domain identifier* (IDI). The AFI identifies what scheme is to follow and the IDI identifies the "authority" responsible for allocating the structure of DSP. The DSP structure is defined by the authority identified by IDI.

As mentioned above, currently three different formats are possible—the DCC, the ICD, and the E.164 format. The formats are basically differentiated by the IDI part of the address. The DCC is defined in accordance with ISO 3166. The ICD identifies an international organization and is maintained by the British Standard Institute (BSI). The E.164 address is basically an international telephone number.

The DSP is composed of the *higher-order DSP* (HO-DSP) and the *end system identifier* (ESI). The HO-DSP is established by the authority identified

14. An *ATM service provider* (ASP) is not required to support any particular format. While this allows maximum flexibility, it also leaves the possibility for interoperability problems.

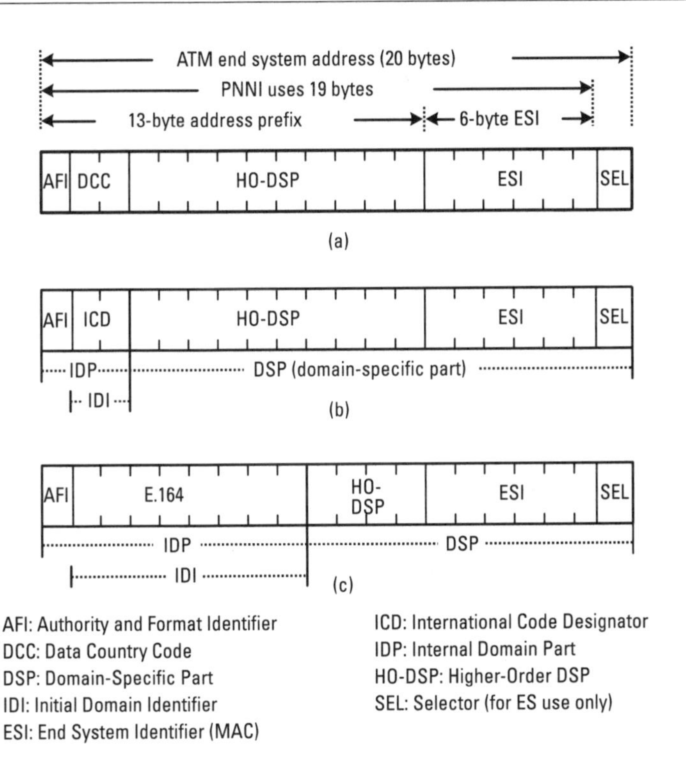

Figure 4.12 ATM end system addresses: (a) DCC format, (b) ICD format, and (c) E.164 format.

in the IDP. This field might contain a hierarchical address such as a routing domain and areas within the domain. The ESI is essentially the 48-bit MAC address format, defined by the IEEE.

The last part of the AESA address, the SEL field, is not used in the ATM network. It usually identifies the protocol entities in the upper layers of the user machine that are to receive the traffic. SEL could contain upper-layer SAPs.[15]

4.2.1.3 Obtaining ATM Addresses

The method of getting ATM addresses assigned to an interface differs depending on the address type used. When E.164 numbers or E.164 AESAs are used, the ITU or the national numbering authority has to be contacted.

15. In the E.164 address there is no field corresponding to the SEL field in the AESA format. Due to the absence of the SEL field, E.164 addresses cannot be used as service identifiers.

When the ICD format is used, BSI, by way of the national registration authority, has the responsibility. When the DCC format is used, the national registration authority for each country has the responsibility.[16]

4.2.2 PNNI Routing

When ATM network equipment was first developed by ITU-T, there were no standards for signaling or routing in ATM networks. Signaling protocols were soon developed, but support of routing was not complete. This forced many vendors to seek proprietary solutions for the support of routing and call setup. Obviously, the problems of intervendor operability became an important issue. PNNI was the solution standardized by the ATM Forum to solve the problems and enable multivendor interoperability, thereby supporting a truly global ATM network.

4.2.2.1 Overview of PNNI

PNNI is composed of two protocols defined together to solve the problems of routing and signaling in ATM networks. By using the routing protocol the network would be able to build up a topological map of the network, while by using the signaling protocol, the user connections could then be set up based on the routes calculated through the use of the routing protocol. Figure 4.13 shows the protocol reference model for PNNI and shows how these two protocol elements fit into the overall protocol reference model.

When work started originally on the PNNI protocol, it was agreed that a two-phase approach would make the most sense—the initial *phase 0* approach enabling the bare minimum of functionality, followed by a *phase 1* with full functionality.

Phase 0 was designed as a short-term solution to solving the basic problem of routing and call setup in ATM networks. It basically consisted of a single protocol, the *interim interswitch signaling protocol* (IISP). IISP used UNI 3.1 signaling (Q.2931/Q.2110) between switches to set up calls. An important aspect of PNNI phase 0 was that there was no real routing protocol. All routing tables and information required manual configuration of static topology and resource tables. Consequently, the solution could be used in small networks but was not scalable to large networks.

Phase 1 was designed from the beginning as the long-term solution for solving the basic problem of routing and call setup in ATM networks. PNNI routing solved the basic routing aspects by introducing a new routing

16. For example, in the United States, the *American National Standard Institute* (ANSI) is responsible for this.

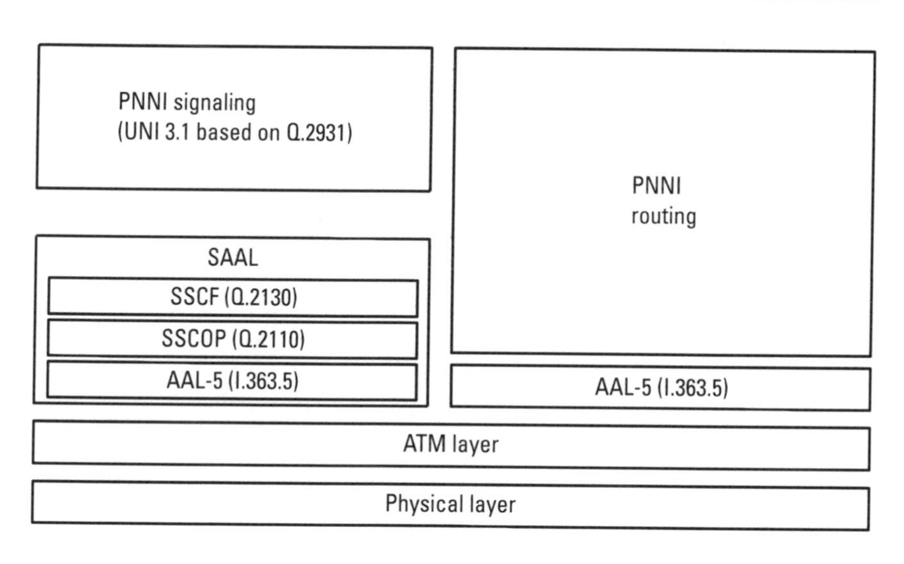

Figure 4.13 PNNI protocol reference model.

protocol based on a hierarchical routing architecture. Its main characteristics include scalability, reachability, QoS-based routing, and support for dynamic topological changes. PNNI signaling was based on a subset of UNI 4.0 signaling, with the addition of other features that were needed to support NNI and routing specifics. Among the new features, the most important are the usage of a *designated transit list* (DTL) and the new functionality of crankback and soft *permanent VPCs* (PVPCs)/*permanent VCCs* (PVCCs).

A quick comparison of PNNI with the routing in TCP/IP networks shows the following points. PNNI routing is based on the use of virtual circuit routing and is designed to support source routing, QoS routing, and link state routing, whereas TCP/IP networks use datagram/hop-by-hop routing, best-effort routing, and various routing protocols including link-state and distance-vector variants.

4.2.2.2 PNNI Routing Protocol

The PNNI routing protocol basically defines a way of distributing the topology information, so that all nodes gain knowledge of the complete topology of the ATM network and consequently make consistent routing decisions.

An important aspect of the PNNI routing is that it uses a hierarchical mechanism. As a result the protocol scales well and can be used for large worldwide ATM networks. Additionally, the PNNI routing and signaling protocols are designed so that all nodes may automatically configure themselves once they are connected to the networks. The PNNI routing protocol uses the 20-byte NSAP ATM address form (E.164, E.191).

PNNI routing is based on source routing. When a call is first set up, the source node decides the complete routing path to the final destination at the start of the connection. This contrasts to the hop-by-hop routing that is used in other packet networks.

Another characteristic of PNNI routing is that it has the ability to choose the "best" path according to the application, current network status, and application constraints. The constraints include how to choose the path that will minimize delay on the path, maximize the throughput, or minimize the cell delay variation.

The PNNI routing protocol is based on the link state routing technique. In this approach the information needed to build and synchronize the topology database is exchanged by means of *protocol topology state packets* (PTSPs). More specifically, neighboring nodes exchange "hello" packets to verify the status of their links and to identify themselves to other nodes, and then the nodes exchange PTSPs. Each PTSP contains multiple *protocol topology state elements* (PTSEs), which are the basic information elements. PTSPs are reliably flooded throughout a single peer group.

The PTSE can be divided into three main types—*nodal state* information, *link state* information, and *reachability* information. Nodal state information consists of node identification and basic nodal capabilities. Link state information consists of link identification, link attributes [such as *allowed cell rate* (ACR) and *cell rate margin* (CRM)], and link metrics [such as *maximum cell transfer delay* (MCTD) and *maximum cell delay variation* (MCDV)]. Reachability information is also made available in the form of specific addresses and address prefixes. When a *peer group* (PG) is first formed, a single *peer group leader* (PGL) is elected to identify the peer group in the higher-level peer groups. Once a peer group is formed, the nodes immediately elect a PGL. The PGL represents the peer group in the next higher-level peer group. Each peer group is represented in the parent peer group by a single node, called a *logical group node* (LGN). LGNs exchange PTSPs with their peer nodes within the parent peer group to inform those nodes of the child group's reachability and attributes. Recursive information obtained by the LGN about the parent group is fed down into the child group. The child nodes can obtain knowledge about the full network hierarchy to construct full-source routes. In this manner a hierarchy of topological information is formed. Figure 4.14 illustrates the hierarchical topology map constructed by the PNNI routing protocol.

4.2.2.3 PNNI Signaling Protocol

The PNNI signaling specifications define the signaling used to set up point-to-point and point-to-multipoint connections throughout an ATM network. Though it is designated as a private NNI signaling protocol, it may be used in public networks as well if the operators agree to it (refer to Section 4.4.1). The basic protocol is based on a subset of the ATM Forum UNI 4.0 signaling but it does not support some UNI 4.0 signaling features such as proxy signaling, *leaf initiated join* (LIJ), and supplementary services.

The following are some of the key concepts on which the PNNI signaling protocol is built: First, it uses the associated signaling method instead of the nonassociated signaling method. Second, it supports complete source routing across each level of hierarchy by using a *designated transit list* (DTL). Third, it supports crankback and alternate path routing. Fourth, it supports soft PVPC/PVCCs.

An end node uses PNNI signaling to request a path (or route) to the destination from PNNI routing. The PNNI routing uses its hierarchical network topology map to decide on one or more paths from the source to the

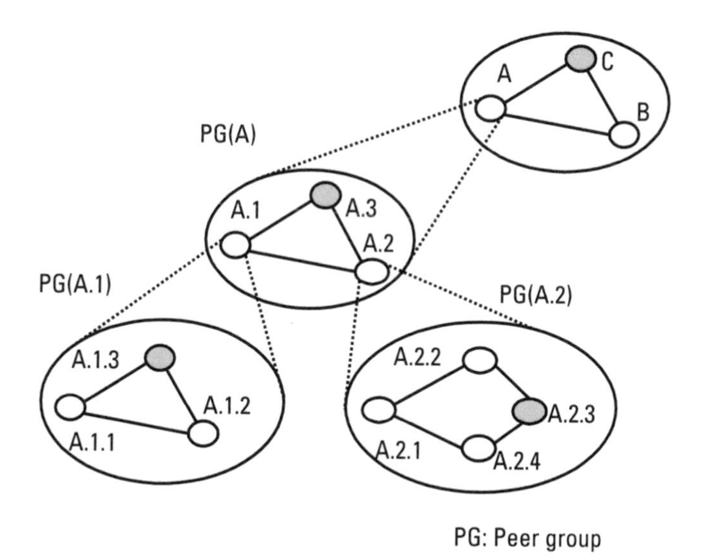

Figure 4.14 Illustration of hierarchical topology map constructed by the PNNI routing protocol.

destination. The PNNI routing function supplies a complete source route in the form of a DTL. The DTLs are saved onto a stack by the PNNI signaling request.

In DTLs, the signaling message is routed through the network reserving resources along the source-routed path. This operation is illustrated in Figure 4.15(a). As the source node A.1 has a detailed knowledge of peer group A, it specifies the source route DTL [A.1, A.2], DTL [A, B, C]. Likewise, the ingress nodes B.1 and C.1 of peer groups B and C, respectively, specify their source routes in a similar manner.

When the call setup with a specified source route fails, the call is "cranked back" to the originator of the top source route DTL, where the originator then generates a new source route, or the originator cranks back the call to the generator of higher-level source route. This operation is illustrated in Figure 4.15(b), which shows how a second DTL is constructed by the ingress node when the initial source route in the peer group B fails.

4.2.2.4 PNNI-Augmented Routing

PNNI-augmented routing (PAR) is an extension of PNNI that facilitates the distribution of information on non-ATM services in the ATM network as part of the PNNI topological database. An example of this would be the distribution of information regarding IP networks and routers connected to a PAR-enabled ATM network. The information is carried in a new type of PTSE, which can carry various *information groups* (IGs) that are the actual containers of the non-ATM information. Currently, specific IGs include OSPF, BGP-4, and DNS. In addition, there is a system capabilities IG that may be used by vendors to carry experimental or proprietary information.[17]

4.3 Multiplexing and Switching

Rapid evolution of technological changes has brought about new switching concepts such as multirate circuit switching, fast circuit switching, and fast packet switching. All these have been considered as possible switching technologies in the B-ISDN environment, and the fast packet switching technology has been finding its embodiment in ATM switching. ATM switching can contribute much toward accommodating versatile services, effectively utilizing resources, and efficiently accommodating bursty services.

---

17. As the prime use of PAR seems to be in supporting TCP/IP traffic over ATM networks, we leave the discussion on PNNI PAR to Chapter 7.

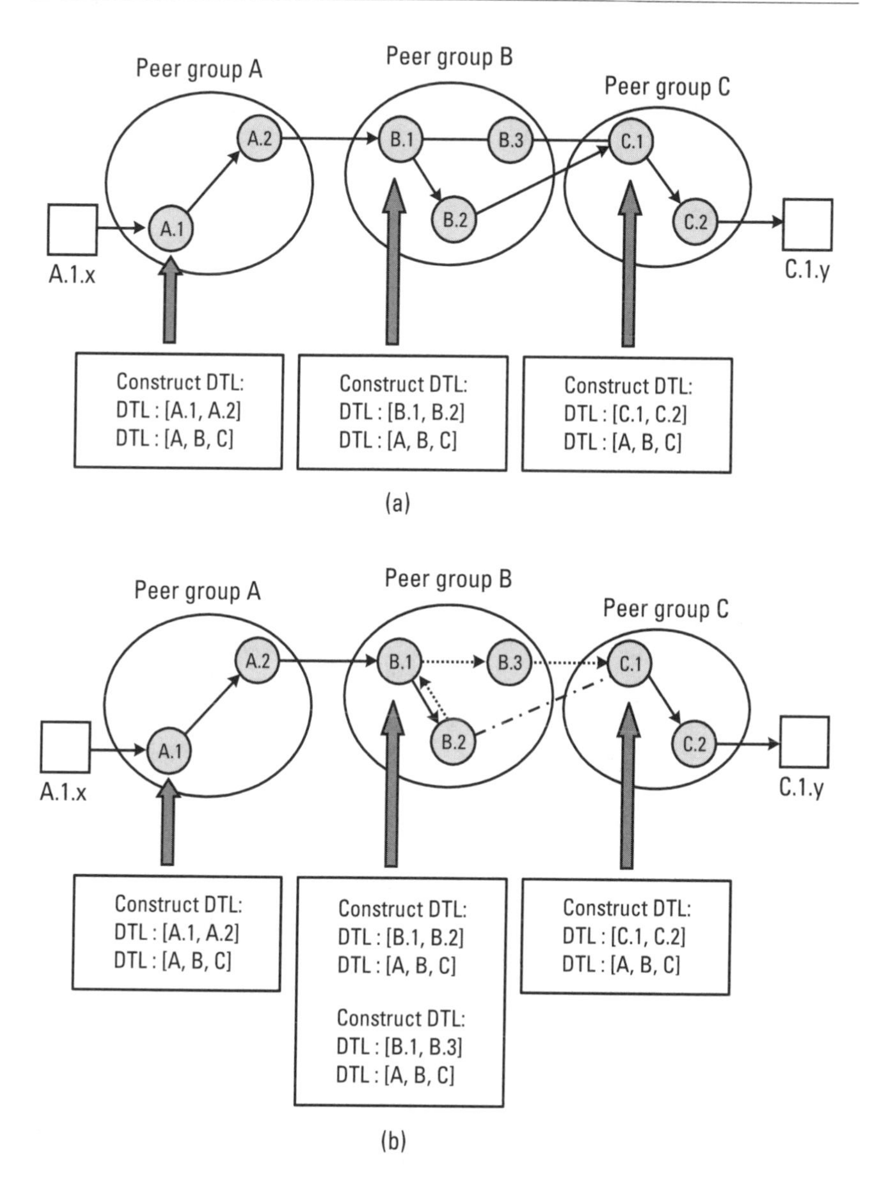

Figure 4.15 Illustration of PNNI routing based on crankback: (a) designated transit list (DTL), and (b) crankback function.

From the functional point of view, an ATM switch is practically the same as the packet switch used in today's computer networks. However, there are two main differences in the incoming packet size and the required switching speed. In ATM switches, the incoming packets (i.e., cells) have afixed size of 53 bytes and are transmitted at or above the 155-Mbps rate. Thus, in contrast to tradional packet switching, ATM switching required new high-speed switching techniques such as parallel processing and self-routing.[18] Also, in contrast to circuit switching, ATM switching requires packet buffering to avoid cell loss caused by the multitude of the cells destined to the same output port. As a consequence, an ATM switch can be considered as a box that provides switching and buffering functions together.

4.3.1 ATM Switch Functions

In general, an ATM switch consists of a set of LIs, a *call processor* (CP), a *signal processor* (SP), and a switching network as depicted in Figure 4.16 [9]. The LI performs optical-to-electrical signal conversion, cell synchronization, VPI/VCI header translation, traffic congestion control, and insertion and extraction of routing information. That is, it performs all the processing required for incoming packets before being transferred to the switching network. The switching network routes incoming packets using the routing information of the packets. The CP and the SP perform the functions concerned with the ATM connection setup and release operations and can be connected to the LIs through two different configurations, *front signal processing* and *rear signal processing*, as shown in Figure 4.16. In front signal processing, signaling packets are transferred through a separate bus to the CP and the SP. In rear signal processing, however, there is no separate bus for signaling packet transfer, but signaling and control information is handled in a unified way together with the user traffic. Therefore, the hardware configuration of the rear signal processing case could be simpler than the front signal processing case.

4.3.2 ATM Switch Architectures

ATM technology was designed from the beginning to support the implementation of high-speed switches. It was always assumed that to become a high-speed switch, much of the processing would have to be done in hardware. Consequently, the development of high-speed switch fabrics that could more or less autonomously route ATM cells was actively studied. This led to

18. Self-routing is a routing scheme in which each incoming packet is transferred to its destined output port based only on its own routing information, without assistance from the centralized processes.

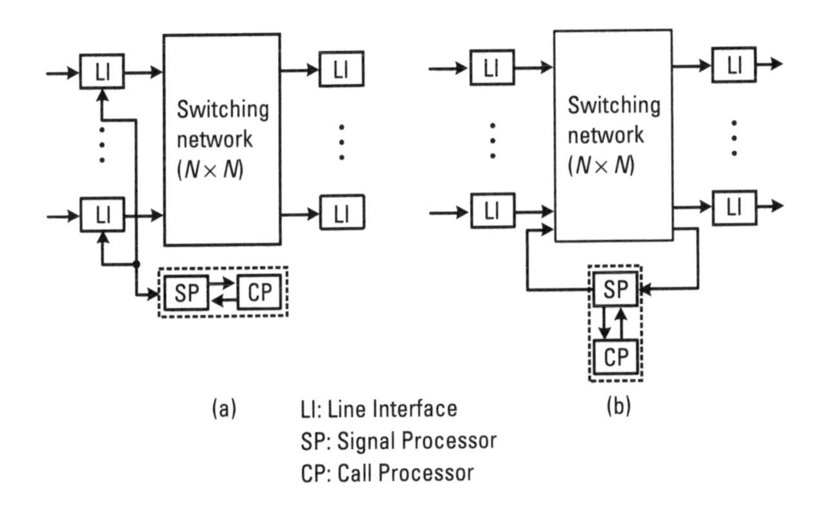

Figure 4.16 Basic structure of ATM switch: (a) front signal processing, and (b) rear signal processing.

an explosion of research in the field of switch fabrics, especially those that supported the self-routing kind.

ATM switches can be classified into *shared-buffer switches*, *shared-medium switches*, and *space-division switches* depending on the switching mechanism employed [10]. The following subsections examine ATM switch architectures according to this categorization.

4.3.2.1 Shared-Buffer Switch

In shared-buffer switches a single buffer memory is shared by all input and output lines as shown in Figure 4.17, where it is used as the central component

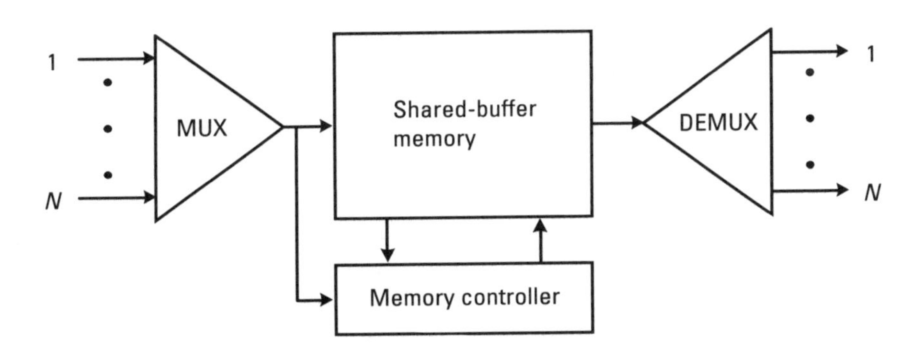

Figure 4.17 Basic structure of a shared-buffer switch.

for the switching operation. Packets arriving at all input lines are multiplexed into a single stream and then fed to the shared memory for storage according to their destined output addresses. At the same time, an output stream of packets is created by retrieving packets in the output queues sequentially. This output stream is then demultiplexed to distribute packets to each individual output line. Accordingly, in shared-buffer switches, high-speed controllers and memories are necessary to process N input packets and N output packets during one slot time. As a consequence, the switch size is determined by the available memory speed and achievable processing speed. Shared-buffer switches have some advantages such as efficient buffer utilization, easy accommodation of point-to-multipoint services, and priority-control-based buffer management.

To perform the required switching function, the shared memory in a shared-buffer switch must operate like N buffers logically. The controller that performs the memory management controls addresses of each output port individually or stores the addresses in the shared buffer in conjunction with the packets in linked list form. It is also possible to simplify buffer management by employing the *content-addressable memory* (CAM) as the shared buffer.

4.3.2.2 Shared-Medium Switch

In shared-medium switches, all packets arriving at the input lines are synchronously multiplexed onto a common high-speed medium of bandwidth equal to N times the rate of a single input line. Each output line is connected to the bus via an interface consisting of an *address filter* (AF) and an output *first-in first-out* (FIFO) queue as shown in Figure 4.18. The AFs determine whether or not the packet observed in the bus is to be written into the FIFO queue. Functionally, this structure is similar to that of the shared-buffer switches, so the point-to-multipoint services can be easily accommodated. However, because independent queues are used at the output lines, efficient buffer utilization is not achieved.

4.3.2.3 Space-Division Switch

Space-division switches can provide a path between an input line and an output in the space-division fashion. As a result, in this type of switch, several independent paths can be set up simultaneously. This is quite different from the case of the shared-buffer switch or the shared-medium switch in which paths are set up sequentially in a time-division fashion. This difference means that the space-division switch may operate at a lower speed than the other

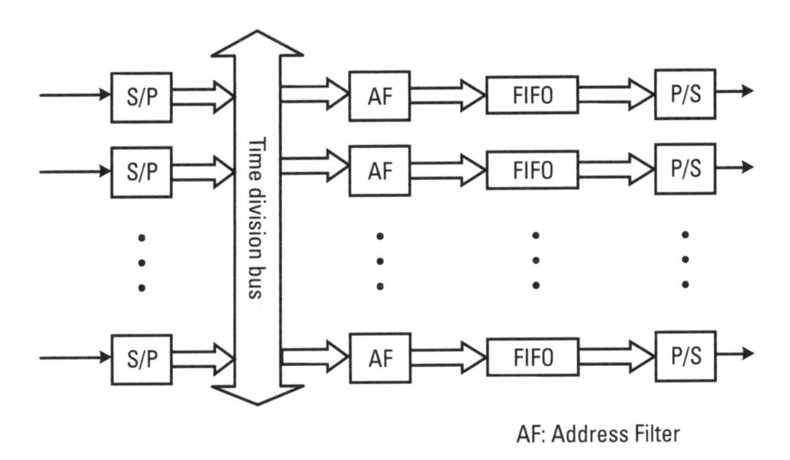

Figure 4.18 Basic structure of shared-medium switch.

types.[19] However, the situation may occur when all incoming packets cannot be transmitted to their desired destination, and, therefore, buffers are necessary to resolve the packet conflicts caused by such resource limitations.

The space-division switches can be classified into four categories depending on the buffer arrangements—*input buffer switch*, *output buffer switch*, *input/output buffer switch*, and *internal buffer switch* (see Section 4.3.2.4). The switching network of space-division switches can be divided into two types—i.e., a *blocking* switching network and *nonblocking* switching network. If a switching network can switch every differently destined input packet to the desired output port then it is a nonblocking switching network; otherwise it is a blocking switching network. In general, blocking switching networks are simpler in structure than, but inferior in performance to, nonblocking switching networks.

The *multistage interconnection network* (MIN) structure is one example of the blocking switching networks. It consists of multiple stages of networks with each stage consisting of a column of 2 × 2 switching elements that performs a fixed pattern of permutation functions. A MIN structure has different names depending on the interconnection pattern between the switching stages, namely, *Banyan network, baseline network, shuffle-exchange network*, and *flip network*. Figure 4.19 depicts the interconnection patterns of these

19. Note that this advantage comes from the constituent hardware, which is much heavier in the space-division switch than in the other switches.

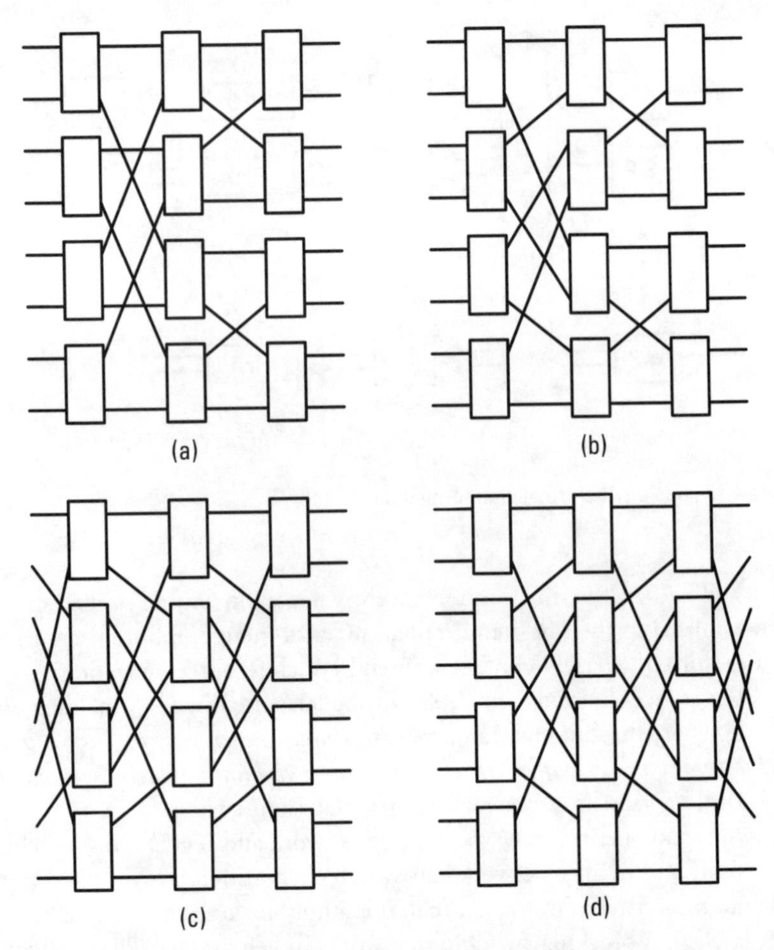

Figure 4.19 Interconnection patterns of multistage interconnection networks: (a) Banyan network, (b) baseline network, (c) shuffle-exchange, and (d) flip network.

four MIN structures. Note that the differences in interconnection patterns do not necessarily lead to any performance difference in the packet-switched environment.

Basically, every MIN has the unique path property that only one path exists from each input to each output. Further, the route that an incoming packet passes within the network can be completely determined by the destination address affixed to the packet. This implies that the connection state of each switching element in the network is completely determined by the destination of the incoming packet. This observation leads to the self-routing scheme, which can be realized by adding a routing header that contains the destination address and other information necessary for the routing to the front end of the packet, and by letting each switching element operate as directed in this header. For example, in the case of the 8 × 8 baseline network shown in Figure 4.20, the three-bit header affixed to each packet determines the route in such a way that the switching elements in the first stage are controlled by the *most significant bit* (MSB) of the routing header, those in the second stage by the next bit, in that order.

Examples of the nonblocking switching network include the crossbar switch, Benes network, and Batcher-Banyan network. The crossbar switch consists of N^2 cross-point switches, one for each input-output pair, so the realizable switch size tends to be limited. In addition, it has the problem that the transit time is not constant over all input/output pairs unless artificial delays are introduced at the inputs and outputs of the switch. The Benes network has a structure in which two baseline networks are cascaded back to back. So, the number of required switching elements is $(2 \log_2 N\text{-}1)N/2$, which is much less than that of the crossbar switch. However, the self-

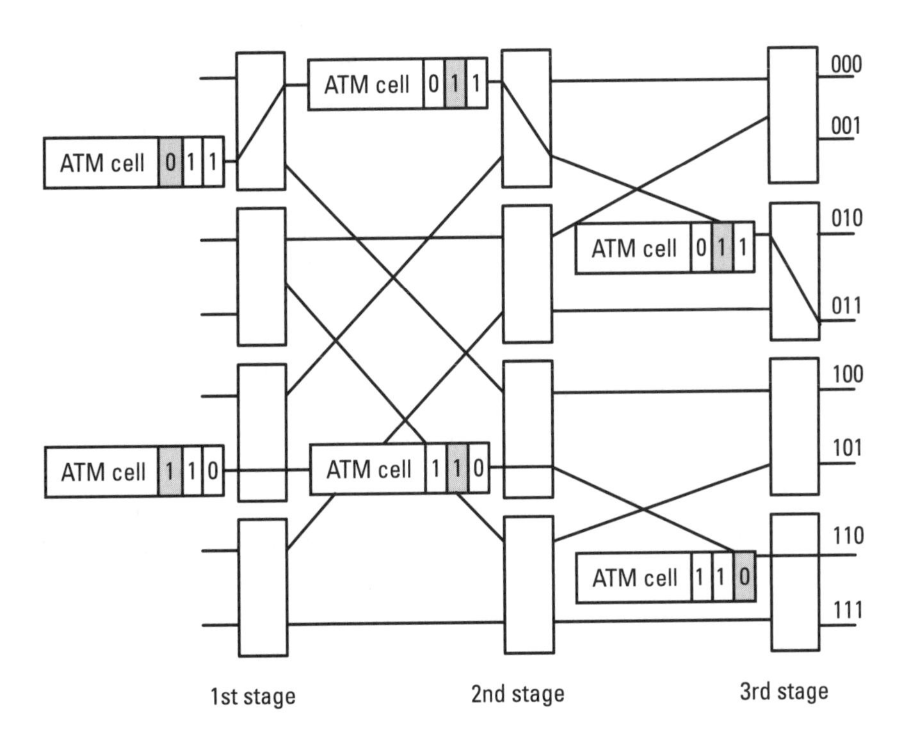

Figure 4.20 A routing example of an 8 × 8 baseline network.

routing capability is not applicable to this network. The Batcher-Banyan network is a nonblocking network that can keep its self-routing capability while overcoming the internal blocking that is the drawback of the Banyan network. The Batcher-Banyan network can avoid internal blocking by sorting the incoming packets based on their destination addresses first and then routing through the Banyan network.

Since the Batcher-Banyan network requires $\log_2 N \{(\log_2 N+1)/2+1\} N/2$ switching elements, it lies in between the crossbar and Benes network as far as hardware complexity is concerned. Figure 4.21 illustrates the basic structure and routing operation of an 8×8 Batcher-Banyan network.

4.3.2.4 Space-Division Switch Architectures

As stated in Section 4.3.2.3, space-division switches can be classified into input buffer switch, output buffer switch, input/output buffer switch, and internal buffer switch depending on the buffer arrangements.

Input buffer switch. In input buffer switches, a nonblocking switching network is used as the switching network and a dedicated buffer is allocated for each input port. Because the switching network can transfer only one packet to each output in each time slot, an arbiter is needed to avoid packet conflicts that may occur in the switching network. The performance of input buffer switches depends on the operation of the input buffers. If the buffers behave like simple FIFO queues, then at the beginning of each time slot only the packets at the *head of line* (HOL) of the buffers contend for access to the switch outputs. If every packet is addressed to a different output, the nonblocking switching network allows each packet to reach the desired output. If k packets at the heads of the N input buffers are addressed to the same output, only one of them is allowed to pass through the switching network, and the other k–1 packets must wait until the next available time slot. While a packet in the HOL is waiting for its turn to access the next available slot, other packets heading for idle output ports must be queued behind it in the buffer. This is known as *HOL blocking.* Due to this blocking the maximum throughput is limited to 0.586 for a uniform traffic[20] and a large N. The contention among the HOL packets can be resolved by employing the ring-reservation scheme. The ring

20. The uniform traffic refers to the traffic whereby the process describing the arrival of packets at an input line is a Bernoulli process, independent of all other input lines, and whereby the requested output port for a packet is uniformly chosen among all output ports, independently of all arriving packets.

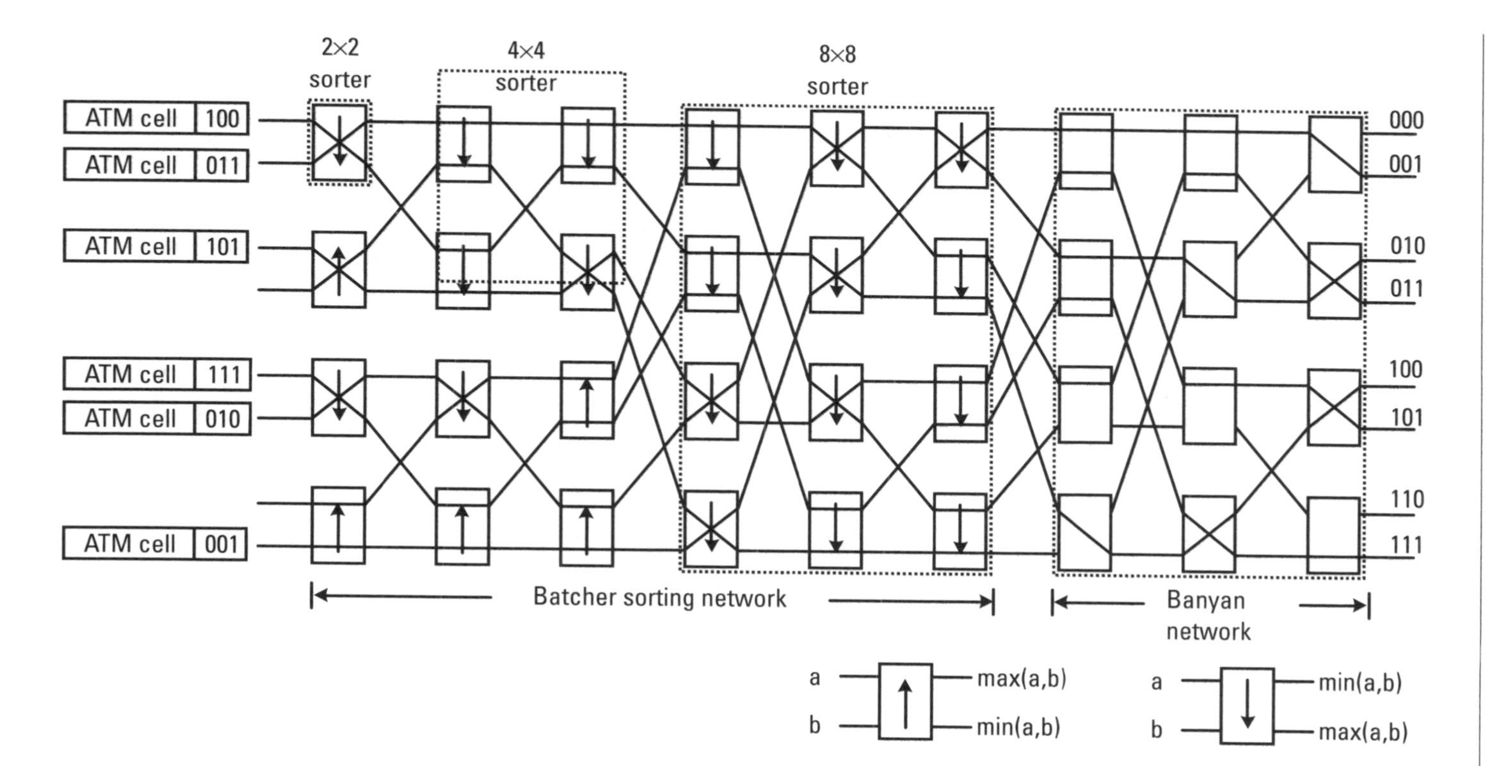

Figure 4.21 Basic structure of an 8 × 8 Batcher-Banyan network.

reservation scheme is actually a token ring scheme in which each HOL packet can make a reservation when it catches the token.

Output buffer switch. In the output buffer switches the interconnection fabric broadcasts each incoming packet to all interface modules simultaneously, and then each interface module selects the packets that are destined to it. That is, a broadcast-and-select mechanism is adopted in the switch. The broadcasting function can be realized by a parallel bus, binary trees, or a high-speed transmission bus. Examples of the output buffer switches include the knockout switch, the sunshine switch, and the tandem Banyan switch.

For example, in the *knockout switch*, N inputs form N buses that are directly connected to each of the N interface modules. This simple structure causes no packet conflicts among the packets bound for different outputs, and brings forth good modularity and broadcast and multicast capabilities. The interface module consists of three major components: packet filter, concentrator, and shared buffer. The packet filter examines the address of every packet on each of the N buses and filters out those addressed to itself. The concentrator then concentrates the input lines to L output lines such that up to L packets can emerge at the output of the concentrator. The shared buffer secures FIFO buffers equivalent to a single queue of L inputs and one output for each interface module to store the concentrator output packets. The knockout switch has low latency and a simple switching mechanism. However, because each bus has a large number of fan-outs, it is difficult to implement large-size knockout switches.

The *tandem Banyan switch* structure consists of Banyan networks and L interface modules. All incoming packets are fed to the first Banyan network. The correctly routed packets at the end of the first Banyan network exit to their respective interface modules. The misrouted packets carry on into the next Banyan network where the same process is repeated. This repeats up to the Lth Banyan network, and the packets that are still misrouted even at the output of the last network are regarded as lost. As the load on successive Banyan networks decreases, so does the likelihood of conflicts. This implies that it is possible to reduce the packet loss rate to the desired low level by attaching a sufficiently large number of Banyan networks.

Various methods exist to resolve conflict if it occurs between two packets in a 2×2 switching element. The random selection method is one simple example. It picks up an arbitrary one of the two conflicted packets and routes it to the correct output port, while misrouting the other packet to the wrong output port with a mark put on its collision bit. Whenever a conflict occurs between a properly routed packet and a misrouted one, the misrouted packet is doomed to lose the conflict. Therefore, it is possible that a packet that gets misrouted at some stage within a Banyan network does not influence the routing of the properly routed packets in that Banyan network. When a misrouted packet begins routing afresh at the next Banyan network, the collision bit is reset to that of a properly routed packet. The packets that arrived at its destined interface module are accepted via the address filters and then stored at its buffer after multiplexing.

Input/output buffer switch. The input buffer switch has a simple switching structure but requires a complex arbiter to obtain the desired switching performance. On the contrary, the output buffer switch does not require an arbiter but its hardware complexity is high. As a compromise, a switch structure that has a simple arbiter and moderate hardware complexity can be devised by introducing buffers both at the input ports and at the output ports. In such an input/output buffer switch, the switching network that can transfer three or four packets to a particular output simultaneously is normally used.

In the input/output buffer switch, it is possible to adopt a back-pressure mechanism between the input buffers and the output buffers so as to overcome packet loss at the output buffers and attain a higher throughput with a quite limited number of output buffers. This requires a large number of buffers installed at the input ports. However, buffering at the input ports costs less, as the input buffer operates at the line speed with a single input line feeding each buffer. On the other hand, an output buffer has to have multiple inputs or run at a speed faster than the output line to be able to accept packets from multiple inputs during the same time slot. Accordingly, it is advantageous to have a small amount of buffering at the output ports and to have a sufficient amount of input buffering to achieve the desired performance.

Internal buffer switch. In the previous three buffering schemes, packet contention within the switching network is avoided by employing an arbitration function outside the switching network, so buffers are located external to the switching network—that is, at the input and/or output ports. In the internal buffer switch, the packet contention resolution functions are distributed internally throughout the switching network and, consequently, the buffering functions are located inside the switching network. The buffered Banyan network is a typical example of the internal buffer switch. In this switch, if two packets at both input ports of a switching element are addressed to the same output at the same time slot, only one is allowed to pass, whereas the other is stored in the internal buffer for one time slot.

4.3.3 Point-to-Multipoint Switching

There are several services today that require multipoint connections—for example, teleconferencing, distributed data processing, LAN bridging, and video distribution. To support such point-to-multipoint services within ATM networks, ATM switches should be capable of providing point-to-multipoint switching. That is, ATM switches should be able to transfer an incoming packet to a set of output port groups simultaneously (i.e., a packet replication function should be added to ATM switching systems).

Addition of the packet replication function to the ATM switch depends largely on the ATM switching system architecture. For example, it can be easily accommodated in the shared-medium switch, because all incoming packets are broadcast to a shared-medium to which all output ports are connected in the switch. However, in the case of ATM switches based on the Banyan network, the replication function cannot be accommodated without altering the switch structure. In this case the packet replication function, in general, can be provided by cascading an additional copy network in the front stage, as shown in Figure 4.22. According to the structure, a point-to-multipoint packet is delivered to its destined outputs through the following three steps: First, the packet is replicated in the front-end copy network. Second, the destined outputs of the replicated packets are inquired after and an appropriate routing field is added to each of them in the *trunk number translators* (TNTs). Third, the replicated packets are routed to their destination ports through the point-to-point switching network.

Because the copy network is in charge of the packet replication function only, it can be easily implemented by using a *broadcast Banyan network* (BBN), which is a Banyan network equipped with the packet replication function. Figure 4.23 shows an example of a copy network based on the BBN. The packet replication process of this network can be divided into the address encoding process and address decoding process. The address encoding process transforms the set of copy numbers specified in the headers of incoming packets into a set of monotone address intervals, which form new packet headers. This process is performed by the concentrator, the *running adder* (RA), and the *dummy adder encoders* (DAEs). The running adder adds the copy numbers of incoming packets sequentially from the top and transfers the results to the DAEs. Then, the DAEs assign a set of monotone output addresses to each packet based on the results. These addresses indicate the outputs to which each packet should be transferred in the following BBN. In the address decoding process, the packet replications are performed at the BBN. Finally, the TNTs determine the destinations of the replicated packets.

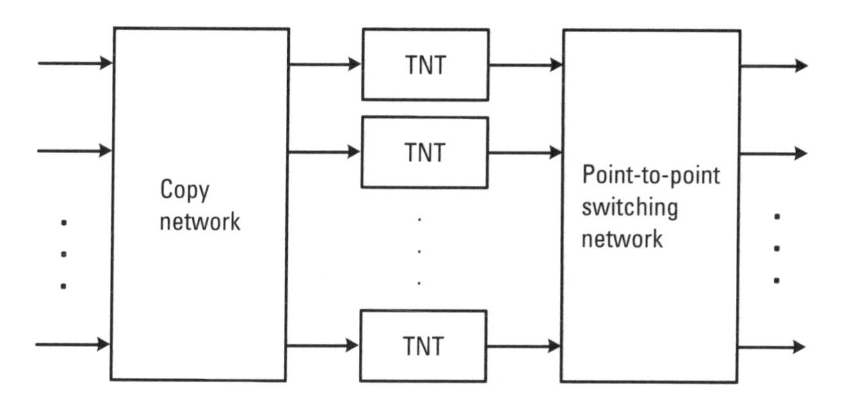

Figure 4.22 Accommodation of point-to-multipoint switching based on a copy network.

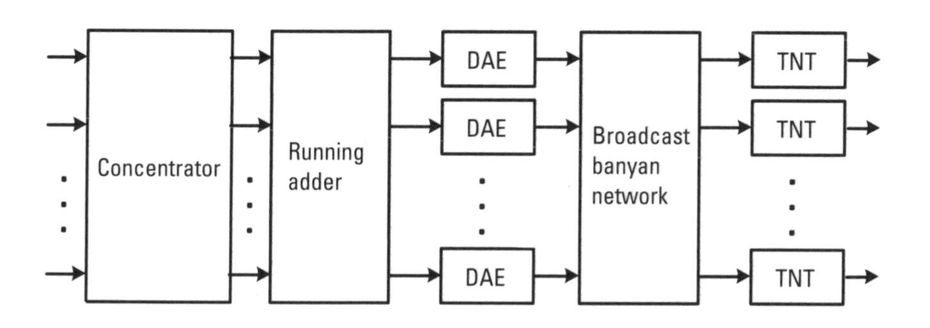

Figure 4.23 Example of a copy network with the BBN [11].

4.4 Network Control and Management

ATM technology was defined from the beginning to have good support for both network control and network management. Signaling is fundamental as ATM employs a virtual circuit model in which connection setup and release are essential functions. At the same time, a reliable and efficient OAM is important as ATM technology is tailored for integrated services of all existing and future networks.

4.4.1 Signaling in ATM Networks

ATM networking has a connection-oriented network paradigm. As such, signaling, which is the generic function of setting up/releasing connections is a basic function of ATM networks. It is based on a type of out-of-band signaling that utilizes separate specific signaling VCs.

4.4.1.1 ATM Signaling Requirements and Standards

The original signaling requirements and functions as defined by the ITU-T were comprised of many features. This was due to the need to support the legacy N-ISDN application services, while at the same time supporting many new service models such as multipoint and multiconnection calls. Realizing the complexity of such an all-inclusive signaling function, the ITU-T took a phased approach to defining the relevant standards. The ATM Forum also took a similar path, trying to minimize the differences between the ITU-T recommendations and the ATM Forum specifications.

Figure 4.24 shows the basic signaling protocol architecture based on the ITU-T model. The UNI side uses Q.2931 while the NNI side uses B-ISUP. The ATM Forum model slightly differs in that the UNI side should use UNI 4.0 while the NNI side may use PNNI or B-ICI. The differences between these models will be explained below.

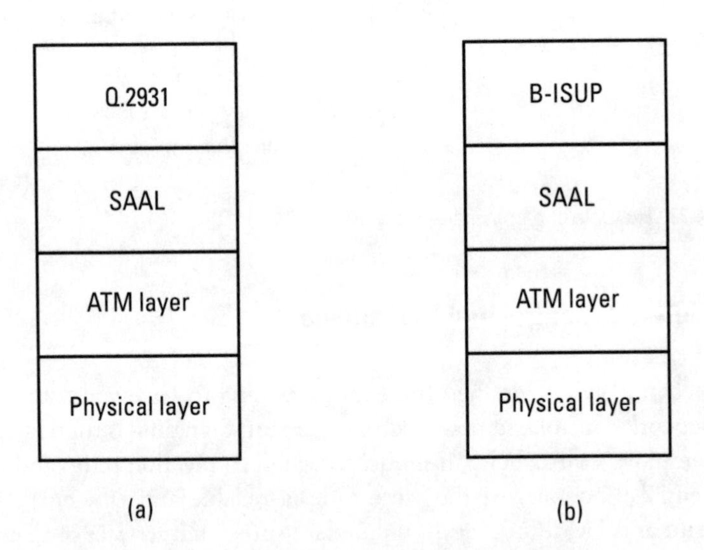

Figure 4.24 Signaling protocol architecture: (a) UNI, and (b) NNI.

There are a number of specifications for the UNI side protocols because both the ATM Forum and the ITU-T have defined a number of versions.[21] As with the UNI, the NNI has a number of specifications defined for it. The currently defined signaling protocols include the following:

- ATM Forum IISP;
- ATM Forum PNNI;
- ATM Forum broadband intercarrier interface (B-ICI);
- ITU-T broadband ISDN services user part (B-ISUP).

The IISP and PNNI signaling interfaces were defined by the ATM Forum to be used in private networks that are composed of multivendor switches (refer to Section 4.2.2.3 for more details on PNNI signaling). The B-ICI and B-ISUP were later defined to be used in public networks between different network operators.

One notable difference between the PNNI solution and the B-ISUP/B-ICI solutions is that B-ISUP and B-ICI do not define any routing protocols. They are purely signaling protocols. In contrast, PNNI defines both routing and signaling protocols. The basic premise is that the operators will use some other routing protocols or methods when using B-ISUP/B-ICI.

4.4.1.2 Signaling at the UNI: Q.2931 and UNI 4.0

As shown in Figure 4.24, in the ITU-T model the higher layer signaling protocol uses Q.2931 at the UNI. The signaling AAL, which lies below the signaling protocol, accomplishes the adaptation function between the signaling protocol and the ATM layer. The Q.2931 protocol is based on the N-ISDN signaling protocol Q.931.

Q.2931 describes in detail the specifications of messages and information elements that are used for signaling procedures between the signaling end points at the UNI. Although based on Q.931, the third layer signaling system of N-ISDN, Q.2931 also includes new features appropriate to

---

21. The ATM Forum UNI 3.0, 3.1 specifications were not originally closely aligned with the corresponding ITU-T specifications of Q.93B, which was a draft version of the Q.2931. Realizing the problems of such an approach, the ATM Forum UNI 4.0 adopted the ITU-T Q.2931 specification. Consequently, the current versions of the standards are now closely aligned in supported functions and features.

B-ISDN whose key technology is ATM. The information elements newly added in Q.2931 include broadband bearer capability, ATM traffic descriptors, AAL parameters, and connection identifiers, which are mainly associated with ATM technology. In addition, the allocation and negotiation of the VPI/VCI, compatibility test at the receiving end, and the interworking procedures between Q.931 and Q.2931 are included. Major signaling procedures described in Q.2931 are the point-to-point call/connection setup procedure, the VPI/VCI allocation and selection procedure, the QoS and traffic parameter selection procedure, the point-to-point call/connection release procedure, the call/connection restart procedure, the error state recovery procedure, the state monitoring procedure, and the interworking procedure with 64-Kbps circuit-mode ISDN.

A user can invoke a call setup procedure by sending a SETUP message after determining the destination address, QoS level, traffic parameters, and so on. As for call release, either the user or the network can invoke the procedure using the RELEASE message. Figure 4.25 shows an example of call setup and release.

The ATM Forum UNI 4.0 specifications are based on the ITU-T Q.2931 specifications. The UNI 4.0 specifications add the following functions to the

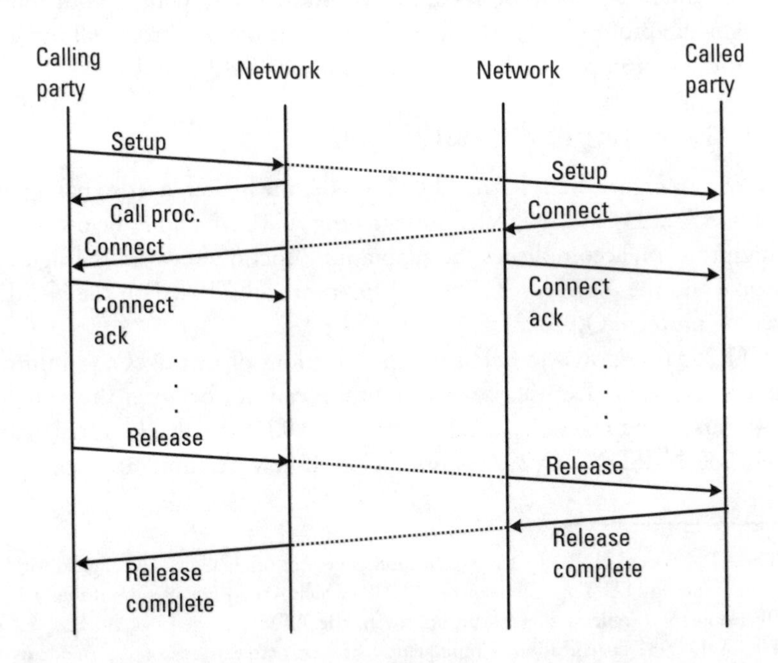

Figure 4.25 Point-to-point call setup and release procedure.

basic ITU-T specifications:[22] group and anycast address support, connection parameter negotiation at setup time, leaf-initiated join point-to-multipoint SVCs, QoS parameters, switched VPCs, proxy signaling, and ABR singling.

4.4.1.3 Signaling Protocol at NNI: B-ICI and B-ISUP[23]

The signaling protocol at the public NNI is B-ISUP, which is specified in Q.2761 through Q.2764. B-ISUP is restructured out of ISUP in such a way that it can fit to the ATM environment and can easily accommodate newly arising services. Consequently, B-ISUP differs from ISUP in both structure and functions.

B-ISUP constructs an *application service element* (ASE) for each function and thus ensures independence among different elements. If a new service is introduced, it can flexibly accommodate it by adding a new ASE. It provides the basic framework of architecture for separate control of call and connection, which is the ultimate goal of the signaling architecture.

B-ISUP interworks with Q.2931 to perform the call setup and release procedures through transparent information transfer between end users and through access to information elements in the network. Call connection control procedures are carried out by exchanging a series of messages between intermediate switches and the switches at the source and destination sides.

B-ICI versions 1.0, 2.0, and 2.1 were defined by the ATM Forum as a way of allowing signaling between carrier networks before the ITU-T specifications were ready. B-ICI version 1.0 initially supported only PVCs with peak rate policing. Later on, B-ICI version 2.0 was developed to support UNI 3.1 SVCs. The basic B-ISUP protocol stack was shared between the B-ICI version 2.0 and B-ISUP. But the two standards are not completely compatible, mainly because the current version of B-ISUP has been extended to support Q.2931 and consequently UNI 4.0 capabilities. In contrast, B-ICI has remained more closely aligned with UNI 3.1.

4.4.1.4 Signaling VC (SigVC)

In B-ISDN, signaling information is transferred separately from the user information through a dedicated channel, an SigVC. There are four types of

22. The main reason for these differences may accrue from the fact that the ATM Forum has been much quicker in adding functions to the standards.
23. We omit the discussions on IISP and PNNI signaling as they are already dealt with in Section 4.2.2.

SigVCs: the *meta-SigVC*, the *general broadcast SigVC*, the *selective broadcast SigVC*, and the *point-to-point SigVC*.

Meta-SigVC is used for establishing, confirming, and releasing the point-to-point SigVCs or selective broadcast SigVCs. The metasignaling procedure, which is a layer management function, is in charge of the transfer of control information only. Meta-SigVC has predetermined VPI/VCI values. Meta-signaling performs the functions of allocating the capacity of SigVC and of associating call setup requests to appropriate service profiles. It also provides a means of resolving simultaneous call requests, which is similar to the procedure that allocates the *terminal equipment identifier* (TEI) in N-ISDN.

Broadcast SigVC is employed to transfer signaling messages to all or some signaling end points. General broadcast SigVC exists in every signaling endpoint regardless of the service profile, and the VPI/VCI values are preassigned. Selective broadcast SigVC is allocated to each service profile of each signaling end point.

Point-to-point SigVC is used for establishing, maintaining, and releasing a VC to convey user information. The VPI/VCI values for the point-to-point SigVCs are also preassigned.

4.4.1.5 Signaling AAL

Signaling AAL (SAAL), outlined in Q.2100, refers to the AAL layer of the control plane, which plays an important role in adapting the signaling protocols such as Q.2931 and B-ISUP to the ATM layer protocol. The standardization for SAAL has been done so that a common protocol can be used for both UNI and NNI to manage the network efficiently. SAAL uses the same SAR and CPCS protocols as AAL-5 does, but it newly defines SSCS protocols suitable for higher layer protocols. This SSCS is composed of the *service-specific connection-oriented protocol* (SSCOP) and the *service-specific coordination function* (SSCF), as depicted in the layered architecture of Figure 4.26.

SSCOP, specified in Q.2110, performs various functions such as sequence integrity, error detection, error recovery through retransmission, receiver-based flow control, assured and nonassured transmissions of the user information, and establishment, release and synchronization of the SSCOP connection. SSCF is for the mapping of service primitives between the higher layer signaling protocol and the SSCOP. It is impossible to use a common SSCF at the UNI and NNI since the higher layer signaling protocols are different from each other. The SSCF at the UNI, which is specified in Q.2130, converts the SSCOP services to the services requisite for Q.2931, and the SSCF at the NNI, specified in Q.2140, converts the SSCOP services to the services requisite for B-ISUP.

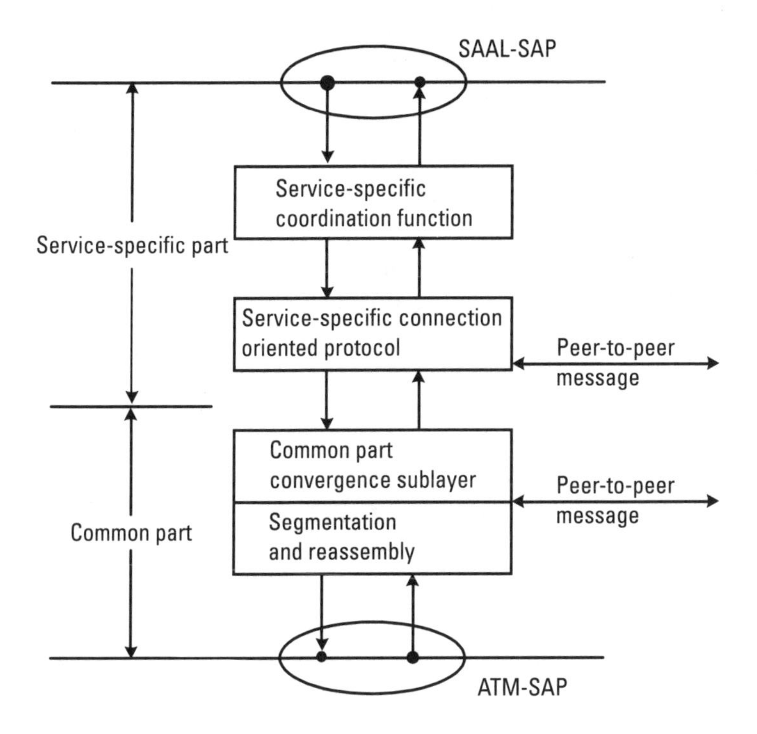

Figure 4.26 SAAL layer protocol structure.

4.4.2 ATM Operations and Management

The OAM functions in ATM networks are required to be capable of such processes as performance monitoring, defect and failure detection, system protection, failure of performance information transfer, and fault locating. As a systematic way to meet these requirements, OAM information flow is assigned and monitored at every network level.

4.4.2.1 OAM Functions

The B-ISDN OAM functions are divided into five types. The first is to monitor, either continuously or periodically, all the entities managed by the network to verify their normal operation. As a result of such performance monitoring, maintenance event information can be generated. The second is to detect failure conditions through a continuous or a periodic inspection. As a result of defect detection, maintenance event information of various alarms can be generated. The third is to minimize the effects of the failure of the managed entity by blocking or replacing it. As a result of such system protection measures, the failed entity is excluded from operation. The fourth is to deliver performance information or impairment information to other management entities. As a result, alarm indications can be delivered to other management planes and a report on the ongoing status can be given. The fifth is to use an internal or an external test system to determine the impaired entity if the given impairment information proves insufficient. As a result of such impairment locating, the impaired entity can be isolated or replaced.

In order to support the above OAM functions, it is necessary to detect loss of frame, loss of cell synchronization and header errors, and monitor the error performance. In addition, it is necessary to generate either an *alarm indication signal* (AIS) or a *remote defect indication* (RDI) signal.

4.4.2.2 OAM Information Flows

ATM networks are layered into an ATM layer network and a physical layer network. The ATM layer network is further divided into *VC level* and *VP level*, and the physical layer network is divided into *transmission path level*, *digital section level*, and *regenerator section level*. The OAM functions are themselves divided into five OAM hierarchical levels in accordance with these five levels. In this arrangement, OAM functions are represented by five information flows, namely, F1, F2, F3, F4, and F5. Figure 4.27 shows the five OAM information flows in contrast to the ATM-layered architecture.

The OAM function associated with each level is independent of that of other levels. For a level to obtain information on performance quality and condition, it has to perform the necessary procedure itself. The result is delivered to the management plane and to the next higher level as the occasion arises. However, the higher-order layer function is not separately required to support the lower-order layer's OAM function.

Physical layer OAM flows. The physical layer encompasses the regenerator section level, the digital section level, and the transmission path level, and the corresponding information flows are defined as F1, F2, and F3, respectively. The method of providing the OAM function required to generate an OAM flow for each case depends on the particular physical layer transmission technique chosen.

In the SDH-based transmission, F1 is conveyed via RSOH, F2 via MSOH, and F3 via *path overhead* (POH) (refer to section 5.2). Parts of F3 are sometimes transported over physical layer OAM cells.[24]

---

24. Refer to Chapter 5 for the acronyms related to SDH/SONET.

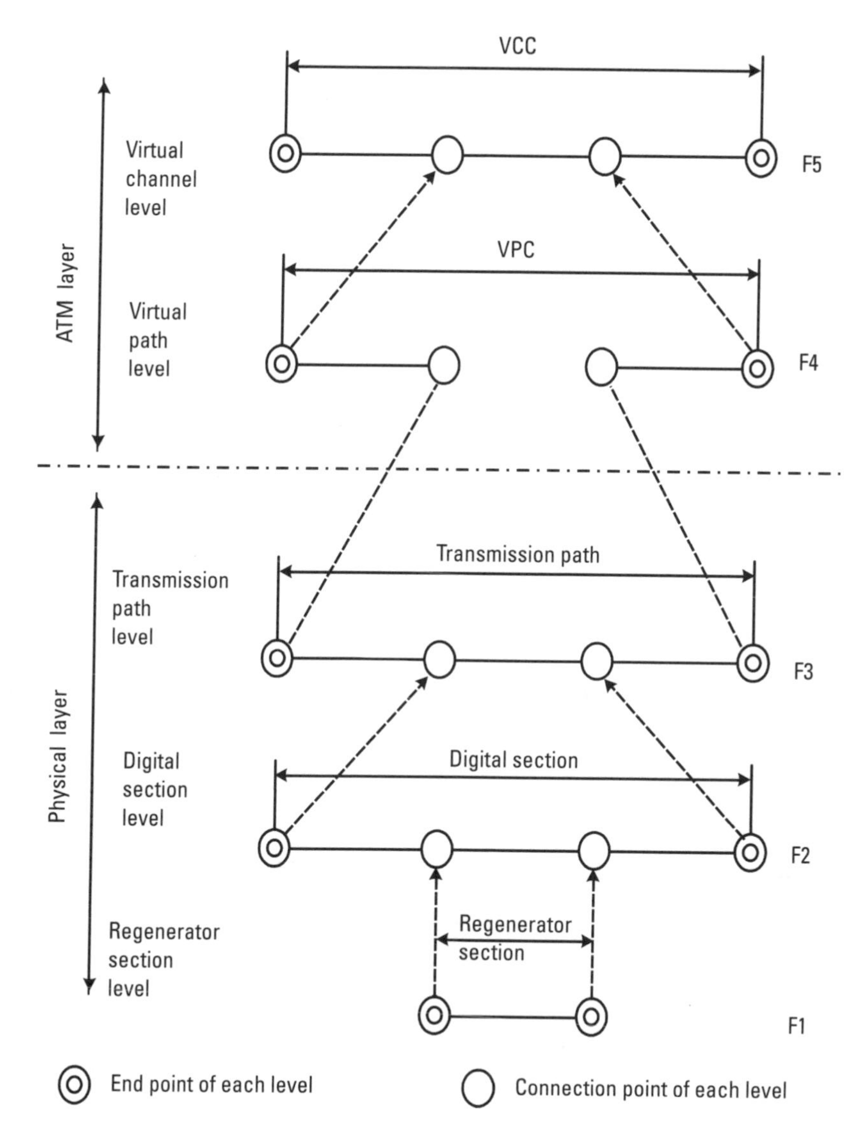

Figure 4.27 ATM-layered architecture and OAM information flows (F1–F5).

In the cell-based case, the multiplexer section is not applicable and hence OAM information flow F2 does not exist. Both F1 and F3 are conveyed via the physical layer's OAM cells, and the headers are assigned with predetermined bit patterns. These physical layer OAM cells are not sent up to the ATM layer.

The physical layer's OAM cells are inserted repeatedly into the ATM cell flow. Here, the insertion of physical layer cells must not hinder the transfer capability of the ATM layer. Consequently, the maximum frequency of physical layer OAM cells allowed is limited to 1 per 27 ATM cells. The minimum frequency possible is 1 physical layer OAM cell per 512 ATM cells.

In G.702's PDH-based transmission, the OAM flow is conveyed through the maintenance function possessed by the system. In this case, the capability to deliver OAM information other than bit messages is extremely limited.

Table 4.4 lists the physical layer OAM functions according to each OAM flow type (refer to ITU-T Recommendation I.610). In the table, entries indicated with "S" and "C" represent SDH-based and cell-based transmissions, respectively. B-NT denotes *broadband network termination* (refer to Figure 4.28 for the location of B-NT1 and B-NT2). Also, PLOAM stands for the *physical layer OAM,* and AU PTR represents the *administrative unit pointer* (refer to Section 5.2.2).

ATM layer OAM flows. The ATM layer encompasses the information flows F4 and F5, which are associated, respectively, with the VP level and the VC level. These information flows are delivered to the VPC and VCC using the cells that are responsible only for the ATM layer OAM function. These cells can be used to achieve communication between peer layers residing in the same management plane of different transmission equipment.

The OAM information flow F4 provides, in support of the VPC OAM function, such capabilities as VPC alarm monitoring, VPC continuity check, and VPC performance monitoring. In case a VPC failure is detected at the VPC point, it sends VP-AIS in the downward direction toward the termination point, and if a VP-AIS or a VPC failure is detected at the VP termination point, it sends VP RDI in the upward direction. Also, if no user information cells have been sent for a fixed duration of time, it creates and sends continuity check cells to verify the continuity of VPC. In addition, information related to error blocking and cell loss/insertion is loaded onto the cells and delivered to the other party for the purpose of end-to-end monitoring.

In support of the VCC OAM function, the OAM information flow F5 provides such capabilities as VCC alarm monitoring, VCC continuity verification, and VCC performance monitoring. The details of their operation are analogous to those of F4. Table 4.5 lists a summary of the ATM layer OAM functions.

Table 4.4
Physical Layer OAM Functions

Level	Function	Flow	Detect/Failure Detection	Alarm/Defect Indication
Regenerator section	Frame alignment (S*)	F1	Cell of frame	Section AIS/RDI
	Section error monitoring (S, C**)		Degraded error performance	Section AIS/RDI
	Section error reporting (C)		Degraded error performance	Section AIS/RDI
	PLOAM cell recognition (C)		Loss of PLOAM cell recognition	Section AIS/RDI
	Cell delineation		Loss of cell sync	Section AIS/RDI
Multiplexer section	Frame alignment (S)	F2	Loss of cell frame	Section AIS/RDI
	Section error monitoring (S)		Degraded error performance	Section AIS/RDI
	Section error reporting (S)		Degraded error performance	Section AIS/RDI
Transmission path	VC-4 offset (S)	F3	Loss of AU PTR	Path AIS/RDI
	Customer network status monitoring (S,C)		Customer network AIS	Path AIS
	Cell delineation (S,C)		Loss of cell sync	Path RDI
	Header error detection / correction (S,C)		Uncorrectable header	(Path management message)
	Header error performance (S,C)		Degraded header error	(Path management message)
	Cell rate decoupling (S,C)		Failure of insertion and suppression of idle cells	(Path management message)
	Path error reporting (S,C)		Degraded error performance	Path AIS/RDI
	Path error monitoring (S,C)		Degraded error performance	Path AIS/RDI
	PLOAM cell recognition (C)		Loss of PLOAM cell recognition	Path RDI

Table 4.5
ATM Layer OAM Functions

Level	Function	Flow	Detect/Failure Detection	Alarm/Defect Indication
VP	Monitoring of path availability	F4	Path not available	–
	Performance monitoring		Degraded performance	–
VC	Performance monitoring	F5	Degraded performance	–

If the OAM flow of the B-ISDN user access is realized in terms of several physical configurations, the result is as shown in Figure 4.28. As can be seen from the figure, F1 terminates at B-NT1 and the regenerator; F2 terminates at B-NT1, B-NT2, and LT; and F3 terminates at B-NT2, ET, and VP-XC. It can also be inferred that F4 terminates at B-NT2 and ET; and F5 at B-NT2 or B-TE. In Figure 4.28, LT denotes *line termination*, NT *network termination*, ET *exchange termination*, XC *cross-connect*, and MUX *multiplexer*.

4.4.2.3 OAM Cell Format

The F4 and F5 OAM flows have a common cell format. The format is shown in Figure 4.29. The VPI field in the F4 OAM cell header is coded according to the VPI value of the VP to be managed. The VCI field is used to differentiate whether the OAM information carried by the cell is for the *link operation* (VCI = 3) or for the *end-to-end operation* (VCI = 4). Other VCI values are reserved for future use. As for F5 OAM flow, both VPI and VCI are used to identify the VC to be managed, and the PT field is used for the discrimination between the *link OAM* (PT = "100") and the *end-to-end OAM* (PT = "101").

The first octet in the payload field of the ATM OAM cell is composed of *OAM type* and *function type*. There are three OAM types: fault management, performance management, and activation/deactivation, and for each OAM type two or three function types are assigned. The *function-specific field* is used to transfer the function-specific OAM information in a predetermined format (refer to ITU-T Recommendation I.610). The octets that are not in use are all stuffed with "01101010."

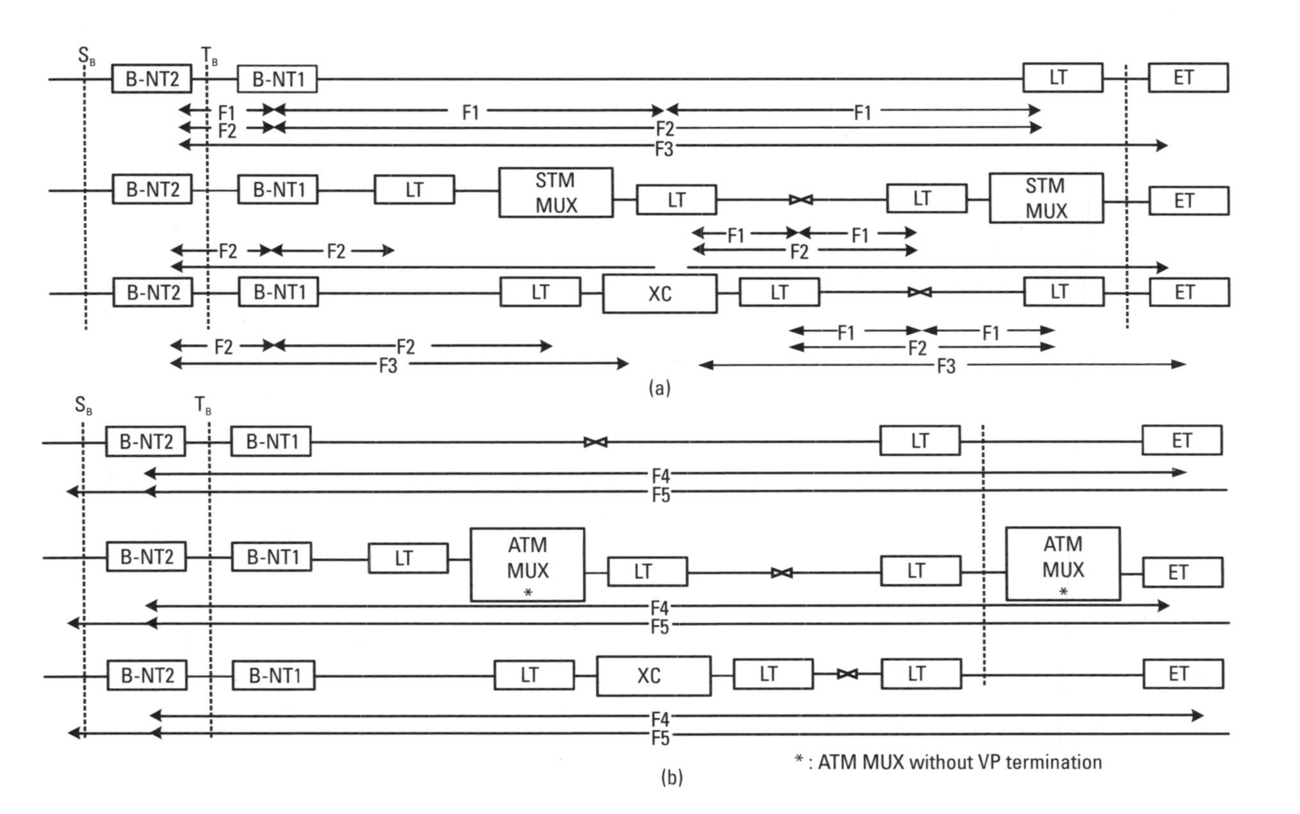

Figure 4.28 Physical structure and OAM flow: (a) OAM flow F1, F2, F3 and (b) OAM flow F4, F5.

Header	OAM type	Function type	Function-specific fields	Reserved	CRC-10

Figure 4.29 ATM OAM cell format.

4.5 Traffic Management

The advantages of ATM lie in the efficient use of network resources and the flexibility to support various services. For these points to be fully exploited, the problems of traffic control and congestion control must be resolved. Many traffic control mechanisms have already been developed for existing packet communication networks. However, these mechanisms are perceived to be minimally effective in controlling congestion in ATM networks, for the following reasons. First, ATM networks should support various services with probably significantly different bit rates. Second, a single connection should accommodate heterogeneous traffic streams with different statistical characteristics and performance objectives. Third, existing and foreseeable services require qualitatively and quantitatively different QoS. Fourth, delay-related performances such as the maximum delay and cell delay variation become important as real-time services emerge.

4.5.1 Basic Concepts

To understand the structure of traffic control and congestion control in the B-ISDN, we need to understand each step involved in setting up, maintaining, and releasing a connection. This section provides a simple overview of these procedures, with an emphasis on traffic control.

When a user wants to set up a connection, first the traffic parameters representing the statistical characteristics of the source and the required level of the QoS are passed to the network. Next, the *connection admission control* (CAC) decides whether the call may be accepted without affecting the QoSs of other connections in progress. Once a connection is admitted, the source may send cells into the network at the rate specified in the traffic contract. While the connection is in use, the *usage parameter control* (UPC) polices the traffic emitted from the source at the UNI to make sure that the source is complying with its traffic contract.

Throughout all these processes, the network may manage network resources using a VP to separate traffic flows according to service characteristics, or may use feedback controls to control the traffic flow injected into the network by the user. Also, the user may generate different priority traffic flows by using the CLP bit so that a congested network element can drop the cell as necessary.

Figure 4.30 shows the reference model for traffic control and resource management as defined by ITU-T. Arrows in the figure indicate the direction of cell flows. We notice that CAC is performed over the entire network, while the UPC and the NPC are located at the UNI and NNI, respectively.[25]

4.5.2 Traffic Management Components

Various mechanisms are used in ATM for traffic management. The main components are the *generic cell rate algorithm* (GCRA) for defining rates and burstiness, traffic descriptors and parameters, CAC methods, UPC, CLP based buffering, traffic shaping, and the newly defined *available bit rate* (ABR) flow control mechanisms.

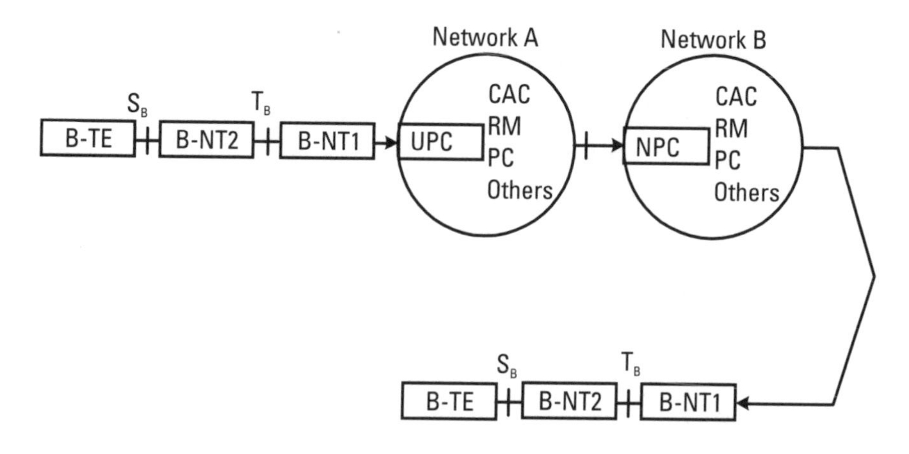

Figure 4.30 Reference configuration for traffic control and congestion control.

25. NPC may apply at some intranetwork NNIs as well.

4.5.2.1 GCRA Algorithm

Before defining any traffic descriptors, it is necessary to first define a method for determining the rate and burstiness of traffic. In ATM networks this method is called the GCRA.

The ITU-T recommendations provide two examples of algorithms —*virtual scheduling algorithm* and *continuous-state leaky bucket algorithm*— that are useful in specifying and monitoring the *peak cell rate* (PCR) in an operational manner, while taking into account a certain *cell delay variation* (CDV) tolerance. These operationally equivalent algorithms are both named the GCRA by the ATM Forum. For each cell arrival, the GCRA determines whether the cell conforms to the traffic contract of the connection, and thus the GCRA is used to provide the formal definition of traffic conformance to the traffic contract.

The GCRA, formally defined in Figure 4.31, depends on two parameters—the *increment* I and the *limit* L. The virtual scheduling algorithm updates the *theoretical arrival time* (TAT) assuming equally spaced cell arrival with an interarrival time of I, and compares it with the actual arrival time t_a. If the actual arrival is too early relative to the TAT, that is, $t_a < TAT - L$, the relevant cell is nonconforming; otherwise the cell is conforming.

In the case of the continuous-state leaky bucket, the value of the leaky bucket counter X increases by increment I and decreases at a continuous rate. If the cell arrives while the value of X is greater than the limit value of L, the cell is nonconforming; otherwise the cell is conforming. The two algorithms in Figure 4.31 are equivalent in the sense that for any cell stream the two algorithms reach the same conclusion as to its traffic conformance to the relevant traffic contract.

4.5.2.2 Traffic Descriptors and Parameters

The first important components of traffic management in ATM networks are traffic descriptors and parameters. A *traffic parameter*, as defined by ITU-T, is a specification of a particular traffic aspect of a source. An ATM *traffic descriptor* is the generic list of traffic parameters that can be used to capture the traffic characteristics of ATM connections, while a *source traffic descriptor* is a subset of traffic parameters belonging to the ATM traffic descriptor. It is used in the connection setup phase to capture the intrinsic traffic characteristics of the connection requested by a particular source. The *connection traffic descriptor*, which specifies the traffic characteristics of

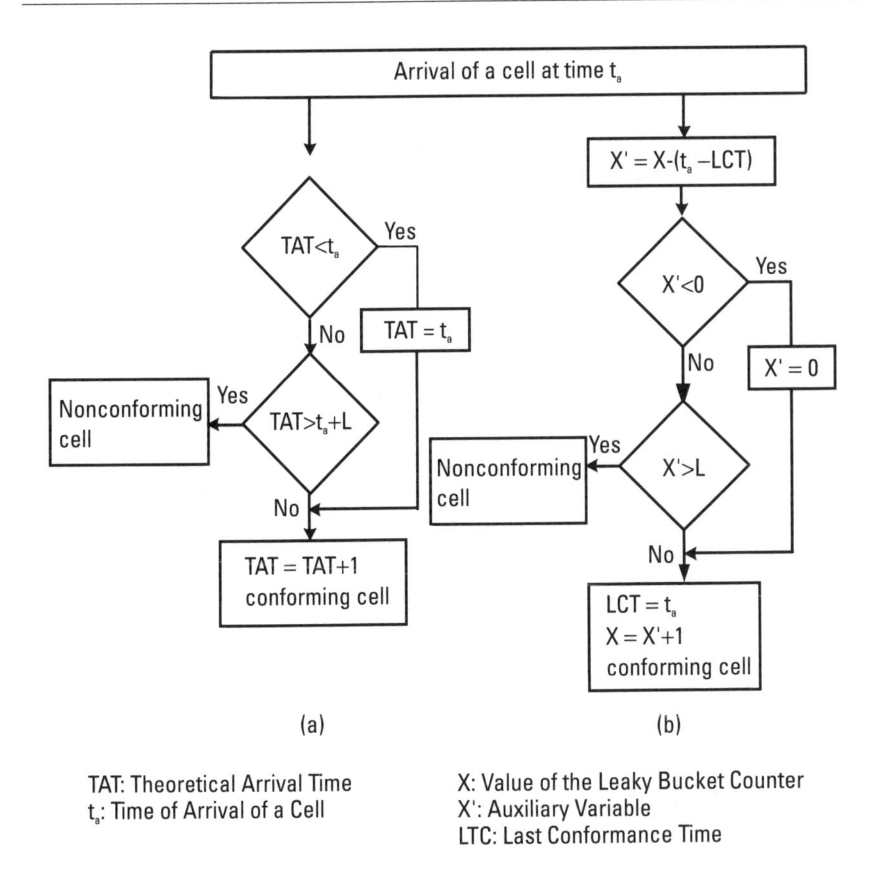

Figure 4.31 Two equivalent GCRA algorithms: (a) virtual scheduling algorithm, and (b) continuous-state leaky bucket algorithm.

the ATM connection at the UNI, includes the cell delay variation, tolerance, and the conformance definition in addition to the source traffic descriptor.

CAC and UPC/NPC procedures require knowledge of the characteristics of the ATM layer connection. This information, called the *traffic contract*, consists of a connection traffic descriptor and a requested QoS class for each direction of the ATM connection.

Traffic parameters. The only traffic parameter originally standardized by ITU-T is the PCR. The PCR in the source traffic descriptor specifies an upper bound on the traffic that can be submitted on an ATM connection. The equivalent terminal configuration for the definition of the PCR is given in Figure 4.32. The PCR is defined at the *physical* (PHY) layer SAP of an equivalent terminal based on the basic event—that is, the request to send an ATM-PDU. The PCR (R_p) of the ATM connection is the inverse of the minimum interarrival time T between the preceding two basic events, where T is called the peak emission interval of the ATM connection.

Due to the need for flexible mapping of the traffic parameters of some existing services such as frame relay, the ATM Forum defined two new parameters, the *sustainable cell rate* (SCR) and *burst tolerance* (BT). The SCR specifies an upper bound on the possible conforming average rate of an ATM connection, whereas the BT together with the SCR and PCR determine the *maximum burst size* (MBS) that may be transmitted at the peak rate and still be conforming. These parameters enable the end-user to describe the traffic characteristics in greater detail than just the PCR. Consequently, the network provider will be able to utilize network resources more efficiently, resulting in possible benefits such as a reduced charge for the end user.

The output of the traffic shaper at the PHY SAP is supposed to conform to GCRA (T, 0). However, due to various operations in the equivalent

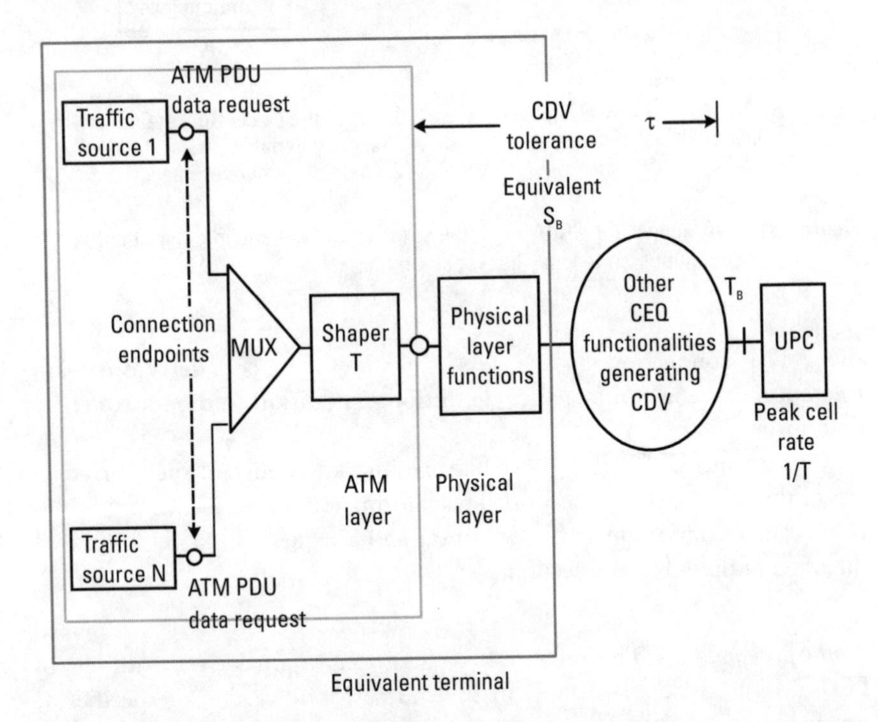

Figure 4.32 Equivalent terminal for the definition of PCR.

terminal and other *customer premise equipment* (CPE), a certain amount of CDV characterized by τ is generated. The value τ is chosen such that the cell flow at the T_B point conforms to GCRA (T, τ). The value τ is called the *CDV tolerance* and represents a bound on the cell clumping phenomenon at the UNI (i.e., T_B point). Users are supposed to select explicitly or implicitly a value for the CDV tolerance at the UNI from a set of values supported by the network.

Connection admission control. One of the main methods for traffic control in ATM networks is CAC, which limits the number of active calls so that QoS requirements for accepted calls can be met. CAC, as defined by ITU-T, refers to the set of actions taken by the network during the call setup phase (or during the call renegotiation phase) to establish whether or not a VC/VP connection request can be accepted. Based on the traffic characteristics, QoS requirements, current network load, and the amount of network resources, the CAC function carries out the following: It decides whether to grant or refuse the connection, determines the traffic parameters for UPC, and allocates network resources.

It has been shown that a so-called *effective bandwidth* exists that is bounded by the peak and the average rate of the source. The most important property of the effective bandwidth is that it is simply additive for several sources.

Another instance of the simple CAC method is the fixed-boundary method, which preallocates a fixed amount of each network resource, such as bandwidth and buffer space, to each class. This CAC algorithm is advantageous in terms of implementation but is likely to suffer from considerably low bandwidth utilization because the unused resources of one class cannot be used by other classes.

In contrast to the above model-based schemes, CAC schemes relying on real-time traffic measurements rather than any user-declared traffic descriptors may be a better solution. The dynamic CAC algorithm is an embodiment of this notion, in which the call acceptance decision is made based on the on-line evaluation of the upper bound of the cell loss probability, which is estimated out of cells arriving in a fixed interval. This algorithm can be used without modeling the input traffic, and thus enables the network to manage its resources tightly using only the peak cell rate.

Policing. The CAC is used to set up a traffic contract between the user and the network. Users may violate the traffic contract negotiated during the call setup phase, either deliberately or due to malfunctioning of the user equipment. As in real life there must be some sort of monitoring of the user to check whether the user stays within the bounds of that contract. This is the role of policing. When the policing action takes place at the edge of the network between customer equipment and the network equipment, it is called UPC. When the action takes place at the boundaries of different networks, it is called NPC.

The UPC/NPC function is used to check the validity of VPI/VCI and monitor cells of a connection to determine whether they conform to the traffic descriptors in the original traffic contract. Depending on the policy of the operator it may discard or tag (to CLP = 1) the nonconforming cells.

A UPC algorithm should be able to detect the violation as quickly as possible and should be transparent to conforming cells. Additionally, it should never be more stringent than the theoretical conformance definition derived from the traffic descriptors and GCRA.

A common example of the UPC/NPC algorithm is the *leaky bucket* (LB) algorithm. The basic form of the leaky bucket is depicted in Figure 4.33(a), and there are a number of variants as well. The most generalized version among them is the buffered leaky bucket shown in Figure 4.33(b), which has both the token pool of size M and the user buffer of size K. It uses a token generator that generates tokens periodically with the period of slots. The generated tokens go into the token pool, and a token is removed from the pool whenever a cell is transmitted. Cells arriving while the token pool is empty are stored in the user buffer until the matching tokens are generated. The LB algorithm is a simple traffic policing algorithm but it can control various traffic parameters by controlling the size of the token pool, buffer size, and token generation rate.

Selective cell discarding and frame discard. As previously mentioned, a congested network element can provide different loss performances among different traffic classes by selectively discarding low-priority cells. This enables the protection of high-priority cell flows and minimizes the spread and duration of congestion. The CLP field in ATM cell header can be used for indicating the cell loss priority such that CLP = 0 for low loss priority and CLP = 1 for high loss priority. In all cases the CLP = 0 cells are guaranteed a CLR no higher than that of the CLP = 1 cells.

The traffic-policing techniques mentioned above enable the detection of cells that do not conform to the agreed-upon traffic contract. The detected

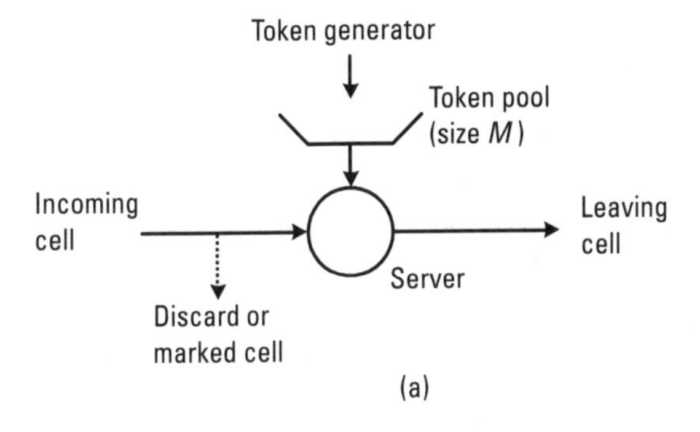

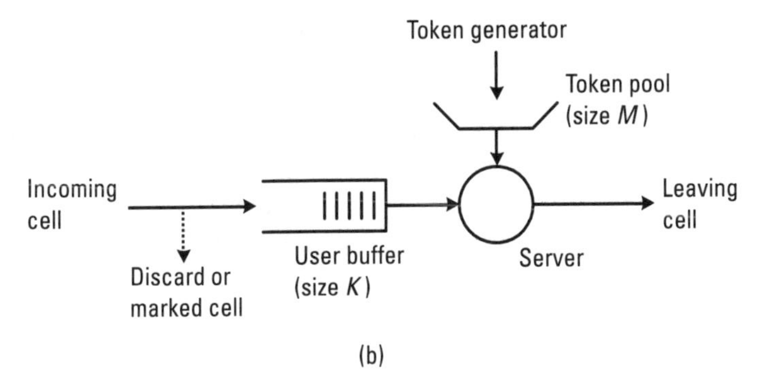

Figure 4.33 LB algorithm: (a) basic LB, and (b) buffered LB.

nonconforming cells are then disposed according to the operation policy. Possible actions of the UPC/NPC function, as defined by ITU-T, are as follows: passing or rescheduling the cells identified by UPC/NPC as conforming and tagging or discarding the nonconforming cells. Another action, which may be optionally taken at the connection level, is to release the noncompliant connection.

Priority control (also known as *selective discard*) is a mechanism by which network elements selectively discard CLP = 1 cells to guarantee a lower CLR to the CLP = 0 cells. Another method is to do tagging—that is, the network only marks nonconformant cells as CLP = 1 rather than immediately discarding them.

A well-known *selective cell discarding* (or space priority) scheme is the *pushout policy* that discards low-priority cells when the queue is full. If a high-priority cell arrives while no buffer space is available, one of the low-priority cells in the queue is pushed out. This policy, even though complicated to implement, has been considered to be optimal with respect to the throughput of low-priority flows while providing a desired level of performance to high-priority flows.

Another popular policy is the *threshold policy*, which accommodates the low-priority cells if and only if the buffer occupancy is below the threshold value. From the implementation's point of view, this scheme is much simpler than the pushout scheme because only the buffer occupancy needs to be monitored in this scheme. There could be many other variations to these policies.

The above discard policies were designed and tested only at the cell traffic level. One of the main findings of the first generation ATM networks was that only optimizing for ATM cell loss was not effective in increasing end user perceived performance. This puzzling result was the consequence of the fact that the ATM networks were in many cases transporting frames of data (e.g., IP packets) that were much larger than the ATM cells themselves. For example, when AAL-5 was used to transfer TCP/IP traffic, it was found that the loss of a single ATM cell would result in the retransmission of large blocks of TCP/IP packets. In some cases the ATM link could be at 100% utilization, but the user throughput would only be 30%. Optimization at the cell transport level did not automatically mean that the frame transport level was optimal.

The main solution to this problem was for the network to discard whole frames rather than individual cells. When AAL-5 is being used, many switches now have the option of employing frame-level discard. To implement frame discard, the network watches for the end of AAL-5 frames and then, if congested, discards the whole next frame rather than individual cells. The end of an AAL-5 frame can be deferred from the use of the bit in the ATM cell header that signifies the end of an AAL-5 frame.

Traffic shaping. Traffic shaping is a mechanism used by the terminal equipment to schedule the entry of cells in the network so they conform to the traffic descriptors at the UPC. Basically the aim of the traffic shaper is to buffer and smooth out any traffic, so that the traffic will better conform to its initial traffic descriptors.

Traffic shaping enables the efficiency of the resource allocation to increase, by introducing a more deterministic traffic pattern. If applied at the egress of the network, traffic shaping cancels the accumulated CDV. Traffic shaping is optional in TM 4.0

Explicit congestion notification. When congestion occurs at an intermediate network node, the node may notify the relevant end nodes of the congestion status so that the end nodes can take some appropriate actions. This congestion control scheme is called the *explicit congestion notification* scheme. Depending on the direction of the notification, this scheme is divided into two types: *explicit backward congestion notification* (EBCN or BECN) and *explicit forward congestion notification* (EFCN or FECN).

In the case of the EBCN, when an intermediate node gets congested, a special cell is generated to notify the status of the congestion at the congested node and is then sent back to all the connected source nodes. In contrast, in the case of the EFCN, the congestion notification is delivered forward to all the destination nodes. Usually, it takes a shorter time to notify backward to the source than to notify forward to the destination. As a result, the EBCN scheme has been widely used in the legacy data networks as a back-pressure mechanism. However, it is not appropriate for ATM networks since it imposes a considerable burden on the intermediate nodes.

In the case of the EFCN, which has been standardized by the ITU-T for use in the ATM network, the destination node is notified of the congestion status over the PT field of the ATM cell. If congestion occurs, the congested node sets the second bit of the PT field in the user information cell to 1 (refer to Section 4.1.3). On recognizing this PT markup, the destination node takes appropriate action by signaling the source node to slow down the upcoming transmission. This EFCN operation is illustrated in Figure 4.34. For this scheme to be effective, end users should cooperate with the network, and the roundtrip delay should be substantially smaller than the expected congestion duration.

Feedback flow control for ABR services. The ABR service defined by the ATM Forum gives the network the opportunity to allocate resources

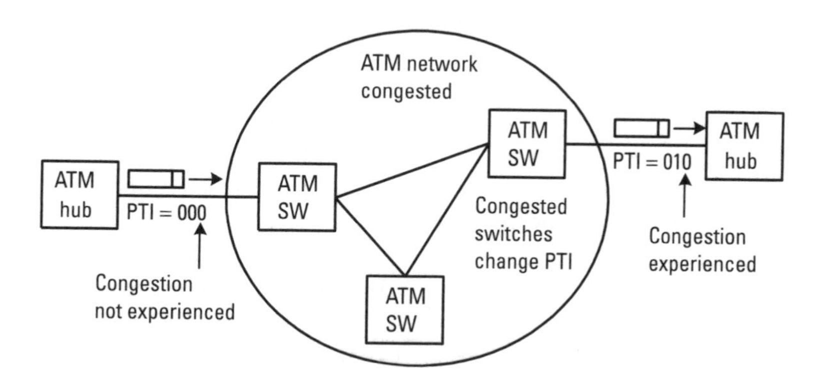

Figure 4.34 Illustration of EFCN operation.

preferentially to high-priority traffic such as the real-time services and share the remaining bandwidth fairly among all ABR connections.

While many data applications are well suited to the original non-real-time service categories, VBR and UBR, some are not. They may need QoS, which is not part of the UBR model. Also their traffic may be dynamic and unpredictable and not fit well into the VBR model. The bandwidth requirement is as much as there is available on an as-needed basis. They may also need some sort of minimum bandwidth to operate correctly. Consequently, ABR service was developed to replace the static VBR traffic constraint with a dynamic ACR based on network feedback. ABR sources that follow the feedback rules get low cell loss.

ABR is essentially a rate-based flow control scheme that tries to match the rate at which the user sends traffic into the network with the current amount of available capacity in the network. Flow control has traditionally played an important role in avoiding congestion collapse in data networks. As pre-TM 4.0 ATM networks have no internal network level flow control, they depend on end-to-end flow control (e.g., TCP) at a higher layer. ABR uses forward and backward explicit feedback notification to enable the end nodes to cooperate and induce an optimal state of transmission rates for all users. It can be proved that the current ABR algorithm will arrive at an optimal rate in all cases.

The aim of ABR is to create an end-to-end controlled flow. At the source end-station, a set of rules called the *source behavior* determine the sending rate, based on control information from the network. At each network element, a set of rules called the *switch behavior* define how feedback may be provided to control the source rate. The decision process for deciding the content of the feedback is not specified by the standard bodies. This decision process or algorithm is viewed as being a key differentiator between different vendor implementations. At the *destination end-station*, a set of rules called the *destination behavior* combines two functions; they define how a destination that receives forward congestion indicators from the network should reflect these back to the source, and they allow a destination to provide its own feedback. Source and destination behaviors are implemented at the end system, for example, in the network interface card.

To provide a mechanism for the network to provide congestion and rate change feedback to the source, the ABR source is required to insert *resource management* (RM) cells periodically within the data flow. The RM cells are turned around by the destination and returned to the source along the return path. Each RM cell contains an *explicit rate* (ER) at which the source wishes to operate, and this rate may be reduced by any network element to the rate it can currently support. Each RM cell also includes binary fields that the source can use to request permission to increase its rate, and that network elements may modify to prevent an increase or force a decrease.

RM cells are sent from source to destination and back in every N_{rm} cells, typically a 3% overhead per direction. The switches and destination modify them to indicate their congestion and the maximum rate they can support.

The key aspect of ABR end node operation is the definition of source behavior and destination behavior. A simplified version of *ABR–source behavior* is as follows:

```
What to do with a returning RM cell:
If CI = 1, decrease rate proportionally
     ACR = (1 - RDF) ACR
else if NI = 0, allow a linear- rate increase
     ACR = ACR + RIF × PCR
Also look at explicit rate feedback:
     ACR = Min (ACR, ER)
Also keep ACR between MCR and PCR
Send a forward RM cell once per Nrm cells, or more
frequently if ACR is low:
When sending a forward RM cell, do two implicit tests
     if source is coming out of idle, reduce ACR to ICR
     if feedback pipe is broken, reduce ACR
     ACR = (1 - CDF) ACR
```

A simplified version of *ABR–destination behavior* is as follows:

```
When a data cell is received save the EFCI indicator
Return saved EFCI bit in the next RM cell returned to
the Source
```

An ABR switch may operate in three different modes. The first mode of operation is the *EFCI binary rate operation*, where the switch marks EFCI bits in the data cells if congested. The second mode of operation is the *relative rate binary operation*, where the ABR switch may mark *congestion indication* (CI), and *nonincrease* (NI) bits in RM cells when congested. The third mode of operation is the *explicit rate operation*, where the switch may explicitly mark the ER field in RM cells to notify the source of the acceptable rate of transmission.

It is permissible to insert *virtual source/virtual destination* (VS/VD) modules into the network so as to protect one segment of the network from another, and to shorten the control loops.

4.6 QoS

The designers of ATM technology had a number of objectives. The first was to enable the efficient use of network equipment by carrying all different types of traffic via the same network elements. This would avoid waste in the existing networks, where separate networks exist for voice and data. Additionally, by using standardized equipment manufactured by multiple vendors, operators could expect corresponding reductions in the cost of equipment and maintenance.

A second objective was to enable optimization by using these resources. This is closely related to the first objective above. These objectives lead to the use of packet network technology, as it would be the best basis for offering a flexible network capable of supporting all types of traffic, while at the same time offering the potential to realize optimal use of the available resources.

This decision leads to another main objective of the design, which was to offer QoS guarantees over such packet networks. The objective was to meet stringent QoS requirements specific to each application and user, while protecting network and other users so as to achieve network performance objectives.

These objectives were some of the main factors that drove the architectural design choices we described in Section 4.1. These choices resulted in a network based on transporting data with small fixed sized cells by way of virtual circuits. As previously mentioned, each of these choices was basically made in some way to support QoS efficiently. Small cells would efficiently carry voice and allow efficient multiplexing of real-time traffic and non-real time traffic. Use of virtual circuits with explicit signaling mechanisms would allow network resources to be allocated to each user as needed.

However, while these basic architectural choices provided the background against which an efficient QoS service could be provided, they were not a total solution by themselves. First, operators would have to come up with a way of defining what level of service was promised to users. This leads to the concept of traffic contracts. Traffic contracts naturally lead to the definition of service categories for different types and levels of user service. Once such an agreement was in place the network operator would need methods for determining whether these agreements were being abided to by the user.

At the same time the operator would have to use the various traffic management tools available to ensure that the user was able to get the contracted level of service.

4.6.1 Traffic Contract and QoS

The first basic component of guaranteeing QoS is the traffic contract between the user and the network. By user we mean the user of network resources from a technological point of view—that is, the application on the PC or mobile telephone, not the actual human user. Of course, this user would have its own traffic contract with the network service provider, but this would then have to be converted into the traffic contract to be used by the network to realize those guarantees.

An ATM user's traffic contract is composed of a service category and traffic descriptor values that define the QoS requirements of the user. This contract is the basis for the allocation of resources in the ATM network. The allocation of these resources should enable the user's traffic to be transported through the network while receiving service at the QoS level demanded in the contract.

The allocation of these resources will also define the cost to the network and ultimately the price to the customer for receiving the service. As the resources allocated is the basis of cost calculation, it is important to ensure that the user only uses network resources in the manner agreed to in the contract—more specifically, that the user's traffic conforms to the traffic descriptors in the contract. This leads to the need for the definition of compliant connections.

As mentioned above, ATM service classes are an essential part of the traffic contract. It defines the context in which the traffic descriptor values that follow should be understood. Both the ATM Forum and the ITU-T have defined ATM service contracts. They are basically similar in structure and content but have some important differences.

The ATM Forum service categories consist of CBR, *real time* (rt) VBR, *non-real-time* (nrt) VBR, *unspecified bit rate* (UBR), and ABR. The ITU-T ATM service categories consist of the *deterministic bit rate* (DBR), which is equivalent to CBR in the ATM Forum standards; *statistical bit rate* (SBR), which is equivalent to VBR in the ATM Forum standards; *available burst transfer with delayed transmission* (ABT/DT); *ABT with immediate transmission* (ABT/IT); ABR; and UBR. In addition, several new service categories have been recently defined by the ATM Forum with the specific aim of supporting TCP/IP traffic and IP-based QoS traffic more efficiently. These are the *guaranteed frame rate* (GFR) and the *differentiated UBR* service classes.

4.6.1.1 DBR (or CBR) Service Category

The DBR (or CBR) is used by connections that request a static volume of bandwidth that is characterized by a PCR. This service category is intended to support CBR applications but is not restricted to these applications. No special cells, such as the RM cells used in the ABR class, are used.

While the guarantee of a rate conceptually maps the service to a traditional CBR service such as PCM voice or circuit emulation, it must be realized that the user may transmit cells at rates less than the agreed-upon PCR. In fact, in many cases the user uses only a fraction of the PCR rate. This can be viewed as another manifestation of the inefficiencies of using circuit switching networks for carrying data traffic.

4.6.1.2 SBR (or VBR) Service Category

The SBR service category is a service category that enables the support of users with variable bit rates but with guarantees that are similar to DBR. As this service tries to give guarantees similar to DBR but for a traffic pattern that is not fixed, it is much harder to implement. The key traffic parameter added to the traffic contract is the *sustained cell rate* (SCR).

The SBR service does not use special cells such as the RM cells used in ABR. The service class may support the tagging option and selective cell discard.

While it is natural to use this service to support the irregular traffic of data networks with non-real-time commitments, the use of SBR for real-time applications (rt-VBR in the ATM-Forum) is a subject that is still unclear. This can be easily seen by considering the fact that some prior knowledge of the user's traffic characteristics must exist for the network to guarantee QoS levels of real-time applications. Without such knowledge, it would be much harder to negotiate the traffic contract with the network due to the inherent statistical properties of the traffic.

4.6.1.3 ABT Capabilities

The ABT service category is a class that only exists in the ITU-T's definition of ATM service categories. In this definition, an ATM block is a group of cells delineated by two RM cells. Reservation of resources in the network is done on the basis of these blocks, essentially realizing a sort of *just-in-time reservation* for the block of cells as it travels through the network. A single CS-PDU may give rise to one ATM block but can also be segmented into several consecutive ATM blocks. Moreover, an ATM block may carry the information contents of several CS-PDUs. Only complete blocks will be delivered at the destination; neither supports the tagging option nor single-cell delivery. In all cases the cell sequence integrity is maintained for both user cells and RM cells.

4.6.1.4 ABR Capabilities

The ABR service category was conceived to support traffic from sources that could adapt to the current state of the network. The transfer characteristics of the ABR service user connection may change subsequent to the connection establishment. In that case the user will be notified of such conditions and will change its traffic generation to fit this. As such, this service class is not designed to support CBR applications.

The ABR service uses *resource management* (RM) cells, but with no commitment on *cell transmission delay* (CTD) and CDV. Traffic parameters used include PCR and *maximum cell rate* (MCR). The ABR connection aims to offer the user a connection with a rate that consists of the MCR promised to the user and an elastic component that depends on the current usage of the network.

Essentially the traffic contract gives guarantees to the user on the quantitative (sometimes qualitative) service that the user can expect. The guarantees can be mainly grouped into three areas—rate guarantees, delay guarantees, and loss guarantees. The parameters related to rate guarantees are PCR, SCR, MCR, and ACR; the parameters related to delay guarantees are those on CTD and CDV; and the parameter related to loss guarantees is *cell loss rate* (CLR).

The above service classes mainly differ in the guarantees received with respect to those parameters. This is shown in Table 4.6.

4.6.1.5 GFR Service Category

Based on recent research results, the ATM Forum defined a new traffic management category called the GFR to more efficiently support TCP/IP traffic. Basically GFR aims to employ a frame discard mechanism while at the same time guaranteeing TCP/IP traffic a minimum data rate. The GFR traffic contract contains PCR, MCR, MBS, and *maximum frame size* (MFS).[26] As

26. The MFS is the size of the AAL-5 that is used. The MCR is usually negotiated to be equal to the long-term average of the connection.

Table 4.6
ATM Layer Service Parameters

Attribute	CBR	rt-VBR	nrt-VBR	UBR	ABR
Traffic parameters					
PCR and CDRV(pcr)	Specified	Specified	Specified	Specified*	Specified†
SCR, MBS, CDTV(scr)	n/a	Specified	Specified	n/a	n/a
MCR	n/a	n/a	n/a	n/a	Specified
QoS parameters					
Peak-peak CDV	Specified	Specified	Unspecified	Unspecified	Unspecified
MaxCTD	Specified	Specified	Unspecified	Unspecified	Unspecified
CLR	Specified	Specified	Specified	Specified	Unspecified±
Flow control					
Closed loop	Unspecified	Unspecified	Unspecified	Unspecified	Specified

* Used either in CAC and UPC or for information purposes only.

† Represents maximum cell rate that ABR source may ever send. The actual maximum cell rate will be determined by network feedback.

‡ CLR is low for ABR sources that adjust cell input rate accordingly to feedback.

the specific reasons and usefulness of this service category can be more easily understood when TCP/IP and ATM interaction is understood, we defer to the detailed discussions on the GFR in Chapter 7.

4.6.1.6 Differentiated UBR Service Category

Recently a great deal of activity in QoS-related work has resulted in the definition of differential services. This has occurred both in the IP layer where it appears as the DiffServ QoS model and in the IEEE 802.1D user priorities work, which aims to introduce QoS to Ethernet networks.

The ATM Forum has recently defined a new refinement that enables the UBR service class to more naturally support this differentiated QoS service traffic. This is done by defining a new attribute to be associated with a

UBR connection. Normally a UBR connection does not offer any service guarantees, basically offering a best-effort service. However, when differentiated UBR is used, a new attribute, the *behavior class* may be associated with the UBR connection. This is done through the use of a *behavior class selector* (BCS) parameter, which is used to indicate the behavior class for the connection when the connection is initially set up. As for the case of GFR, differentiated UBR can be better understood when TCP/IP and ATM interaction is understood, so we defer detailed discussions to Chapter 7.

4.6.2 Conformance and Compliance

The key to enabling QoS guarantees based on traffic contracts is in checking the conformance of the user traffic to the agreed upon traffic contract. In all cases, the guarantees offered by the network are only valid if the traffic conforms to the negotiated traffic contract.

For CBR, VBR, and UBR, conformance is defined by the GCRA. For ABR, conformance is defined by the source and destination behavior, but a dynamic GCRA is still a useful example of how to test conformance.

There are various causes for the nonconformance including excessive rate, excessive clumping, and excessive burst. As mentioned before, nonconforming cells may be discarded or, when permitted, tagged with CLP = 1 for lower priority.

An important topic is the idea of GCRA-based conformance. Essentially, this is a conformance test that is based on the GCRA algorithm. The conformance test differs for each service class. CBR service must conform to *GCRA (1/PCR, CDVT)*. VBR service must conform to both: *GCRA (1/PCR, CDVT)* and *GCRA (1/SCR, CDVT+BT)*. ABR service must conform to *dynamic GCRA (1/ACR, t)* where ACR is the network-specified rate (PCR>ACR>MCR) and *t* is a dynamically variable burst tolerance. For the case of UBR services, as the UBR rate is unspecified, whether a UBR connection must conform to test is not defined in the standards. Network operators are free to apply the conformance test *GCRA (1/PCR, CDVT)* on a UBR connection to ensure that the peak rate is not violated.

It is important to understand the difference in meaning between conformance and compliance. *Conformance* is an attribute of the cell, and is judged at the cell level. *Compliance* is an attribute of the connection. QoS is guaranteed for all conforming cells of a compliant connection. A connection is defined as compliant when the number of nonconforming cells is below a threshold set by the network provider. The network provider specifies the threshold for compliance in the traffic contract.

4.6.3 Miscellaneous Considerations

Overall, ATM was built from the beginning to support QoS. The decision to use small fixed-size cells transported over a virtual connection topology was driven by the desire to support QoS. In addition, the later addition of the ABR rate-control mechanism was technically an optimal solution to the problem of sharing network resources between many cooperative users.

Today, the use of data packets has clearly outstripped the need for voice traffic–oriented network design, so the use of small cells may no longer be important. Also, due to the use of high-speed integrated ASICs and new algorithms for fast packet switching, the use of fixed-sized cells may also not be considered as being essential in building high-speed switches. In fact, recent problems in building any ATM SAR chips that work beyond the STM-16/OC-48c (refer to Section 5.1.2) speeds indicate that use of "small" cells may even deter high-speed network development. Nevertheless, it is clear that the use of a connection-oriented scheme is still a viable and important method for supporting QoS in packet-based networks. This can be clearly evidenced by the TCP/IP networks that are evolving to support some form of signaling. The most relevant example is the recently developed RSVP protocol.

References

[1] Lee, B. G., M. Kang, and J. Lee, *Broadband Telecommunications Technology,* 2nd ed., Norwood, MA: Artech House, 1996.

[2] ITU-T Rec. I.321 (04/91) "B-ISDN Protocol Reference Model and Its Application."

[3] ITU-T Rec. I.361 (02/99) "B-ISDN ATM Layer Specification."

[4] ITU-T Rec. I.362 (03/93) "B-ISDN ATM Adaptation Layer (AAL) Functional Description."

[5] ITU-T Rec. I.363 (03/93) "B-ISDN ATM Adaptation Layer (AAL) Specification."

[6] ITU-T Rec. I.363.1 (08/96) "Type 1 AAL."

[7] ITU-T Rec. I.363.2 (11/2000) "B-ISDN ATM Adaptation Layer Specification: Type 2 AAL."

[8] ITU-T Rec. I.363.3 (08/96) "Type 3/4 AAL."

[9] Eng, K. Y., M. J. Karol, and Y. S. Yeh, "A Growable Packet (ATM) Switch Architecture: Design Principles and Applications," *IEEE Transactions on Communications,* Vol. 40, No. 2, February 1992, pp. 423–439.

[10] Guizani, M., and A. Rayes, *Designing ATM Switching Networks,* New York: McGraw-Hill, 1998.

[11] Lee, T. H., "Design and Analysis of a New Self-Routing Network," *IEEE Transactions on Communications,* Vol. 40, No. 1, January 1992, pp. 171–177.

Selected Bibliography

Armbruster, H., and G. Arndt, "Broadband Communication and Its Realization with Broadband ISDN," *IEEE Communications Magazine,* Vol. 25, No. 11, November 1987, pp. 8–19.

ATM Forum, "ATM User-Network Interface Specification Version 2.0," 1992.

ATM Forum, "Network Compatible ATM for Local Network Applications," Phase 1, Version 1.0, 1992.

Basch, B. E., et al., "VISTAnet: A BISDN Field Trial," *IEEE LTS Magazine,* Vol. 2, No. 3, August 1991, pp. 22–30.

Bauch, H., "Transmission Systems for the B-ISDN," *IEEE LTS Magazine,* Vol. 2, No. 3, August 1991, pp. 31–36.

Boudec, J. Y. L., "The Asynchronous Transfer Mode: A Tutorial," *Computer Networks and ISDN Systems,* Vol. 24, 1992, pp. 279–309.

Byrne, W. R., et al., "Evolution of Metropolitan Public Network and Switch Architecture," *IEEE Communications Magazine,* Vol. 29, No. 1, January 1991, pp. 69–82.

Chao, H. J., and J. S. Park, "Centralized Contention Resolution Schemes for a Large-Capacity Optical ATM Switch," *Proceedings of IEEE ATM Workshop,* May 1998.

Cidon, I., I. Gopal, and R. Guerin, "Bandwidth Management and Congestion Control in Planet," *IEEE Communications Magazine,* Vol. 30, No. 10, October 1991, pp. 54–65.

Cooper, I. R., and M. A. Bramhall, "ATM Passive Optical Networks and Integrated VDSL," *IEEE Communications Magazine,* Vol. 38, No. 3, March 2000, pp. 174–179.

Coudreuse, J. P., "Network Evolution Towards B-ISDN," *IEEE LTS Magazine,* Vol. 2, No. 3, August 1991, pp. 66–70.

Fonsenca, N. L. S., and M. Zukerman, "ATM Dimensioning and Traffic Management and Modeling Tutorial," *Proceedings of IEEE GLOBECOM'97,* November 1997.

Georgatsos, P., et al., "Technology Interoperation in ATM Networks: The REFORM System," *IEEE Communications Magazine,* Vol. 37, No. 5, May 1999, pp. 112–18.

Handel, R., and M. N. Huber, *Integrated Broadband Networks: An Introduction ATM-Based Networks,* Reading, MA: Addison Wesley Publishing Co., 1991.

Hung, A., G. Kesidis, and N. McKeown, "ATM Input-Buffered Switches with Guaranteed-Rate Property," *Proceedings of IEEE ISCC'98,* July 1998, pp. 331–35.

Iida, K., et al., "Delay Analysis for CBR Traffic in Static-Priority Scheduling: Single-Node and Heterogeneous CBR Traffic Case," *Proceedings of IEEE GLOBECOM'98,* November 1998, pp. 1256–1263.

Iida, K., et al., "Performance Analysis of Flow Aggregation of Constant Bit Rate Type Traffic at Ingress Router," *Proceedings of IEEE GLOBECOM'99,* December 1999, pp. 92–99.

ITU-T Rec. I.311 (08/96) "B-ISDN General Network Aspects."

ITU-T Rec. I.311 Amendment 1 (03/2000) "B-ISDN General Network Aspects."

ITU-T Rec. I.363.5 (08/96) "Type 5 AAL."

ITU-T Rec. I.366.1 (06/98) "Segmentation and Reassembly Service Specific Convergence Sublayer for the AAL Type 2."

ITU-T Rec. I.366.2 (11/2000) "AAL Type 2 Service Specific Convergence Sublayer for Narrowband Services."

ITU-T Rec. I.371 (03/2000) "Traffic Control and Congestion Control in B-ISDN."

ITU-T Rec. I.371.1 (11/2000) "Guaranteed Frame Rate ATM Transfer Capability."

ITU-T Rec. I.381 (03/2001) "ATM Adaptation Layer (AAL) Performance."

ITU-T Rec. I.630 (02/99) "ATM Protection Switching."

ITU-T Rec. I.630 Amendment 1 (03/2000) "ATM Protection Switching."

ITU-T Rec. I.630 Corrigendum 1 (03/2000) "ATM Protection Switching."

ITU-T Rec. I.731 (10/2000) "Types and General Characteristics of ATM Equipment."

ITU-T Rec. I.732 (10/2000) "Functional Characteristics of ATM Equipment."

ITU-T Rec. I.741 (07/99) "Interworking and Interconnection Between ATM and Switched Telephone Networks for the Transmission of Speech, Voiceband Data and Audio Signals."

ITU-T Rec. I.751 (03/96) "Asynchronous Transfer Mode Management of the Network Element View."

ITU-T Rec. I.761 (03/2000) "Inverse Multiplexing for ATM (IMA)."

ITU-T Rec. I.762 (03/2000) "ATM Over Fractional Physical Links."

Lee, T. T., "Nonblocking Copy Networks for Multistage Packet Switching," *IEEE Journal on Selected Areas in Communications,* Vol. 6, No. 9, December 1988, pp. 1455–1467.

Li, S. Q., "Performance of a Nonblocking Space-Division Switch with Correlated Input Traffic," *IEEE Transactions on Communications,* Vol. 40, No. 1, January 1992, pp. 97–108.

Minzer, S. E., "Broadband ISDN and Asynchronous Transfer Mode (ATM)," *IEEE Communications Magazine,* Vol. 27, No. 9, September 1989, pp. 6–14.

Minzer, S. E., and D. R. Spears, "New Directions in Signaling for Broadband ISDN," *IEEE Communications Magazine,* Vol. 27, No. 2, February 1989, pp. 6–14.

Mir, N. F., "An Efficient Multicast Approach in an ATM Switching Network for Multimedia Applications," *Journal of Network and Computer Applications,* Vol. 21, January 1998, pp. 31–39.

Mir, N. F., "Evaluation of an ATM LAN Constructed with a Cyclic Deflection-Routing Network," *Journal of Computer Communications,* Vol. 21, No. 7, June 1998, pp. 661–69.

Mitra, N., and S. D. Usikin, "Relationship of the Signaling System No. 7 Protocol Architecture to the OSI Reference Model," *IEEE Communications Magazine,* Vol. 5, No. 1, January 1991, pp. 26–37.

Nong, G., J. K. Muppala, and M. Hamdi, "Analysis of Nonblocking ATM Switches with Multiple Input Queues," *IEEE/ACM Transactions on Networking,* Vol. 7, No. 1, February 1999, pp. 60–74.

Pazos, C. M., M. R. Kotelba, and A. G. Malis, "Real-Time Multimedia Over ATM: RMOA," *IEEE Communications Magazine,* Vol. 38, No. 4, April 2000, pp. 82–87.

Shiomoto, K., et al., "Scalable Multi-QoS IP+ATM Switch Router Architecture," *IEEE Communications Magazine,* Vol. 38, No. 12, December 2000, pp. 86–92.

Smouts, M., *Packet Switching Evolution From Narrowband to Broadband ISDN,* Norwood, MA: Artech House, 1992.

Special issue, "Architecture and Protocols for Integrated Broadband Switching," *IEEE Journal on Selected Areas in Communications,* Vol. 9, No. 9, December 1991.

Special issue, "B-ISDN: High Performance Transport," *IEEE Communications Magazine,* Vol. 29, No. 9, September 1991.

Special issue, "Congestion Control in High-Speed Networks," *IEEE Communications Magazine,* Vol. 29, No. 10, October 1991.

Special issue, "Congestion Control in High-Speed Packet Switched Networks," *IEEE Journal on Selected Areas in Communications,* Vol. 9, No. 7, September 1991.

Special issue, "Gigabit Networks," *IEEE Communications Magazine,* Vol. 30, No. 4, April 1992.

Special issue, "Large Scale ATM Switching Systems for B-ISDN," *IEEE Journal on Selected Areas in Communications,* Vol. 9, No. 7, September 1991.

Special issue, "Teletraffic Analysis of ATM Systems," *IEEE Journal on Selected Areas in Communications,* Vol. 9, No. 3, April 1991.

Stallings, W., *Advances in ISDN and Broadband ISDN,* Los Alamos, CA: IEEE Computer Society Press, 1992.

Yashiro, Z., T. Tanaka, and Y. Doi, "Flexible ATM Switching Architecture for Multimedia Communications," *Proceedings of IEEE BSS'97,* 1997, pp. 58–64.

5

SDH/SONET

In contrast to the upper-layer network technologies such as TCP/IP and ATM that handle application services using "fine fingers," SDH/SONET and WDM/optics are lower-layer network technologies that provide transport services using "big hands." Consequently, the points of discussion shift from the channel-level issues such as switching, routing, traffic management, and QoS to transport-related issues such as add/drop multiplexing and cross-connection, synchronization, network operation and management, and network survivability.

The SDH and SONET are digital transport structures that operate by appropriately arranging the payloads and transporting them through synchronous transmission networks. Before the advent of the SDH and SONET, the most common digital hierarchy in use was *plesiochronous digital hierarchy* (PDH), which is still widely used in the form of North American and European tributaries.[1] These PDH tributaries are multiplexed through the SDH/SONET's synchronous multiplexing structure to form the *synchronous transport module level n* (STM-*n*)/*optical carrier level m* (OC-*m*) signal

1. The existing digital hierarchy is called *plesiochronous* as the bit rate of each tributary is within the specified tolerance of the nominal rate. North American PDH tributaries include DS-1, DS-1C, DS-2, and DS-3 signals, and European PDH tributaries include DS-1E, DS-2E, DS-3E, and DS-4E signals. Refer to Figure 5.5 for the relations among them.

for transmission.[2] When compared with the PDH, the SDH has an extremely simple structure and operation but the process of synchronous multiplexing that maps PDH tributaries into STM-*n*/OC-*m* signals is no trivial matter.

Though the transport of the PDH tributaries was the first target of services of SDH/SONET in the beginning, ATM has become another important object to transport. As SDH/SONET was tailored for PDH tributaries, the payload space does not neatly fit to the ATM cells. Nevertheless, SDH/SONET can effectively accommodate and transport ATM cells over the synchronous network. Furthermore, SDH/SONET can provide an efficient transport means to TCP/IP packets directly in the payload space or with encapsulation.

The term *synchronous* in the SDH and SONET originates from the fact that the process of multiplexing the PDH tributaries into STM-*n*/OC-*m*, respectively, adopts a *synchronous multiplexing* structure. A synchronous multiplexing structure is advantageous over the plesiochronous one for the PDH in various aspects. It has a simplified multiplexing/demultiplexing process, direct access to low-rate tributaries without demultiplexing/multiplexing all the intermediate signals, enhanced operation and management capabilities, and easy transition to higher bit rate signals yet to come in step with the evolution of the transmission technology. As such, the synchronous multiplexing structure plays the most essential role of SDH/SONET.

Synchronous multiplexing plays the central role in SDH/SONET systems, so the advantages of synchronous multiplexing above are directly tied with the strengths of SDH/SONET, which renders a powerful and effective means of connecting the low-end customer channels to high-end optical fibers in a modular and reliable way. The layered architecture it adopts enables an organized handling of various different types of constituent signals, thereby lending itself to an adaptation function that coordinates the IP or ATM layer above with the WDM layer below. SDH/SONET can accommodate not only digital hierarchical tributaries but also ATM cells and, further, can easily extend the mapping pattern to accommodate IP packets as well. The use of affluent overhead bytes enables a flexible and resilient operation and management of the synchronous network in harmony with the existing telecommunication management network.

2. Synchronous transport module at level *n* (STM-*n*) and optical carrier at level *m* (OC-*m*) are the transport signals of the SDH and SONET respectively. OC-*m* is an optical signal whose electrical counterpart is *synchronous transport signal at level-m* (STS-*m*).

This chapter provides an overview of SDH/SONET, as a key technological component of broadband networks that is tailored for optical transmission. To this end, we first examine the concept and the overall architecture of SDH/SONET in relation to the layering concept. Then we discuss the multiplexing and synchronization issues in conjunction with the associated mapping and pointer processing. On this basis, we discuss the add/drop and cross-connect functions and the network operation and management issues, with stress put on the reliability/survivability aspect.[3] In describing all these subjects we shall adopt the SDH terminology as it has a wider coverage, while indicating its SONET counterpart whenever needed.

5.1 Concept and Architecture

SDH/SONET provides a transport means to the PDH tributaries and the ATM cells through synchronous optical networks. In this aspect, SDH/SONET may be regarded as a lower-layer entity that provides services to the upper-layer entity PDH. However, SDH/SONET stands at the same level as PDH in the capacity that SDH/SONET is another form of digital hierarchy that transports all existing user information even including those readily contained in the PDH tributaries, based on conceptually different mechanisms. Among various differences that SDH/SONET has, the layering concept is one of the most important facets that accompanied many other distinctive features of SDH/SONET. The layering concept triggered restructuring of the conventional type of frame formats to a modern one that consists of layer-dedicated space allocation. It also brought forth the concept of "layer networks."

5.1.1 Layering Concept and Frame Structure

The synchronous multiplexing of the SDH/SONET allocates the requisite bit space in a systematic manner, in accordance with the layering concept. Consequently, the resulting SDH/SONET frame structure becomes an orderly embodiment of the layering concept [2, 3].

5.1.1.1 Layering Concept

The layering concept in the SDH/SONET is closely associated with the hierarchical processing of digital transmission in which user information is

3. We organize the chapter rather concisely as SDH/SONET is a comparatively well-established and widely known technology that has reached a steady state with minimal evolution occurring. The reader is advised to refer to Chapter 3 of [1].

transmitted via the path, multiplex, regenerator, and physical medium, as illustrated in Figure 5.1. Each element of this transmission procedure can be viewed as a *layer*, and hence the entire digital transmission processing is divided into *path layer*, *multiplex section layer*, *regenerator section layer*, and *physical medium layer*, accordingly. These comprise the SDH terminology, whose SONET counterparts are *path layer*, *line layer*, *section layer*, and *physical medium layer*, respectively. The path layer can be subdivided into *higher-order path layer* and *lower-order path layer* depending on the level of the tributary it carries. Figure 5.2 shows the resulting layer architecture in stack.

The layering concept in digital transmission divides the overall functionality into a disjointed set of layered functions and is implemented in such a way that the overheads governing each layer function are segregated out of the user information space and put together in a particular location of the frame. This arrangement renders an efficient means for transmission processing, as the layer entity of a particular layer can accomplish its own functionality without affecting other layers: The overheads relevant to a layer can be accessed without opening the overhead space of other layers.

5.1.1.2 Frame Structure

Figure 5.3 illustrates how the layering concept is imbedded in the STM-*n* [and *synchronous transport signal level m* (STS-*m*) inside the bracket] frame structure. The STM-*n* frame is partitioned into five regions, four of which

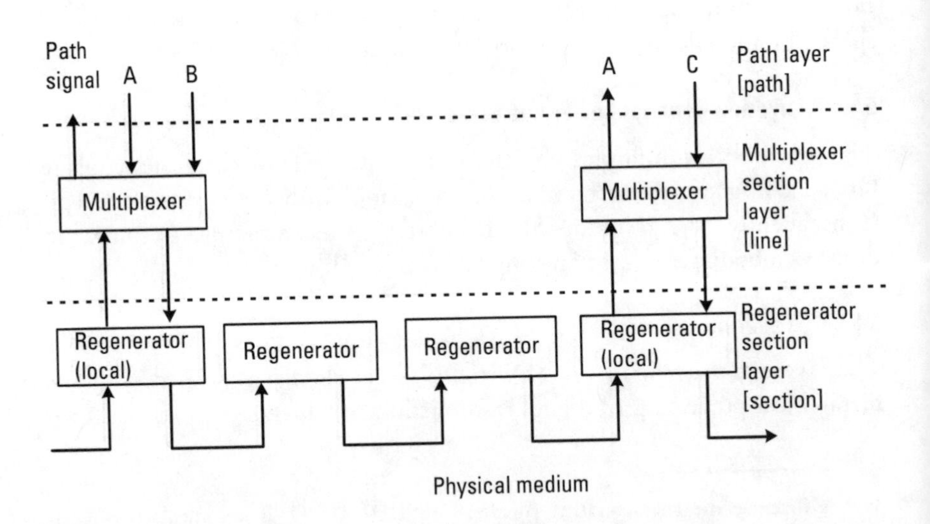

Figure 5.1 Illustration of the layering concept in SDH/SONET transmission (unidirectional).

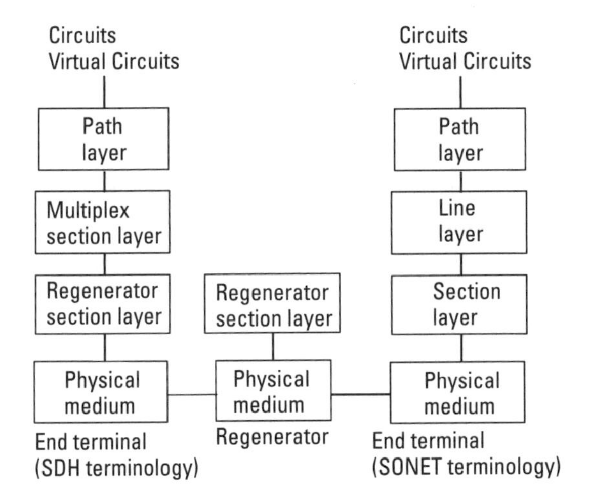

Figure 5.2 Layer architecture of SDH/SONET.

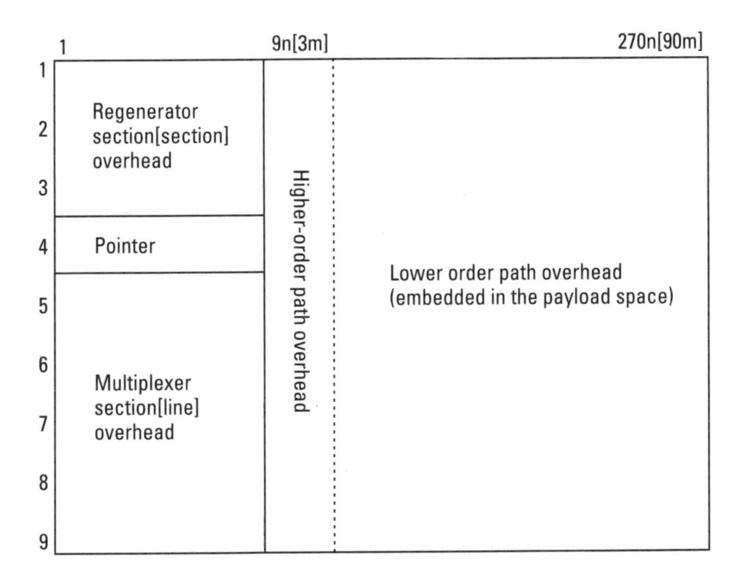

Figure 5.3 Layering concept embedded in the SDH/SONET frame structure.

are providing overhead functions for different layers and the fifth region is assigned for the *pointers* (PTRs). The *regenerator section overhead* (RSOH) and the *multiplexer section overhead* (MSOH) belong to the regenerator section layer and the multiplexer section layer, respectively, and are collectively referred to as *section overheads* (SOHs). The *path overhead* (POH) belongs to the path layer and is divided into *higher-order POH* (HO-POH) and *lower-order POH* (LO-POH). The POH is located in the overhead of the *virtual container* (VC) that carries the corresponding tributary signal. The POH shown in the frame structure is the HO-POH, and the LO-POH is imbedded in the STM-*n* payload space that carries the lower-order VCs.[4]

The SOH is used for block framing, maintenance, performance monitoring, and other operational functions in the regenerators and multiplexers and the POH is used for the end-to-end communication between the nodes where the relevant VC is terminated. The SOH bytes are inserted in the last step of constructing an STM-*n*, while the POH bytes are inserted every time a VC is constructed. Accordingly, the SOH bytes are extracted in the first step of receiving an STM-*n*, and the POH bytes at every time the pertinent VC is terminated.

Figure 5.4 shows the frame structure of the STM-1 (and STS-1 inside the bracket), the basic-level end product of the layered synchronous multiplexing procedure in SDH (and SONET), with the pointer processing

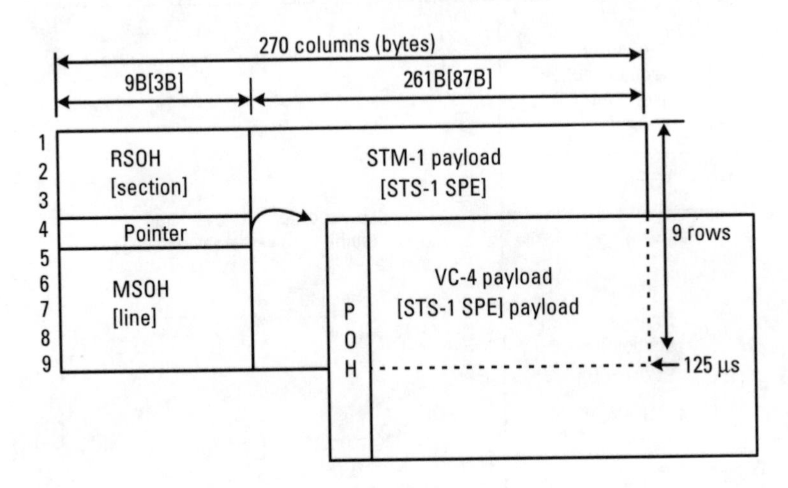

Figure 5.4 Frame structure of STM-1 [STS-1].

4. The LO-POH is contained in the overhead space of the VCs of the lower bit-rate PDH tributaries such as DS-1, DS-1E, DS-2 and DS-3, and the HO-PDH in the VCs of the higher bit-rate ones such as DS-3 and DS-4E. The VCs are named after the corresponding tributary levels, namely, VC-11 (for DS-1), VC-12 (for DS-1E), VC-2 (for DS-2), VC-3 (for DS-3), and VC-4E (for DS-4). Refer to Section 5.2.2 for more details on VCs.

incorporated together. The STM-1 frame structure occupies a 9B × 270 space over 125 μs, which yields a 155.520-Mbps bit rate. The SOH occupies 3 × 9B (for RSOH) and 5 × 9B (for MSOH) partitions of space. The PTR, which indicates the starting point of the virtual container contained within the STM-1 payload space, takes the 1 × 9B space in the middle, while the remaining 9B × 261 space is reserved for the STM-1 payload. The STM-*n* frame has the same basic structure but the width is expanded by *n* times. It has the structure of 9B × 270 × *n* and the bit rate of 155.520 × *n* Mbps. A similar relation holds for SONET: STS-*m* (or OC-*m*) frame has the same basic structure as STS-1 (or OC-1) but the width is expanded by *m* times. (Refer to Section 5.2.1 for more details.)

5.1.2 PDH, SDH, and SONET

PDH is the conventional digital hierarchy relying on asynchronous multiplexing, and SDH is its outgrowth based on synchronous multiplexing. The relation between PDH and SONET is similar except that the coverage of SONET is limited to the North American PDH tributaries. SDH and SONET are equal-footed but SDH has different lower-layer signal formats and a broader multiplexing coverage.

5.1.2.1 PDH Versus SDH

The PDH is divided into North American and European designations, as shown in Figure 5.5. North American PDH in Figure 5.5(a) consists of DS-1 (1.544 Mbps), DS-1C (3.152 Mbps), DS-2 (6.312 Mbps), DS-3 (44.736 Mbps), and DS-4E (139.264 Mbps); and the European PDH in Figure 5.5(b) accommodates DS-1E (2.048 Mbps), DS-2E (8.448 Mbps), DS-3E (34.368 Mbps), DS-4E (139.264 Mbps), and DS-5E (564.992 Mbps). Each multiplexing step is plesiochronous, and synchronization is achieved with the use of bit stuffing and positive justification [4–7].

The SDH is composed of STM-*n* signals, as shown in Figure 5.6(a). For the number *n*, which indicates the number of multiples of the base STM-1 contained in the STM-*n*, 1, 4, 16, and 64 are of major interest, whose corresponding bit rates are 155.52, 622.08, 2,488.32, and 9,953.28 Mbps, respectively. The STM-*n* signal is formed by synchronously multiplexing the DS-1, DS-1E, DS-2, DS-3, and DS-4E tributaries. The DS-1C and DS-5E signals are excluded in this synchronous multiplexing. The STM-*n* signal is formed by *byte-interleaved multiplexing* (BIM) *n* of the STM-1 signals. Likewise, SONET is composed of STS-*m* (or OC-*m*) signals as shown in Figure 5.6(b), and is formed in a similar manner.

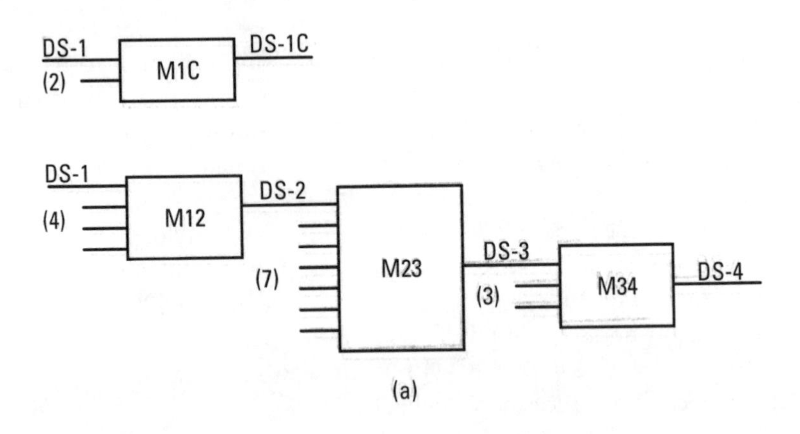

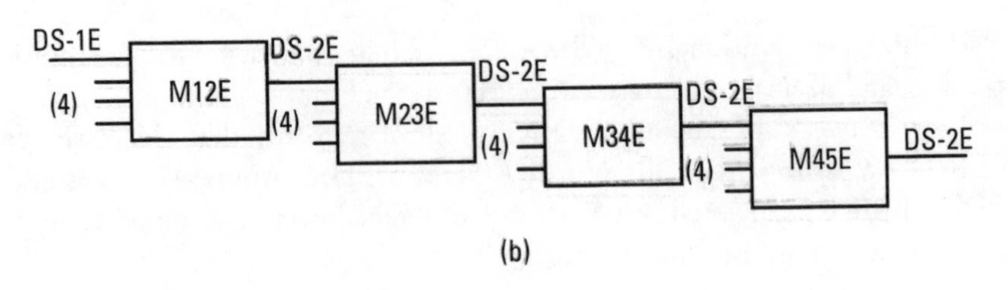

Figure 5.5 PDHs: (a) North American, and (b) European.

From Figures 5.5 and 5.6, it is apparent that the SDH/SONET has a much simpler structure than the PDH. In the SDH, North American and European tributaries go through only a single stage of multiplexing. In the PDH, *asynchronous multiplexing* is used when a tributary is multiplexed into a tributary of higher bit rate but, in the SDH, *synchronous multiplexing* is used instead. Also, in the PDH, DS-k is a higher tributary of the DS-(k-1) signal, but in the SDH, all the DS signals are equal in status.

In addition, SDH/SONET has various distinctive features over the PDH. First, the frame structure of SDH/SONET has a time duration of 125 μs, which enables access to low-level signals, especially DS-0, directly from high-level signals. Second, SDH/SONET integrates all existing PDH signals into one STM-n/STS-m (or OC-m) structure, thus enabling one-step multiplexing, which renders add/drop and cross-connect simple and economical. Third, it employs the layering concept, taking advantage of the "divide-and-conquer" approach in the transmission process. Fourth, it systematically divides and maximally utilizes the overhead space for enhancing

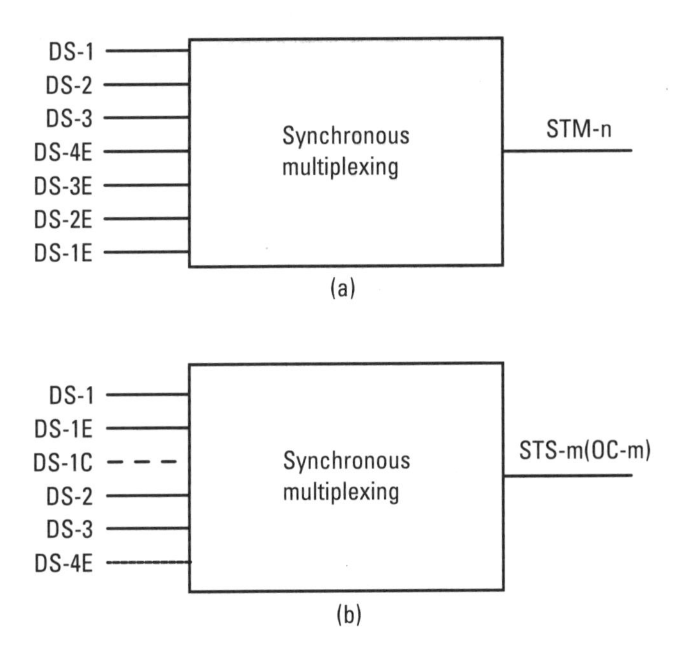

Figure 5.6 Synchronous digital hierarchies: (a) SDH, and (b) SONET.

the communication operation and maintenance capabilities. Fifth, it employs the pointer-processing technique for flexible and dynamic manipulation of synchronization. Sixth, it enables a global transport network through repeated use of pointers and unification of the North American and European hierarchies.

5.1.2.2 SDH Versus SONET

SDH is identical to SONET in concept and operation, wherein SDH becomes a *threefold expansion* of SONET and has a broader coverage. That is, the *concatenation* of three OC-1 (i.e., OC-3c) is equivalent to STM-1 at 155.52 Mbps, even though the use of the overhead bytes may be a little different. A main difference between SDH and SONET is that the base rate of SDH is 155.52 Mbps (for STM-1), while that of SONET is 51.84 Mbps (for STS-1 or OC-1). This implies that SDH employs DS-4E as the highest-level tributary to accommodate lower-order signals, whereas SONET employs DS-3 as the highest-level tributary.

SDH and SONET differ slightly in granularity and diversity of the transmission rates and frame format. However, the rates of primary interest are the same: The STM signals of primary interest are STM-1 (155.52 Mbps), STM-4 (622.08 Mbps), STM-16 (2,488.32 Mbps), and STM-64 (9,953.28 Mbps), which respectively correspond to OC-3, OC-12, OC-48, and OC-192. In general, the rate of STM-*n* is three times the rate of OC-*n*—that is, STM-*n* is the same as OC-3*n* in bit rate. This relation also applies to frame formats. The frame format of STM-*n* is a three-fold expansion of that of OC-*n*; that is, the frame format of STM-*n* is identical to the concatenated signal OC-3*nc*.

SDH and SONET differ in the synchronous multiplexing structure. In the case of SDH the construction of STM-1 requires systematic multiplexing of all the tributaries from DS-1 to DS-4, while OC-1 needs to multiplex just five of the tributaries, DS-1, DS-1E, DS-1C (at 3.152 Mbps), DS-2, and DS-3. As a result, the construction of STM-1 necessitates the establishment of various intermediate signal units such as *container* (C), *virtual container* (VC), *tributary unit* (TU), *tributary unit group* (TUG), *administrative unit* (AU), and *administrative unit group* (AUG) (refer to Section 5.2.2 and Figure 5.13).[5] In contrast, OC-1 requires only one type of intermediate signal, called the *virtual tributary* (VT). The VT is equivalent to the VC of SDH, and the VTs that correspond to VC-11, VC-12, and VC-2 are called VT1.5, VT2, and VT6, respectively. In addition, VT3 is used for the DS-1C signal.

As SDH and SONET have different sets of intermediate signal units, so they have different multiplexing procedure. As shown in Figure 5.7, SDH requires a systematic multiplexing structure that links C, VC, TU, TUG, AU, AUG, and STM-*n*, while the multiplexing structure of SONET only has to link DS-*k*, VT, and STS-1 (or OC-1) *synchronous payload envelope* (SPE), which is much simpler. Here, the mapping of tributaries into VT1.5, VT2, and VT6 is equivalent to their mapping into VC-11, VC-12, and VC-2, respectively. The mapping of these VTs into STS-1 (or OC-1) SPE is analogous to the mapping of the related VCs into VC-3 via TUG-2. The mapping of DS-3 into STS-1 SPE is the same as the mapping of DS-3 into VC-3. (Refer to Section 5.2 for details.)

From the terminology point of view, VT1.5, VT2, and VT6, respectively, correspond to VC-11, VC-12, and VC-2; STS-1 (or OC-1) SPE corresponds to VC-3; and OC-3c to STM-1. Also, with respect to the terms

5. Note that there are four different types of VCs, namely, VC-11, VC-12, VC-2, and VC-3, whose capacities are equivalent to those of the tributaries DS-1, DS-1E, DS-2, and DS-3, respectively.

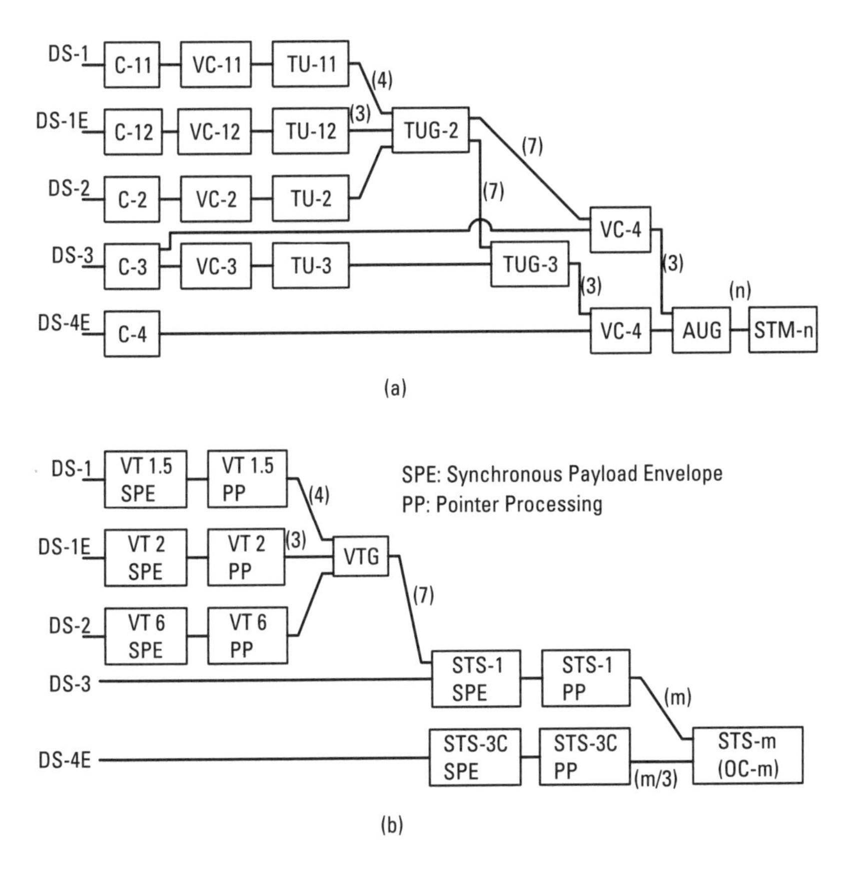

Figure 5.7 Synchronous multiplexing procedure for (a) SDH, and (b) SONET.

related to layering, the physical medium, regenerator section, multiplexer section, and path layers of SDH are respectively called *photonic, section, line,* and *path* layers in SONET. The respective terms related to mapping, multiplexing, overhead, and synchronization are mostly identical.

Table 5.1 compares SDH and SONET. As such, SDH and SONET are akin to each other in every aspect. They slightly differ in terminology and coverage, and the multiplexing procedure differs accordingly. However, more fundamentally, they are identical in format and rate, when STM-1 and OC-3c are compared. As a consequence, they may be handled collectively without individual differentiation. In this respect, therefore, we will make all discussions in this chapter in terms of SDH, which has a broader coverage, unless particular distinction is needed. (Refer to [2, 3, 8–10] for more details of the discussions on the SDH/SONET comparison.)

Table 5.1
Comparison of SDH and SONET

	SDH	SONET
Transport signals	STM-1 (155.52 Mbps)	STS-3/OC-3
	STM-4 (622.08 Mbps)	STS-12/OC-12
	STM-16 (2,488.32 Mbps)	STS-48/OC-48
	STM-64 (9,953.28 Mbps)	STS-192/OC-192
Base-rate signal and format	STM-1 (155.52 Mbps)	STS-1/OC-1 (52.84 Mbps)
	9B × 270	9B × 90
Entry-level tributaries	DS-1, DS-2, DS-3	DS-1, DS-1C, DS-2, DS-3
	DS-1E, DS-4E, (DS-2E, DS-3E)	DS-1E, (DS-4E)
Intermediate signals	C, VC	VT
	TU, TUG, AU, AUG	VT-PP
	STM-1 payload space	STS-1 SPE
	VC-11, -12, -2, -3, -4	VT1.5, 2, 6, (3)
Layering	Path	Path
	Multiplex section	Line
	Regenerator section	Section
	Physical medium	Photonic

5.1.3 Layer Networks

As observed above in Section 5.1.1, the layering concept forms the basic foundation for synchronous digital transmission. The transmission process becomes layered, which means that it is divided into the path layer and the section layer. This is reflected in the STM-*n* structure in the form of POH and SOH. The path layer is further divided into the higher-order path layer associated with the VC-3/VC-4 and the lower-order path layer associated with the VC-1/VC-2/VC-3. Likewise, the section layer is further divided into the multiplex section layer and the regenerator section layer.

Table 5.2 represents the layering concept in three different resolutions. In the table, the circuit layer is placed on top of the path layer, with the physical medium layer put below the section layer. The circuit layer represents the layer for various kinds of services that are transported through a

Table 5.2
Layering Concepts in SDH/SONET

Coarse Layering	Medium Layering	Fine Layering
Circuit layer	Circuit layer	Circuit layer
Path layer	Path layer	Lower-order path layer
		Higher-order path layer
Transmission medium layer	Section layer	Multiplex section layer [line]
		Regenerator section layer [section]
	Physical medium layer	Physical medium layer

common path layer. The physical medium layer forms the lowest part of the transmission process and represents optical or radio media. The physical medium layer and section layer can be combined and labeled as the *transmission medium layer.*

Figure 5.8 illustrates the layering concept in conjunction with the synchronous multiplexing processes. The figure shows that the circuit layer includes a 64-Kbps circuit-switched service, packet-switched service, leased-line circuit service, and ATM service. It also shows that inside the path layer the lower-order path layer consists of VC-11, VC-12, VC-2, and VC-3, while the higher-order path layer consists of VC-3 and VC-4. In addition, the AUG and MSOH processing falls in the multiplexer section layer and the RSOH processing in the regenerator section layer.

Based on the layering concept, a communication network can be divided into multiple layer networks, namely, the *circuit layer network, path layer network,* and *transmission medium layer network.* Here, each pair of adjacent upper and lower layer networks maintains a server/client relationship, and each layer network possesses its own OAM capability.

The *circuit layer network* provides communication services to users through the circuit layer access points. As stated above, its target services include circuit-switched service, packet-switched service, leased-line service, and ATM service. The configuration of the circuit layer network varies depending on the kinds of services a specific network can provide. The circuit layer network, in principle, operates independently of the path layer network.

The *path layer network* delivers information to the path layer access points in support of the circuit layer network. The path layer network

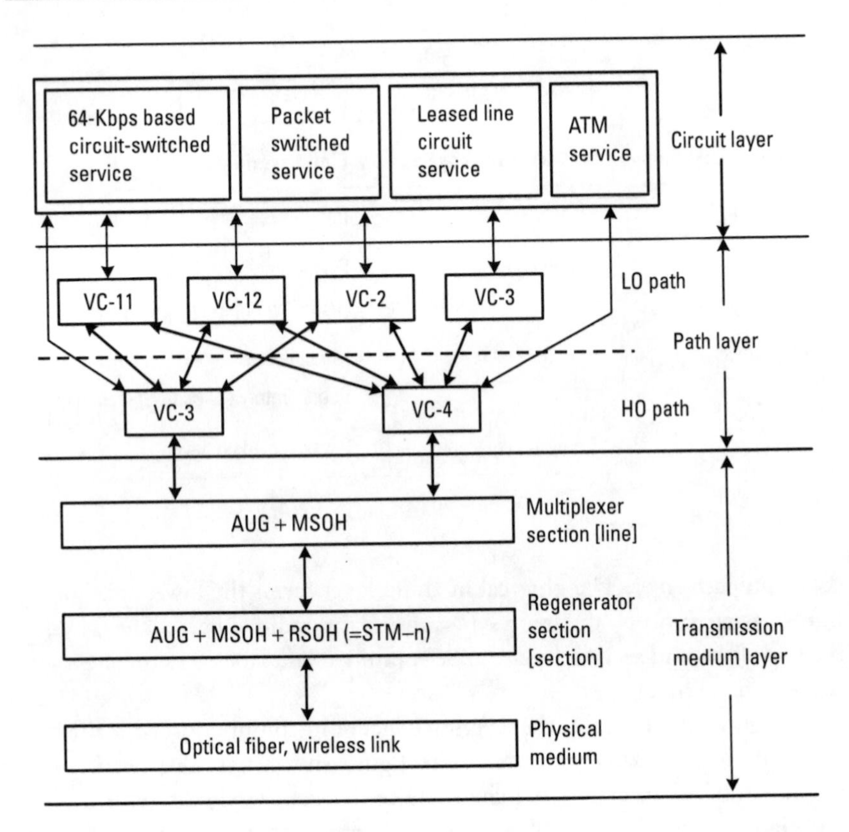

Figure 5.8 Organization of layers for SDH/SONET transmission.

functions as a lower-order layer network that can be shared by different sets of services. The path layer network can be divided into the lower-order path layer and the higher-order path layer and is independent of the transmission medium network.

The *transmission medium layer network* supports the path layer by transporting information from one path layer access point to another. The transmission medium layer network is dependent on the actual physical medium used, such as optical fiber and radio link. The internal layers within the transmission network consist of the multiplexer section layer, the regenerator section layer, and the physical medium layer.

Figure 5.9 illustrates the layered networks for an actual transmission network. In the figure the circuit layer is the network connecting service transport termination points, while the path layer network forms its sublayer that connects path layer access points. Here, the respective circuit from each

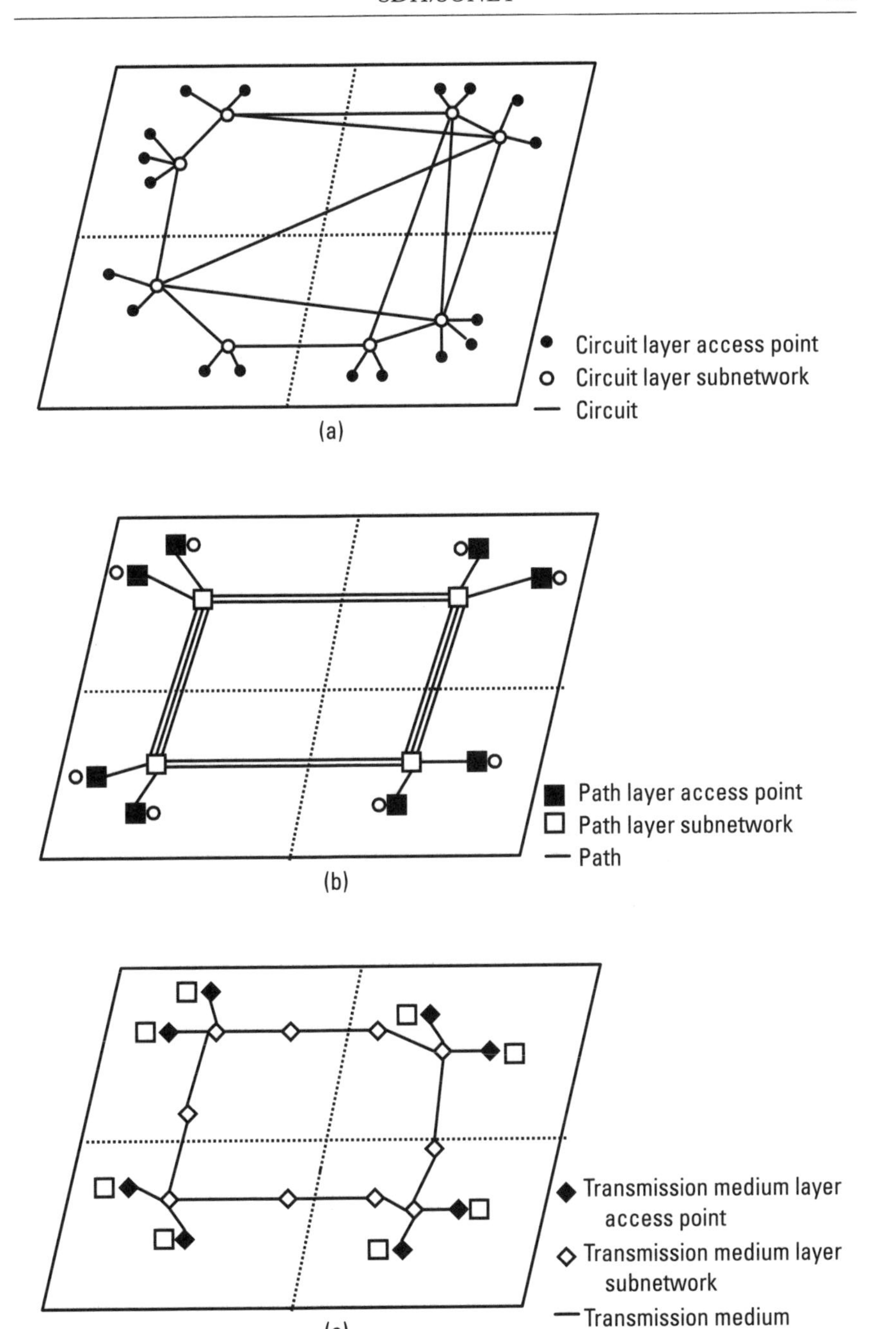

Figure 5.9 Illustration of layer networks: (a) circuit layer network, (b) path layer network, and (c) transmission medium layer network.

circuit layer network follows one unique path in the path layer network. The transmission medium layer network becomes a physical layer that is established in support of the path layer network.

For example, if we consider the case of conveying TCP/IP packets over an SDH/SONET optical network at STM-64 rate, the circuit layer network will provide a packet-switched service, presumably through the DS-1 tributary. The path layer network then provides a transport service to carry the DS-1 tributary (in a virtual container VC-4, for example) and the corresponding access point lies in an SDH/SONET multiplexer. The transmission medium layer network provides a physical transmission pipe (i.e., STM-64) to convey this path layer signal, with the access point lying in a cross-connect system. The transmission medium layer network in this case may possibly contain WDMs and optical fibers.

5.2 Multiplexing

Due to the diversity in the bit rates of the traffic signals to carry, SDH/SONET is equipped with various frames of different size in the low end, coming up with a fixed-sized STM-*n*/OC-*m* frame at the high end. To accommodate the diversity of bit rates in a systematic manner, it employs the synchronous multiplexing mechanism, which consists of a diverse set of intermediate signals of different sizes in the horizontal direction and a layered architecture in the vertical direction. In this synchronous multiplexing process, a comparatively large number of overhead bits are squeezed into the containers, and pointer-based processing is employed for synchronizing the payloads inside the containers (refer to [9–11] for details on synchronous mulitplexing).

5.2.1 Organization of STM-*n* Frame

The STM-*n* frame is the core of the SDH, as the distinct characteristics of the SDH such as the layering concept, a transmission rate of 155.52 Mbps, the visibility of the DS-0 signal at the VC-1 level, and synchronization via pointers are all incorporated in the STM-*n* frame. To understand the SDH, therefore, it is essential to understand the organization of the STM-*n* frame. Noting that the STM-*n* frame is a simple *n*-fold extension of the STM-1 frame, we focus on examining the STM-1 frame below.

5.2.1.1 SOH

SOH is intended for reliable transmission of STM-*n* through regenerators and multiplexers, and is housed in the 3×9B and 5×9B space in the upper and lower parts of the AU PTR. The RSOH in the upper part is used to maintain the transmission reliability between regenerators. Each regenerator is supposed to look at the RSOH only and ignore the information in the rest of the frame. The MSOH in the lower part is used to support the multiplexer section. When a multiplexer receives an STM-1 signal, it checks and examines this MSOH part, tossing the contained payload contents to the destined path entities.

An SOH is added on the multiplexed framework of n (n = 1, 4, 16, 64) AUGs, to form the STM-n signal. The formats of the SOHs in the STM-n for n = 1, 4, 16, and 64 are closely related one another. Figure 5.10 shows the SOH structure of the STM-1, which corresponds to an n:1 reduction of the SOH of the STM-n. Specifically, the sizes of Al, A2, B2, C1, X, and Z0 are reduced by a factor of n, and the remaining parts are fixed at one byte each. Note that in the case of SONET, the SOH structure again corresponds to a 3:1 reduction of the SOH of the STM-1.

RSOH. The RSOH consists of the bytes A1, A2, B1, C1 (or J0, Z0), D1, D2, D3, E1, and F1. A1 and A2 bytes are frame alignment words for delineating the STM-n boundary. B1 is a *bit-interleaved parity* (BIP) byte of period 8 used for the error monitoring function, with the ith (i = 1, 2,..., 8) bit of B1 containing the even parity of all the ith bits of the STM-n bytes. C1 contains

	1	2	3	4	5	6	7	8	9
1	A1	A1	A1	A2	A2	A2	J0 C1	X*	X*
2	B1			E1			F1		
3	D1			D2			D3		
4	AU pointer								
5	B2	B2	B2	K1			K2		
6	D4			D5			D6		
7	D7			D8			D9		
8	D10			D11			D12		
9	S1					M1	E2	X	X

* Unscrambled bytes

Figure 5.10 Structure of the section overhead of STM-1.

the identification number of the corresponding STM-1 within the STM-*n*, which is used for aligning the STM-*n* frame or for extracting a single STM-1 frame. The C1/J0 byte becomes C1 when used for identification of STM signals and becomes J0 when used for tracing the transmission signals in the regenerator sections. J0 is the regenerator section trace field for the continuity check of connections within the regenerator sections. D1, D2, and D3 are DCC used by the regenerator sections, which jointly forms one single channel of 192-Kbps (3 × 64 Kbps) capacity. E1 is an orderwire channel used by the regenerator section for voice communication. F1 is a user channel reserved for communication network operators. The X and Z0 bytes are assigned respectively for national and international uses.[6]

MSOH. The MSOH consists of the bytes B2, D4 to D12, E2, K1, K2, M1, and S1. Three B2's are BIP-24 parity check bytes used for multiplexer section error monitoring. In the case of the STM-*n*, B2 consists of 3*n* bytes and hence the parity checking is done in BIP-24*n* format. The D4 to D12 bytes form an MSOH DCC, which forms one channel of 576-Kbps capacity. E2 is the multiplexer section orderwire, which can be used for voice communication. K1 and K2 form the *automatic protection switching* (APS) channel that carries the APS signaling for switching from the working channel to a protection channel. Part of K2 carries the AIS or the RDI signal. M1 is the *remote error indication* (REI) byte, which notifies the far end of the BIP error count in the B2 byte. S1 is used to indicate the level of synchronization or the quality of synchronization of the multiplexer section. Table 5.3 lists a summary of various SOH bytes.

5.2.1.2 Payload and POH Space

The payload space of the STM-1 frame of 9B × 261 size is allowed to carry one VC-4 or three VC-3 signals. Loading of one VC-4 (or three VC-3) onto the STM-1 payload space is done in a *floating mode*, with the AU PTR indicating the location of its first byte in the VC. The location in this case refers to the address within the STM-1 payload space, which is numbered from 0 to 782, in units of 3 bytes, starting immediately after the AU PTR (see Figure 5.11). This addressing scheme is applicable to VC-4 as well as VC-3, in which case three sets of 783 addresses are used. Therefore, 10 bits are

6. In Figure 5.10, the X-bytes with asterisks (*) are the bytes that are left unscrambled. This is due to the fact that the first row of the SOH does not go through the scrambling process.

Table 5.3
Functions of Section Overheads

Overhead	Function	Note
A1, A2	Frame alignment	"11110110", "001010000"
B1	Regenerator section error monitoring	BIP-8
B2	Multiplexer section error monitoring	BIP-24
C1	STM-1 identifier	
D1-D3	Regenerator section data communication	
D4-D12	Mutiplex section data communication	
E1, E2	Orderwires	
F1	User channel	Network operator
H1, H2	AU-4 PTR/path AIS	/"11111111"
H3	Pointer action	Negative justification
J0	Regenerator section path trace	
K1, K2	Automatic protection switching	
K2 (bits 6–8)	Section AIS/section RDI	"111"/"110"
M1	Section REI	B2 error count
S1	Synchronization state indication	
X	Reserved byte for national use	
Z0	Reserved byte for international use	

required for addressing the whole STM-1 payload space. The address thus assigned indicates the degree of offset of each address location from the pointer location.

The STM-1 payload space consists of the POH space and the VC-3/VC-4 payload space. POH is designed for the control of the constituent VCs. A POH is inserted at the point a VC is originated and removed at the destination of the VC. There are two different types of POHs—HO-POH for the higher-order VC's, namely VC-3 and VC-4, and LO-POH for the lower-order VC's, namely VC-11, VC-12, VC-2, and VC-3. The HO-POH bytes are located in the front column of the VC-3/VC-4, while the LO-POH byte is positioned inside the VC-3/VC-4 payload space. Figure 5.12 shows the POH bytes in the higher-order and lower-order VCs.

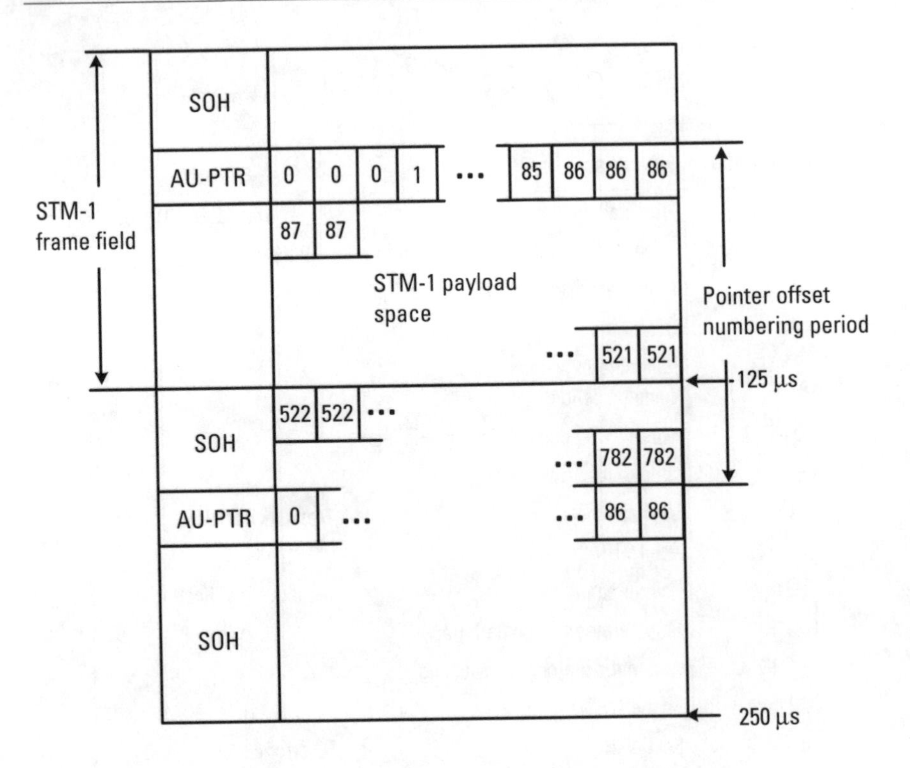

Figure 5.11 Addressing (or pointer-offset numbering) of the STM-1 payload space.

HO-POH. The HO-POH is made up of the J1, B3, C2, G1 F2, H4, F3, K3, and N1 bytes, as shown in Figure 5.12(a). J1 is a path trace byte that checks the continuity of connection between path terminations. B3 is a BIP-8 check byte used for the path error-monitoring function. C2 is a signal label byte for indicating the contents of the VC-3/VC-4. The F2 byte is allocated for path user communication purposes between equipment. G1 is a channel used by the receiver to feed back the transmitter the path condition and performance. The first 4 bits of G1 are for REI and bits 5~7 are for RDI. The H4 byte is a time reference indicator for multiframe arrangement of specific payloads. In case the VC-4 carries an ATM cell, which does not require a multiframe indicator, H4 is used to denote the starting position of the ATM cell. K3 is an APS byte (bits 1~4) for path protection of the higher order path. N1 is used for tandem connection monitoring.

LO-POH. The LO-POH is composed of V5, J2, N2, and K4 bytes as shown in Figure 5.12(b). As illustrated in the figure, V5 includes bits for BIP check

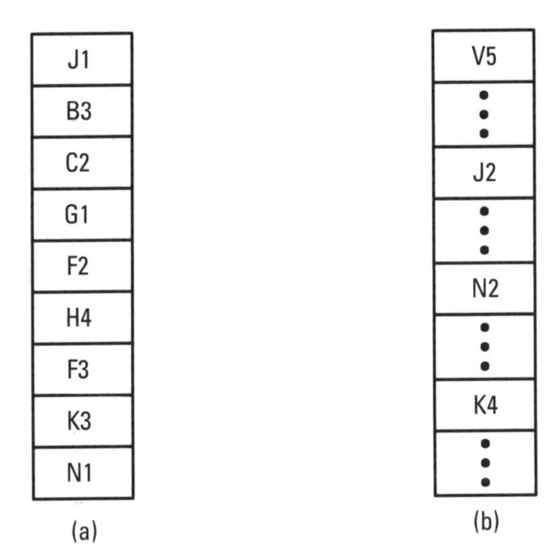

Figure 5.12 Structure of the path overheads: (a) HO-POH, and (b) LO-POH.

(i.e., BIP-2), REI, *remote failure indication* (RFI), signal labels (L1, L2, L3), and RDI. The RDI bit notifies the far end if a path AIS or signal defect is received. RFI notifies the far end if, after a path failure, recovery is not made until the maximum allowed time for protection switching. A detailed indication on the RDI-related defect can be made by sending an RDI code over bits 5~7 of K4 byte. The J2, N2, and K4 (bits 1~4) are used for the same purpose as J1, N1, and K3 (bits 1~4), respectively. Table 5.4 summarizes various POH bytes.

5.2.1.3 Pointer Space

The PTR (more specifically, AU PTR) space is located on the fourth row of the STM-1 overhead. Pointer is employed to keep track of the shifting location of the starting point of the VCs within the corresponding AU or TU. The 9-byte AU PTR, which applies to the VC-3 and VC-4, consists of three triplets of H1, H2, and H3 (see Figure 5.22 in Section 5.3). In case the STM-1 carries a VC-4, only the first triplet of H1, H2, and H3 is used to indicate the starting point of the VC-4. However, in the case of carrying three VC-3s, each of the H1, H2, and H3 triplets independently keeps track of the address of each VC-3. Among the 24 bits contained in the H1, H2, and H3 triplet, 10 bits are used to indicate the addresses from 0 to 782, and the remaining 14 bits are used for other purposes. In the case of TU PTR, the three TU PTR bytes are designated as V1, V2, and V3 and are contained

Table 5.4
Functions of POHs

Overhead	Function	Note
B3	Path error monitoring	BIP-8
C2	Path signal label	
F2, F3	User channel	Path user
G1 (bits 1–4)	Path REI	B3 error count
G1 (bits 5–7)	Path RDI	
H4	Multiframe indication	ATM cell offset (bits 3–8)
J1, J2	Path trace	
K3, K4	Path APS, RDI	
N1, N2	Tandem path monitoring	
V5	Lower-order POH	BIP-2, REI, RFI, signal label, RDI

in the corresponding TU. (Refer to Section 5.3 for a detailed discussion on pointer and pointer processing.)

5.2.2 Intermediate Signal Elements

Synchronous multiplexing is carried out through a multistage process, and a number of intermediate signal elements are involved in this process. PDH tributaries are converted or mapped into *container* (C), *virtual container* (VC), *tributary unit* (TU), *tributary unit group* (TUG), *administrative unit* (AU), and *administrative unit group* (AUG) before being multiplexed into the STM-*n* (refer to Figure 5.13). ATM cells are also multiplexed into the STM-*n* via those intermediate signals, and other forms of user data can join the synchronous multiplexing as well. In the following sections, we review the intermediate signal elements, one by one, based on the synchronous multiplexing procedure depicted in Figure 5.13.

5.2.2.1 Tributaries

All North American or European PDH tributaries are multiplexed into the STM-*n* in one form or another. In the case of North American tributaries, the DS-1, DS-2, and DS-3 signals can be mapped into the corresponding

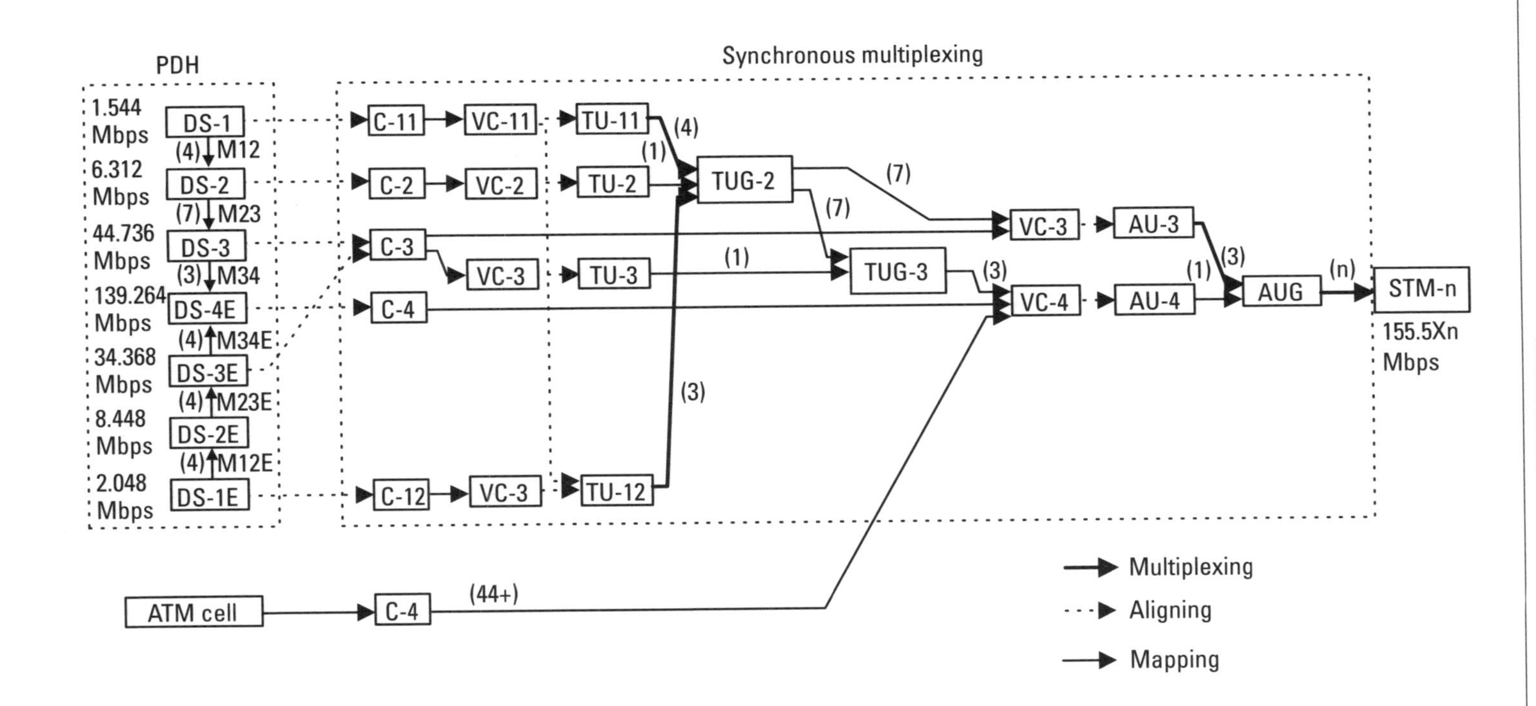

Figure 5.13 Synchronous multiplexing procedure.

container directly or into an upper-level container after being multiplexed. For example, the DS-1 signal can be either loaded directly onto the C-1 or multiplexed to the DS-2 first and then mapped into the C-2. The same scenario applies to DS-2 and DS-3. In the case of the European tributaries, the synchronous multiplexing is less complete in the mapping path than its North American equivalent. We observe from Figure 5.13 that direct mapping to container is not available in the case of DS-2E. It has to be asynchronously multiplexed to DS-3E or DS-4E first and then mapped into the corresponding container.

5.2.2.2 Container (C)

The container is the most fundamental unit of the synchronous multiplexing structure in the sense that all of the North American and European PDH tributaries have to be mapped into the respective containers as the first-step processing toward the synchronous multiplexing. Containers are categorized into classes C-1, C-2, C-3, and C-4, with the number denoting the corresponding digital hierarchical levels. C-1 can further be categorized into C-11 and C-12, with C-11 accommodating the North American DS-1, and C-12 accommodating the European DS-1E. C-4 can carry either a DS-4E from the PDH, or the ATM cells.

5.2.2.3 Virtual Container (VC)

The function of a VC is to support the connections between the path layers in synchronous transmission. The VC consists of the POH and the payload that carries the information data. The payload portion corresponds to a container, and the whole VC frame is repeated every 125 or 500 μs. The four classes of VC, namely, VC-1, VC-2, VC-3, and VC-4, are equivalent to C-1, C-2, C-3, and C-4, respectively, in terms of the contained information. Similar to C-1, VC-1 can be further categorized into VC-11 at the DS-1 rate and VC-12 at the DS-1E rate. VC-1 and VC-2 are called the *lower-order* VCs, and VC-3 and VC-4 are called the *higher-order* VCs. The POH for the lower-order VCs is named *V5* and the POH for the higher-order VCs is called *VC-3 POH* or *VC-4 POH.*

5.2.2.4 Tributary Unit (TU)

The TU is designed to provide adaptability between higher-order and lower-order path layers. For instance, lower-order VCs can be mapped into higher-order VCs through a TU or a TUG. A TU is created if a TU PTR is attached to a lower-order VC. The pointer indicates the degree of offset of the lower-order VC relative to the starting position of the higher-order VC's frame. The TU is categorized into TU-1, TU-2, and TU-3. TU-1 is further categorized into TU-11 and TU-12, depending on the type of VC it contains.

5.2.2.5 Tributary Unit Group (TUG)

The role of the TUG is to collect one or more TUs and load them onto a fixed location on the payload of a higher-order VC. No overhead is added when a TUG is formed from the TUs. There are two classes of TUG: TUG-2 and TUG-3. TUG-2 is formed by assembling a homogeneous group of TU-1s or by a direct mapping of a single TU-2. Similarly, TUG-3 could be an assembly of multiple TU-2s or a single TU-3.

5.2.2.6 Administrative Unit (AU)

The AU functions as an adapter between the higher-order path layer and the multiplexer section layer. As in the case of TU, AU consists of payload and a pointer. The payload carries a higher-order VC, and the AU PTR indicates the relative offset between the starting positions of the AU payload and the frame of the multiplexer section layer. Specifically, the two types of AU, namely, AU-3 and AU-4, carry VC-3 and VC-4, respectively, and the AU PTR indicates the degree of offset of VC-3 or VC-4 with respect to the STM-*n* frame.

5.2.2.7 Administrative Unit Group (AUG)

One or more AUs occupying a fixed location on an STM payload is called an AUG. An AUG can consist of three AU-3s or a single AU-4.

5.2.2.8 Synchronous Transport Module (STM)

The STM is the final product of the synchronous multiplexing process and is the signal that is actually transmitted over the SDH networks. The STM-*n* signal is formed by byte-interleaving *n* AUGs and then adding an SOH.

5.2.3 Synchronous Multiplexing Procedures

The STM-*n* signal is obtained by synchronously multiplexing the incoming PDH tributaries, DS-1 to DS-4E, following the procedures depicted in Figure 5.13. The figure describes the procedures based on the intermediate signal elements introduced in Section 5.2.2. We find that the procedures consist of 11 different multiplexing paths. As it is tedious to follow each individual path, we instead consider a high-level sketch of the procedures in this section.[7]

5.2.3.1 Overall Structure

In the first stage of synchronous multiplexing, each tributary is mapped into the corresponding container C. If a POH is attached to the container, it becomes a VC, and if a pointer is added as well, it becomes a TU. When a VC is mapped into an STM-1 without going through other VCs, then the corresponding TU becomes an AU. Assembling a number of TUs creates a TUG, which can then be multiplexed into the next higher level of VC. If a PTR is added, a corresponding TU or AU is obtained. In this process, the number (k = 1, 2, 3, 4) attached to all signal units indicates that the corresponding signal unit is equivalent to the DS-k tributary in terms of rates. In the final stage, one or more of AUs are multiplexed into STM-n.

A notable feature in this synchronous multiplexing is that a tributary DS-k directly enters the synchronous multiplexing processing via VC-k and TU-k without being multiplexed into the next higher tributary DS-(k+1). For example, DS-1 gets into the synchronous multiplexing procedure via VC-11, TU- 11, TUG, and so on, without passing through the DS-2, DS-3, and DS-4 multiplexing paths, finally reaching the destination of STM-n. This is called *one-step multiplexing.*

The synchronous multiplexing can be categorized into the *lower-order multiplexing paths,* which involve VC-1 and VC-2, and the *higher-order multiplexing paths,* which involve VC-3 and VC-4 (refer to Figure 5.8). The lower-order case involves the multiplexing of the low-level tributaries DS-1, DS-1E, and DS-2, which are first aligned into C-11, C-12, and C-2, and then mapped into VC-11, VC-12, and VC-2, respectively. As an exceptional case, the DS-3 and DS-3E tributaries are also regarded as undergoing lower-order multiplexing path if they pass through the C-3, VC-3, and TU-3 path, before being multiplexed into VC-4. The higher-order case involves the multiplexing of the DS-3 and DS-4 tributaries, as well as the above lower-order VCs, via VC-3 and VC-4.

In the case of ATM cells, they are mapped to VC-4 first, then aligned into AU-4, and finally multiplexed into STM-n. As the payload space in VC-4 is not an integral multiple of the ATM cell size, the starting point of

7. For details of the synchronous multiplexing procedures, refer to Section 3.3 of [1].

ATM cells vary in every VC-4. The starting point of the first ATM cell in the sixth row of the VC-4 is optionally indicated in the H4 field of the POH.

5.2.3.2 Tributary Mappings to VCs

The first step of synchronous multiplexing is tributary mapping. The PDH tributaries are mapped into containers, Cs, or, equivalently, to virtual containers, VCs, thereby entering the synchronous domain. Hence, synchronization is the central function of the tributary mappings. Tributary mapping includes the mapping of DS-1, DS-1E, DS-2, DS-3, and DS-4E signals into VC-11, VC-12, VC-2, VC-3, and VC-4, respectively.

The mapping of a tributary into a VC can be classified as asynchronous and synchronous, with the latter subclassified into *bit-synchronous* and *byte-synchronous* cases. Asynchronous mapping refers to the case when the clock used for generating a VC or VT is not identical to the clock used for generating the tributary contained in the VC, while synchronous mapping refers to the case when the clocks are identical. The synchronous case is called byte-synchronous if the byte boundary is visible externally, and bit-synchronous, otherwise.

In the asynchronous mapping case, synchronizing process is needed when mapping a tributary clock to C or VC. In this process *positive-zero-negative* (P/Z/N) justification via *bit stuffing* is used for DS-1, DS-1E, DS-2, and DS-3E signals, and positive justification with bit stuffing is used for DS-3 and DS-4E signals. In general, P/Z/N justification is used when the bit rate of a VC payload is the same as the nominal bit rate of the corresponding tributary. In normal state, no justification occurs (*zero justification*), but when the tributary bit rate decreases below the bit rate of the VC payload, a null bit (or a garbage bit) is sent in place of the effective information data (*positive justification*). Conversely, if the tributary's bit rate increases above the VC bit rate, a spare bit in the payload is used to absorb the offset (*negative justification*).

Figure 5.14 shows the format for asynchronous mapping from DS-1 to C-11, as a typical example of the tributary mappings. The C-11 frame has the size of 104B which is contained in 500-μs multiframe, consisting of four subframes of 125 μs each. The P/Z/N justification function is performed through the justification opportunity bits J_1, J_2, and the execution status is indicated through the use of the justification control bits C_1 and C_2.. In the figure, R stands for reserved bits; I, information bit; I*, information byte; and O, overhead bit.

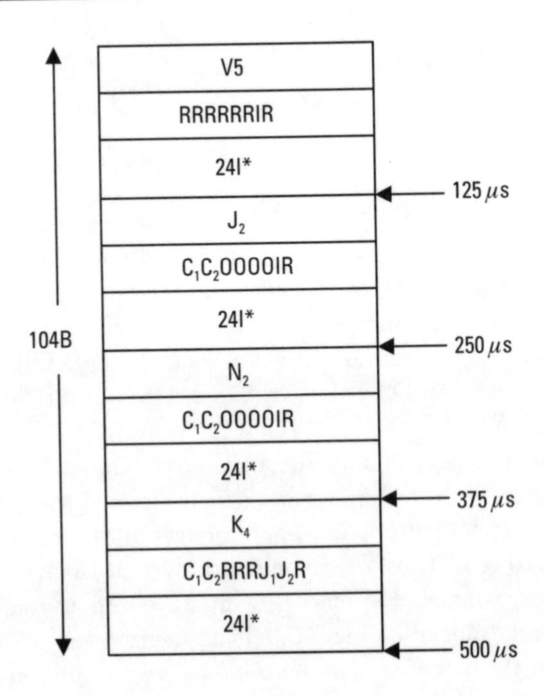

Figure 5.14 Channel frame for mapping from DS-1 into C-11 (floating mode).

5.2.3.3 Multiplexing of Lower-Order VCs

The lower-order virtual containers, VC-11, VC-12, and VC-2, are composed of 104B, 140B, and 428B, respectively, in a multiframe of 500 μs, which is a concatenation of four 125-μs frames of 26B, 35B, and 107B each. The first byte in each VC, designated as V5, is the POH of the VC and the remainder is the space for carrying the container C-11, C-12, or C-2, and the relevant overheads. This is depicted in the left-hand side of Figure 5.15.

Four pointer bytes, designated V1, V2, V3, and V4, are attached to the VC, 1 byte per 125-μs frame, to form a TU. Hence, TU-11, TU-12, and TU-2 occupy 27B × 4, 36B × 4, and 108B × 4, respectively, over a 500-μs multiframe, with the 26B × 4, 35B × 4, and 107B × 4 portion respectively taken by the corresponding VC. The four pointer bytes indicate the relative position of the payload within the TU multiframe. The right-hand side of Figure 5.15 shows the organization of the TUs.

In terms of the base 125-μs frame unit, TU-11, TU-12, and TU-2 can be respectively aligned in the form of 9B × 3, 9B × 4, and 9B × 12 formats, that is, 3, 4, and 12 columns of 9B stacks. Hence, if four TU-11s or three TU-12s are joined through byte-interleaved multiplexing, 12 columns are

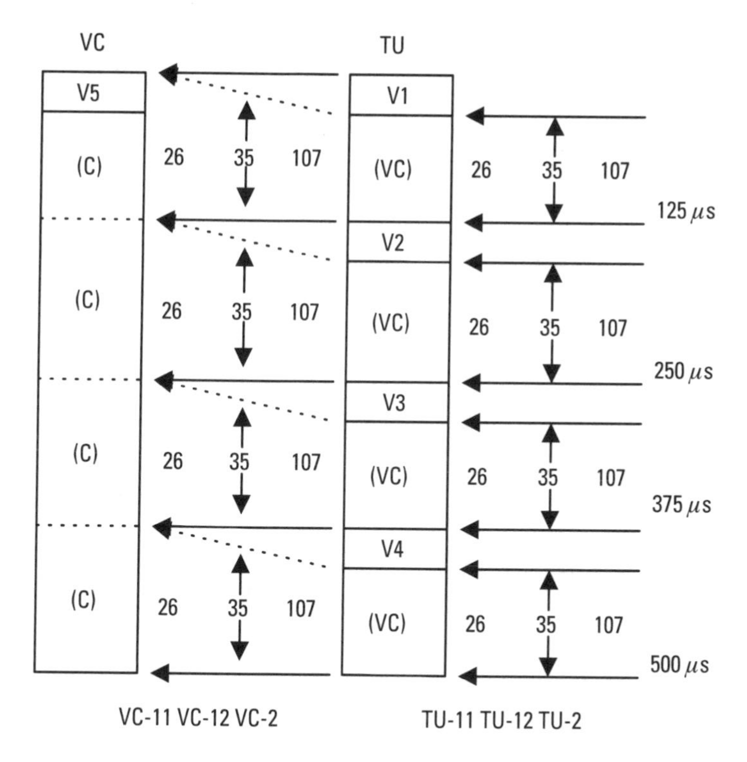

Figure 5.15 Composition of low-order VCs and TUs.

produced, which is equivalent to a single TU-2. These 12 columns can be used to create a single TUG-2. This relation is depicted in Figure 5.16.

5.2.3.4 Multiplexing of Higher-Order VCs

Among the higher-order VCs, the VC-3 consists of 85 columns of 9B stacks. The first column is used for the POH, and the remaining 84 columns correspond to the payload, which can carry a C-3 or seven TUG-2s. TUG-3 consists of 86 columns of 9B stacks, allowing the mapping of one VC-3 or TU-3, or seven TUG-2s into its payload. The TU-3 is formed if three pointer bytes H1, H2, and H3 are added to the VC-3, and the TUG-3 is obtained if 6 fixed-stuff bytes are further added in the same column. Figure 5.17 shows this process.

In the case of the VC-4, it is composed of 261 columns of 9B stacks. The first column is used for the POH, and the remaining 260 columns for the payload, to which a C-4 or three TUG-3s can be mapped. An AU-3 is formed by adding two 9B columns of fixed-stuff bytes to a VC-3, thus acquiring 87

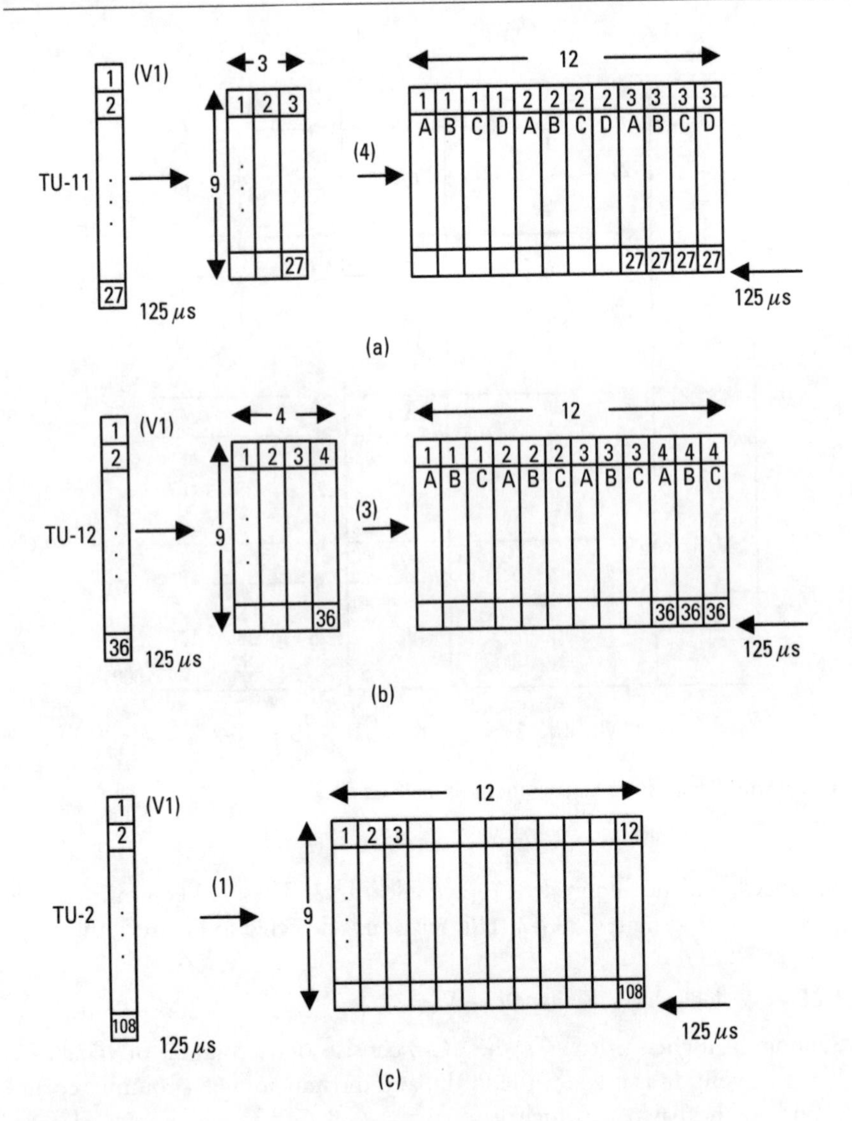

Figure 5.16 Formation of TUG-2 based on (a) TU-11, (b) TU-12, and (c) TU-2.

columns in all, and then adding a 3B pointer. An AU-4 is formed if a 9B pointer is added to a VC-4. Figure 5.18 illustrates this process.

An AUG is composed of 261 columns of 9B stacks plus a 9B row for pointers. An AUG can be formed either from three AU-3s or from a single AU-4. An AU-4 signal, which is equivalent to the AUG, is produced if an AU-4 PTR is added to the VC-4 signal. The AUG is also produced if three

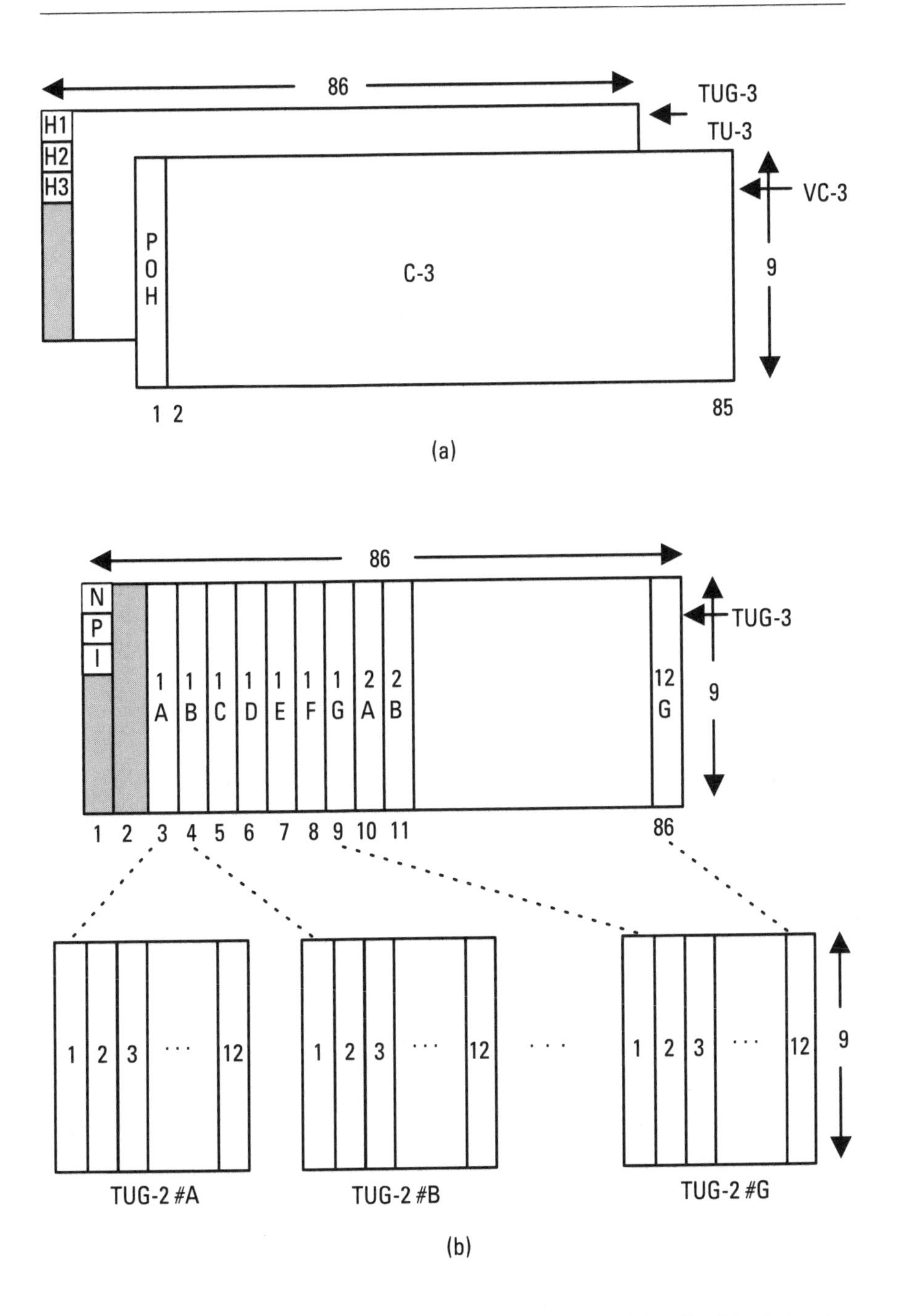

Figure 5.17 Multiplexing to TUG-3 based on (a) VC-3/TU-3, and (b) TUG-2 (shaded region for fixed-stuff bytes).

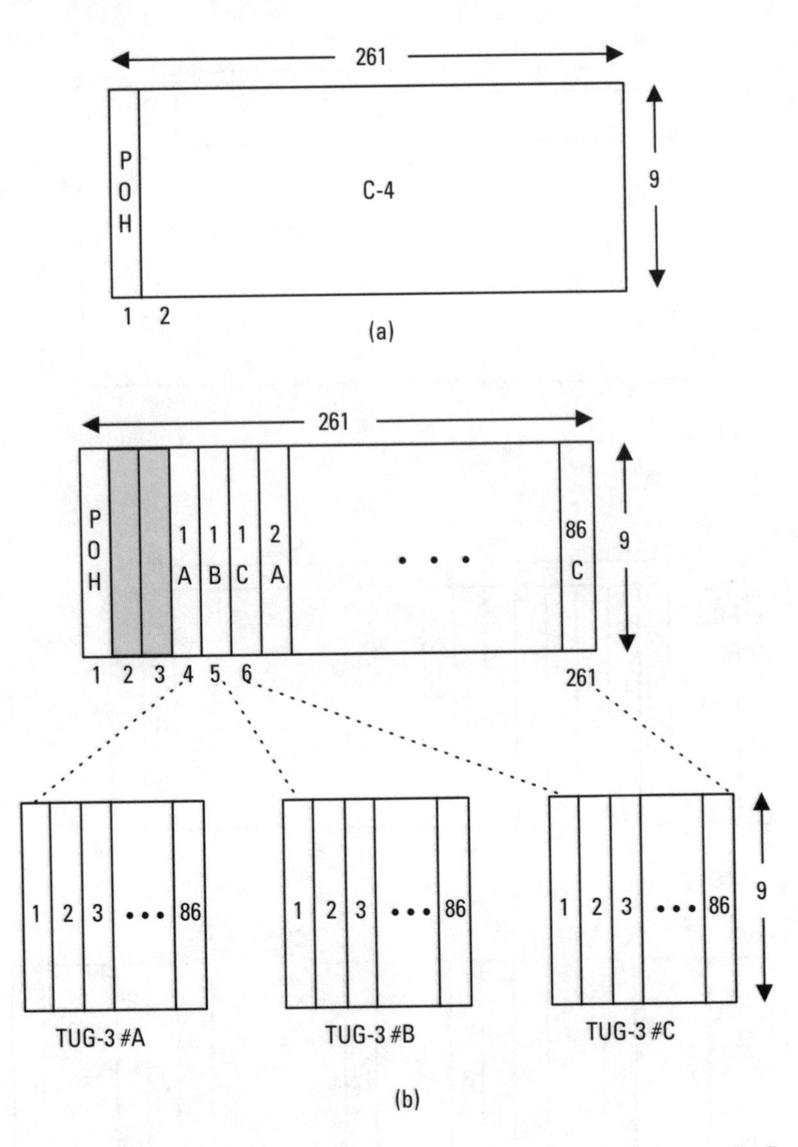

Figure 5.18 Mapping into VC-4 based on (a) C-4, and (b) TUG-3 (shaded region for fixed-stuff bytes).

AU-3s are *byte-interleaved multiplexed*, as shown in Figure 5.19(a). If *n* AUGs are byte-interleaved multiplexed again and an SOH is attached, the STM-*n* signal is generated, as shown in Figure 5.19(b).

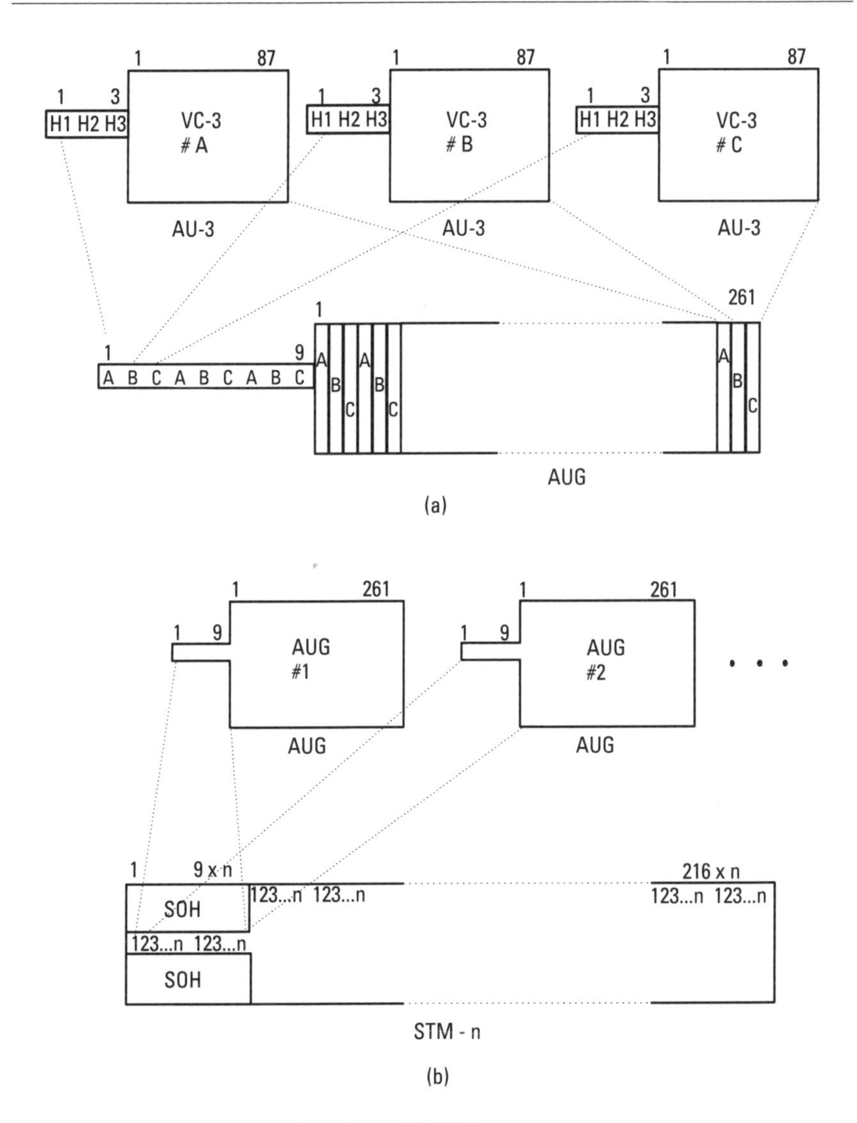

Figure 5.19 Organization of (a) AUG, and (b) STM-*n*.

5.2.3.5 Mapping of the ATM Cell

The ATM cell is composed of 5 bytes of header and 48 bytes of information data, with the fifth header byte, the HEC byte, used for detecting the cell boundary. ATM cells are transported over the SDH/SONET network after being mapped into the VC-4 of the STM-*n*. As the size of the VC-4, 2,340B, is not a multiple of the ATM cell size, 53B, there are always one or more

ATM cells that cross the boundary of the VC-4 and, consequently, the starting location of an ATM cell within the VC-4 varies from one VC-4 to another. So, the H4 byte in the VC-4 POH is used, optionally, to indicate the starting point of the first ATM cell in the same row as H4 itself.[8] Figure 5.20 illustrates the mapping of ATM cells in the VC-4.

The receiver can detect the ATM cell boundary in two possible ways. First, after searching through the STM-*n* frame, it locates the VC-4 starting location through the AU-4 PTR, and then finds the ATM cell boundary by reading out the address written in H4. Second, it can apply the HEC based delineation process instead of H4 reading-out (refer to Section 4.1.3.1).

5.2.3.6 Concatenation of AU-4

If the payload to be transmitted is greater than the transmission capacity of a single VC-4, AU-4-*x*c, which is a concatenation of *x* AU-4s, can be used instead. Externally AU-4-*x*c looks identical to the AU-4*x* or AUG-*x* but the internal payload forms a single byte stream at *x* times the base rate, not a multiplexed stream of *x* base-rate signals. The first of the resulting *x* pointer sets maintains a normal pointer function for the concatenated stream but the remaining *x*-1 pointer sets denote the concatenation status. In the case of

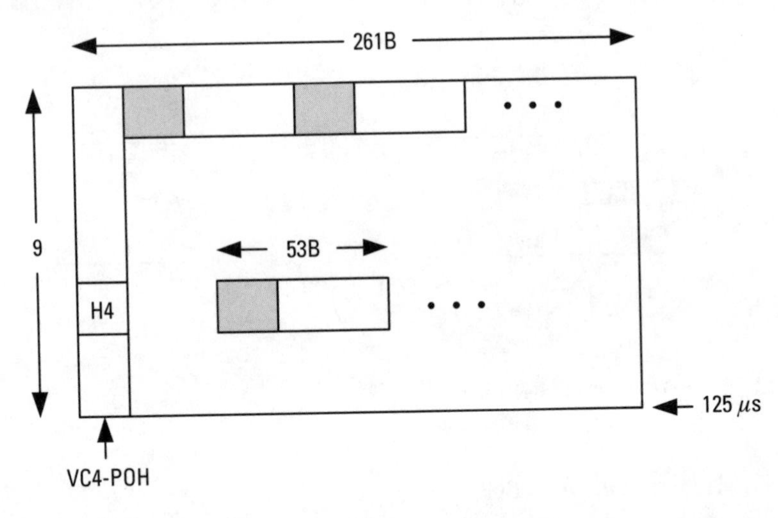

Figure 5.20 Mapping of ATM cells into VC-4.

8. H4 is usually employed to furnish the multiframe phase information when the lower-order tributaries are mapped into the VC. However, in the case of ATM cell mapping, such a function is not required and H4 is free to use for other purposes.

SONET, concatenation of three STS-1s, namely STS-3c, yields a frame structure identical to an STM-1.

5.2.3.7 An Illustration of the Synchronous Multiplexing Procedure

Among the 11 different multiplexing paths connecting DS-*k* to STM-*n* shown in Figure 5.13, the path that emerges as the most economical is the C-1/VC-1/TU-1/TUG-2/TUG-3/VC-4/AUG path. Since TUG-2 and TUG-3 merely do an assembly function, this path yields a minimum-processing multiplexing path from DS-l/DS-1E into STM-*n*.

Figure 5.21 illustrates the synchronous multiplexing procedure through the DS-1/C-11/VC-11/TU-11/TUG-2/TUG-3/VC-4/AU-4/AUG/STM-*n* path. The DS-1 signal is first mapped into C-11, which becomes VC-11 if a VC-11 POH is added. A TU-11 is formed out of VC-11 if a TU-11 PTR is attached, and a TUG-2 is produced if four TU-11s are multiplexed.

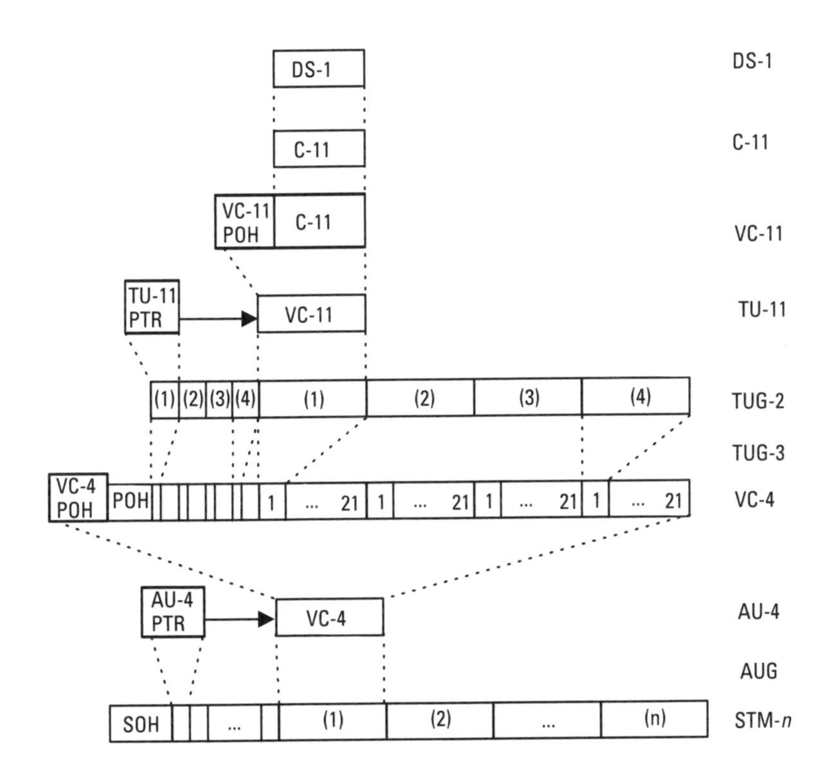

Figure 5.21 Example of multiplexing procedure for the multiplexing path DS-1/C-11/VC-11/TU-11/TUG-2/TUG-3/VC-4/AU-4/AUG/STM-*n*.

A TUG-3 is obtained if seven TUG-2s are multiplexed and a *fixed overhead* (FOH) is attached, and a VC-4 is produced if three TUG-3s are multiplexed and another FOH and VC-4 POH are attached. Therefore, the VC-4 signal is equivalent to the signal constructed by multiplexing 21 TUG-2s and attaching VC-4 POH and FOH to the head of its frame.

As a result of this highly coordinated multiplexing procedure, each of the constituent 84 TU-11 signals can be accessed separately on a VC-4. In this case, the FOH is just an overhead to match the size of a VC-4 frame to that of the VC-3. A VC-4 together with the AU-4 PTR produces an AU-4, which is equivalent to an AUG. Multiplexing *n* AUGs and adding an SOH produces an STM-*n*.

5.3 Synchronization

Synchronization is an essential feature of the SDH and SONET systems, as is well signified by the initial "S" of the names of both systems. In fact, synchronization takes place in all digital transport networks whenever two or more signals meet for multiplexing. In the conventional asynchronous digital transmission systems, synchronization was usually achieved among the participating plesiochronous tributaries through bit stuffing. In the SDH/SONET systems, in contrast, synchronization is achieved through the use of pointers that identify the location of the information payload; an AU/TU pointer is allocated to each AU/TU to convey the information regarding the starting point of the constituent VC (refer to [12]).

5.3.1 Synchronization via Pointer Processing

In general, the VCs that are aligned within the AU or TU are created from a different source, which has a different clock from the one that produces the AU/TU. Of course, these two clocks have a plesiochronous relationship, and the degree of discrepancy between the two is small and within the specified tolerance. Nevertheless, since they are not perfectly synchronized either, a synchronization procedure is always needed. This can be achieved by using pointers (refer to [9, 11]).

5.3.1.1 Composition of Pointers

Pointers can be grouped into higher-order pointers, such as AU-4 PTR, AU-3 PTR, and TU-3 PTR, and lower-order pointers, such as TU-11 PTR, TU-12 PTR, and TU-2 PTR.

The higher-order pointer is contained in bytes H1, H2, and H3, which appear once in every 125-μs frame (see Figure 5.4). Among the three bytes, H1 and H2 are used as the address indicator for the starting location of the corresponding VC, and H3 is used for the execution of negative justification.

The lower-order pointer is contained in the bytes V1, V2, and V3, which appear once in every 500-μs TU frame (see Figure 5.15). These three bytes are located at the beginning of the four 125-μs partitions of the 500-μs frame, and are identical in functions to H1, H2, and H3.

Figure 5.22 shows the composition of the pointer bytes. The 16 bits of H1 [or V1] and H2 [or V2] are divided into three sections: the first 4 bits carry the *new data flag* (NDF); the next 2 bits indicate the signal type of the frame; and the last 10 bits are address bits, which consist of 5 *increment* (I) bits and 5 *decrement* (D) bits in an interleaved format. The address bits indicate the starting location of the VC in the regular mode of operation but "blink" momentarily whenever the address change takes place. The change is reflected through the inversion of I- or D-bits in such a manner that all the I-bits are inverted when the VC is to be shifted up and all the D-bits are inverted when the VC is to be shifted down.

5.3.1.2 Pointer-Based Synchronization

The general principle of the synchronization procedure via pointers is as follows. First of all, a VC does not get "locked" into the AU/TU frame but is allowed to "float" within the payload space in this frame, with the address of

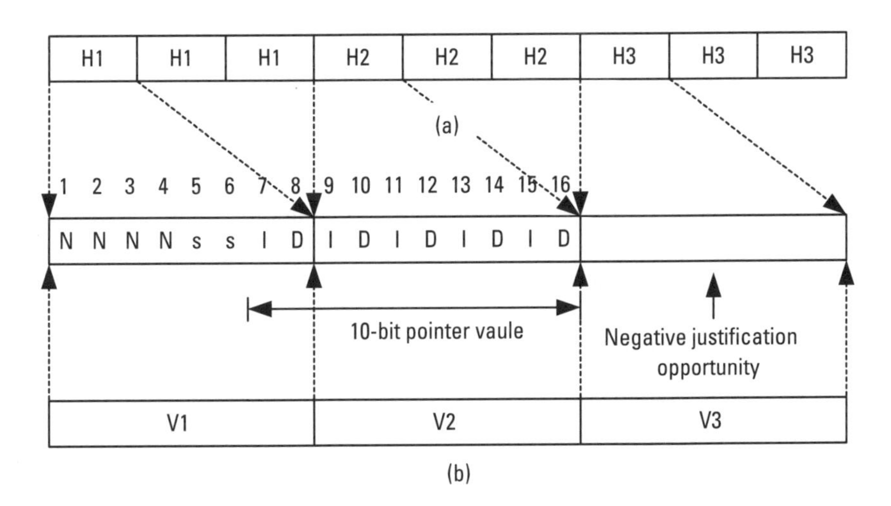

Figure 5.22 Pointer structures of (a) AU PTR, and (b) TU PTR.

the first byte of the VC recorded in the pointer. We assume the case when the bit rate of the VC is lower compared to that of the AU/TU, which implies that the incoming bit rate is lower than the processing bit rate of the multiplexing node. If the accumulated data offset becomes 1-byte long (3 null bytes in the case of AU-4), then 1 null byte (3 null bytes in the case of AU-4)[9] is inserted into the payload. At this time, the starting address of the VC is shifted up by one, with the altered address recorded in the pointer. Hence, the clock discrepancy between the VC and AU/TU—that is, the discrepancy between the incoming and system clocks, has been effectively resolved. Such a process is called *positive justification.*

Conversely, if the bit rate of the VC is high compared with that of the AU/TU, an opposite treatment called *negative justification* is used to resolve the clock offset. To elaborate, when the data offset becomes 1-byte long (3 bytes for AU-4), 1 spare byte is used to convey the extra byte of the VC. At this time, the address of the VC is shifted down by one, with the altered address recorded in the pointer. Here, H3 (or V3) is the spare byte that conveys the extra data.

5.3.1.3 Illustration of Pointer Processing

Figure 5.23 depicts a conceptual structure of a frame with pointer, which is common to all different types of pointers. The pointer portion in the shaded region corresponds to the H1, H2, and H3 (or V1, V2, and V3). The address ranges from 0 to N-1, with N-1 being 782, 764, 427, 139, and 103 respectively for AU-4/AU-3, TU-3, TU-2, TU-12, and TU-11. In Figure 5.23, the address location 0 is positioned in the middle of the frame, which in practice comes right after the H3 byte in the case of AU-4/AU-3 and after the V2 in the case of TU-2, TU-12, and TU-11. NJ and PJ represent the *negative justification* and *positive justification* opportunity bytes, respectively. The actual position of the PJ execution byte immediately follows H3 (or V3). N, S, and I/D are for NDF, signal type, and increment/decrement indications, as mentioned above.

First, we consider the execution procedure of positive justification based on Figure 5.24, where the number N-1 is taken to be 8. We assume that, just before the positive justification process is initiated, the starting-point address of the VC was recorded as 2 in the address (i.e., I/D-bit) field and the entire payload space was filled with the VC data [Figure 5.24(a)]. Once the positive

---

9. To be more accurate, it should be written, "when the accumulated offset data becomes one address location long." This length is equivalent to the size of H3, which is 3 bytes in the case of AU-4, and 1 byte for the others.

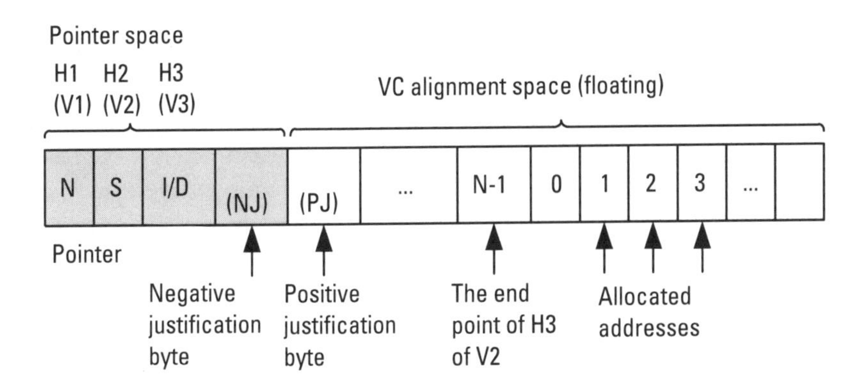

Figure 5.23 Conceptually reorganized AU/TU structure.

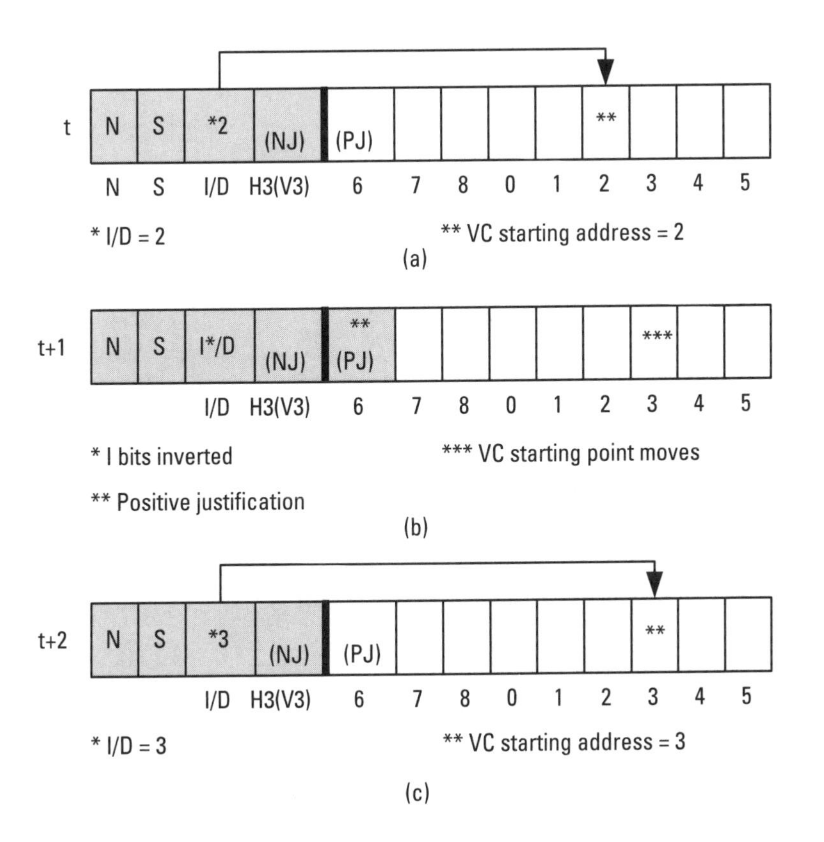

Figure 5.24 Illustration of positive justification: (a) before the justification, (b) during the justification, and (c) after the justification.

justification starts, all 5 I-bits in the I/D-bit field are inverted, the PJ execution byte is loaded with 1 null byte (or left as a blank space), and the VC data get loaded only onto the remaining VC alignment space. In that case, the starting-point address of the VC gets incremented by one [Figure 5.24(b)]. After the termination of the positive justification procedure, the new starting-point address, 3, is recorded onto the I/D-bit field, and the payload space gets filled with the effective VC data [Figure 5.24(c)].

The negative justification procedure is basically equivalent to that of positive justification except for the direction of execution. We examine its operation using Figure 5.25, assuming the identical initial status to the positive justification case [Figure 5.25(a)]. When the negative justification is executed, the five D-bits are inverted, and the payload data gets loaded to the NJ

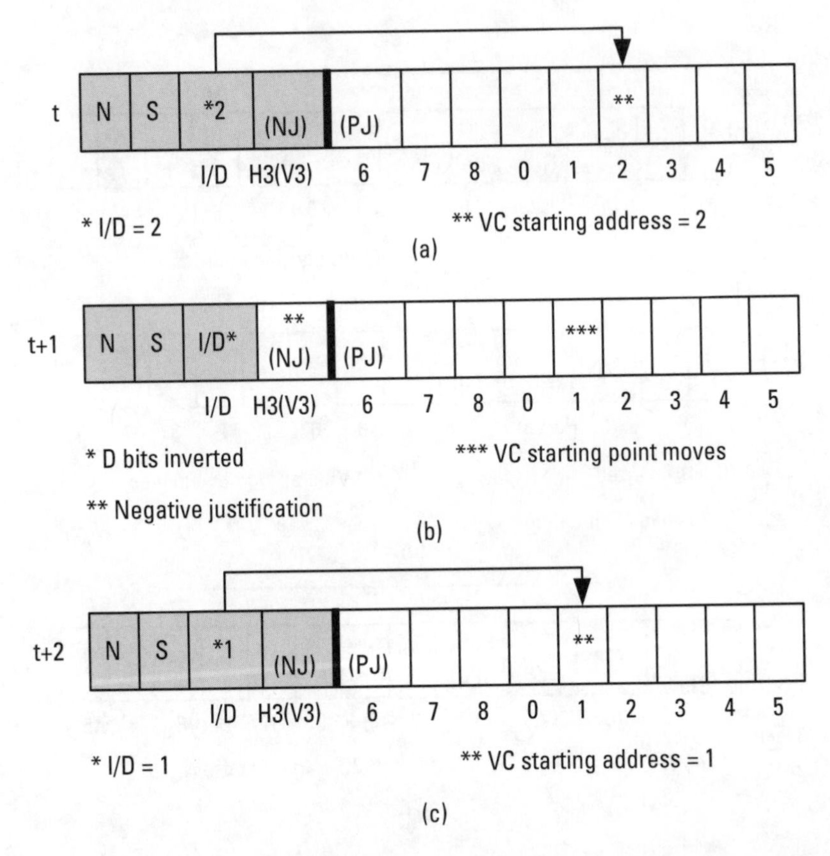

Figure 5.25 Illustration of negative justification: (a) before the justification, (b) during the justification, and (c) after the justification.

byte as well as the entire payload space. In that case, the starting-point address of the VC decreases by one [Figure 5.25(b)]. On termination of the negative justification, the new starting-point address, 1, is recorded onto the I/D-bit field, and the payload data is then filled in the payload space only [Figure 5.25(c)].

5.3.2 Pointer Processing Jitter

The pointer technique is one of the most distinct features of SDH/SONET transmission. The pointer-based synchronization method can achieve synchronization without repetitive frame search procedures, and it can also cope with a plesiochronous environment with small elastic store, making synchronization possible over wide area networking. However, because the pointer technique is linked with the 125-μs duration frame, it generates low-frequency high-amplitude jitters. This is because the justification ratio corresponds to 0.5, which is the worst case as far as the waiting time jitter is concerned, and the justification is performed on a byte basis.

To resolve the jitter problem, various jitter reduction techniques have been devised. Included among them are the direct reduction scheme, which utilizes narrowband *phase-locked loops* (PLLs) for desynchronization, and the bit-leaking control scheme, which splits the processing interval of byte size into bit size or a fraction of a bit.[10]

In fact, pointer-based synchronization mechanism is what enables the operation of SDH/SONET in plesiochronous networks. Jitter and wander are important performance indicators for the SDH/SONET network synchronization. The jitter and wander performance of the SDH/SONET system is determined by the internal and external clock performances of the transmission network, the output wander at the network interface, and the jitter and wander of the synchronous line system.

The tolerance range of jitter and wander in the SDH/SONET network is regulated for the G.702 interface and the STM-*n* signal, which is categorized into multiplexer-related jitter and line-system-related jitter. If optical transmission is employed, the jitter related to the transmission line system is of no concern. Thus, in the SDH/SONET network, waiting time jitter becomes crucial to resolve due to the pointer-based synchronization employed.

10. Refer to [13, 14], as well as ITU-T Recommendations G.823, G.824, G.743, G.752, G.783, and G.958 for the requirement on jitter and wander in the SDH network.

5.3.3 Network Synchronization and Timing

In the SDH/SONET system, the external synchronization signal is physically interfaced to the internal timing source through *multiplexer timing physical interface* (MTPI). The internal timing source, *multiplexer timing source* (MTS), provides a timing standard to each functional element in the SDH/SONET system.

There are several timing methods depending on the timing source—external timing, line timing, loop timing, through timing, and internal timing. *External timing* synchronizes the system clock to the externally received clock; *line timing* takes the system clock out of the received STM-*n* signal and uses it for transmission in both directions; *loop timing* uses the received clock for transmission in the backward direction; *through timing* uses the received clock for transmission in the forward direction; and *internal timing* takes the clock from the locally installed oscillator, using it for independent synchronization or for timing replacement in the case of system clock failure.

Clock references can be extracted from three different sources. The first is the G.702 external synchronization interface, the second is the G.702 tributary interface, and the third is the STM-*n* interface. The 2.048-kHz external timing signal is furnished by the G.702 synchronization interface; and the timing signal for line timing, loop timing, or through timing is taken from the signal received through the G.702 interface or the STM-*n* interface.

Synchronous transmission equipment exchanges synchronization state messages to achieve network synchronization in an effective and consistent manner. This message is delivered through bits 5~8 of the S1 byte in the multiplexer section overhead. The synchronization state message specifies how to choose a synchronization source for each section of the network and helps to make synchronization strategy compatible among all timing methods. It also helps to maintain the best possible synchronization quality by indicating the quality of each timing source, and further protects from clock looping, which could occur when each end system takes the timing source from the received signal.

5.4 Add/Drop and Cross-Connect

The functions of the SDH/SONET networks that correspond to switching and routing of the ATM and IP networks, respectively, are add/drop and cross-connect. Add/drop and cross-connect provide bulk data processing for the rearrangement of the constituent high-rate transmission signals, while switching and routing provide connection rearrangement among low-rate user signals. The ADM and *digital cross-connect system* (DCS) that perform the add/drop and cross-connect functions, respectively, deal with the STM-*n* signals for interfacing, and the PDH tributaries and ATM signals for internal processing. At the edge of the SDH/SONET networks that contain ADMs and DCSs, there is *line terminal equipment* (LTE) for terminating and accommodating the PDH tributaries and ATM signals.

5.4.1 Line Termination

The line terminal is a piece of basic SDH/SONET equipment that does multiplexing on various PDH tributaries to form the STM-*n* signal and its reverse function. It contains a multiplexer and demultiplexer pair that performs the synchronous multiplexing and demultiplexing processes respectively. There can exist various different types of line terminals depending on the PDH tributaries to accommodate, and the STM-*n* signals to produce. For efficient organization of the system, a module-based arrangement is desirable for constructing the line terminal as well as other SDH/SONET systems.

If we construct the line terminal according to the synchronous multiplexing procedures in Figure 5.13, we get the module configuration shown in Figure 5.26.[11] In the figure, the VC-1 and VC-2 modules respectively map the PDH tributaries DS-1 and DS-2 to the corresponding VCs, and then align them into the corresponding TUs, finally forming TUGs. Likewise, the VC-3/4 module maps the PDH tributary DS-3 or DS-4 to the corresponding VC, then align them into AU-3 or AU-4, finally forming AUGs. The STM-1, STM-4, STM-16, and STM-64 modules multiplex 1, 4, 16, and 64 AUGs, respectively, then affix the corresponding SOHs, finally transmitting the resulting STM-*n* signals at 155.520, 622.080, 2,488.320, and 9,953.280 Mbps, respectively. Among the modules, the PDH interface modules can be protected by board switching, and the VC-3/4 adaptation module and the STM-*n* module can be protected by 1:*n* or 1+1 manner (refer to Section 5.6.2).

Other modules included in Figure 5.26 are the *message communication function* (MCF) module which involves utilizing the *data communication*

11. This is the module configuration of the SDH line terminal equipment. The module configuration of the SONET case can be drawn in a similar form based on the synchronous multiplexing procedures shown in Figure 5.6(b).

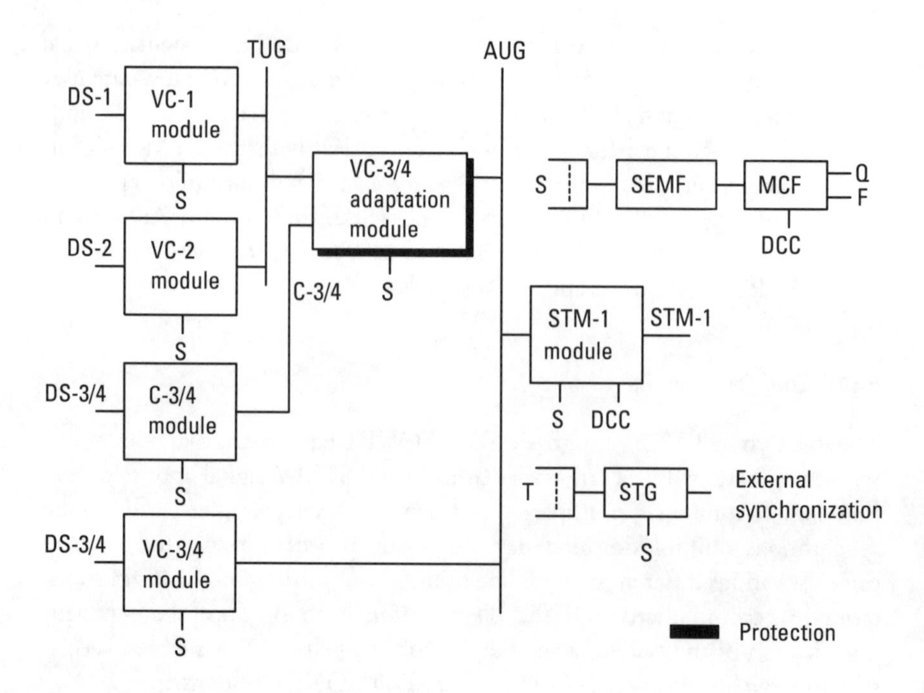

Figure 5.26 Module configuration of line terminal equipment.

channels (DCCs), and the *synchronous equipment management function* (SEMF) module which involves the OAM-related functions and SDH system management. In addition, the *synchronous timing generation* (STG) module provides timing-related functions such as interfacing external timing sources and generating timing signals for distribution to other modules.

Among the interface signals in the block diagram, T represents the timing signal, S the supervisory (i.e., monitoring, alarm, and control) signal, and Q and F are the message interfaces to MCF, with Q being the MCF access point to the TMN. (Refer to [15, 16] for functional blocks and related timing.)

5.4.2 ADM

Add/drop refers to the function of dropping a small number of lower-rate signals contained in a higher-rate input signal, then adding new input signals to rebuild an output signal that has the same constitution as the original higher-rate input signal. It is equivalent, in terms of functionality, to the pair of demultiplexing and multiplexing functions that are put back-to-back. In this respect, add/drop is a cost-effective way of passing through most of the lower-rate signals without applying further upper-layer processing as we did in PDH systems.

An ADM is the system that realizes the add/drop function. It is formed when two functional modules—TU and AU connection modules—are added to the line terminal above. That is, the TU exchange function is added to the VC-3/4 adaptation module, and the AU connection module is additionally attached. The AU and TU connection modules are to reconfigure the AU- and TU-level connections, respectively, to aid selected add/drop of the constituent signals. The resulting system configuration is as shown in Figure 5.27.

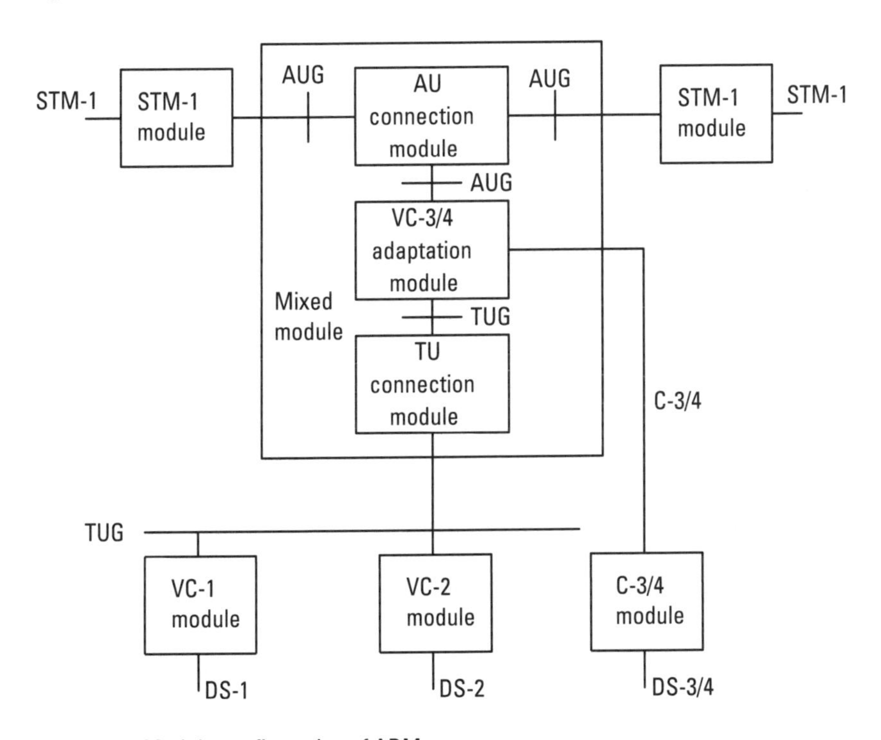

Figure 5.27 Module configuration of ADM.

In the ADM system in Figure 5.27, a part of the VCs in the original STM-1 signal are extracted while passing through the AU and TU connection modules and then get dropped as the corresponding PDH tributaries. During the foregoing process, some other PDH tributaries are inserted through a reversed process to fill the blank positions in the STM-1. Among the modules in the ADM system, the mixed module of the AU connection, VC-3/4 adaptation, and TU connection functions, as well as the STM-*n* module, can be protected by 1+1 protection switching.

We observe from Figure 5.27 that if there is no channel passing through during the process, an ADM becomes a simple multiplexer. In the case of STM-*n* ADM, it is also possible to arrange the functional blocks in such a way that the dropped/added signals are in the STM-1 unit. In this case the ADM is called *synchronous ADM.*

The ADM may be used in any node of a network, but it is very effective when used for engineering an *SDH/SONET ring.* The ring in this case forms an STM-*n* ring, and PDH tributaries or STM-1 signals can be dropped off and added on at each ADM node in the ring. Such an SDH/SONET ring forms a cost-effective, reliable transport network as will be discussed in Section 5.6.3.

5.4.3 DCS

Cross-connect is the function that reshuffles the constitution of the lower-rate upper-layer signals residing in the STM-*n* signals. An SDH/SONET DCS, or *broadband DCS* (BDCS), is the system that realizes the cross-connect function, in the lower-order path and higher-order path levels, among multiple STM-*n* (n = 4, 16, 64) signals. Some of those upper-layer signals can be accessed externally for test purpose, or can be added on or dropped off in the cross-connecting process.

Figure 5.28 shows the module configuration of the DCS. Apparently, it is similar to that of the ADM in module configuration. However, the DCS distinguishes itself in that the switching operation in the AU/TU connection modules is the major function of the DCS, which, in the case of the ADM, is merely an auxiliary function that helps to select the desired higher-order signals to drop off.

The broadband DCS is usually located at a highly traffic-concentrating point of a synchronous mesh network, so its network management capability is very important. For this reason, a standard network management capability is normally installed in the DCS, along with the performance and failure monitoring capability for the tandem connection links, and the automatic failure recovery capability for the mesh networks and ring networks.

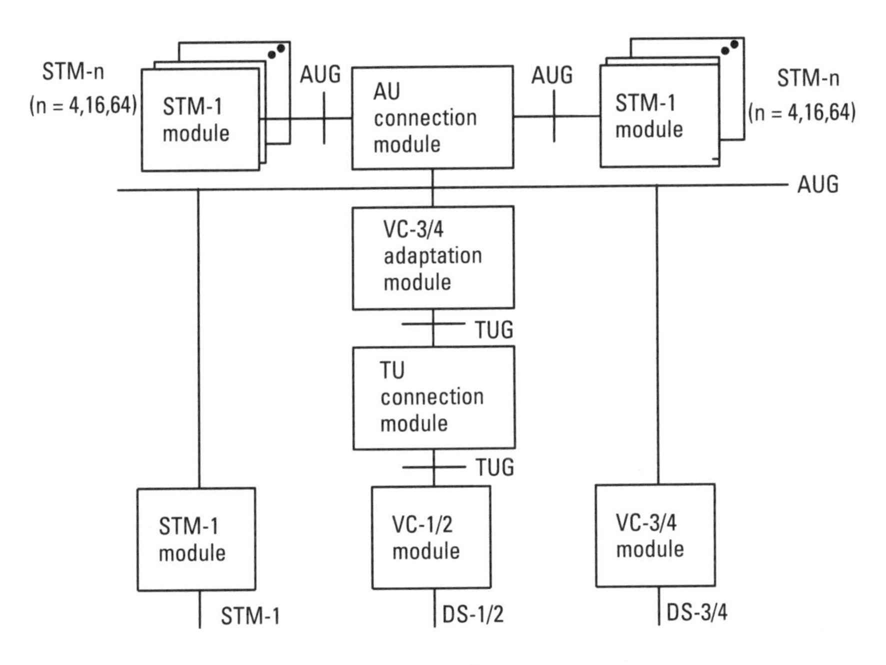

Figure 5.28 Module configuration of broadband cross-connect system.

The TU/AU connection module in the DCS enables, through VC-*n* (*n* = 1, 2, 3, 4) level switching, to reconfigure dynamically the connections for distribution, add/drop, path switching, test access, and broadcasting. This capability helps to make the *digital system cross-connect* (DSX) functions of the *main distribution frame* (MDF) more automatic and electronically processed. It also helps to improve the transmission service quality by consolidating various types of transmission facilities into a single DCS.

Figure 5.29 shows an example of the SDH/SONET network that contains synchronous multiplexes (S-MUXs, or equivalently, LTEs), ADMs, DCSs, and other network devices. The network accommodates both PDH tributaries and ATM cell streams, which are respectively indicated by solid lines (PDH/STM) and dashed lines (ATM/STM). In Figure 5.29, *circuit switch* (C-SW) involves the PDH signals, while *broadband network termination* (B-NT) and *ATM switch* (A-SW) involve the ATM streams.

The S-MUXs are installed at all user interfaces where those PDH tributaries and ATM cell streams meet the SDH/SONET network. ADMs can be installed at any node in the SDH/SONET network but are especially useful when put in the ring architecture. ADM rings well match the ring-based protection schemes as will be discussed in Section 5.6.3. In the case of the DCSs,

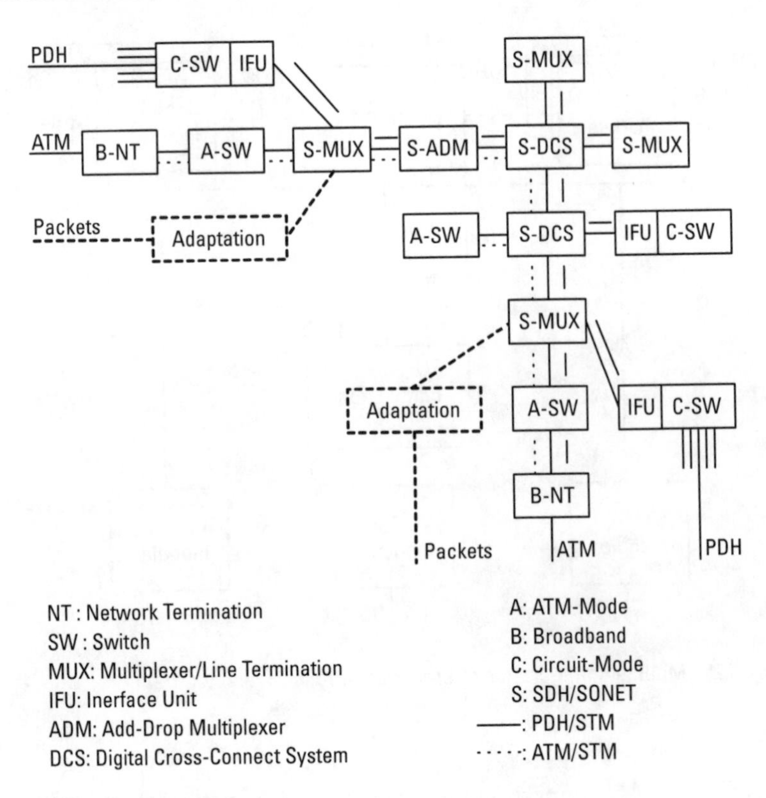

Figure 5.29 Illustration of signal flow in SDH/SONET network.

it is more appropriate to install them at the nodes in the mesh networks. Traffic grooming can be effectively performed in those DCS-equipped nodes.

The notion of mapping ATM cells into the STM-*n* signal to form the ATM/STM signal can be equally extended to map packets into the STM-*n* signal. The resulting "packet/STM" signal is typically matching with an IP over SDH/SONET network arrangement. The "packet/STM" signal thus obtained is also compatible with the PDH/STM signal in the transmission layer, and therefore PDH/STM and ATM/STM signals may coexist in the SDH/SONET network.

5.5 Network Management

As stated in Chapter 2, network management in general is composed of five components, namely fault management, accounting management, configuration management, performance management, and security management.

From SDH/SONET's point of view, the most important of these are fault management and performance management, which are the key elements of the OAM function.[12]

A comprehensive OAM function is a distinguished feature of the SDH/SONET networks. OAM contains such functions as detection, confirmation, location, isolation, and recovery of failures. Within the SDH/SONET frame, a significantly large bandwidth is allocated to section and path overheads to carry the OAM-related information within the SDH/SONET network, or in connection to external dedicated networks for network operations and management. The SOH and POH are arranged in layered architecture, thereby supporting systematic use of the OAM channels.

Section 5.5.1 concentrates on the OAM functions, discussing the OAM signals and their applications, and defers the discussions of the network survivability aspect to Section 5.6.

5.5.1 OAM

In the SDH/SONET network, network operation is monitored and network failure is detected all the time. There are various different OAM signals that indicate the status of operation and failure. The OAM signals are carried over the section and path overheads to the far end and network operations and management centers for due countermeasures.

5.5.1.1 Failures and OAM Signals

Failures are divided into *hard failure* and *soft failure;* hard failure causes service failure due to malfunctioning of the hardware and/or software of the system, and soft failure causes performance degradation. In general, a failure is regarded as a hard failure if it yields a BER of 10^{-3} or higher; it is regarded as a soft failure otherwise. In case a hard failure occurs, an alarm signal is generated and protection switching is triggered, whereas in the case of a soft failure, performance is monitored more cautiously with optional protection switching (refer to Section 5.6 for discussions on protection switching).

Failures in the section or path are detected by checking the state of the received signal. Failures in the SDH/SONET network include transmitter

12. Refer to [17–20] as well as ANSI Standards T1.119.01, T1.204, T1. 208, T1.214, T1.215, and T1.229 for detailed descriptions of network management in general and OAM functions.

failure, receiver failure, *loss of signal* (LOS), *loss of frame* (LOF), *loss of pointer* (LOP), *signal label mismatch* (SLM), signal *unequipped* (UNEQ), *trace identification mismatch* (TIM), and failure of protection switching. LOS is declared when the physical layer signal is lost for 100 μs or longer; LOF is declared when a severely errored frame persists 3 ms or longer; and LOP is declared when the pointer failure occurs.

The OAM signals indicating the state of failure include an *alarm indication signal* (AIS), which notifies of failure in the forward direction; and RDI, which notifies of failure in the backward direction. In addition, there is REI, which reports the error monitoring results to the far-end, and RFI, which indicates the receipt of AIS to upstream network elements of the same peer layer (refer to [21]).

5.5.1.2 OAM Channels

OAM channels in the SDH/SONET network include the performance monitoring channels, maintenance signals, orderwire channels, user channels, and data communication channels. Performance monitoring channels carry bit-interleaved parity, LOS, LOF, LOP, and so on. Maintenance signals include AIS, RDI, and RFI. Orderwire channels are for voice communications connecting section or path terminations; user channels are for network operators' use; and data communication channels are for regenerator-to-regenerator, or multiplexer-to-multiplexer data communication.

The OAM channels for synchronous transmission exist in the POH and SOH. OAM channels in POH include those channels indicating the AIS, LOP, RDI, REI, RFI, and error checking of the path. In the SOH there are also OAM channels for indicating the AIS, LOS, LOF, RDI, REI, RFI, and error checking of the section. In addition, SOH supports the *network operator maintenance channel* (NOMC), which includes orderwire, the user channel, and DCC. Also, for management purposes, TMN can be additionally employed.

In terms of POH and SOH bytes, AIS is generated by setting some particular bits (such as bits in K2 in STM-*n*, H1 and H2 in AU, and V1 and V2 in TU) to the "1" state or by replacing all lost frames with "all 1" data. RDI message is delivered through some particular bits (such as K2 in STM-*n*, G1 in VC-3/4, V5 in VC-1/2) in the normal frame traveling backward to the remote end. LOP is relevant to the pointer bytes H1 and H2, and LOF to the frame alignment bytes A1 and A2. Error monitoring is done by counting the errors occurring in the B1, B2, and B3 bytes, and the monitoring result is sent in the form of REI through the M1 or G1 byte (refer to Tables 5.3 and

5.4). Orderwire bytes, E1 and E2, are reserved for voice communications; user channels F1–F3 are reserved for the network operator's use; and DCCs D1–D3 (regenerator section) and D4–D12 (multiplex section) are used to carry OAM information.

5.5.2 Application of OAM Signals

There are two different ways to monitor performances and alarms in the synchronous transmission network: in-service-monitoring and out-of-service monitoring. The former is used when monitoring performance and alarms while providing services continuously and the latter is applied when locating failures through loopback or test access.

5.5.2.1 Propagation of OAM Signals

In the synchronous network, maintenance is done systematically in a layered approach, based on regenerator section, multiplexer section, higher-order path and lower-order path, as illustrated in Figure 5.30. If a failure occurs in the regenerator section and, consequently, the eastward signal in the STM-*n* gets lost, then the *regenerator section termination* (RST) detects this failure and sends out the "all 1" signal (or MS-AIS) in the forward direction. Detecting this MS-AIS, the *multiplexer section termination* (MST) in the east issues the AIS in the forward direction by replacing the AU-3/4 signal in the failed STM-*n* with "all 1" data and, at the same time, sends out MS-RDI backward to the MST in the west. The higher-order and the lower-order *path termination equipment* (PTE) in the east detects the "all 1" AIS, and then transmits the RDI to the PTE in the west. As such, AIS enables the detection of the upward link failure and RDI the downward link failure in the remote end in a systematic manner. Once a failure is detected, each piece of equipment delivers alarms to the relevant SEMF and the recovery procedure is activated.

5.5.2.2 Detection and Generation of OAM Signals

The detection of OAM signals and the corresponding countermeasures are shown in Figure 5.31. In the figure, the relevant actions are categorized into the section layer, the higher-order path layer, and the lower-order path layer, with the generation and the detection of the signals indicated by the filled circles and the countermeasures by blank circles. In the section layer, in the event of the detection of LOF, LOS, section AIS in the received signal, or excessive bit error (B2), the far-end is notified of the condition through RDI

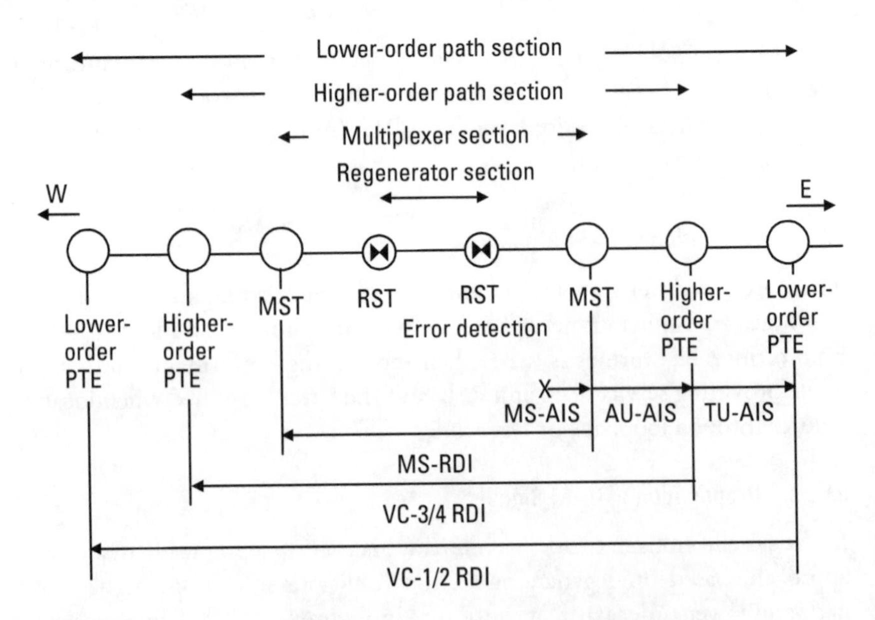

Figure 5.30 Propagation of OAM signals in SDH/SONET network.

and a VC-3/4 path AIS is sent to the higher-order path. In the higher-order path, if higher-order path AIS or higher-order path LOP is detected, it is reported to the far-end through higher-order path RDI, and lower-order path AIS is sent to the lower-order path. Also, if any bit error (B3) is detected, it is confirmed and notified to the far-end through REI. In the lower-order path layer, the far-end is notified if the lower-order path AIS or the lower-order path LOP is detected, and AIS is sent to the associated containers. Also, any bit error is confirmed and reported to the far-end through REI in V5.

If a hard failure is detected in the line system, protection switching is triggered. Here, the line system designates the multiplex section that starts at a multiplexer termination, goes through the regenerator, and ends at the far-end multiplexer. Protection switching can be performed in the 1+1, 1:1, or 1:*n* format (refer to Section 5.6.2), and the employment of the protection action can be unidirectional or bidirectional and may also be revertive or nonrevertive. (Refer to Section 5.6 for discussions on protection switching.)

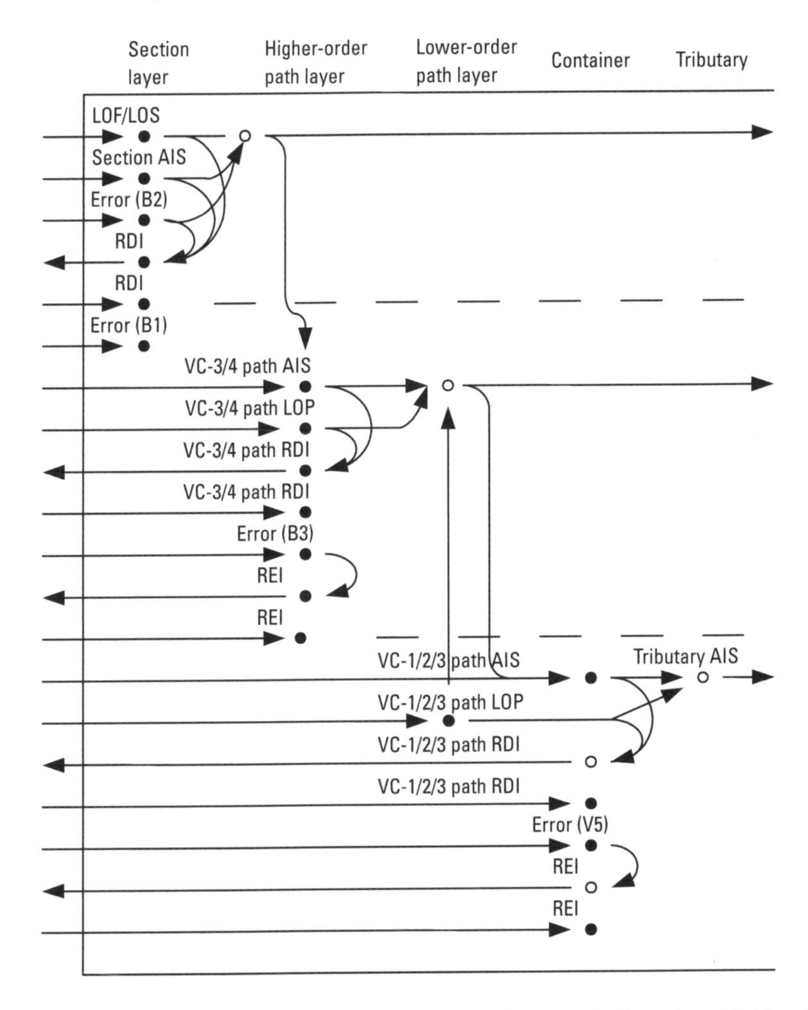

Figure 5.31 Detection and generation of OAM signals (refer to Tables 5.3 and 5.4 for the related overheads).

5.5.3 Synchronous Management Network

The SDH/SONET network is equipped with overhead channels for various maintenance functions and data communication channels for centralized network management. The 192-Kbps channel formed by channels D1~D3 of the RSOH and the 576-Kbps channel formed by D4~D12 of the MSOH provide the DCCs for transmission of OAM information. Such DCCs render the physical link to operate the *embedded control channel* (ECC), which enables the construction of an SMN among *network elements* (NEs). The SEMF represents the function of accepting the OAM signals through regenerators or multiplexers to be sent to the MCF, or its reverse. The MCF receives and transmits messages through the DCC, or delivers it to the TMN, and it can also function in the reverse direction (refer to [22]).

The SMN is a network that manages NEs in the synchronous transmission network, and it may be regarded as part of the TMN. The SMN is composed of management application function, network element function, and MCF that contributes to the generation, termination, collection, and transfer of the TMN messages. The MCF is carried out by operating the ECC through the DCC. The operation of the ECC provides access to the DCN and the *local communication network* (LCN), which are required in the TMN. Consequently, the LCN of the TMN can be easily accessed if synchronous transmission equipment is installed in every central office and, further, the DCN can be constructed through the SMN if the *operating system* (OS) for the network management is deployed at some central offices.

Therefore, the network management function of the SDH/SONET network may be considered in the context of the TMN access function. The most important objective of network management through the TMN is the survivability of network services. For network survivability, or for an uninterrupted service provision, a real-time management capability is necessary that can react to network environment changes actively. For example, it is important to have the capability of identifying the degraded network elements and setting up detouring paths. In addition, it needs to be capable of evaluating the performance quality of a network element by measuring the failure occurrences and then taking appropriate reactive actions, and to be able to distribute properly the network facility in response to service traffic changes.

In the PDH network environment, the network management system had to be newly defined whenever a new system was developed, and the OAM information gathered from different network equipment had to be transformed to the data formats adequate for an integrated network management. To improve such inefficiency and equipment dependency of network management, an integrated management information system is necessary. The SMN meets this necessity in the SDH/SONET network environment.

5.6 Network Survivability

The synchronous transmission network is a large-capacity transmission system with transmission rates ranging from several hundred megabits per second to tens of gigabits per second, so any failure in transmission line or equipment can lead to catastrophic service failures. In this aspect, it is important to arrange the network so that services can be provided without interruption even when failure occurs in the network. Such network survivability can be considered at different layer networks, namely, the ATM network level (i.e., virtual channel and virtual path layers), the SDH/SONET network level (i.e., transmission path and section layers), or the WDM/optical level (i.e., physical medium layer). In this section we consider the network survivability at the SDH/SONET network level.

Network survivability in the SDH/SONET network can be further enhanced because ring-type or mesh-type architecture is also available in place of the conventional star-type architecture. In general, network survivability is acquired in different ways depending on the network architecture. In the case of the point-to-point architecture APS can be applied, but in the case of the ring networks consisting of ADM a *self-healing ring* (SHR) arrangement can be employed. In addition, in the case of the mesh network including broadband DCS, detouring paths can be established within the network. (Refer to [23–28] for more detailed discussions on network survivability.)

5.6.1 Protection and Restoration

There are various different ways of protection and restoration schemes that may be applied to secure network survivability. Here *protection* refers to the arrangement that switches a working channel in operation to some protection channel in standby when the working channel fails, whereas *restoration* refers to the arrangement to utilize spare channel capacity to establish detouring paths when failure occurs in the operating network. In protection, network survivability totally relies on the redundant capacity in the protection line, while, in restoration it relies on the spare capacity in the alternative path that is determined by the reconfigurable cross-connects residing in the nodes.

5.6.1.1 Point-to-Point, Ring, and Mesh Networks

In the point-to-point network, protection channels are deployed separately from the operation channels. When an operation channel fails it is automatically switched to the protection channel. Since the span between two adjacent nodes is switched to each other, the APS in this case is also referred to as *span switching*. The protection channels for APS may lie in parallel with the operation channels along the same line, or may be separately placed to take a detouring path as shown in Figure 5.32(a).

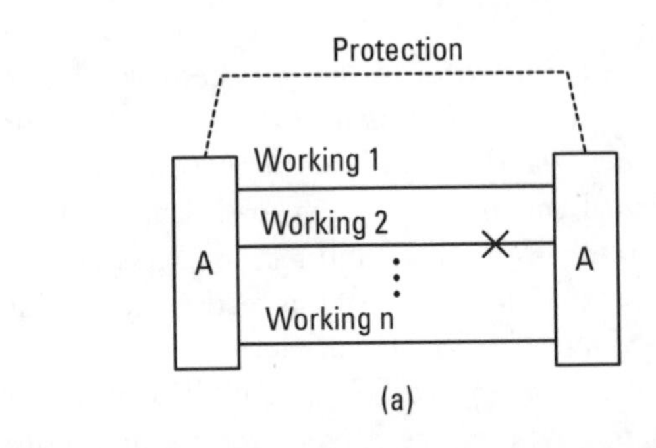

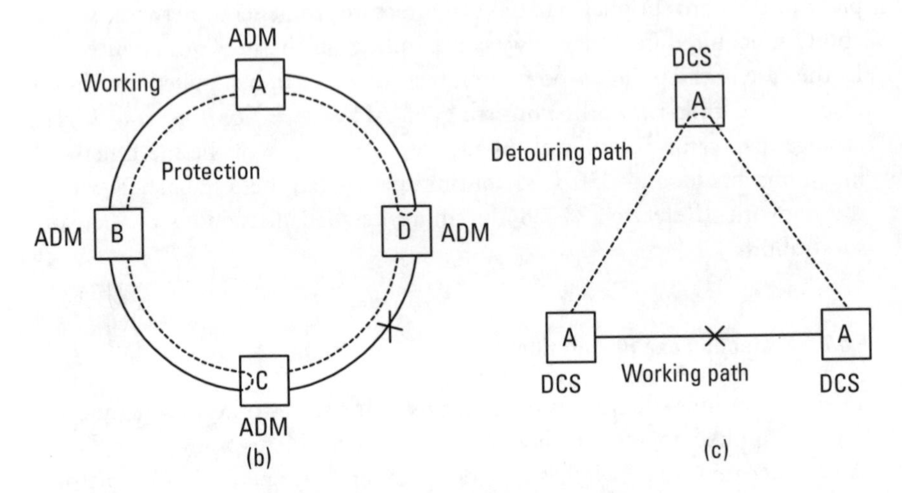

Figure 5.32 Survivability of point-to-point, ring, and mesh networks: (a) automatic protection switching point-to-point network, (b) self-healing ring network, and (c) path-detouring mesh network.

In the case of the ring network containing ADMs, a protection ring is deployed in addition to the operation ring, as shown in Figure 5.32(b). If failure occurs in a span within the operation ring, the neighboring nodes take APS procedures to activate the protection ring. If a ring network has the capability to "heal" the failure itself in this manner, it is called an SHR. The operation channel in the SHR is protected 1:1 or 1+1, and thus has a fast failure-recovery capability as in the case of the APS-based point-to-point network.

In the case of the mesh network containing DCSs, in contrast to the APS-based point-to-point network or the SHR network, protection channels are not arranged separately. Instead, once a link is failed, it is restored by utilizing the spare capacity in a detouring path in the mesh network. For example, if the link connecting nodes A and B fails in the mesh network in Figure 5.32(c), then the DCSs in nodes A, B, and C are activated in such a way that a detouring path can be established via node C using the spare capacity in the pertaining links.

5.6.1.2 Dedicated Protection and Shared Protection

For protection in point-to-point and ring networks, either dedicated protection or shared protection may be applied. *Dedicated protection* refers to the protection scheme that assigns a dedicated channel to protect total traffic capacity in the operation channel, whereas *shared protection* refers to the scheme to make multiple operation channels share one or multiple protection channels. The *1+1 protection* and the *1:n protection* for the point-to-point network are the typical examples of the dedicated and shared protection schemes, respectively.

On the other side, protection switching can be operated in the *revertive* switching mode in which the operation channel returns to the operational stage after failure recovery or in the *nonrevertive* switching mode in which the protection channel stays in operation even after failure recovery. Dedicated protection may be operated in either mode, but shared protection must be done in the revertive switching mode.

5.6.1.3 Operation of APS

In the case of dedicated protection, traffic is transmitted over both the working and the protection lines, among which the receiver selects the line that carries a better quality of traffic. As a result, the fastest possible protection is done in this case, as the receiving node can make the decision for itself without communicating with the transmitting node. In the case of the shared protection, however, the protection line does not carry a duplicate traffic in normal state, so if a failure occurs the transmitter and the receiver both have to switch connection from the working line to the protection line simultaneously. However, the failure is detected only by the receiver in this case, so the failure information must to be sent to the transmitter over the APS data channel. Upon receiving the APS data, the transmitter can then switch the failed line to the protection line.

For the multiplexer section, line protection is done by the *multiplexer section protection* (MSP) functional block, and the APS data are carried over the K1 and K2 bytes in the protection channel. For path protection, the APS data is also carried over the K3 byte (in the case of the higher-order path) and the K4 byte (in the case of the lower-order path).

5.6.2 Point-to-Point Link Protection

In the point-to-point links, protection is normally done through 1+1, 1:1, or 1:*n* protection switching. In the case of *1+1 protection*, or dedicated protection, the transmitter loads the traffic data on both the working and the protection lines simultaneously. The receiver listens to both lines in normal operation, taking the data from the working fiber, but switches to the protection line on detecting any failure. This protection switching takes place very fast as the receiver can make decision on its own without contacting any other nodes. As the price, however, it requires two lines for the transmission of one-line capacity. Consequently, the channel efficiency is low in the case of 1+1 protection. Such 1+1 protection switching is normally done in nonrevertive manner.

In the case of *1:1 protection*, the traffic data is loaded only on one of the two lines, with the protection line staying in standby. If failure occurs in the line or the node equipment, both the transmitter and receiver switch to the protection line simultaneously. As the failure is detected by the receiver, the failure information is sent to the transmitter over the APS protocol. Due to this transmitter-receiver communication, 1:1 protection cannot operate as fast as 1+1 protection. However, 1:1 protection has the advantage that the protection line, in normal operation, is free to use for carrying low-priority traffic data, as long as it can be interrupted whenever protection switching takes place.

In the case of *1:n protection*, or shared protection, one protection line is shared among *n* working lines. This increases the channel utilization significantly. If failure occurs in a line or a piece of node equipment, the switches at both ends of the line simultaneously switch to the protection line. In support of this, the transmitter and receiver communicate through the APS protocol, as for the 1:1 protection case. Such 1:*n* protection switching is done in revertive manner. Figure 5.33 illustrates those three protection mechanisms.

On the other hand, the protection switching in the point-to-point links can also be divided into *line protection* and *path protection* depending on the location of adaptation points to protect. In the case of the line protection, the line (or the multiplex section) is protected, whereas the path (i.e., the higher-order path) is protected in the case of the path protection.

5.6.3 SHR Network

A ring network is inherently resilient to failures because it is two-connected in its most simple form of structure. If protection lines are deployed in addition, the survivability increases substantially. Much of the SDH/SONET

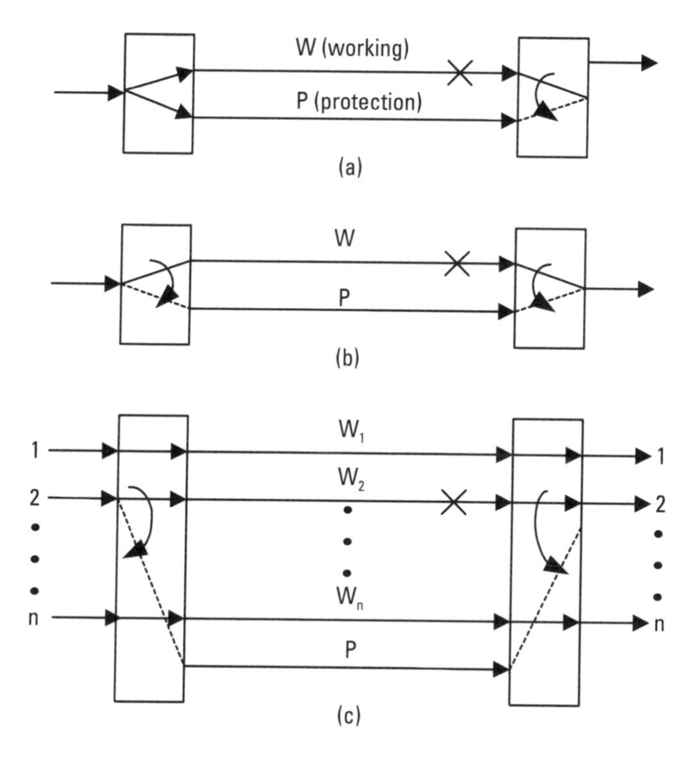

Figure 5.33 Protection switching mechanisms in the point-to-point links: (a) 1+1 protection, (b) 1:1 protection, and (c) 1: *n* protection.

networks today deploy ring architecture. They are called *self-healing* if they are equipped with the protection capability that detects failures and reroutes traffic promptly, resuming the communication rapidly. That is, SHR is a network that connects the constituent nodes in physical ring topology with bandwidth sharing and a self-healing capability. In general, an ADM is installed in each node of the self-healing ring to support the self-healing capability.

In SDH/SONET SHRs, service must be restored within 60 ms after a failure. This includes the failure detection time, the hold-off time, and the propagation delay in the system. This time duration is set by the requirement that the lower-speed streams such as DS-1 and DS-3 that are multiplexed into the STM-*n*/OC-*m* streams must not lose frame synchronization at their receivers while protection switching takes place.

The types of SHR differ depending on the path traveled by a duplex communication channel between each node pairs. Depending on the direction of the travel paths, SHR architecture is divided into *unidirectional SHR* (USHR) and *bidirectional SHR* (BSHR). In the case of the USHR, each duplex channel connecting a node pair travels unidirectionally over different paths, whereas each duplex channel travels bidirectionally over the same path in the case of BSHR. BSHR is subdivided into *four-fiber BSHR* (BSHR/4) and *two-fiber BSHR* (BSHR/2) depending on the number of fibers in the ring. Figure 5.34 illustrates the resulting three different types of SHRs. The figure shows that the USHR in (a) has the duplex channel connecting node A and node C, for example, running unidirectionally over different paths, one via node B and the other via node D. In contrast, the duplex channel connecting node A and node C in the BSHR/4 or BSHR/2, in (b) or (c), runs in the same path, bidirectionally, via node B (or, equivalently, via node D).

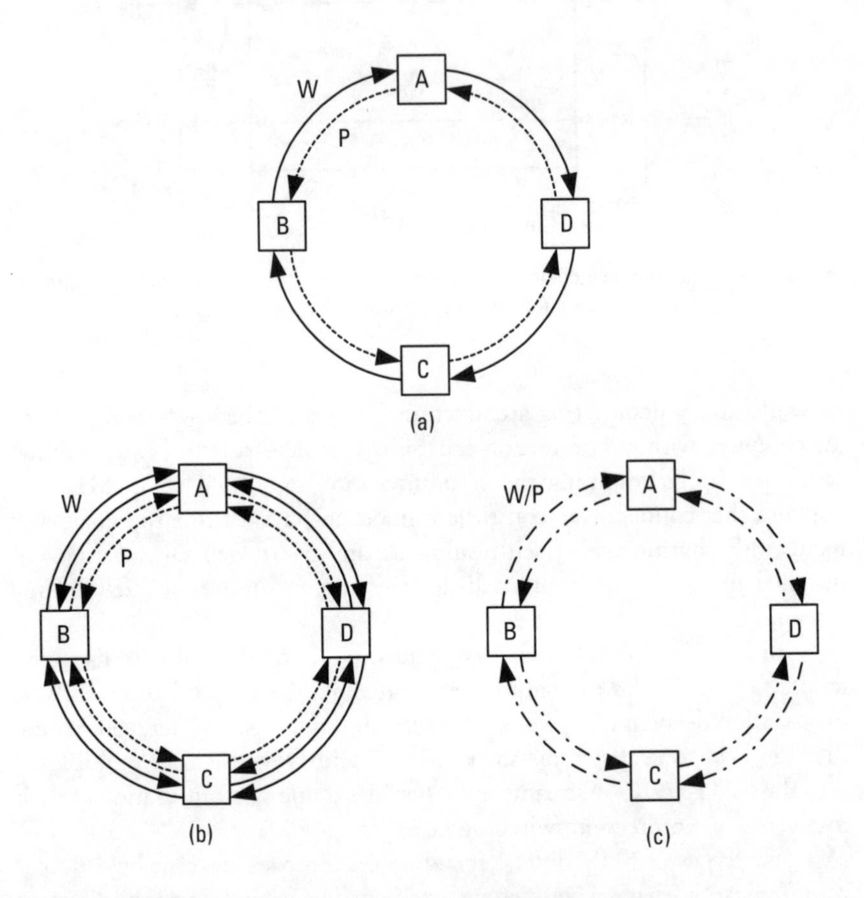

Figure 5.34 Self-healing rings: (a) USHR, (b) BSHR/4, and (c) BSHR/2.

There are two possible SDH/SONET self-healing control schemes applicable to SHRs—*section* (or *line*) *protection* switching and *path protection* switching. The section/line protection switching scheme uses SDH/SONET section/line overhead for protection switching and restores the desired section/line from the failed facility. In contrast, the path protection switching scheme uses path overhead and restores the desired VC path from the failed facility. A USHR, which has only one working ring, usually relies on path protection switching, whereas a BSHR uses both path protection switching and section/line protection switching. For this reason, a USHR is also called a *unidirectional path-switched ring* (UPSR), and a BSHR is called a *bidirectional line-switched ring* (BLSR).

5.6.3.1 USHR

A USHR is equipped with two fiber rings—one for working and the other for protection—and one ADM in each node. The duplex channel formed by the two-fiber rings travel in the opposite direction. The USHR architecture, with path protection, is based on the concept of signal dual-feed, whose operation is similar to that of 1+1 protection in point-to-point links. So the maximum capacity of a USHR is the same as a single ring.

In normal operation, an identical STM-*n*/OC-*m* signal is transmitted in both the clockwise and the counterclockwise directions on the two rings. This pair of signals, one primary and one secondary, are both monitored at the destination node, and the primary signal is selected for service. The primary signal is supposed to be on the working ring but could be on the protection ring as well when the USHR is operating nonrevertive. If a failure occurs, due to fiber-cut or a hardware failure, the destination node switches the connection to the secondary input, thereby resuming normal operation immediately. This protection switching happens very quickly as the destination node keeps monitoring both incoming signals and takes action on detecting a failure. Figure 5.35 illustrates the protection operation of the USHR.

5.6.3.2 BSHR/4

A BSHR/4 operates over four-fiber rings and a protected ADM in each node. The four fiber rings form two sets of opposite-directed fiber rings—one set for working channels and the other set for protection channels. Protection capability of the BSHR/4 is achieved by using APS for a loop-back function.

In normal operation, two working sets of fiber rings carry the traffic in the opposite directions. The working traffic in the BSHR/4 travels

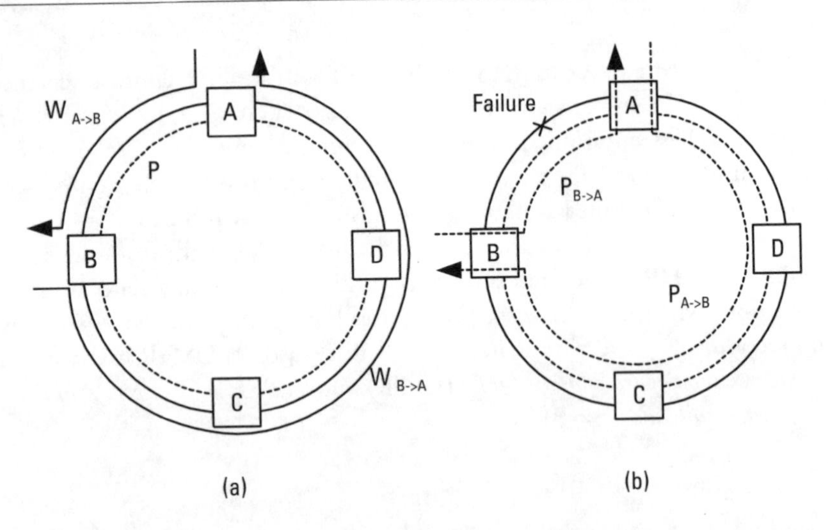

Figure 5.35 Operation of USHR: (a) normal, and (b) protection.

bidirectionally on one pair of working fiber rings, with the other pair serving as backup. In the case of failure, the traffic is intercepted in the next node and rerouted back to the destination on the protection fiber ring. The BSHR/4 architecture requires a protection ADM for each working ADM and a 1:1 nonrevertive lower-speed electronic protection switch at each node. It is also possible to implement it so that regenerators are taken as the protection components instead of the duplicated ADMs. In this case the control mechanism becomes relatively complex as the regenerators have no intelligence on determining whether to drop off the signals. Figure 5.36 illustrates the protection operation of the BSHR/4.

5.6.3.3 BSHR/2

A BLSR/2 is equipped with two fiber rings and an ADM at each office. The two fiber rings are oppositely directed, with half the total capacity in each ring allotted for the operation channels and the other half for the protection channels. When a failure occurs in a ring, the operation channel is switched to the protection channel in the other ring.

In normal operation, traffic is evenly split into the inner ring and the outer ring, by filling even or odd numbers of time slots or by filling the first or second half of time slots. In case a failure occurs in the operation fiber ring or node equipment, the traffic is switched into the vacant time slots in the opposite direction. It is possible, for a simple implementation, to load traffic to a half of the STM-*n*/OC-*m* carried by the operation ring in normal operation,

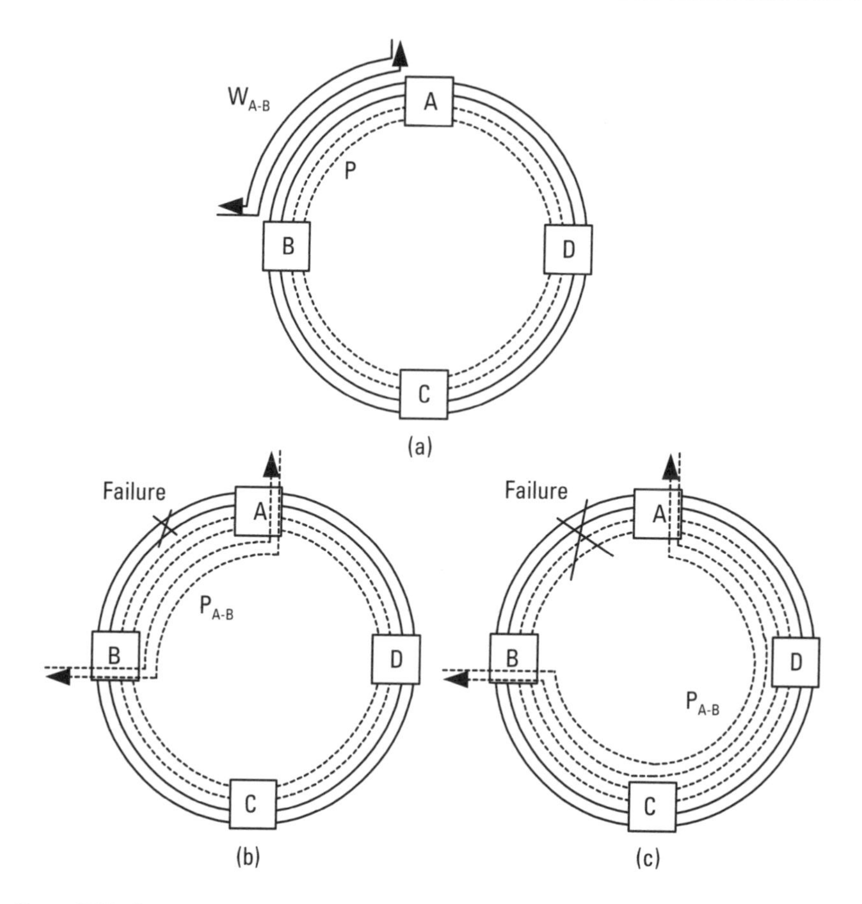

Figure 5.36 Operation of BSHR/4: (a) normal, (b) span protection, and (c) line protection.

switching it to the other vacant half in the opposite-directed ring. Consequently, the loading of traffic must be kept below 50% of the line capacity in normal operation, and the channel efficiency becomes less than a half.

It is also possible to operate BSHR/2 in shared mode, with all but one constituent STM-*n*/OC-*m* allocated to carry traffic in normal operation. In this case, a dual-ended switching scheme is applied along with the ring APS protocol over the APS channel. Channel efficiency increases above a half but the operation becomes complicated. Figure 5.37 depicts the protection operation of the BSHR/2 in dedicated form.

5.6.3.4 Comparisons

In the networks where the traffic dominantly converges at one node along the ring, USHR will serve best. A typical example of this case is the local exchange

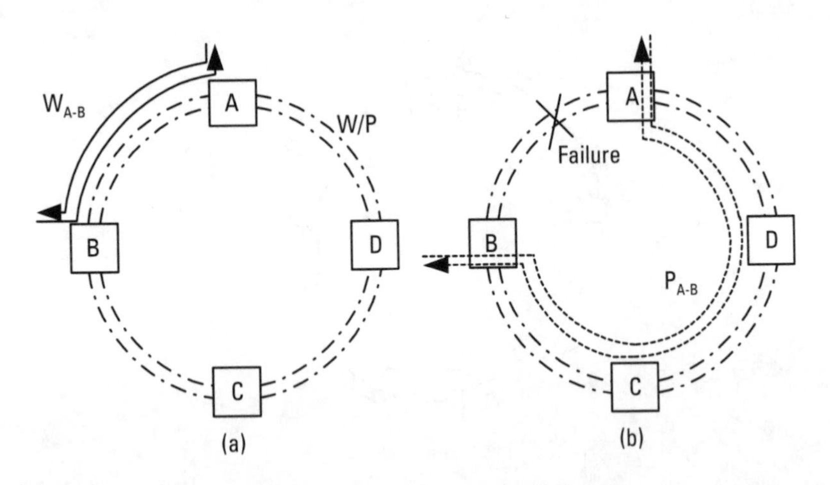

Figure 5.37 Operation of BHSR/2: (a) normal, and (b) line protection.

carrier access networks. In contrast, in the networks where the traffic is more evenly distributed, for example, in the case of the local carrier interoffice networks or interexchange carrier networks, BSHR can be more diversely utilized.

The protection/restoration time is shorter for USHR than BSHR, even if both may meet the 60-ms requirement. USHR does not require the processing/communication time for the APS action, which is mandatory for BSHR. The switching time of USHR is not affected by the ring size or the number of ADMs in the ring.

If USHR and BSHR/4 are compared, a BSHR/4 has higher capacity than a USHR. However, a BSHR/4 requires twice as much equipment and facilities, such as fiber and ADMs, than a USHR does. If USHR and BSHR/2 are compared, the situation is similar. That is, the equipment and facility may be twice as high for a BSHR/2 as for a USHR.

Between BSHR/4 and BSHR/2, a BSHR/4 can handle more failures, and consequently performs more robustly than a BSHR/2, as a BSHR/4 can simultaneously handle one transmitter failure on each span in the ring. Also a BSHR/4 is easier to service than a BSHR/2 since it can service multiple spans independently. Instead, ring management is more complicated in a BSHR/4 than in a BSHR/2 because it requires coordination among multiple protection mechanisms.

5.6.4 Mesh Network Restoration

In the mesh network, most nodes are connected to multiple other nodes, so there are multiple transmission paths between each pair of nodes. It is more complicated, in general, to coordinate the protection switching control among the multiple nodes in the mesh network than in the point-to-point link or ring networks. As a result, the protection switching time could become longer in the mesh network than in other networks unless the network is configured in a special form.[13] Moreover, the dedicated spare capacity secured in every connection in the mesh network for protection switching degrades channel efficiency significantly.

Therefore in the mesh network, in general, restoration is the preferred choice of achieving network survivability over protection; it is more efficient, or more economical, to restore failures by utilizing the spare capacity than to secure dedicated protection channels or fibers. As the price, however, it requires protocols for coordinating the failure restoration among the related nodes such that optimal detour paths can be configured using the spare capacity. As a consequence, the restoration processing becomes complicated and the restoration time becomes comparatively long.

Failures in the mesh network can be restored either in the form of *line restoration* or in *path restoration*, which are illustrated in Figure 5.38. In Figure 5.38, the sender S and the chooser C indicate nodes adjacent to the failed links or equipment that initiate collaborative works among the relevant nodes to establish detouring paths. In the case of line restoration, detouring paths are established with respect to the two adjacent nodes where the failed line terminates, by utilizing the spare channels in all the links that connect the two nodes. The dashed lines in Figure 5.38(a) indicate the paths where the spare channels are available. In contrast, in the case of path restoration,

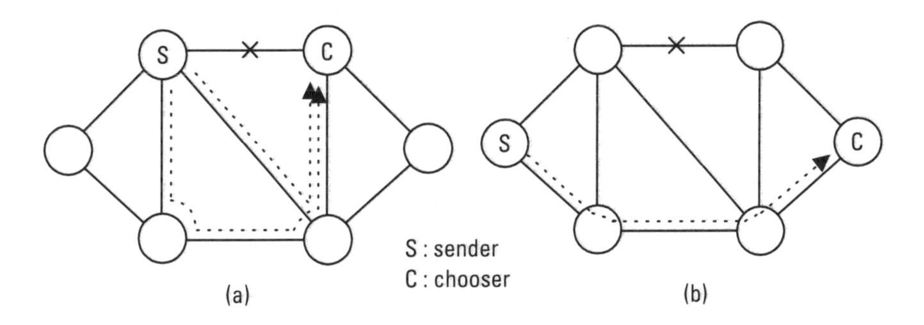

Figure 5.38 Restoration in mesh network: (a) line restoration, and (b) path restoration.

13. A network configuration that enables fast protection switching in the meshed WDM network is discussed in Section 6.5.3.

detouring paths are established with respect to the virtual channels carried over the failed line, independently of the failed line itself. In Figure 5.38(b), S and C are respectively the sender and the chooser for the restoration of the virtual channels carried over the failed line.

There are two ways to control the mesh network restoration—*centralized control* and *distributed control*. In centralized control, the operating system that manages the overall network elements directly controls the network elements to configure detour paths for restoration. In contrast, in distributed control, the nodes that are adjacent to the failed links or equipment (i.e., the sender and the chooser) initiate collaborative works among the relevant nodes. For this, the sender that detects the failure notifies adjacent nodes of the failure state, and the adjacent nodes do the same to their adjacent nodes, and the process is repeated until the failure message gets propagated to the chooser. The chooser finally returns the message to the sender.

5.6.5 Survivability in ATM Over SDH/SONET

For the case in which the SDH/SONET network forms the physical transport infrastructure for the ATM network, another dimension can be added to the network survivability. As the interest of the ATM network is in recovering the VCs and VPs, network survivability can be achieved not only by recovering the relevant paths in the SDH/SONET layer but also by establishing detouring paths in the ATM layer.[14]

In such multilayer arrangement, a failure in the lower network layer usually causes failures in multiple connections in the upper network layer. Figure 5.39 illustrates this on the ATM and SDH/SONET layers; a failure that occurred on a link in the SDH/SONET layer can lead to the failure of multiple VPs in the ATM layer. Apparently, in this situation, recovering the failed link in the SDH/SONET layer has the effect of recovering multiple VP failures at the same time. This example demonstrates that harmonization of two adjacent network layers can diversify network survivability schemes.

Figure 5.40 illustrates two different ways of recovering the VP in the ATM layer for the failure occurred in the SDH/SONET layer. Figure 5.40(a) shows how to restore the failure by setting a detour path in

---

14. A similar relation holds for the SDH/SONET over WDM networks. The interest of the SDH/SONET network is in recovering the transport signals STM-*n*/OC-*m*, so that network survivability can be achieved not only in the STM-*n*/OC-*m* level but also in the optical layer level of the WDM network.

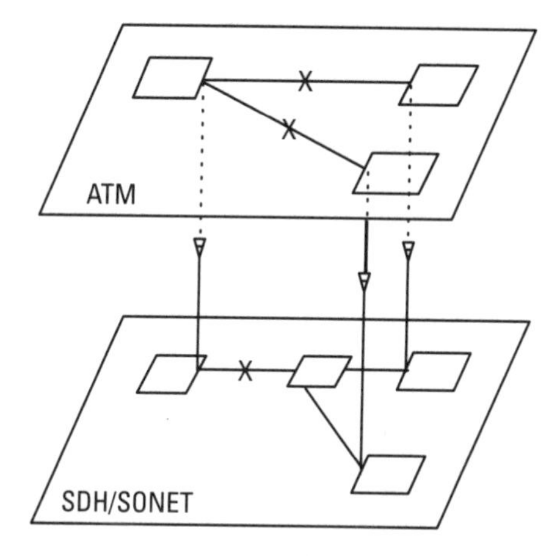

Figure 5.39 Relation of the SDH/SONET layer and ATM layer with respect to failure occurrence.

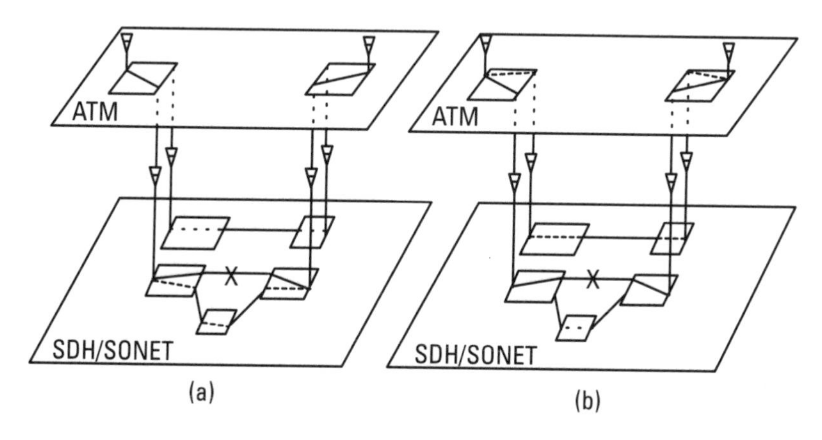

Figure 5.40 Restoration of the VP (a) in the SDH/SONET layer, and (b) in the ATM layer.

the SDH/SONET layer, whereas Figure 5.40(b) illustrates how to restore the failure by reconfiguring the VP in the ATM layer. Such an interlayer collaborative restoration, in which the paths or virtual paths are restored through collaborative work among different layers, is called *escalation.* Escalation enables numerous ways to restore the network failure and thus helps to enhance network survivability by adopting the most economical and effective means from among those available. This, however, will cost increased complexity in restoration management.

References

[1] Lee, B. G., M. Kang, and J. Lee, *Broadband Telecommuniations Technology, 2nd ed.*, Norwood, MA: Artech House Publishers, 1996.

[2] ITU-T Rec. G.708 (07/99) "Sub STM-0 Network Node Interface for the Synchronous Digital Hierarchy (SDH)."

[3] ANSI T1.105-1995, "Synchronous Optical Network (SONET)—Basic Description Including Multiplex Structure, Rates and Formats," 1995.

[4] ITU-T Rec. G.702 (11/98) "Digital Hierarchy Bit Rates."

[5] ITU-T Rec. G.703 (10/98) "Physical/Electrical Characteristics of Hierarchical Digital Interfaces."

[6] ANSI T1.101-1999, "Synchronization Interface Standard," 1999.

[7] ANSI T1.102-1993(R1999), "Digital Hierarchy—Electrical Interfaces," 1993 (R1999).

[8] Siller, Jr., C. A., and M. Shafi, SONET/SDH: *A Sourcebook of Synchronous Networks,* IEEE Press, 1996.

[9] ITU-T Rec. G. 709, "Synchronus Multiplexing Structure," 1993.

[10] ANSI T1.105.02-1995, "Synchronous Optical Network (SONET)—Payload Mappings," 1995.

[11] ANSI T1.107-1995, "Digital Hierarchy—Formats Specifications," 1995.

[12] Klein, J.K., and R. Urbansky, "Network Synchronization—A Challenge for SDH/SONET?" *IEEE Communications Magazine*, Vol. 31, No. 9, September 1993, pp. 42–50.

[13] ITU-T Rec. G.825 (03/2000) "The Control of Jitter and Wander Within Digital Networks Which Are Based on the Synchronous Digital Hierarchy (SDH)."

[14] ANSI T1.105.03-1994, "Synchronous Optical Network (SONET)—Jitter at Network Interfaces," 1994.

[15] ITU-T Rec. G.782 (01/94) "Types and General Characteristics of Synchronous Digital Hierarchy (SDH) Equipment."

[16] ANSI T1.105.09-1996, "Synchronous Optical Network (SONET)—Network Element Timing and Synchronization," 1996.

[17] ITU-T Rec. G.784 (07/99) "Synchronous Digital Hierarchy (SDH) Management."

[18] ITU-T Rec. M.3010 "Principals for a Telecommunications Management Network, 1991."

[19] ANSI T1.119-1994, "Information Systems—Synchronous Optical Network (SONET)—Operations, Administration Maintenance, and Provisioning (OAM&P)," 1994.

[20] Holter, R. F., "SONET: A Network Management Viewpoint," *IEEE Magazine of Lightwave Communication Systems*, Vol. 1, No. 4 November 1990, pp. 4, 7–13.

[21] ITU-T Rec. G.775 (10/98) "Loss of Signal (LOS), Alarm Indication Signal (AIS) and Remote Defect Indication (RDI) Defect Detection and Clearance Criteria for PDH Signals."

[22] ANSI T1.105.04-1995, "Synchronous Optical Network (SONET)—Data Communication Channel Protocol and Architectures," 1995.

[23] ITU-T Rec. G.841 (10/98) "Types and Characteristics of SDH Network Protection Architectures."

[24] ANSI T1.105.01-1998, "Synchronous Optical Network (SONET)—Automatic Protection," 1998.

[25] Hague, I., W. Kremer, and K. Chaudhuri, "Self-Healing Rings in a Synchronous Environment," IEEE Press, 1996, pp. 131–139 of [8].

[26] Wu, T. H., *Fiber Network Service Survivability*, Norwood, MA: Artech House Publishers, 1992.

[27] Wu, T. H. and R. C. Rau, "A Class of Self-Healing Ring Architectures for SONET Networks Applications," *IEEE Transactions on Communications*, Vol. 40, No. 11, November 1992, pp. 1746–1756.

[28] Wu, T. -H., "Emerging Technologies for Fiber Network Survivability," *IEEE Communications Magazine*, Vol. 33, No. 2, 62–74, February 1995, pp. 58–59.

Selected Bibliography

Alexander, G., and S. Alexander, "Architecture for Restorable Call Allocation and Fast VP Restoration in Mesh ATM Networks," *IEEE Transactions on Communications*, Vol. 47, No. 3, March 1999, pp. 397–403.

Anelli, P., and M. Soto, "Evaluation of the APS Protocol for SDH Rings Reconfiguration," *IEEE Transactions on Communications*, Vol. 47, No. 9, September 1999, pp. 1386–1393.

ANSI T1.102.01-1996, "Digital Hierarchy—VT1.5 Electrical Interfaces," 1996.

ANSI T1.105.03a-1995, "1995 Supplement ANSI T1.105.03a-1995," 1995.

ANSI T1.105.03b-1997, "Supplement ANSI T1.105.03b-1997," 1997.

ANSI T1.105.05-1994, "Synchronous Optical Network (SONET)—Tandem Connection Maintenance," 1994.

ANSI T1.105.06-1996, "Synchronous Optical Network (SONET)—Physical Layer Specification (Revision of ANSI T1.106-1988)," 1996.

ANSI T1.105.07-1996, "Synchronous Optical Network (SONET)—Sub-STS-1 Interface Rates and Formats Specification," 1996.

ANSI T1.105.07a-1997, "Supplement ANSI T1.105.07a-1997," 1997.

ANSI T1.117-1991(R1997), "Digital Hierarchy Optical Interface Specifications (Short Reach)," 1991.

ANSI T1.119.01-1995, "SONET: OAM&P-Communications-Protection Switching Fragment," 1995.

ANSI T1.119.02-1998, "SONET: OAM&P-Communications-Performance Management Fragment," 1998.

ANSI T1.231-1997, "Digital Hierarchy—Layer 1 in-Service Digital Transmission Performance Monitoring," 1997.

Bellamy, J. C., "Digital Network Synchronization," *IEEE Communications Magazine,* Vol. 33, No. 4, April 1995, pp. 70–83.

Black, U. D., and S. Waters, *SONET & T1: Architectures for Digital Transport Networks,* Englewood Cliffs, NJ: Prentice Hall, 1997.

Bregni, S., M. D'Agrosa, and L. Valtriani, "Jitter Testing Technique and Results at VC-4 Desynchronizer Output of SDH Equipment," *IEEE Transactions on Instrumentation and Measurement,* Vol. 44, No. 3, June 1995, pp. 675–678.

Cavendish, D., "Evolution of Optical Transport Technologies: From SONET/SDH to WDM," *IEEE Communications Magazine,* Vol. 38, No. 6, June 2000, pp. 164–172.

Chow, M. -C., *Understanding SONET/SDH: Standards & Applications,* Holmdel, NJ: Andan Publishers, 1995.

Chung, S. -H., et al., "Cost-Minimizing Construction of a Unidirectional SHR with Diverse Protection," *IEEE/ACM Transactions on Networking,* Vol. 4, No. 6, December 1996, pp. 921–928.

Davidson, R. P., *SONET/SDH: Foundation for Modern Broadband Networks,* New York: IEEE, 1997.

Fatato, M., "Modeling Telecommunications Networks' Transmission Systems," *IEEE Communications Magazine,* Vol. 34, No. 3, March 1996, pp. 40–47.

George, D., and G. Don, *ATM & SONET Basics,* Fuguay-Varina, NC: APDG Publishing, 1999.

Gersht, A., S. Kheradpir, and A. Shulman, "Dynamic Bandwidth-Allocation and Path-Restoration in SONET Self-Healing Networks," *IEEE Transactions on Reliability,* Vol. 45, No. 2, June 1996, pp. 321–331.

Gilbert, H., *High-Speed Digital Transmission Networking: Covering T/E-Carrier Multiplexing* 2nd ed., New York: Wiley, 1999.

Goralski, W. J., *SONET,* New York: McGraw Hill, 2000.

Grover, W. D., "High Availability Path Design in Ring-Based Optical Networks," *IEEE/ACM Transactions on Networking,* Vol. 7, No. 4, August 1999, pp. 558–574.

Imaoka, A., and M. Kihara, "Time Signal Distribution in Communication Networks Based on Synchronous Digital Hierarchy," *IEEE Transactions on Communications,* Vol. 45, No. 2, February 1997, pp. 247–253.

ITU-T Rec. G. 691 (10/2000) Optical Interfaces for Single-Channel STM-64, STM-256 and Other SDH Systems with Optical Amplifiers.

ITU-T Rec. G.707/Y.1322 (10/2000) "Network Node Interface for the Synchronous Digital Hierarchy (SDH)."

ITU-T Rec. G.707 Corrigendum 1 (03/2001) "Network Node Interface for The Synchronous Digital Hierarchy (SDH)."

ITU-T Rec. G.774.1 (02/2001) "Synchronous Digital Hierarchy (SDH) Bidirectional Performance Monitoring for the Network Element View."

ITU-T Rec. G.774.2 (02/2001) "Synchronous Digital Hierarchy (SDH) Configuration of the Payload Structure for the Network Element View."

ITU-T Rec. G.774.3 (02/2001) "Synchronous Digital Hierarchy (SDH) Management of Multiplex-Section Protection for the Network Element View."

ITU-T Rec. G.774.4 (02/2001) "Synchronous Digital Hierarchy (SDH) Management of the Subnetwork Connection Protection for the Network Element View."

ITU-T Rec. G.774.5 (02/2001) "Synchronous Digital Hierarchy (SDH) Management of Connection Supervision Functionality (HCS/LCS) for the Network Element View."

ITU-T Rec. G.774.6 (02/2001) "Synchronous Digital Hierarchy (SDH) Unidirectional Performance Monitoring for the Network Element View."

ITU-T Rec. G.774.7 (02/2001) "Synchronous Digital Hierarchy (SDH) Management of Lower Order Path Trace and Interface Labeling for the Network Element View."

ITU-T Rec. G.774.8 (02/2001) "Synchronous Digital Hierarchy (SDH) Management of Radio-Relay Systems for the Network Element View."

ITU-T Rec. G.774.9 (02/2001) "Synchronous Digital Hierarchy (SDH) Configuration of Linear Multiplex Section Protection for the Network Element View."

ITU-T Rec. G.774.10 (02/2001) "Synchronous Digital Hierarchy (SDH) Multiplex Section (MS) Shared Protection Ring Management for the Network Element View."

ITU-T Rec. G.780 (07/99) "Vocabulary of Terms for Synchronous Digital Hierarchy (SDH) Networks and Equipment."

ITU-T Rec. G.781 (07/99) "Synchronization Layer Functions."

ITU-T Rec. G.783 (10/2000) "Characteristics of Synchronous Digital Hierarchy (SDH) Equipment Functional Blocks."

ITU-T Rec. G.783 Corrigendum (03/2001) "Characteristics of Synchronous Digital Hierarchy (SDH) Equipment Functional Blocks."

ITU-T Rec. G.785 (11/96) "Characteristics of a Flexible Multiplexer in a Synchronous Digital Hierarchy Environment."

ITU-T Rec. G.804 (02/98) "ATM Cell Mapping Into Plesiochronous Digital Hierarchy (PDH)."

ITU-T Rec. G.813 (08/96) "Timing Characteristics of SDH Equipment Slave Clocks (SEC)."

ITU-T Rec. G.803 (03/2000) "Architecture of Transport Newworks Based on the Synchronous Digital Hierarchy (SDH).

ITU-T Rec. G.825 Erratum 1 (08/2001) "The Control of Jitter and Wander Within Digital Networks Which Are Based on the Synchronous Digital Hierarchy (SDH)."

ITU-T Rec. G 829 (03/2000) "Error Performance Events for SDH Multiplex and Regenerator Sections."

ITU-T Rec. G.831 (03/2000) "Management Capabilities of Transport Networks Based on the Synchronous Digital Hierarchy (SDH)."

ITU-T Rec. G.832 (10/98) "Transport of SDH Elements on PDH Networks—Frame and Multiplexing Structures."

ITU-T Rec. G.842 (04/97) "Interworking of SDH Network Protection Architectures."

ITU-T Rec. G.957 (07/99) "Optical Interfaces for Equipments and Systems Relating to the Synchronous Digital Hierarchy."

ITU-T Rec. G.958 (11/94) "Digital Line Systems Based on the Synchronous Digital Hierarchy for Use on Optical Fiber Cables."

ITU-T Rec. G.981 (01/94) "PDH Optical Line Systems for the Local Network."

Johnston, C. A., "Architecture and Performance of HIPPI-ATM-SONET Terminal Adapters," *IEEE Communications Magazine,* Vol. 33, No. 4, April 1995, pp. 46–51.

Kartalopoulos, S. V., *Understanding SONET/SDH and ATM: Communications Networks for the Next Millennium,* New York: IEEE, 1999.

Kiefer, R., and J. Nutley, *Test Solutions for Digital Networks: Basic Principles & Measurement Techniques for PDH, SDH, ISDN & ATM,* San Francisco, CA: dpunkt. verlag/Morgan Kaufmann Publishers, 1999.

Lee, B. G., and S. C. Kim, "Low-Rate Parallel Scrambling Techniques for Today's Lightwave Transmission," *IEEE Communications Magazine,* Vol. 33, No. 4, April 1995, pp. 84–95.

Manchester, J., et al., "IP over SONET," *IEEE Communications Magazine,* Vol. 36, No. 5, May 1998, pp. 136–142.

Miura, H., K. Maki, and K. Nishihata, "SDH Network Evolution in Japan," *IEEE Communications Magazine,* Vol. 33, No. 2, February 1995, pp. 86–92.

Saman, S. A., A. Cantoni, and V. Sreeram, "A Comprehensive Analysis of Stuff Threshold Modulation Used in Clock-Rate Adaptation Schemes," *IEEE Transactions on Communications,* Vol. 46, No. 8, August 1998, pp. 1088–1096.

Serizawa, Y., "Additive Time Synchronous System in Existing SDH Networks," *IEEE Aerospace and Electronics Systems Magazine,* Vol. 14, No. 2, February 1999, pp. 19–28.

Serizawa, Y., et al., "SDH-Based Time Synchronous System for Power System Communications," *IEEE Transactions on Power Delivery,* Vol. 13, No. 1, January 1998, pp. 59–65.

Sexton, M., and A. Reid, *Broadband Networking,* Norwood, MA: Artech House, 1997.

SR-NOTES-SERIES-01, "Telcordia Notes on the Synchronous Optical Network (SONET)," Telcordia, December 1999.

SR-NWT-001756, "Automatic Protection Switching for SONET," Telcordia, October 1990.

SR-NWT-002439, "Interface Functions and Information Model for Initial Support of SONET Operations Using OSI Tools," Telcordia, December 1992.

SR-OPT-002104, "TIRKS Time Slot Numbering Schemes for SONET Add/Drop Multiplex Equipment (ADM)," Telcordia, November 1991.

SR-TSV-002387, "SONET Network and Operations Plan: Feature Functions, and Support," Telcordia, August 1992.

TA-TSY-000842, "Generic Requirements for SONET Compatible Digital Radio," Telcordia, July 1988.

TA-TSY-001040, "SONET Test Sets for Acceptance and Maintenance Testing: Generic Criteria," Telcordia, July 1990.

Takatori, M., et al., "A High Performance Switch for OC-12 SONET Self-Healing Ring Networks," *IEEE Journal on Selected Areas in Communications,* Vol. 14, No. 2, February 1996, pp. 353–361.

TR-NWT-000233, "Wideband and Broadband Digital Cross-Connect Systems Generic Criteria," Telcordia, November 1993.

TR-NWT-000917, "SONET Regenerator (SONET RGTR) Equipment Generic Criteria," Telcordia, December 1990.

Yamagishi, K., N. Sasaki, and K. Morino, "An Implementation of a TMN-Based SDH Management System in Japan," *IEEE Communications Magazine,* Vol. 33, No. 3, March 1995, pp. 80– 85.

Zhang, X., and C. Qiao, "An Effective and Comprehensive Approach for Traffic Grooming and Wavelength Assignment in SONET/WDM Rings," *IEEE/ACM Transactions on Networking,* Vol. 8, No. 5, October 2000, pp. 608–617.

6

WDM/Optics

WDM/optics, together with SDH/SONET, provides a lower-layer network technology that supports the upper-layer network technologies such as TCP/IP and ATM. In particular, WDM/optics makes available the bottom layer of all communication layer stacks, providing physical interfaces by means of the optical transmission medium. Optics alone would be limited to a physical medium function unless WDM adds an additional dimension of multiplexing function on it. Owing to the multiplexing functionality of the WDM, WDM/optics retains a wider scope of coverage, including multiplexing, layering, ADM and cross-connect, and network operation and management.

WDM was originally intended to expand the transmission bandwidth. With all the optical cables readily deployed, there were only two promising ways to meet the demand for bandwidth increase: one was to increase the multiplexing speed and the other was to increase the number of multiplexing wavelengths. As the WDM and the related optical technologies have matured, other types of applications have been added to the WDM networks. A typical example is the wavelength-routing networks (WRNs) in which the wavelength management opens a new layer for network routing. On the other hand, the WDM networks are diversifying the service targets; in the past, the main service target was SDH/SONET but it is now being directed toward upper-layer services such as TCP/IP and ATM. This brings significant changes to communication networks design, as WDM-added optics is no longer a passive point-to-point physical medium but an active network solution.

This chapter aims to provide an overall picture of WDM/optics. Specifically, where the *first-generation optical network* rooted on the SDH/SONET standards was the subject of Chapter 5, this chapter discusses the *second-generation optical network* containing WDM and wavelength-conversion capabilities.[1] To this end, we organize the discussions in this chapter in the following manner: First, we examine the concept and the layering of the WDM/optics by reviewing the evolution of the optical networks, and then we provide a concise but comprehensive description of the enabling technologies. Subsequently, we discuss optical network architectures, ADM and cross-connect functions, and finally network management issues.

6.1 Concept and Layering

Optical communication is utilizing the numerous advantages of optical fiber over copper wire. Fiber is a flexible, lightweight, hair-thin medium whose propagation loss is very low and whose transmission bandwidth is extremely large, enabling high-speed, broad-bandwidth, long-distance transmission. Technologies for fiber-optic communications have continuously improved since the early 1970s, gradually converting copper cable networks into optical fiber networks. Optical fiber was deployed in long-haul trunk in the early stage and then diffused into subscriber networks in recent years. In the 1990s, the optical communications based on the SDH/SONET standards have spread worldwide, replacing the legacy proprietary optical communication systems.

6.1.1 Optical Network Evolution

In the 1980s, the main thrust of optical communication was in developing technologies to transmit user data at higher bit rates over longer distances. As a result, the transmission medium was replaced with optical fiber but the processing for switching and transmission all remained in the electronic domain. There was no standard for optical transmission, and consequently all the optical transmission systems were proprietary ones. We may call the communication networks including such nonstandard optical transmission

1. Refer to Sections 6.1.1.1 and 6.1.1.2 for the definition of the first- and second-generation optical networks.

systems the *primitive* optical network. Typical examples of optical transmission systems in this category were the nonstandard 405-Mbps, 417-Mbps, 565-Mbps, and 1.7-Gbps systems that prevailed before the advent of SDH/SONET systems. The Metrobus optical transmission system[2] developed in the mid 1980s made a turning point to those point-to-point optical systems, thereby opening a new era for the network-oriented synchronous optical systems that have led to SDH/SONET systems.

6.1.1.1 First-Generation Optical Network

The standardization works of the 1980s gave birth to the SONET systems in North America and the SDH systems for the rest of the globe. This standardization produced two different global standard optical systems that are closely compatible. The standard optical systems were built based on the point-to-multipoint electrical network concept, by employing electrical ADM and cross-connect functions. This new generation of optical networks that has been built throughout the 1990s may be referred to as the *first-generation* optical network. The SDH/SONET-based first-generation optical network distinguishes itself from the primitive nonstandard optical network in that it is network-oriented, wherein the ADM and cross-connect functions are realized efficiently and reliably. The first-generation networks have formed the core of the telecommunications infrastructure in Europe, Asia, and North America.

As the demand for wider bandwidth has been increasing continuously, solutions have been sought in two different directions—increasing the transmission bit rate and increasing the number of optical wavelengths into which to multiplex user signals. In the first solution, multiple lower-rate data streams are multiplexed into a higher-speed stream by means of electrical TDM. TDM can be done in optical domain as well, in which case it is called *optical TDM* (OTDM). In the second solution, multiple wavelengths of optical carriers are optically multiplexed together, with each wavelength carrying independent bit streams. It is called WDM. As WDM is independent of TDM, it is common to combine the two multiplexing technologies to maximize the transmission capacity.

6.1.1.2 Second-Generation Optical Network

The optical networks containing TDM, WDM, or their combination may be still categorized into the first-generation optical network, as those

2. Description of the Metrobus optical system is available in [1].

multiplexing means merely contribute to increasing the transmission capacity. However, if the optical WDM networks retain the added functionality of wavelength-routing, they may be classified into the *second-generation* optical networks. The *optical ADM* (OADM) or *wavelength ADM* (WADM), which provides wavelength-based ADM function, and the *optical cross-connect system* (OXC) or *wavelength cross-connect system* (WXC), which provides wavelength-based cross-connect function, are the key optical systems realizing this second-generation optical network. Of course, OTDM or WDM can be employed for a transmission capacity increase in the second-generation network, and a wavelength converter can also be employed for efficient wavelength utilization in ADM and cross-connect operations.

As the transmission speed becomes faster, the relevant switching and routing processes must become faster as well. The required processing speed can be attained by employing faster electronics but can also be achieved by adopting optical processing. In reality, it is advantageous to adopt optical switching and routing means over the electronic counterparts, as they are more amenable at higher bit rates and do not require repeated optical-to-electrical and electrical-to-optical conversions. In addition, optical switching may be further expanded to optical packet or burst switching. The optical packet networks with the switching and routing functions done in optical domain may be referred to as the *third-generation* optical network. In optical networks, the control plane can be managed either electronically or optically depending on the technology advancement.

While the first-generation optical network was the subject of Chapter 5, the second-generation optical network is the subject of this chapter. We will view the optical WDM network in the capacity of a wavelength-routing-capable network and consider the supporting systems such as OADM and OXC as its important building blocks. The interaction/integration issues among different layers of network technologies to be discussed in later chapters, namely IP over ATM over SDH over WDM, IP over SDH over WDM, or IP over WDM, are all based on this second-generation network. The third-generation network is still in the research stage in the aspects of the enabling technologies, and so it is excluded from discussion in this chapter.

6.1.2 WDM

Optical fiber made of silica manifests low loss in two comparatively wide-wavelength regions in the vicinity of 1.3 μm and 1.55 μm. These low-loss wavelength bands can be used to carry multiple numbers of different wavelengths in a single optical fiber. The process of combining multiple optical signals in different wavelengths into one optical signal is called WDM and the reverse process is called *wavelength-division demultipexing*. As each wavelength carries a high-bit-rate TDM signal, the WDM that aggregates the TDM signals can yield an extremely high-bit-rate optical transmission. For example, if 40 of 10-Gbps STM-64/OC-192 TDM signals are WDM-multiplexed at 40 different wavelengths, then the aggregate bandwidth of 400 Gbps is carried by one optical fiber!

WDM is essentially the same as the FDM that has been used in radio communications over a century to multiplex different data streams at different carrier frequencies simultaneously. Only the term *wavelength* is used in replacement of the term *frequency* in the case of optical communications. Optical signals in different wavelengths normally interfere with each other if the wavelengths are closely spaced, producing various undesirable effects. So when performing WDM, the wavelengths should be arranged sufficiently far apart. In earlier WDM technology, the number of wavelengths was limited to about 10, but now *dense WDM* (DWDM) technology is available with a larger number of wavelengths, usually 40 or more, by packaging wavelengths more densely [2].

WDM has been made possible owing to the enabling optical component technologies. The most fundamental ones are the wavelength-division multiplexer and demultiplexer. They can be constructed out of several optical components to be discussed in Section 6.1.3, among which are *arrayed waveguide grating* (AWG), the Mach-Zehnder interferometer, Bragg ratings, and fiber gratings. The AWG is also used in building the static wavelength router, in which multiple different wavelengths in multiple incoming optical signals are reshuffled in the outgoing multiple optical signals. Likewise, the Bragg gratings and fiber gratings are useful in building WDM ADMs in which signals in one or more wavelengths are dropped off and their replacements are added on.

Figure 6.1 shows a block diagram of a WDM multiplexer and demultiplexer connected via a WDM network. In front of the multiplexer and right after the demultiplexer, a group of *wavelength converters* (WCs), or *transponders,* are attached for wavelength matching. Such wavelength conversion-and-matching is often needed before multiplexing because the optical signals that come into the WDM multiplexer, in general, are generated by different pieces of SDH/SONET LTE located in geographically different places. WDM optical networks are supposed to include OADMs (or WADMs) and OXCs (or WXCs) that respectively add/drop and cross-connect optical signals in the level of wavelength.

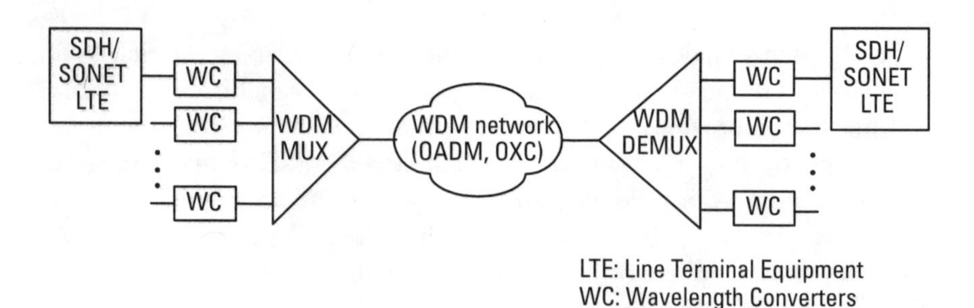

Figure 6.1 Wavelength division multiplexer and demultiplexer connected via a WDM optical network.

6.1.3 Optical Layer

According to the ITU-T G.872 specifications [3], the *optical channel* (OCh) *trail,* OMS, and OTS form the basis for an optical transport network; an OCh trail or *lightpath* refers to a point-to-point optical layer connection between two access points in an optical network. An OMS layer provides the transport of the optical channels. The information contained in this layer is a data stream comprising a set of optical channels, which have a defined aggregate bandwidth. An OTS (or *optical amplifier section*) layer provides functionality for amplification and transmission of the optical signal on optical media of different types.

In the case of the first-generation optical network, the lightpath was merely a hard-wired physical medium that carried the SDH/SONET multiplexed signals. However, in the case of the second-generation optical network, the embedded WDM function creates an additional layer, the OMS layer. The OMS layer supports point-to-point WDM links and provides related multiplexing and protection functions. The wavelength-level exchange functions such as ADM and cross-connect also take place in this layer.

In the first-generation optical networks, the SDH/SONET layer contains four sublayers—the *path layer, multiplex section* (or *line*) *layer, regenerator section* (or *section*) *layer,* and *physical medium layer.* Given a channel (or a circuit) for service, a path is established between the corresponding node pair for end-to-end connection. The multiplex section layer then gets involved for multiplexing multiples of such path connections and provides protection switching function. The regenerator section layer does the regeneration and the related performance monitoring functions. The physical medium layer takes the responsibility of conveying the SDH/SONET-multiplexed data stream in the form of optical signal.

In the case of the second-generation optical network, the counterpart of the SDH/SONET layer is the *optical layer,* which comprises the three layers previously mentioned. The OCh layer takes care of the end-to-end routing of the lightpaths for transmission. The OMS layer takes a role that is similar to the multiplex section layer of the SDH/SONET, providing the wavelength multiplexing and related protection functions. The OMS layer represents a point-to-point link connecting WDM nodes along the lightpath, and OTS layer is a subsection of the OMS layer where optical amplification takes place.

Figure 6.2 shows the layer architecture of the optical layer in comparison with the SDH/SONET layer. If the WDM optical network is going to carry the SDH/SONET-multiplexed signal, the three sublayers of the optical layer should be put underneath the four sublayers of the SDH/SONET layer. However, as far as the functionality is concerned, the OCh, OMS, and

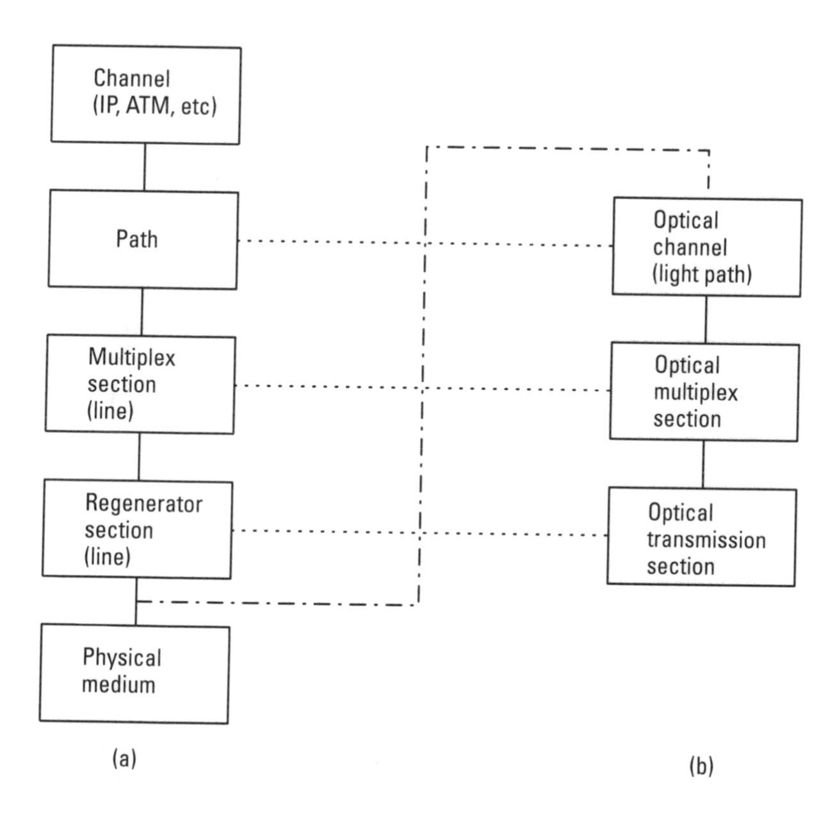

Figure 6.2 Comparison of layer architectures: (a) SDH/SONET layer, and (b) optical layer.

OTS layers have similarities with path, multiplex section, and regenerator section layers, respectively. The latter three sublayers respectively perform the path, multiplexing and regeneration functions for the electrical path signal, while the former three sublayers respectively do the path, multiplexing (or WDM), and amplification (or a simple form of regeneration) functions for the optical path (i.e., lightpath) signal.

In the case of the SDH/SONET over WDM services above, the number of sublayers grows to seven, making the overall layering look complicated or duplicated. However the layer functions cannot be eliminated as one set is in electrical domain and the other in the optical or wavelength domain. Savings may be better sought for in the electrical domain in case the SDH/SONET carries other electrical forms of signals like IP or ATM. The two layer stacks differ not only in signal representation but also in bit-rate granularity. For example, the SDH/SONET layer can more efficiently handle lower rate signals (e.g., up to STM-4/OC-12 or STM-16/OC-48 level) carried by one wavelength, while the optical layer is more efficient for processing higher rate signals in plurality of wavelengths. The former is fine-fingered while the latter has big hands. A similar distinction exists in handling protection; some higher levels of failures can be better protected at the SDH/SONET layer while lower-level failures, like fiber cutting, can be better handled at the optical layer.

Figure 6.3 shows the layer architecture related to the *optical line termination* (OLT), OADM, and OXC. OLT gets involved with all three sublayer functions in the optical layer, but OADM and OXC get involved with the OMS and OTS layers only. This block diagram is similar to the SDH/SONET layer case, in which the three layers are replaced respectively with path, multiplex section, and regenerator section layers. The SDH/SONET counterparts of OLT, OADM, and OXC are then respectively LTE, ADM, and DCS. Likewise, the SDH/SONET regenerator takes the place of the optical amplifier.

6.2 Enabling Technologies

To get a thorough grasp of the characteristics of optical communications and networks, especially of WDM optical networks, it is important to understand the enabling technologies. Optical fiber and optical components are the essential enabling building blocks of the optical systems and networks. Optical fiber provides the propagation path to communication signals in optical form. Optical components perform various optical signal processing

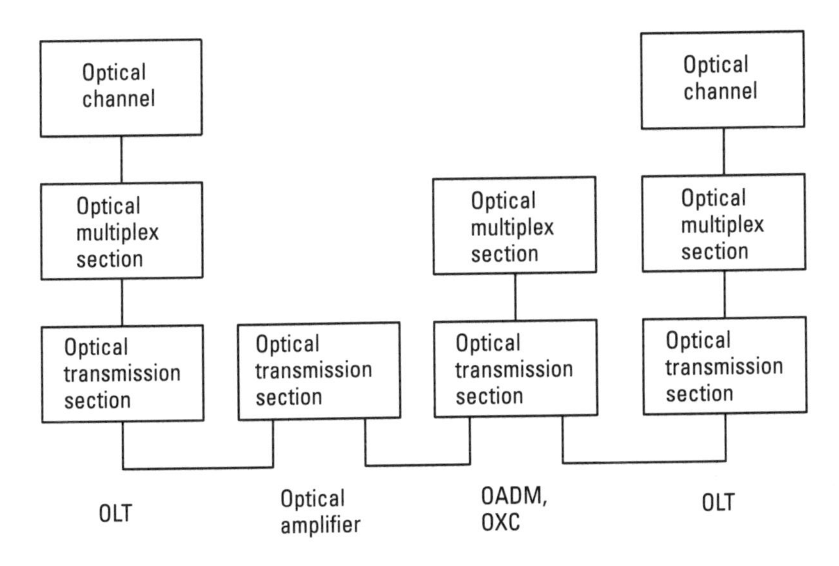

Figure 6.3 Layered processing at OLT, optical amplifier, and OADM/OXC.

functions from the generation of the light signals to the detection of the light signals, which include amplifying light signals, combining and splitting of light signals, switching among different light signals, converting to different wavelengths, and multiplexing and filtering different wavelengths. The optical components that perform these functions are lasers, optical detectors, optical amplifiers, couplers, switches, WCs, wavelength multiplexers, and filters, respectively.

6.2.1 Optical Fiber

Optical fiber, which is a slender strand of silica, consists of the center portion, called the *core,* and the outer portion, called the *cladding.* These components are protected by external sheaths as shown in Figure 6.4. The core and cladding are designed such that the refractive index of the core is slightly greater than that of the cladding. This slight disparity of refraction index contributes to the light propagation in the core through total internal reflection [4].

6.2.1.1 Multimode and Single-Mode Fibers

An optical fiber can be categorized into three different types according to its refractive index profile and the resulting light propagation characteristics (see Figure 6.5). In the case of the *step-index multimode* fiber, the refractive index

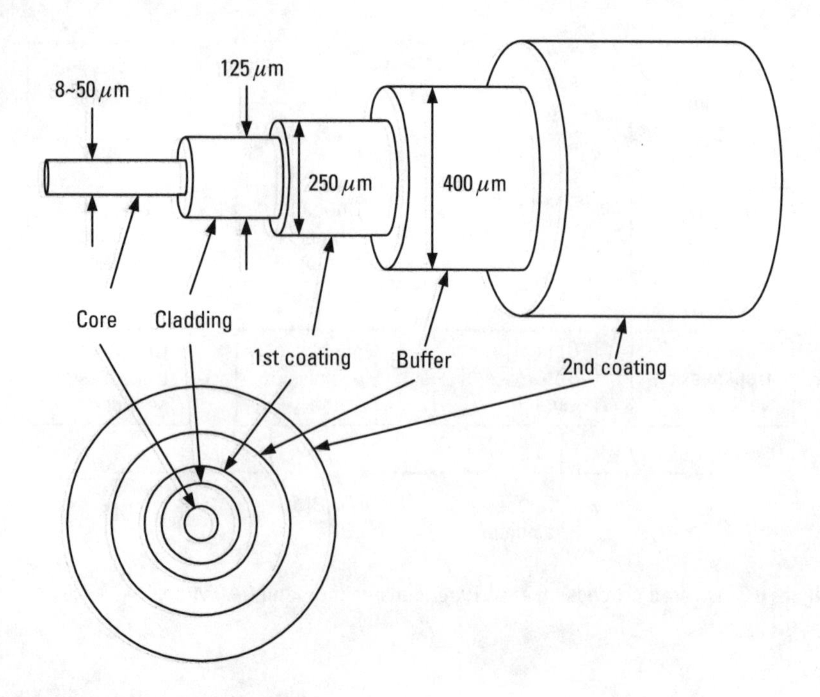

Figure 6.4 Structure of optical fiber.

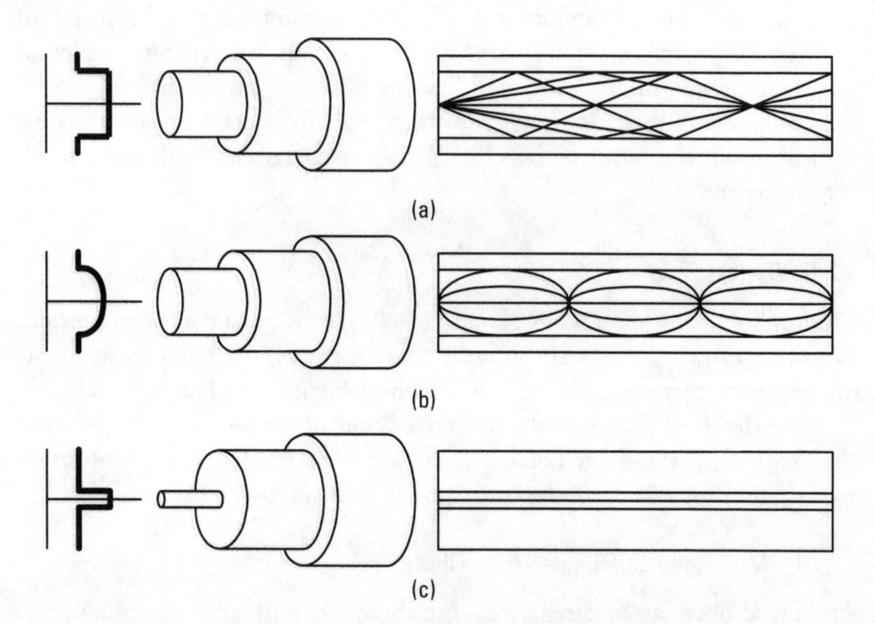

Figure 6.5 Refractive index and light propagation characteristics: (a) step-index multimode, (b) graded-index multimode, and (c) single mode.

changes abruptly at the core-cladding boundary, while the change becomes gradual in the case of the *graded-index multimode* fiber [5]. These two belong to the case where the diameter of the core is in the neighborhood of 50 μm, which is large enough to allow for multiple light modes to exist inside. Unlike these two cases, the *single-mode* fiber has a small diameter in the vicinity of 8 μm, which allows for only a single light mode to exist. The diameter of cladding is 125 μm in all the three cases [6].

The propagation of light in the fiber can be explained from the ray optics' point of view as follows: In the case of step-index multimode fiber, for example, when the light ray reaches the boundary of the core and cladding, each having a different refractive index, a part of the light ray is transmitted through while the rest is reflected. When the incident angle of the light from the core (which is a denser medium) to the cladding becomes less than the critical angle, then light is totally reflected internally. This phenomenon, called *total internal reflection,* governs the light propagation in such a way that a ray of light entering the core at less than the critical angle experiences total reflection at the core-cladding boundary. Consequently, if optical fiber is bent abruptly, the incident angle in the bent region may become larger than the critical angle, causing light to stray to the cladding portion and thus experiencing loss of light power.[3]

In the case of step-index multimode fiber, there can exist multiple light rays of different incident angles, all falling within the total internal reflection angle, that take different paths of propagation within the fiber. They are called multiple modes, or *multimodes.* Due to the difference of propagation paths, the multiple modes of light rays that belong to an identical light pulse can travel different lengths of distance, and this causes broadening of the light pulse at the receiving end. This phenomenon is called *modal dispersion.* Modal dispersion causes a transmission capacity limit on the order of tens of megahertz per kilometer. To alleviate this capacity limit problem the graded-index fiber was designed. The gradually changing profile of the refractive index in the core makes the light paths bend as they travel away from the center, making the bent light rays arrive at the receiving end at almost the same time as the straight beam. This helps to increase transmission capacity by up to a few gigahertz per kilometer. In the case of the single-mode fiber, the diameter of the core is drastically reduced such

3. Such a reflection-based description of light propagation is called ray theory or geometrical optics approach. There is another approach called wave theory, which regards the light as a wave and applies the electromagnetic wave theory to it. The term "mode" in multimode and single mode originates from the wave theoretic approach.

that only one mode, or the straight propagation path, can exist. As a result, there is no modal dispersion in the case of single-mode fiber and thus the transmission bandwidth can go above 100 GHz-km.

Among the three different types of optical fibers, single-mode fibers are used in all long-haul applications. In the case of the optical subscriber network, multimode optical fibers were widely used in the early field trials, but single-mode fibers have been predominantly used later in commercial applications. Since most subscribers are distributed within a few kilometers from the central office, multimode fiber may have sufficient channel capacity if just the initial services are considered. Multimode fiber is also advantageous in connecting and splicing fibers. However, the almost unlimited bandwidth of single-mode fiber should be attractive in preparation for future expansion.

6.2.1.2 Loss and Dispersion

The transmission performance of the optical fiber is characterized by *transmission loss,* or *attenuation,* which limits repeater spacing, and *light dispersion,* which limits channel capacity. Figures 6.6 and 6.7, respectively, show the loss and dispersion characteristics as functions of the optical wavelength.

Loss. The main loss mechanisms in optical fiber are *material absorption* and *Rayleigh scattering.* Material absorption refers to the absorption by silica and

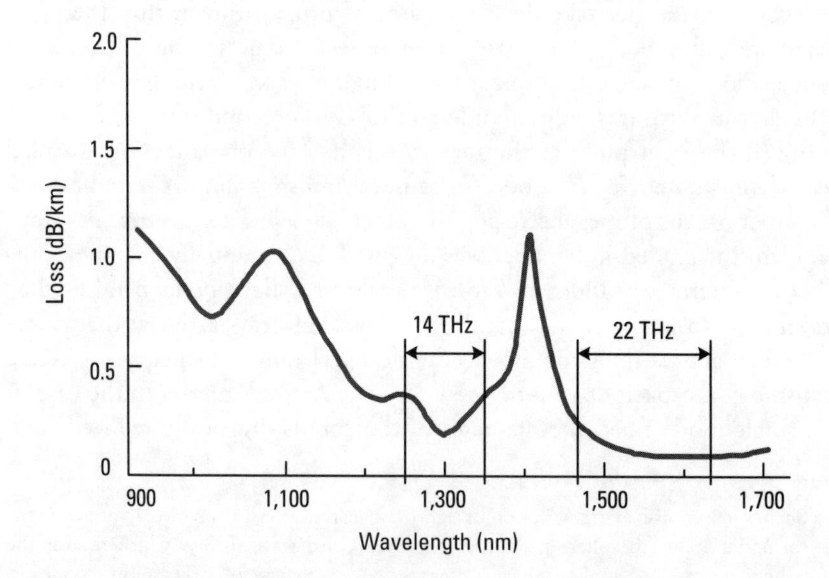

Figure 6.6 Attenuation characteristics of optical fiber.

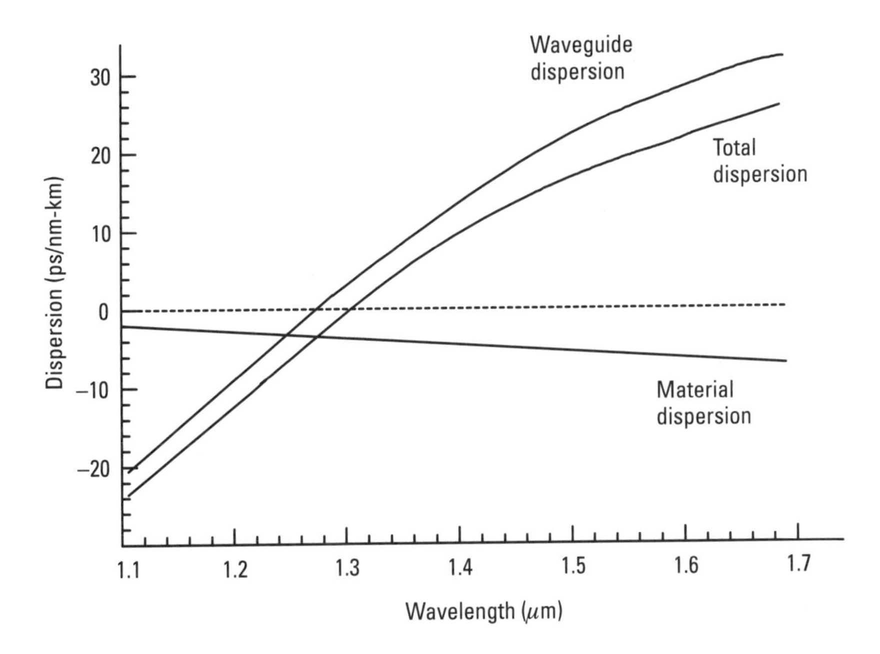

Figure 6.7 Dispersion characteristics of optical fiber.

the impurities in the fiber. The material absorption by pure silica is negligible, and the major source of material absorption is the impurities. The material absorption by impurities was the major factor that limited the loss performance in the past, but it has dropped down to a negligibly low level today due to matured fiber manufacturing technology. Therefore the loss due to material absorption, overall, is negligible in the wavelength range of practical interest, except for the loss peak in the vicinity of 1.4 μm, which is caused by hydroxyl group, or residual water vapor in silica. Rayleigh scattering occurs because of fluctuations in the density of the medium at the microscopic level. The loss due to Rayleigh scattering is inversely proportional to a fourth power of the wavelength, so it decreases rapidly with the wavelength increase. Rayleigh scattering sets a floor of loss in the wavelength range of practical use, which becomes the dominant component of fiber loss, today.

From Figure 6.6, we observe that the loss characteristics of optical fiber have local minima at the wavelengths 1.3 μm and 1.55 μm. These are the wavelengths commonly used in optical communication systems. In the early stage of optical transmission, the wavelength at 0.8 μm was used. The typical transmission losses at 0.8, 1.3, and 1.55 μm are 2.5, 0.4, and 0.25 dB/km, respectively. Noting that the loss budget is about 30 dB per span, the loss of 0.4 dB/km corresponds to a repeaterless transmission of 75 km, and the loss of 0.25 dB/km to a repeaterless transmission of 120 km.

The frequency bandwidth available in optical transmission can be measured in terms of Δf, for the frequency f having the relation $f = c/\lambda$ for the speed of light c and the wavelength λ. By taking the derivative of the equation and then applying the available width of wavelengths, $\Delta\lambda$, which are about 80 nm at the 1.3-μm band and 180 nm at the 1.55-μm band, we get frequency bandwidths of about 14 THz and 22 THz, respectively, at the two bands. Therefore the total bandwidth available amounts to 36 THz.

Dispersion. Dispersion refers to the phenomenon in which different components of the transmitted light pulse travel at different velocities, thereby arriving at the receiving end at different times. So the light pulse, when transported along the optical fiber, gets distorted in shape, with the pulse width widened as it propagates over long distances. Dispersion can cause adjacent pulses to overlap and consequently limit the transmission capacity of optical fiber. Dispersion can be categorized into modal dispersion and chromatic dispersion.

Modal dispersion occurs in multimode fibers in which multiple modes of lightwaves, or multiple rays of light, coexist. Due to the difference of propagation paths, the multiple light rays that belong to an identical light pulse can travel different lengths of distance and the light pulse gets broadened at the receiving end. In the case of single-mode fiber, it is free from modal dispersion, so chromatic dispersion is of primary concern.

Chromatic dispersion occurs because different spectral components in the light ray travel at different velocities. Chromatic dispersion has two different components—material dispersion and waveguide dispersion. *Material dispersion* is the principal component of chromatic dispersion that happens because the refractive index of silica is wavelength dependent. *Waveguide dispersion* is a secondary component that happens because the effective value of refractive index of fiber changes depending on the proportion of power contained in the core and cladding, which is a function of the wavelength. Figure 6.7 plots the material dispersion and waveguide dispersion with respect to the wavelength of the light ray. We observe that the total dispersion becomes zero at the wavelength 1.3 μm.

Considering the loss characteristics and dispersion together, we find that optical cable at the 1.3-μm wavelength exhibits a good cost-performance compromise as the loss and dispersion are both low. Optical cable at the 1.55-μm wavelength has an even smaller transmission loss but its dispersion is comparatively high. So *dispersion-shifted fiber* that shifts the zero-dispersion point to the 1.55-μm wavelength is required in this case. Wavelengths up to 1.55 μm are considered useful for direct information carry as well as WDM applications of the future.

6.2.1.3 Nonlinear Effects and Soliton

Linear characteristics may be assumed on the optical systems up to a moderate level of transmission rate and transmission power. However, as the rate or power goes up higher, nonlinear characteristic becomes dominant and it becomes more serious in the WDM systems. The nonlinear characteristics usually bring about undesired effects that are hard to analyze or utilize. However, in some special situations the nonlinear effects can contribute constructively by helping to overcome the pulse broadening effects, which culminates in the example of the soliton pulses [7].

Nonlinear effects. Nonlinear effects in optical systems occur due to the dependence of the refractive index on the intensity of the applied electric field. There are several different nonlinear effects such as *self-phase modulation* (SPM), *four-wave mixing* (FWM), *stimulated Raman scattering* (SRS), and *stimulated Brillouin scattering* (SBS), among which SPM is most dominant.

SPM happens because the refractive index of the fiber has an intensity-dependent component. Such refractive index causes induced phase shift, which increases as the intensity of the pulse increases. As a result, different parts of the pulse undergo different phase shifts, thereby causing pulse chirping. *Chirping* here refers to the phenomenon that the frequency of the light pulse changes with time. Pulse chirping happens in such a way that it can compensate for the pulse broadening effects by dispersion, and it becomes stronger as the transmitted signal power increases. Therefore the SPM-induced chirping affects the systems with high bit rates that are subject to dispersion limitations and the systems that use high transmission powers.

If the wavelength of the light pulse is below the zero-dispersion point of the fiber then spectral broadening causes temporal broadening of the pulse while it propagates. On the other hand, if the wavelength is above the zero-dispersion wavelength then chromatic dispersion and SPM can interact to the opposite direction, thereby reducing temporal broadening.

Soliton. For a particular type of light pulse, with a short duration and high peak power spectrum, spectral broadening due to nonlinear effects of and the dispersion of the fiber can compensate each other. Soliton is a typical example of such light pulses. Most light pulses undergo broadening due to group velocity dispersion while propagating through optical fiber, but the soliton pulses take advantage of the nonlinear effects, specifically SPM, to overcome the pulse-broadening effects of group velocity dispersion. For soliton pulses, the balance between pulse compression and broadening occurs in such a balanced way that the pulse neither changes in shape nor undergoes any periodic changes of envelope. The former types of solitons are called fundamental solitons, and the latter types are called higher-order solitons.

The solitons are very important for optical communication as they render a means to overcome the chromatic dispersion problem completely. Thus, the solitons make it possible to propagate light pulses over long distances without changing the pulse shape. Therefore, if solitons are used in conjunction with optical amplifiers, which can compensate for attenuation, then high bit-rate, repeaterless data transmission becomes possible over an extremely long distance. It was demonstrated in a laboratory that such a combination can provide repeaterless transmission of 20-Gbps data over a distance of 7,150 km, which amounts to 143-Tbps-km product.

6.2.2 Lasers and Detectors

Light sources and detectors are two fundamental components that make optical communication possible. They generate and detect light signals over which communication data are carried. In this respect, all other optical components may be said to be auxiliary devices that help to modulate/demodulate communication data on/out of the light signal, amplify/regenerate the modulated light signal for elongated transport, or process/modify light signals for various other purposes.

The major light sources for communications are *light-emitting diodes* (LEDs) and lasers. An LED is a *p-n* junction diode operating in forward bias mode in which the injected minority carriers recombine with majority carriers through the *spontaneous emission* process, thereby producing light. Since the light generation relies on spontaneous emission, in contrast to the *stimulated emission* for a semiconductor laser (see Section 6.2.2.1), the spectrum of an LED is much broader than that of a laser. Moreover, the light power generated by an LED is usually much lower than that of a laser. Therefore, laser is a preferred choice of light source over the LED in optical communications, even though the LED is a low-priced choice for low-speed, low-power-budgeted applications such as local area uses.

Photodetector is the key component in the receiver of optical communication systems. The receiver is composed of a photodetector followed by a front-end amplifier, and an optical amplifier may be optionally put in front.

The photodetector and the front-end amplifier are usually combined together in one integrated device, called an *optoelectronic integrated circuit* (OEIC).

6.2.2.1 Semiconductor Laser

Laser is an acronym for *light amplification by stimulated emission of radiation.* As the acronym indicates, laser is essentially an optical amplifier that generates light by the stimulated emission process in the absence of input signal. *Stimulated emission* refers to the process by which the transition of atoms occurs from a higher energy band, E_2, to a lower energy band, E_1, emitting photons of energy hf_c, for plank's constant h and frequency f_c. This increases the number of photons of energy hf_c, thereby yielding an amplification of the optical signal of that frequency. Semiconductor lasers use semiconductor as the substrate for light generation, while fiber lasers typically use erbium-doped fiber.

Figure 6.8(a) shows a simplified structure of a semiconductor laser that operates based on the Fabry-Perot principle. It consists of *p*-type and *n*-type semiconductor materials, and partially reflecting mirrors are put on both side edges. The edges may be carefully carved for partial reflection, instead. When forward bias is applied to the device, an active region is formed in the *p-n* junction, where light is generated through radiative recombination of the injected minority carriers. Within the cavity, the light is reflected back and forth between the mirrors and certain wavelength components are reinforced by constructive interference in this process. The wavelengths for which such resonance occurs are those that are the submultiple of the roundtrip distance between the mirrors. The resonance occurs when the amplification gain that is combined with the mirror reflectivity grows sufficiently large beyond the lasing threshold. The device then works as an oscillator, or a laser.

In practice, the structures of Fabry-Perot laser in practical use are much more complex in geometry than that in Figure 6.8(a). Buried *double-heterostucture* (DH) shown in Figure 6.8(b) is a typical example of laser structure for practical communications use. In this case the light is confined to a narrow region, which reduces the required threshold current and also reduces the light-emitting area to a small spot.

While the Fabry-Perot laser generates light of *multiple longitudinal mode* (MLM), which has as wide a spectral width as approximately 10 nm, there are several other types of lasers that generate light of SLM. This narrow-spectrum light of SLM lasers helps to reduce the chromatic dispersion of the optical fiber, so is very important for broadband optical communication. *Distributed*

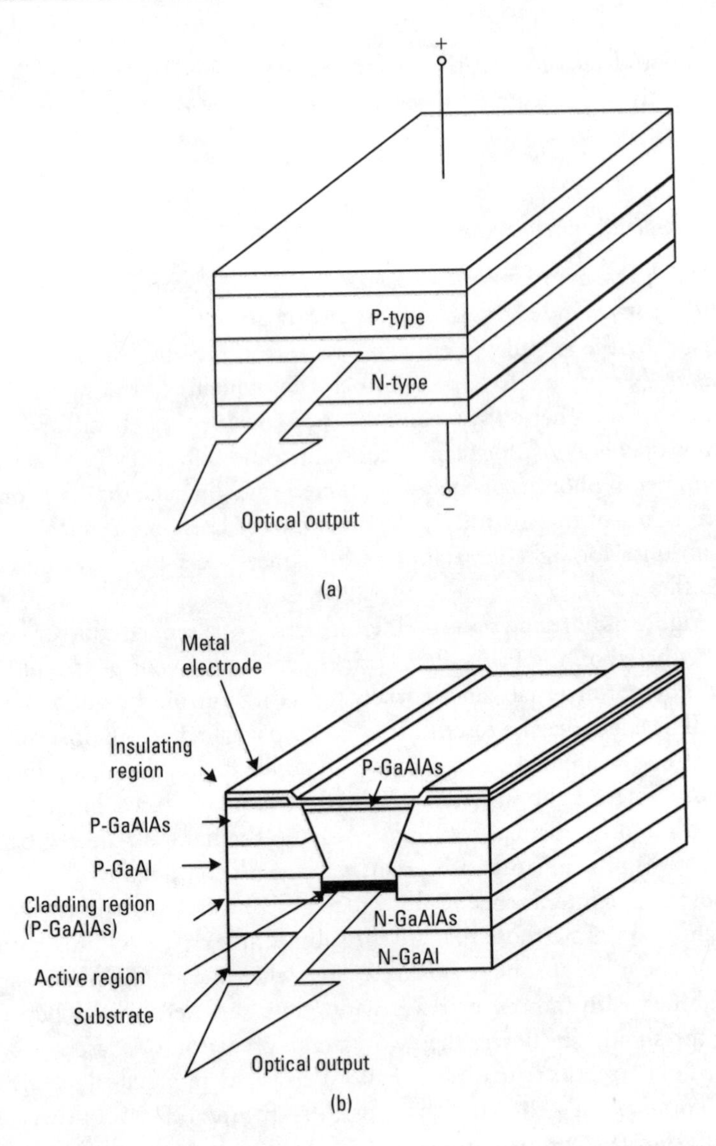

Figure 6.8 Structure of lasers: (a) simplified structure of Fabry-Perot laser, and (b) buried double-heterostructure (DH).

feedback (DFB) laser and *distributed Bragg reflector* (DBR) laser are most typical examples of single longitudinal mode (SLM) lasers, both of which adopt corrugated waveguides to achieve SLM operation. The corrugation contributes to single-mode lasing and to the precise adjustment of wavelength.

In multiwavelength optical networks such as the wavelength-routing network or WDM network, *tunable lasers* whose wavelengths are adjustable are needed. A tunable semiconductor laser can be realized by applying the phenomenon that the refractive index of the semiconductor changes with the injected current density. Accordingly, it is possible to adjust the wavelength by adjusting the injected current. However, the change of injected current accompanies the change of the output light power as well. So it is necessary to decouple the two effects. An example of such a decoupling mechanism can be found in using two separate currents for independent control, which is the case in *two-section DBR lasers.*

6.2.2.2 Optical Detector

The light signal conveyed over the optical fiber is first optical-amplified, which is optional, then photodetected, and finally electrical-amplified, before being fed to the decision circuit. Photodetectors that convert light signals to electrical signals play the central role in this series of processes.

A photodetector is basically a *p-n* junction diode. A *p-n* junction diode has two bands of electron energy levels, namely the valence band and the conduction band, having the boundary energies E_1 and E_2 respectively. Photons incident on this diode generate electron-hole pairs in the valence band, pushing up the electrons to the conduction band. In this case only the photons that have energy larger than the band gap can make this happen. If an external voltage is applied the electron-hole pairs drift to form electric current. To improve the photon-to-electron conversion efficiency, a very lightly doped *intrinsic* semiconductor is inserted between *p*-type and *n*-type semiconductors. Efficiency increases in this configuration since the depletion region where electrons are generated by absorption of photons extends across the intrinsic semiconductor. This type of photodiode is called the *pin photodiode*, with *pin* standing for *p*-type, *i*ntrinsic, and *n*-type semiconductors. On the other hand, the *avalanche photodiode* that takes the advantage of avalanche multiplication of electrons in a strong electric field can further increase the output current.

As for the front-end electrical amplifier that follows the photodetector, it should be designed to minimize the effect of thermal noise and to maximize the dynamic range as well. There are two different types of front-end amplifiers: the high-input impedance voltage amplifier and transimpedance amplifier. The *transimpedance amplifier* is actually a current-to-voltage converter and is most frequently used in existing systems as it outperforms the high-input impedance amplifier in terms of the thermal noise and dynamic range. Figure 6.9 shows the equivalent circuit of a transimpedance front-end amplifier in conjunction with the photodetector.

As optical transmission speed reaches tens of gigahertz, implementation of the relevant optical devices and electrical circuits becomes a challenging issue. OEIC is the technology that has been introduced as a means of increasing device speeds by reducing the size of the devices and the influence of external conditions. OEICs integrate both optical devices and electrical circuits onto a single chip, thereby reducing parasitic reactance and operation noise. OEICs are useful devices for optical fiber submarine systems that require high-speed laser diode modulation and high receiver sensitivity, as well as for optical subscriber systems that require a low cost and enhanced reliability.

6.2.3 Optical Amplifiers

As discussed earlier, optical fiber has two limiting factors in long-distance light propagation—loss and dispersion. Light signals get attenuated in amplitude and distorted in shape while propagating through the optical fiber over distance. In the existing optical communication systems, such limitation has been resolved by employing regenerators in electrical domain; after propagating over certain distance which is dictated by optical power budget, the optical signal was converted into electrical domain, then regenerated in electrical form, and finally converted back to optical domain again. In this process, optical detectors, optical transmitters, and electrical regenerators that perform

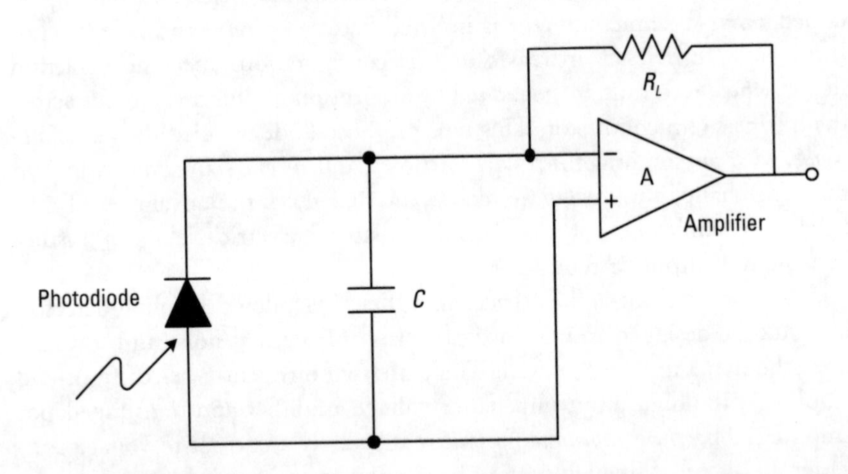

Figure 6.9 Electronic circuit equivalent to a transimpedance amplifier attached to photodetector [7].

amplifying, reshaping, and retiming functions are needed. Consequently the processing overhead for the optical signal regeneration was significantly high. In the case of the optical signal carrying WDM data, the overhead increases even more as the regeneration should be done wavelength based.

Once the limitation due to dispersion is lifted partially by using single-mode fiber or dispersion-shifted fiber, or completely by using soliton, the major limitation that remains then is loss. As there is supposed to be no signal distortion, the main concern is how to amplify the optical signal itself. In this situation, it is desirable to find the solution in a purely optical domain that does not require any electrical conversion or electrical processing. To this end optical amplifiers that amplify optical signals directly in the optical domain have been strenuously developed.

There are several different types of optical amplifiers, depending on how the amplification is achieved [8]. The most popular among them are semiconductor optical amplifiers and fiber optical amplifiers. The structure and operation of them are in principle identical to those of semiconductor lasers and fiber lasers; they are both based on stimulated emission of radiation by atoms in the presence of electromagnetic field. They differ in the mechanism of increasing the population of atoms at the higher energy band; the former relies on electrical current and the latter on optical pumping signal. There are other types of optical amplifiers, such as fiber Raman and fiber Brillouin amplifiers, whose operation is based on nonlinear effects of the optical fiber, not by stimulated emission.

6.2.3.1 Semiconductor Optical Amplifiers

Figure 6.10 shows the block diagram of a *semiconductor optical amplifier* (SOA). It is essentially a *p-n* junction diode, in which the depletion region formed in the junction provides the active region for optical amplification. Optical signal is amplified through stimulated emission when it passes through the active region. An *antireflection* (AR) coating is put on both edges of the device to eliminate ripples in amplifier gain over different wavelengths.

When forward-biased, or current pumping is done, the *p-n* junction diode generates more electrons transiting from the conduction band to the valence band by stimulated emission. In this process, the optical signal of frequencies f_c for which the energy hf_c exceeds the bandgap $E_2 - E_1$ are amplified. As the forward bias increases, the injected electrons occupy higher energy levels and the range of wavelengths for amplification becomes wider. It is even possible to amplify optical signals in the 1.3-μm and 1.55-μm wavelength bands simultaneously.

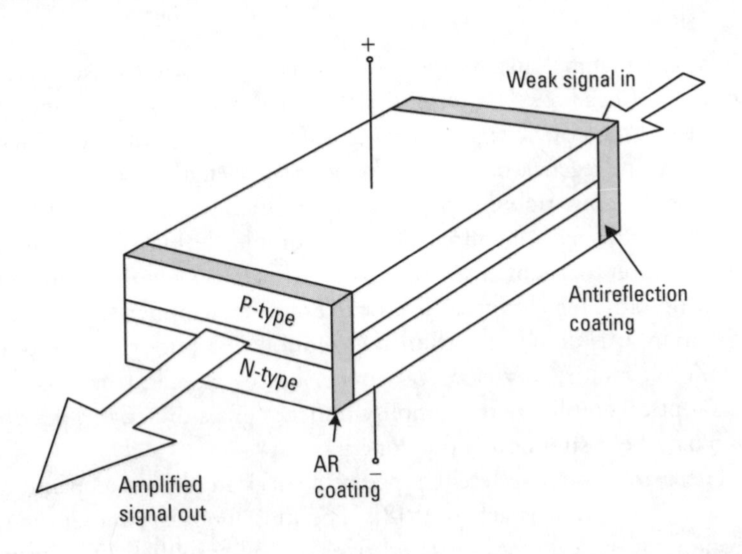

Figure 6.10 Simplified structure of an SOA.

The SOA devices in practical use have more complex structures than the one in Figure 6.10; they normally take a heterostructure, which has a thin layer of different semiconductor material inserted between the *p*-type and *n*-type semiconductors. This inserted semiconductor material forms the active region and helps to confine the carriers injected to the active region as well as the optical signal for amplification.

SOAs, in general, have a large bandwidth and relatively high gain. However they are not used as widely as the fiber amplifiers for several reasons. SOAs introduce severe crosstalk among simultaneously amplified signals, which becomes a serious problem when they are used in WDM systems. Losses due to coupling and polarity-dependency for SOAs are higher than the fiber amplifier case. Nevertheless, SOAs are used in the applications where they can be combined with other semiconductor components into OEIC. They find applications as a preamplifier of optical receivers and as a power amplifier of laser transmitters. They also find applications in switches and wavelength converters.

6.2.3.2 Erbium-Doped Fiber Amplifier

Figure 6.11 shows a block diagram of an erbium-doped fiber amplifier (EDFA) [9]. It consists of a pumping laser, a wavelength selective coupler, isolators, and an erbium-doped fiber. The pumping laser provides energy

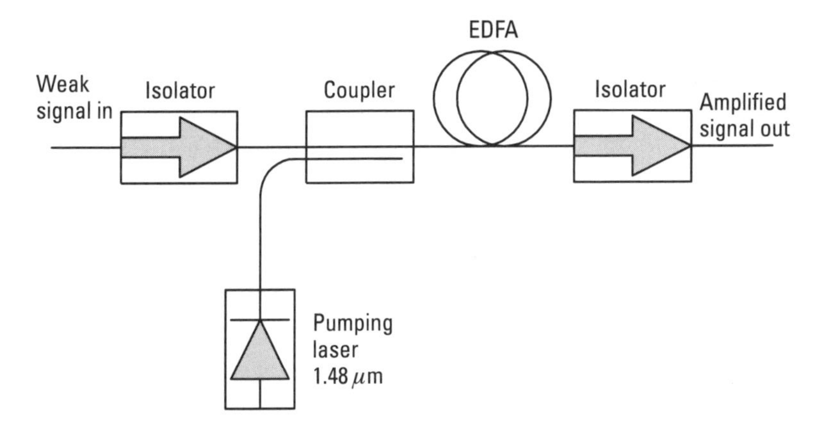

Figure 6.11 Structure of forward-pumping EDFA.

for light amplification, and the semiconductor lasers at the resonance wavelength of 0.98 μm or 1.48 μm are used as the pumping laser. The wavelength-selective coupler couples or decouples the pumping light and the input light, and the isolator isolates the reflected lights. The pumping shown in Figure 6.11 is a forward-pumping mode in which the pumping light travels toward the same direction as the input light. However, a backward-pumping mode is also available in which the pumping light travels in the reverse direction (see Figure 6.12).

If the light having a wavelength in the neighborhood of 0.98 μm or 1.48 μm is pumped into the erbium-doped fiber, atoms in the fiber get excited by absorbing the energy of the pumped light and move up to the energy band E_3 which is above the E_2 level. Atoms that have been moved up to level E_3 quickly transit to level E_2 by the spontaneous emission process. The atoms in level E_2 then drop to level E_1, emitting photons of energy hf_c by the stimulated emission process. The frequencies that can be amplified by this stimulated emission process corresponds to the wavelength range 1.525 to 1.570 μm. Therefore, if input light at a 1.55-μm wavelength passes through the EDFA, the excited atoms transfer the energy to the input light, thus amplifying the input light. It takes about 10 ms for the excited electrons to return to the ground state after emitting their energy into the 1.55-μm light, which is long enough to ensure that sufficient energy is accumulated for the light amplification.

In practice, two-stage design of EDFA shown in Figure 6.12 is more commonly used than the single-stage one in Figure 6.11. In this case the first

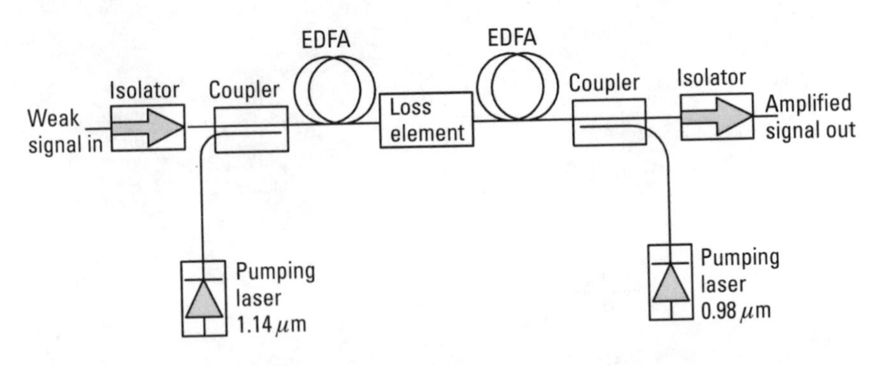

Figure 6.12 Structure of two-stage EDFA.

stage is designed to provide high gain and low noise, while the second stage is designed to provide high output power. The two-stage design consequently yields a high-performance optical amplifier with low noise and high output power. The two-stage arrangement also renders redundancy in pumping; even if one pumping laser fails, the EDFA still functions correctly, due to the other pumping laser.

EDFAs can be installed in the transmitting end, in the receiving end, or in the repeaters. The EDFA installed in the transmitter works as a power amplifier, whereas the EDFA in the receiver takes the role of a preamplifier, which increases the receiver sensitivity by amplifying weak received signals. In general, an EDFA is a preferred optical amplifier over an SOA. There are several factors that make this happen: It is an all-fiber device that is polarization-independent and easy to couple light in and out of the fiber. It is simple to manufacture, and compact and reliable high-power semiconductor pump lasers are available for this. It can directly and simultaneously amplify a wide wavelength region with a relatively flat gain. It does not introduce crosstalk when amplifying WDM signals, so is adequate for WDM use. However, it has room for improvements in making flat gain over different wavelength channels. Also, it is not a small device that can be adequately integrated to other semiconductor devices.

6.2.4 Couplers and Isolators

The light signal in the optical fiber contrasts to the electrical signal in the copper wire in combining and splitting the carried signals. When combining two different electrical signals in two different copper wires, a multiplexing process is needed in a time-division or frequency-division manner. On the other hand, when splitting (i.e., copying, not demultiplexing) electrical signals into two different copper wires, the two wires may be directly connected to the given wire. In contrast, when handling two different wavelengths in two different optical fibers, a coupler is needed to perform the necessary functions in both cases.

6.2.4.1 Couplers

A *coupler* or a *directional coupler* is a general term for the devices that combine multiple input light signals into one output and/or split a light signal into multiple outputs. If a coupler only splits the optical signal of one input into two or more outputs, it is called a *splitter*. The most common form of splitter is 1×2 splitter, and the most common splitting ratio is half-to-half, although other ratios are also possible. If a coupler does the reverse function of the splitter, combining multiple input optical signals into one output, it is called a *combiner*.

Figure 6.13 shows a typical structure of a 2×2 directional coupler. Coupling takes place in the midregion, where part of the input power flows into the other output port. The amount of power flow depends on the length of the coupling region. If a 2×2 directional coupler has a half-and-half power split, it is called a *3-dB coupler*. In general, it is possible to realize an *$n \times n$ coupler* such that each input optical signal is divided into n different outputs with the power equally split among them.

In couplers, two different types of loss mechanisms occur. While most input power is divided into multiple output ports, a small amount of the power is reflected in the opposite direction back to the input port. This is called *return loss*. In addition, part of the input power is lost while the input

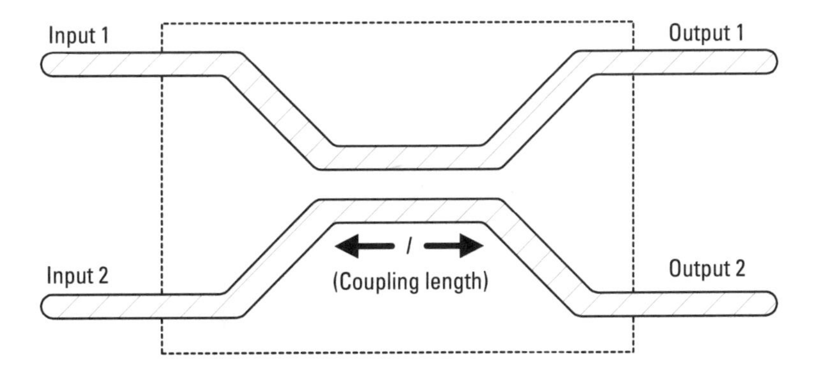

Figure 6.13 Structure of a 2×2 directional coupler [7].

optical signal is launched into the coupler device. This is called *insertion loss.* It is desirable to design couplers such that both return loss and insertion loss get minimized.

Couplers are used as building blocks for several other optical devices. They are principal components in building optical switches. They are also important in constructing optical filters and multiplexers/demultiplexers. Couplers that combine different wavelength signals are used in building WDM devices and EDFAs.

6.2.4.2 Isolator

An *isolator* is a nonreciprocal device that blocks the reflection of light signals. It is used in systems that necessitate one-direction propagation of light signals, blocking all reflections in the backward direction. Common examples of its use can be found in optical amplifiers and lasers. In these cases, isolators are used primarily to prevent reflected signals from entering those devices, as otherwise they would degrade the performance.

An isolator typically consists of a front-polarizer, an intermediate Faraday rotator, and a rear-polarizer as shown in Figure 6.14. The front-polarizer passes the light signal of only one state of polarization, (e.g., vertical component); the Faraday rotator rotates the state of polarization by 45 degrees (e.g., clockwise), regardless of the direction of propagation; and the rear-polarizer passes only the light component whose state of polarization is rotated by 45 degrees. A reflected light signal in this case will have 45-degree rotated

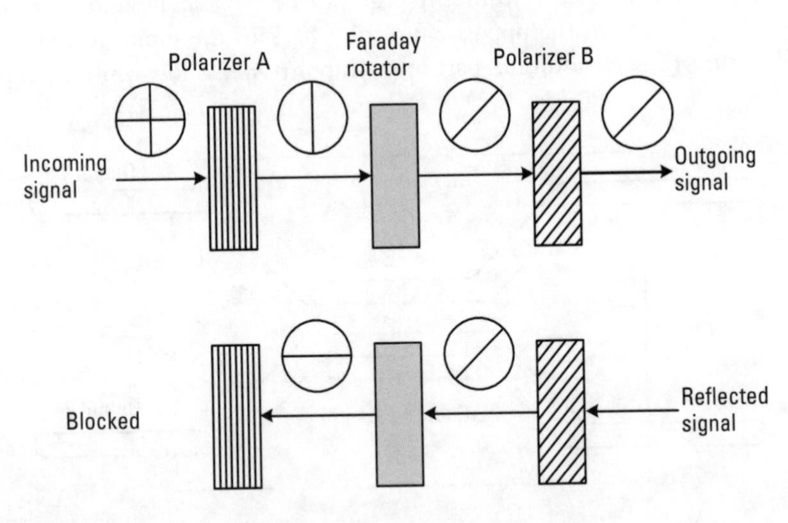

Figure 6.14 Block diagram of optical isolator.

state of polarization, and will become 90-degree-rotated (i.e., horizontal component) after passing through the Faraday rotator, therefore getting blocked at the front-polarizer when reflected back.

In isolators, two parameters are important as their figure of merit: insertion loss and isolation. *Insertion loss* is the loss in the forward direction, and *isolation* is the loss in the backward direction. As for the isolator, it is desirable to have a minimal insertion loss and a maximal isolation.

Isolator is a two-port device, in which the signal coming into port 2 is blocked internally. If a third port, port 3, is added to the device and the internal structure is reorganized in such a way that the incoming signal at port 2 is transmitted to port 3 and the incoming signal at port 3 is transmitted to port 1, then a *circulator* is resulted. Likewise, it is possible to add port 4 in the same manner to produce a four-port circulator.

6.2.5 Switches and Wavelength Converters

A switch is a node device that provides connection between multiple input ports and output ports, in a time- or space-shared manner. A switching system is located at a network center where a number of transmission links are interconnected, and it processes the aggregated traffic. Consequently, the processing speed is normally very high, going up even higher in the case of optical networks. In most existing optical networks, switching has been mostly done in the electrical domain; the optically transmitted signals are first converted into electrical signals, switched into the electrical domain, and then converted back to the optical domain. The burden of such repeated conversions as well as the burden of high-speed operation can be relieved if the switching can be done in optical domain. To this end, optical switching technology has been continuously developed for decades. As the optical amplification technology is matured to enable all-optical links for node-to-node connections, the optical switching technology, provided that it reaches practical level of maturity, would render the optical networks all-optical [10].

6.2.5.1 Switching Elements

Two-input two-output (i.e., 2×2) switch is the unit element of switch. Larger switches are constructed by cascading multiple switching elements. It is desirable if a switching element has high *extinction ratio* (or large difference of power in the on and off states), small insertion loss, low crosstalk (or low leakage of power to undesired output port or from undesired input port) and low polarization-dependent loss. In addition, it is desirable to have a short switching time, or a fast switching operation.

There are various ways of designing a switching element—mechanical, thermo-optic, electro-optic, and electronic. Mechanical switch operates mechanically by moving the mirrors residing inside. Tiny mirrors are installed using a *micro-electro-mechanical system* (MEMS) technology, which passes through or reflects the incoming light rays, thus performing the switching function. Mechanical switch exhibits good performance in terms of extinction ratio, insertion loss, crosstalk- and polarization-dependent loss, but the operation speed is too low, in the range of 10 ms. Accordingly, mechanical switches are better fit to slowly operating cross-connects than to data switching. The thermo-optic switch is essentially a 2 × 2 Mach-Zehnder interferometer built on waveguide material whose refractive index changes according to the temperature change. Its operation is slow also, on the order of few milliseconds.

The electro-optic switch is a voltage-controlled 2 × 2 directional coupler as shown in Figure 6.15. Used in the coupling region is a material whose refractive index changes according to the applied voltage, such as lithium niobate ($LiNbO_3$). The electro-optic switch operates at extremely high speeds, in the subnanosecond range, and is adequate to integrate into larger-size switches. The SOA can be used as a switching element. In this case, the bias voltage that adjusts the generation of minority carriers for light amplification does the on-off control function. Amplification function

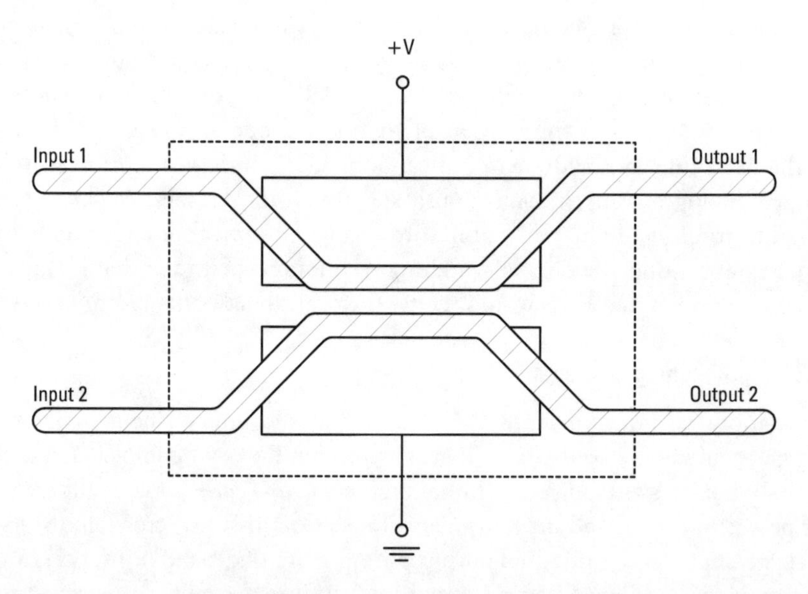

Figure 6.15 An electro-optic switch based on voltage-controlled 2 × 2 directional coupler.

becomes active in the on state and extinct in the off state, so the extinction ratio becomes very large. The SOA switch has very fast switching speed, on the order of nanoseconds, and can be effectively integrated into large switches.

6.2.5.2 Switches

Larger switches are constructed by interconnecting multiple-stage arrays of 2×2 switch elements. There are several aspects to look into when constructing large switches. First, it is desirable if it can be designed using fewer switching elements. It is also important that the loss profile is kept uniform for different types of input signal patterns. As a large switch is likely to be built on the same substrate, it is important to minimize or nullify the number of line *crossovers*. Crossovers in optical switches cause power loss as well as cross-talk. It is also desirable to design it to be nonblocking.

The most primitive form of $n \times n$ switch is a *cross-bar switch*, which is comprised of $n \times n$ matrix of 2×2 switch elements. It is a nonblocking switch, requiring as many as n^2 switching elements but no crossover. The *Benes switch* is a modified form of cross-bar switch that requires a minimal number of switch elements but an increased number of crossovers. The number of required switch elements is $(n/2)(2\log_2 n - 1)$. It can be made nonblocking by rearranging the routing pattern. Figure 6.16 shows the architecture of an 8×8 *Spanke-Benes* switch. The Spanke-Benes switch stands in the midway of the above two switches such that the number of switch elements is medium and no crossover take place. The number of required switch

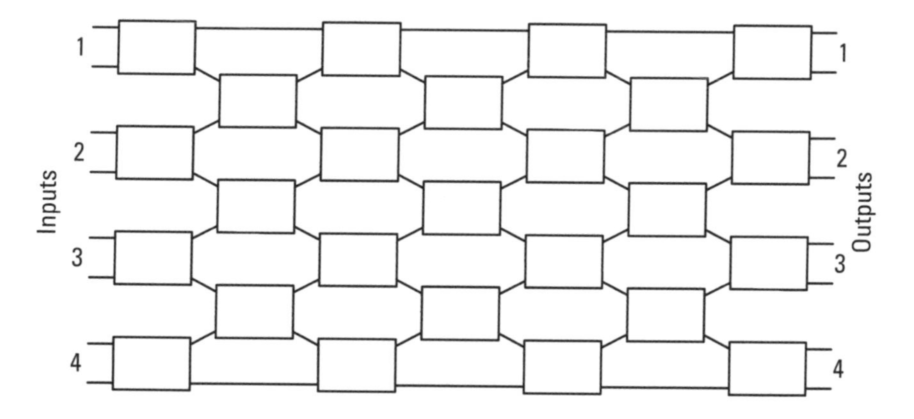

Figure 6.16 Example of an 8×8 Spanke-Benes switch [7].

elements is $n(n-1)/2$. However, it is not nonblocking in a wide sense and the loss is not uniform. If we compare the three architectures for an 8×8 switch, the number of switch elements is 64 for the crossover switch, 20 for the Benes switch, and 28 for the Spanke-Benes switch.

6.2.5.3 Wavelength Converters

In optical networks, especially in multiwavelength WDM networks with wavelength-routing capability, each lightpath is set for data transport between a node pair. In this case, a wavelength is allocated to each lightpath, with an identical wavelength serving every link in the same lightpath. This wavelength-continuity constrains the flexibility of the network in such a way that a lightpath cannot be added to certain node pair even if bandwidth is available. This limits the flexibility and efficiency of the network. If it is made possible to allocate different wavelengths to different links in the same lightpath, wavelengths can be flexibly reused in different links of the same network, thereby increasing the wavelength efficiency significantly. The device that enables this capability is the *wavelength converter*. A wavelength converter converts the wavelength of the incoming light signal to a different wavelength, yet maintains the contents of the data stream intact.

There are basically two different ways of implementing wavelength converters—*optoelectronic* and *all-optical*. Optoelectronic conversion is a back-to-back conversion in which an optical signal in one wavelength is converted to an electrical signal, regenerated in the electrical domain, and then converted back to an optical signal of a different wavelength. The regeneration in this case can be done in three different levels—analog amplification only, regeneration with reshaping, and regeneration with retiming. There are efficiency-complexity trade-offs among the three.

All-optical wavelength conversion relies on the nonlinear effects in the optical devices—namely wave mixing and cross modulation. *Wave mixing* is a phenomenon that occurs owing to the nonlinearity in the transmission medium. In particular, the four-wave mixing phenomenon generates a new frequency component at the frequency $f_1 + f_2 - f_3$ out of the three incoming frequencies f_1, f_2 and f_3. *Cross modulation* utilizes the nonlinear phenomenon of the semiconductor optical devices such as SOA. The intensity-modulated input signal modulates the gain in the SOA due to gain saturation. A continuous-wave signal at the desired output wavelength is modulated by this gain variation in such a way that it carries the same information as the original input signal. This is called *cross-gain modulation*. Wavelength conversion can be achieved through *cross-phase modulation* also.

6.2.5.4 Optical Packet Switches

In the case of the *optical transport packet networks* (OPTNs) where optical processing is done in the packet level, optical packet switches are basically needed. Figures 6.17 through 6.19 show three distinct examples of the optical packet switches that operate based on fixed packet sizes [11].

First, the optical packet switch in Figure 6.17 uses a broadcast-and-select switch architecture, in which input signals are all combined together and broadcast to all the output ports and then selected by the wavelength selectors. *Fixed-wavelength converters* (FWCs) are used to convert incoming wavelengths to a fixed wavelength. The combined signal is passed through a splitter to K different *optical delay lines* (ODLs), with the ith ODL having a delay of i timeslots. The output of each ODL is sent to a two-stage selector for each output port. The first selector selects one of the K ODL outputs and the second selector chooses one of the N wavelengths from the combined outputs.

Figure 6.18 shows an optical packet switch with space switch fabric, which assumes that all packets are fixed sized and have a delay of T. The packet encoder divides the input packets into N different wavelengths. Each signal is then put into a *tunable wavelength converter* (TWC) that changes the wavelength of the input packet to a wavelength that is free at the destined

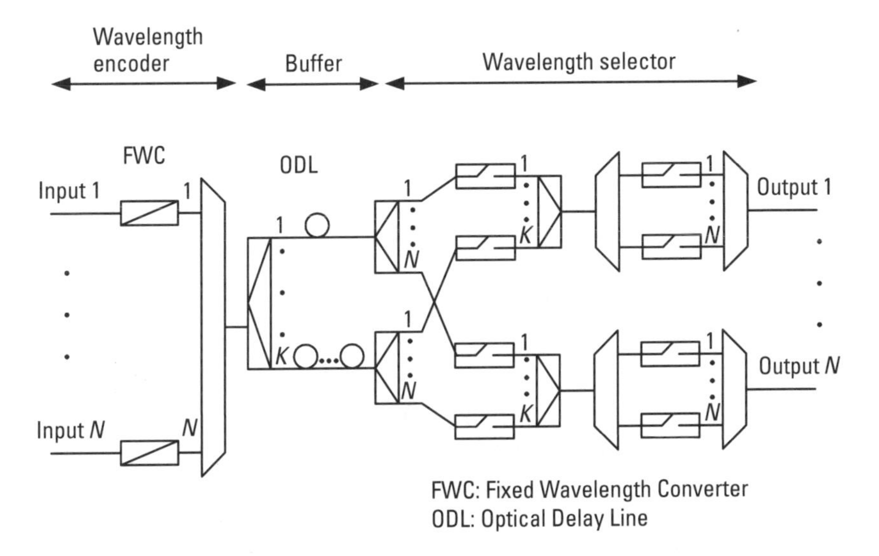

Figure 6.17 Optical packet switch architecture with broadcast-and-select switch fabric [11].

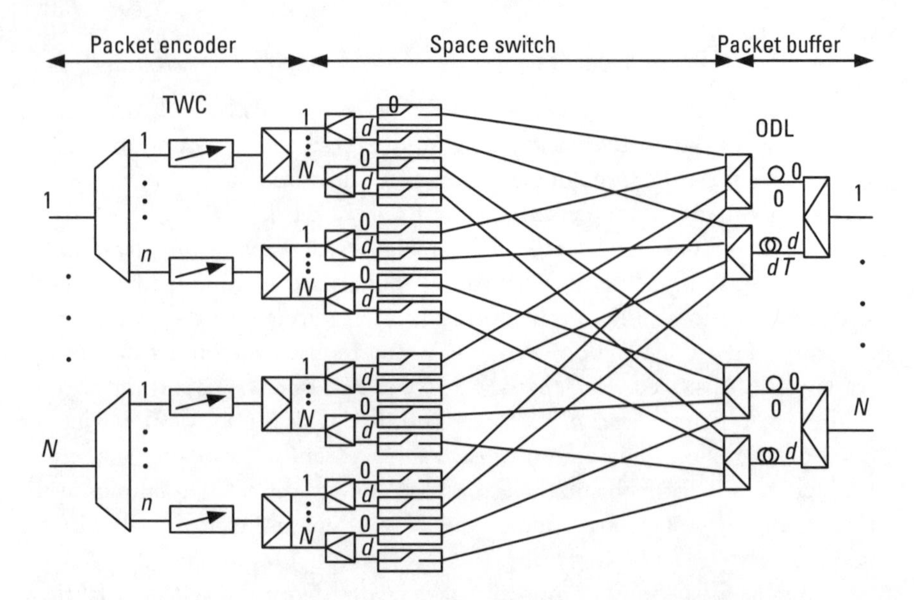

Figure 6.18 Optical packet switching architecture with space switch fabric [11].

output port. Each output port has ODLs that act as buffers of 0 to *d* timeslot sizes. By appropriately setting each TWC and the optical gates in the middle space switch stage it is possible to ensure that the packets arriving at the same port simultaneously are sent to different ODLs on the same output port, thereby solving the contention resolution problem.

Figure 6.19 shows an optical packet switch that uses wavelength-routing switching fabric. It employs a two-stage, feedback structure, where the first part solves the contention resolution problem while the second part routes the packet to the destined output port. The first part is an AWG that can route optical signals from different input ports to different output ports based on the wavelength. By setting the TWC in front of the AWG, it is possible to route the packets to the ODL buffers or directly to the second stage. Note that TWCs are also inserted on the outputs of the ODL buffers that are connected to the input of the AWG. Then, by setting the space switch of the second stage appropriately, it is possible to route the signal to the desired output port.

There are some basic problems that must be addressed in realizing optical packet switches. The first problem is the problem of packet coding. Various solutions have been studied including the bit serial, bit parallel, and out-of-band solutions. In the case of the bit serial solutions, the signals are transmitted serially using *optical code division multiplexing* (OCDM), optical

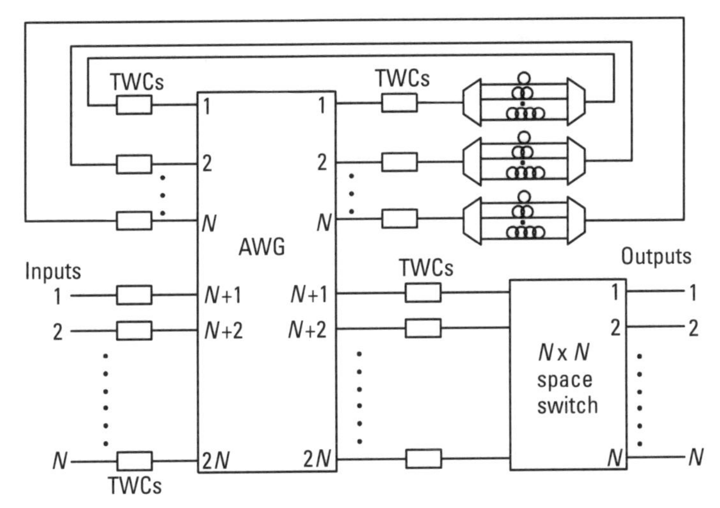

Figure 6.19 Feedback optical packet switching architecture incorporating TWCs and AWGs [11].

pulse interval, and mixed rates. The OCDM systems carry the routing information in the code, while both optical pulse and mixed-rate systems send the routing information in the packet header in front of the packet payload. In the case of the bit parallel solutions, multiple bits are sent simultaneously on different wavelengths. In the case of the out-of-band solutions the packet header is sent on a different channel from that on which the packet payload is sent over.

The second problem is how to resolve contention in the optical switches. Contention occurs at the output ports as packets simultaneously arrive at the same port for transmission. As in the case of classical packet switching, there could be a number of solutions including buffering and deflection routing. Currently, however, there are limitations on using optical buffering as it is still in a rather primitive stage. It basically relies on the use of ODLs, and so cannot be used to buffer a large number of packets for a large number of times. In fact, the unavailability of optical buffers is the main problem for many optical systems. A deflection routing scheme minimizes the use of optical buffers and ODLs but has the disadvantage that the deflected route will most probably result in a longer path to the destination than the original path. Additional solutions would become possible by utilizing wavelength converters, which would then depend on the number of available wavelengths and the availability of wavelength converters.

6.2.6 Wavelength Multiplexers and Filters

Wavelength selective optical components play important roles in optical transmission systems, especially those including WDM operations. Optical filters may be most essential among all frequency selective optical components as others could be constructed out of optical filters. However, wavelength multiplexers and wavelength routers are more directly applicable components. The three components can be compared in the following manner: An *optical filter* takes one wavelength at the output among multiple, different input wavelengths. A *wavelength multiplexer* combines all the input wavelengths into one output light signal. Finally, a *wavelength router* provides reshuffled wavelengths at the output as illustrated in Figure 6.20—that is, two different input ports may be switched to the other outputs port independently of other wavelength components.

There are various different technologies that help to realize the optical filters and wavelength multiplexers/demultiplexers. Recently emerged key technologies among them are *multilayer interference* (MI) filters, *fiber Bragg gratings* (FBG), AWGs, and *acousto-optic tunable filters* (AOTF).

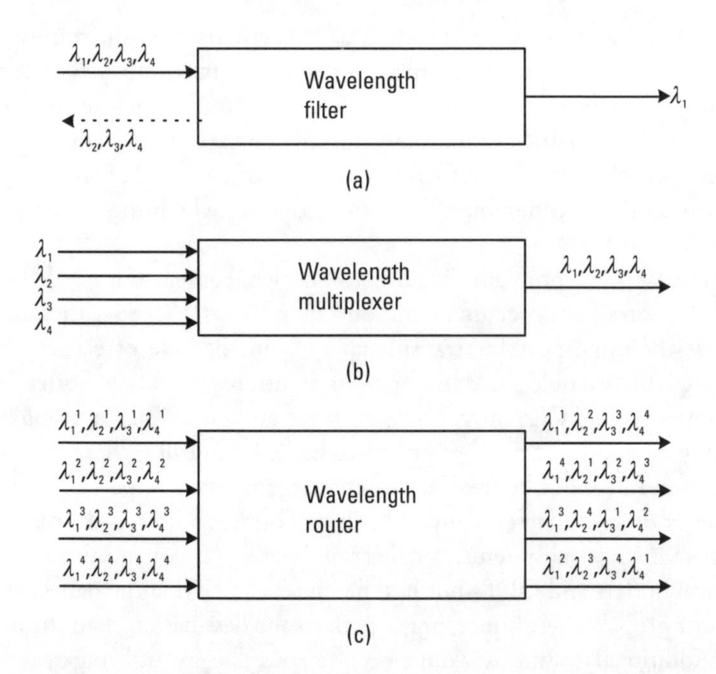

Figure 6.20 Wavelength filters and multiplexers: (a) wavelength filter, (b) wavelength multiplexer, and (c) wavelength router [7].

6.2.6.1 MI Filter

The MI filter consists of a glass substrate on which multiple layers of dielectric thin films are deposited. The thickness and dielectric constants of those multiple layers are arranged such that the interference caused by multiple reflections between the adjacent layers allows a selected wavelength to be transmitted and all others reflected back through the glass. It is a reciprocal device that does the reverse function when input and output are reversed.

Figure 6.21(a) illustrates the operation of the MI filters. When multiple wavelengths are put as the input light signal to the MI filter, one wavelength is propagated through the filter and then transmitted while all the other wavelength components are reflected back. This is a filtering or demultiplexing mode of operation. Conversely, in a multiplexing type of operation, input wavelengths are installed as shown in Figure 6.21(b), with their multiplexed signal taken at the left upper corner.

The filter characteristics of the MI filter can be adjusted by choosing appropriate layering geometry, and it is possible to realize a nearly ideal optical frequency response, having flat passband and steep sideband slopes.

6.2.6.2 FBG

Gratings, in general, refer to the devices whose operation involves interference among multiple components of an identical light source but that have different relative phase shifts. In optics, gratings have been widely used for a separating a light signal into its constituent wavelength components. The same operation can provide wavelength-demultiplexing function if used for separating different wavelength signals out of a multiplexed light signal.

Grating originally means a grating plane on which multiple narrow slits are put equally spaced. Light incident on one side of the grating is transmitted through the slits, and due to diffraction, light transmitted through each

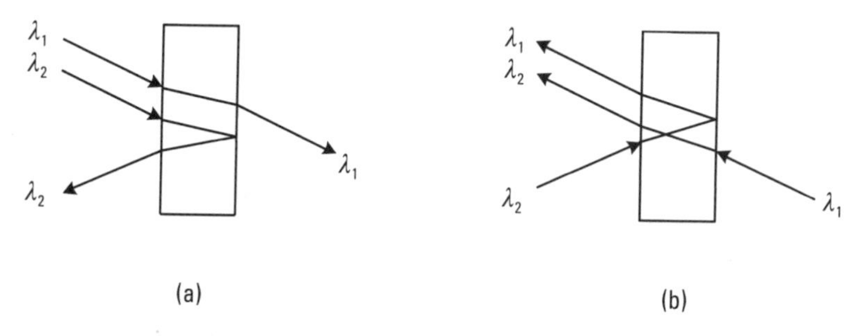

Figure 6.21 Operation of MI filter: (a) multiplexing, and (b) demultiplexing [12].

slit spreads out in all directions. The spread light components from different slits interfere one another, and at some point inside the grating plane the interference occurs constructively, with the constructive interference points differing among different wavelength components. This can lead to wavelength filtering or wavelength demultiplexing, depending on the subsequent optical processing.

Bragg grating refers to the grating caused by periodic perturbation in the propagating medium. The most popular form of perturbation is a periodic variation of the refractive index. Bragg grating, in principle, does the same operation as the conventional grating, segregating different wavelength component signals. If the Bragg gratings are "written" on the optical fiber, they are called fiber Bragg gratings [13].

In FBG, gratings are "written" in the fiber by using photosensitivity of certain type of fibers such as silica fiber doped with germanium. If such a fiber is exposed to *ultraviolet* (UV) light, the refractive index of the fiber core changes. So, grating can be "written" on the fiber by exposing its core to two interfering UV beams, as it causes the radiation intensity to vary periodically along the length of the fiber. The radiation intensity variation is directly connected to the refractive index variation in such a way that the index increases at the point where the intensity is high.

FBG has several advantages—low loss, low temperature coefficient, easy installation, insensitivity to polarization, and low cost. FBG is used for a variety of applications, including optical filters and optical ADM functions. Figure 6.22 illustrates how the FBG can be used in conjunction with the circulator to provide the ADM functions.

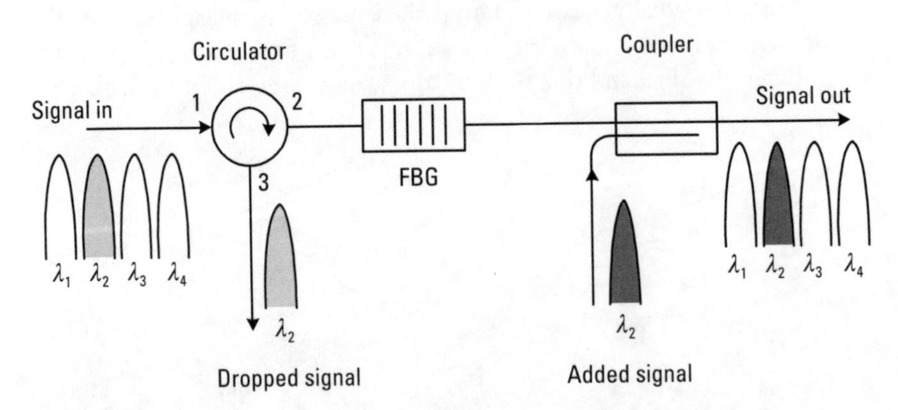

Figure 6.22 Illustration of ADM functions based on the FBG and circulator [7].

6.2.6.3 AWGs

If two 3-dB directional couplers are put in cascade as shown in Figure 6.23, with the length of the two connection arms differing by ΔL, then it can be used for filtering and multiplexing/demultiplexing. This device is called a *Mach-Zehnder interferometer* (MZI).

For the input signal applied to the upper input port, the phase of the optical signal coming out at the lower output port of the first 3-dB coupler lags that at the upper output port by $\pi/2$. So, after passing through the second 3-dB couplers and the connection arm, the phase difference between two signal components (i.e., one via the upper arm and the other via lower arm) grows to $\pi + \beta\Delta L$ at the upper output port and $\beta\Delta L$ at the lower output port for the propagation constant β. Therefore, signal cancellation occurs between the two components such that the two signal components having the wavelengths for which $\beta\Delta L$ becomes an even multiple of π cancel out at the upper output port; and those for which $\beta\Delta L$ becomes an odd multiple of π cancel out at the lower output port.

AWG is a generalization of the MZI, where the number of input-output ports increased from 2×2 to $n \times n$. An MZI is viewed as a device in which two different copies of the same signal, differing in phase shifts, are added together, thereby getting selective signal cancellation. Likewise, an AWG may be viewed as a device in which n different copies of the same signal, differing in phase shifts, are added together. The operation of AWG is therefore similar to that of MZI.

AWG may be used in different ways. It may be used as an $n \times 1$ multiplexer in which n different input signals are combined into one output. Conversely, it can be used as an $1 \times n$ demultiplexer, in which the multiplexed

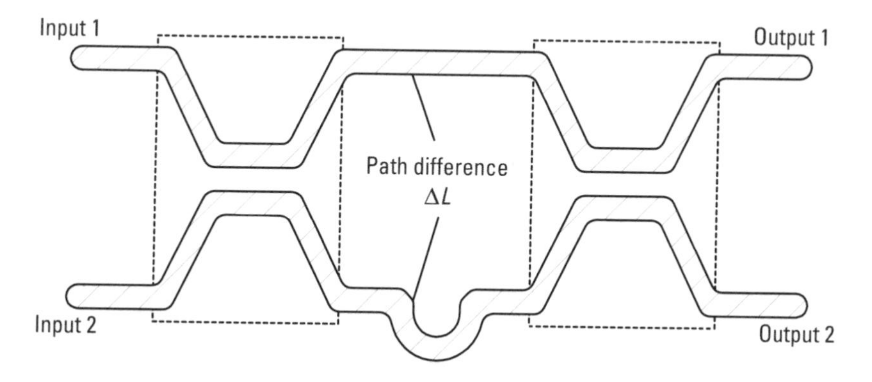

Figure 6.23 Structure of an MZI [7].

signal in one input port is demultiplexed into n output streams at the output. While it is also possible to realize such a multiplexer/demultiplexer using a chain of MZIs, AUG is the preferred choice as it has a lower loss, a flatter passband, and easier implementation on an integrated-optic substrate. AWG can also be used as a router shown in Figure 6.20 or its $n \times n$ extension.

6.2.6.4 AOTF

An AOTF is a tunable filter that can select multiple wavelengths simultaneously. It is constructed by taking advantage of the interaction of sound and light. AOTF is composed of three parts as shown in the integrated-optics implementation in Figure 6.24, namely the front-end coupler, the rear-end coupler, and the acousto-optic polarization region.

The front-end coupler functions as the input polarization beam splitter and the rear-end coupler as the output polarization splitter. A polarization beam splitter splits the *transverse electric* and *transverse magnetic* components of the incident signal as indicated in Figure 6.24. The TE and TM components correspond to the vertical and horizontal components with respect to the substrate surface.

The acousto-optic polarization flipping region is the place where the sound-light interaction takes place. The acoustic transducer in this region, which is driven by a radio frequency source, generates a surface-acoustic wave that propagates in the same direction as the light signal. Then the acousto-

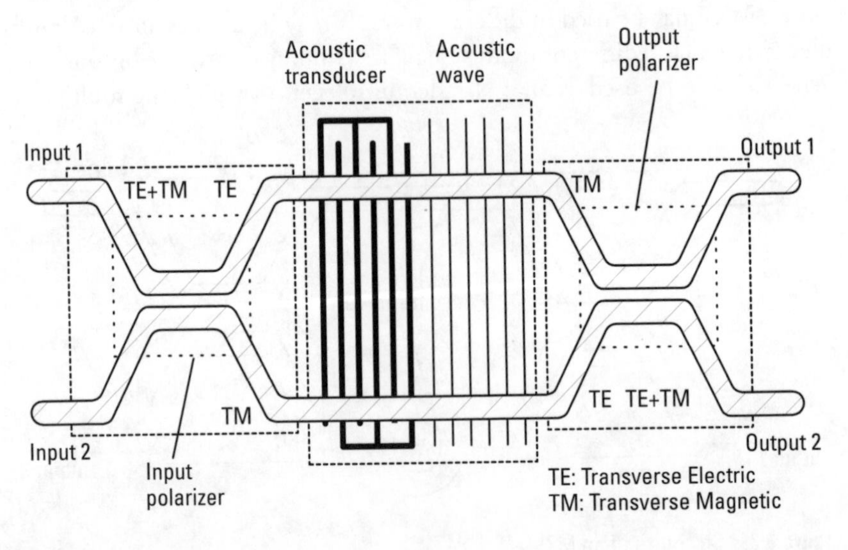

Figure 6.24 Structure of an integrated AOTF [7].

optic interaction process in the material forms a moving grating. This process can be arranged to be phase-matched to the light signal at a selected wavelength by choosing the right frequency of the radio-frequency excitation. A signal that is phase-matched, or "selected," is then flipped from the TM to TE mode, which is then directed to the lower output port by the output polarization beam splitter. The "unselected" signal stays at the upper output. This polarization-flipping-selecting process also takes place for the TE component in a similar manner. As a result, the device sets the cross state for the selected wavelength. Therefore it is possible, by superimposing excitations of multiple radio frequencies, to select multiple wavelengths simultaneously.

As the switch states of AOTF can be controlled independently for multiple wavelengths, an AOTF can function as a 2×2 wavelength selective filter, or a wavelength router. An AOTF renders a general form of 2×2 dynamic router, as the routing pattern can be easily controlled by varying the frequencies of the acoustic waves launched in the device, or the radio frequency applied externally. An $n \times n$ dynamic router can be built by suitably cascading multiple stages of AOTFs.

AOTF is a remarkable device that can "switch" multiple wavelengths simultaneously. The acoustic wave based tuning can take place over a wide range of wavelengths and the switching time falls on the order of microseconds. However, its sideband suppression capability needs improvements.

6.3 Network Architectures

WDM networks can be categorized into two different architectures depending on the ways the wavelengths are utilized. In the local or metropolitan areas, which used to be dominated by bus- and ring-based broadcast networks for copper-wired data communication, wavelength-broadcasting schemes are better fitted. Each station is allocated with a particular wavelength that is distinguished from all other stations in the same network. The different wavelengths from different nodes are collected together at a central node, which are then combined by optical couplers and broadcast back to all the nodes. Finally each node selects using tunable optical filters the wavelength of the destination station out of the aggregate wavelengths. The WDM network architecture of such an arrangement is called *broadcast-and-select architecture.*

On the other hand, wide area networks used to be dominated by mesh networks conventionally, and the same architecture is applied to WDM optical networks. In this case, different wavelengths are allocated to different lightpaths, so wavelengths can be reused in different lightpaths as long as no geographical overlap occurs. Consequently, routing in the network can be wavelength-based. The WDM network architecture having such wavelength-based routing capability is called *wavelength-routing architecture.* The reuse factor of wavelengths in the wavelength-routing network increases if each node has the added capability of wavelength conversion. In this case, a lightpath may be composed of different wavelengths, link by link, so a new lightpath can be created as long as any wavelength is available for every link of the target lightpath.

With the above two network architectures being the two basic but extreme cases, there arrive variations in architecture in the midway. The variation in which the network nodes are capable of routing but the routing is independent of wavelength is called *linear lightwave* architecture. Such an architecture makes sense as the wavelength-independent routers are easier to build, and thus less expensive, than the wavelength-routing ones.

Whereas the above three WDM network architectures are all based on physical topology that relies on physical interconnection of optical fibers, their lightpath-level network architectures may take a much different shape depending on the assignment of wavelengths in the optical layer. In contrast to the term physical topology, the latter is called *virtual topology* (or *logical topology*). For practical use of WDM networks it is important to design the virtual topology in an efficient manner, thereby maximizing the utilization but minimizing the congestion and the cost of the WDM network.

The optical networking revolution brought about by WDM technology has also caused various technical standards organizations to consider defining network architectures that can fully utilize the new technologies. Along this line, the ITU-T has developed the G.872/G.709 Recommendations to define the future *optical transport network* (OTN) based on WDM technology. Specifically, G.872 handles the architecture of the OTN and G.709 handles the NNI for the OTN. As many subjects of these OTN standards are still under development, we place discussions of the OTN in Section 6.6, as an addendum to the chapter.

6.3.1 Broadcast-and-Select Network

In the WDM optical networks employing broadcast-and-select architecture, the source and destination node pairs are connected through a primitive way of broadcasting and selecting. The network collects all the optical signals from all senders in different wavelengths and sends them to all the nodes in the network. The destination node then selects the desired optical signal by tuning the receiver to the corresponding wavelength. It is essentially a single-hop network, so no routing function is involved in this process. The processing is as primitive as in all legacy local area networks.

6.3.1.1 Topologies

As the *broadcast-and-select network* (BSN) shares the common nature of broadcasting and selecting operation with LANs and MANs, the network topologies are also similar. The most popular topologies are the star and bus topologies, which are also called *broadcast-star* and *folded-bus* in the BSN case. Figure 6.25 illustrates the two topologies. In the case of the broadcast-star, the star coupler, which forms the internal core, can be reorganized as shown in Figure 6.26. Broadcast-star and folded-bus networks are both built out of transceivers and 2 × 2, 3-dB couplers. In the case of the folded bus, one out of the four arms of the 3-dB coupler is not used. Each transceiver is assigned with a particular wavelength to transmit signals and is capable of receiving all different wavelengths.

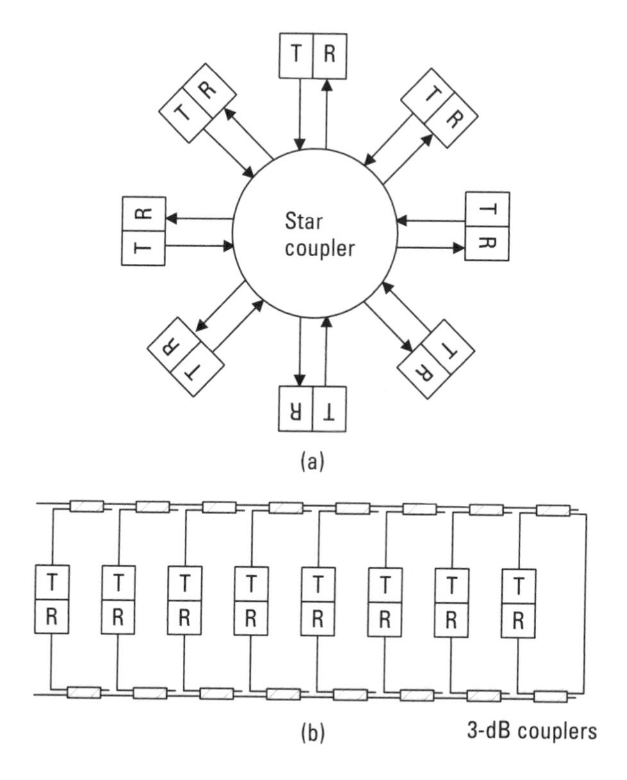

Figure 6.25 Broadcast-and-select network topologies: (a) broadcast-star, and (b) folded-bus [7].

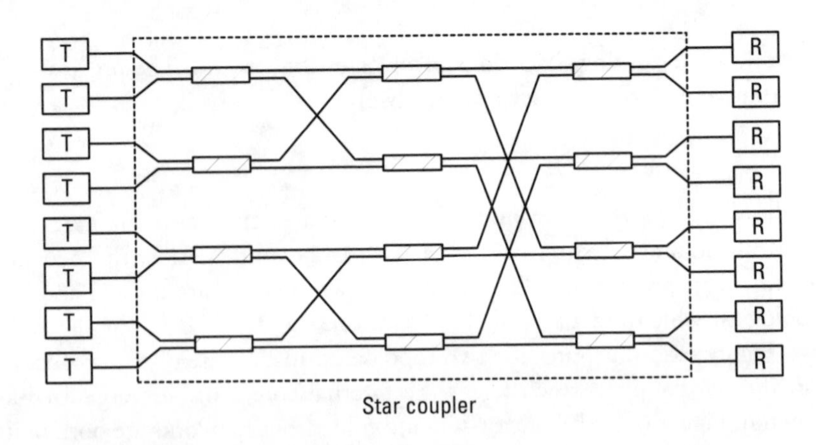

Figure 6.26 Rearranged broadcast-star topology [7].

As can be observed in Figures 6.25 and 6.26, broadcast-star and folded-bus topologies contrast with each other in the number of couplers to install in the network and the number of couplers to pass through for a light signal. For example, in the case of a broadcast-and-select network having n stations, the number of required couplers is $(n/2)\log_2 n$ for the broadcast-star topology and is n for the folded-bus topology. On the other hand, the number of couplers that a light signal has to pass through is $\log_2 n$ for the broadcast-star topology and is n on average and $2n$ at maximum for the folded-bus topology. It is not desirable that the number of couplers changes depending on the light signal, as each additional pass-through implies an additional loss of 3 dB in power.

It is desirable to use one broadcast-star network to accommodate all the stations as much as possible but in the case of large network it does not become feasible (see Section 6.3.1.4). In this case it is desirable to arrange multiple broadcast-star networks to be interconnected through wavelength routers. For example, the 4 × 4 wavelength router shown in Figure 6.20(c) can help interconnect four broadcast-star networks. Let the four broadcast-star networks, N_1 through N_4, each having the wavelengths $\lambda_{n,1}$ through $\lambda_{n,4}$, for $n = 1, 2, 3, 4$, be connected to the left-hand side of the wavelength router, and the four output optical signals connected back to the respective broadcast-star networks. Then broadcast-star network N_1 can be connected to N_2 via $\lambda_{1,2}$, to N_3 via $\lambda_{1,3}$, and to N_4 via $\lambda_{1,4}$. In a circular fashion, each network can be connected to the other networks. So a signal in wavelength $\lambda_{1,1}$ in broadcast-star network N_1, for example, can thus be sent to a port in N_2 by first being transmitted to $\lambda_{1,2}$. A set of wavelengths may be reused in different broadcast-stars for in-star communications as long as the wavelengths are different from those used for inter-star communication.

6.3.1.2 Medium Access

As in the case of LANs and MANs where source nodes broadcast packets on the shared medium and destination nodes select the desired packets among them, the broadcast-and-select optical networks also need MAC protocols for efficient sharing of the medium and maximized throughput. In the case of broadcast-and-select networks, we assume that each station is assigned with a wavelength, and the bandwidth available on a single wavelength can be shared among multiple applications of the same source or destination node, or even of different source and destination nodes. In most cases this sharing is done based on time division multiplexing, with each wavelength channel divided in units of time slots. Then data is generated and transmitted in the form of packets, which are sized to fit into one or multiple of the time slots.

Unlike the LAN or MAN cases, the MAC protocols in the BSN are to be designed considering the available transceivers. The MAC protocol design changes depending whether or not tunable transmitters and receivers are available in the network. Since tunable transmitters and receivers are much more costly than fixed-tuned ones in current technology, it is practical to design MAC protocols to operate with some components being fixed-tuned. Such time-division multiplexed packet processing requires the transmitter or receiver to tune to the desired wavelength in a small fraction of the time slot, which is normally very short. It is technologically challenging to build such fast components, so such optical packet networks are not commercially available to date.

In addition to the component availability issue, synchronization is another important issue in designing packet-based BSNs. As optical buffers are not available in general, and, further, the amount of data carried by a wavelength is enormously large even in a small duration of time, data processing cannot be done properly without keeping the whole network in keen synchronization. As a result, every node needs a suitable time reference that is tuned to the network in such a way that the data transmitted in different time slots can be duly serviced without collision. In the case of the broadcast-star topology network, for example, we keep synchronization of the network by measuring the propagation delay from each station to the star-coupler in the center in advance. Depending on the amount of delay, each station rigorously controls the time to transmit and receive data.

There are several different ways to realize such synchronization. For example, it is possible to set a synchronization node, which issues periodic synchronization pulses, taking into account the distances to each node. In general, the synchronization pulse interval forms a frame and the frame is divided into an integer number of time slots. Once frame synchronization is done, each node can divide it into time slots and determine which slot to load the desired packets.

6.3.1.3 Multiple Access

Among various MAC protocols for multiwavelength optical networks, we first consider the *time-wavelength division multiple access* (T-WDMA) protocol. The T-WDMA protocol assumes that the number of stations n is equal to the number of wavelengths W and that there exists a control wavelength. It requires that each station is equipped with a fixed-tuned data transmitter, a tunable data receiver and, in addition, a fixed-tuned control transmitter and receiver operating at the control wavelength. Time is divided into slots, both for the data channels as well as the control channel, with the size of a data slot being n times the size of a control slot. Each data slot is associated with the corresponding control slot in such a way that for each data transmission a control packet is transmitted first in the control slot, which is immediately followed by a data packet transmission in the data slot. Since data packets of different stations are transmitted on different wavelengths, no collision of data packets occurs. Likewise, there occurs no collision in control packets either, since control packets of different stations are transmitted on different control slots. However contention occurs when two or more nodes intend to transmit data packets simultaneously. In this case, since all the stations monitor the control channel, the transmitting nodes become aware of the contention at the same time as the receiving node does. The throughput of the T-WDMA protocol is much higher than that of the slotted Aloha protocol, which is a primitive MAC protocol.

The T-WDMA techniques are effective means for achieving high degree of connectivity and high throughput. However the station complexity increases rapidly as the number of connections increases. A high degree of connectivity requires many wavelengths, rapidly tunable transceivers, and very high-speed channels, each of which is very costly. A low-cost, low-complexity replacement can be found in *subcarrier multiple access* (SCMA) techniques, in which optical processing in the multiple access is replaced with electronic processing. At each SCMA transmitter, operating at a preassigned optical wavelength, each user data in multiplexed form passes through a frequency multiplexing process on a preassigned subcarrier frequency before being loaded on the optical wavelength. The star coupler in the BSN combines the optical wavelengths coming from all the transmitters and distributes it to all SCMA receivers. Then each receiver receives the aggregated optical signal using a wide-band fixed-tuned receiver such that all wavelengths are included as well as their corresponding subcarrier frequencies. Finally, the receiver detects the desired user data by demultiplexing the corresponding subcarrier frequency, electronically.

Unlike orthogonal techniques such as T-WDMA and SCMA, *optical code division multiple access* (O-CDMA) renders a quasi-orthogonal scheme in the sense that different channels are not completely, but approximately, separable at the receiver. In the O-CDMA technique, each channel is allocated with a unique code, by which the signal spectrum is spread over a wider spectrum. The spread spectrum is overlapped by the spread signals of all constituent channels and delivered to the receivers. Each receiver then despreads the received aggregate signal using the code of the desired channel, thereby detecting the desired signal. In this process the signals of all other channels whose codes do not match turn into interference signals.

6.3.1.4 Limitations

In the BSNs, connectivity is dictated solely by the access stations. Multiplexing and multiple access procedures contribute to realizing high connectivity, but heavily rely on resource sharing in the stations and on the fibers.

The broadcast-and-select architecture in broadcast-star or folded-bus topologies has limitations in scaling up to large networks. There are several factors that limit the scalability. First, it is not possible to reuse optical spectrum because all transmissions share the same fibers. Consequently, the required number of wavelengths grows in proportion to the number of transmitting stations. Second, synchronization, signaling overhead, time delays, and processing complexity all increase rapidly as the number of stations and the number of connections increase. Third, the system has no mechanism for survivability, that is, failure at a critical point (e.g., the star-coupler of a broadcast-star or the head-end of a folded-bus) can break down the whole network.

As a result, the number of stations in a broadcast-star topology is limited to the order of 100. Nevertheless, it is possible to build large networks using broadcast-and-select networks as basic building blocks. In this case, those building blocks are interconnected through wavelength routers, as discussed above.

6.3.2 WRN

In the WDM optical networks employing wavelength-routing architecture, the connectivity between source-destination node pairs is provided by lightpaths, each of which utilizes a single wavelength. Lightpaths are formed by choosing a subset of the links that connect end to end. Wavelength conversion capability may be included in the network such that the series of links constituting a lightpath are comprised of different wavelengths. It is the optical layer that provides such lightpaths to upper layers.

The wavelength-routing capability of the WRNs is in fact implemented in the nodes. The nodes of wavelength-routing networks, in general, contain OLTs, OADMs, and/or OXC systems. An OLT provides an interface point between upper layer signals and WDM signals; OADM provides ADM functionality of the upper layer signals contained in the WDM signals; and OXC reconfigures the constitution of the upper layer signals contained in multiple WDM signals. These node systems are detailed in the following section.

6.3.2.1 Topologies

Most WANs today belong to WRNs and form, in general, mesh networks consisting of nodes and interconnecting links. There are two different kinds of WRNs: static and reconfigurable. The two differ in that static networks use fixed (or static) wavelength converters but not any dynamic wavelength converters or switches inside the OXC nodes, while reconfigurable networks use dynamic wavelength converters or switches. This difference, in effect, is tied with the difference in lightpath reconfigurability; the set of lightpaths established among users is fixed for a static network but can be changed for a reconfigurable network. The reconfiguration is done by changing the states of switches or wavelength converters at the OXC nodes.

A static network can be built out of passive components, so is less costly but more reliable than a reconfigurable network that contains active components. In static networks, wavelengths are tied with the physical network ports, and the wavelengths and ports are determined at the time the network is constructed. Therefore, a static network design is appropriate when a fixed set of lightpath requests exists. In practice, however, such a set of requests is not known in advance or changes with time, so the applicability of static network design is quite limited. In general, a static network requires a much larger number of wavelengths than a reconfigurable network does even when the traffic pattern changes with small variation.

In the case of static networks, the major interest lies in how to reuse the same wavelengths and how many wavelengths are needed to provide a permutation routing among all the nodes. For the former, it is apparent that a wavelength can be reused in geographically different links as long as no wavelength overlap occurs. As to the number of wavelengths needed, it is theoretically known to be possible to construct a static network that is capable of permutation routing by using the wavelength that is about half the user number [14].

In the case of reconfigurable networks, the major interest lies in the minimum required number of switching states or, equivalently, the number of wavelengths. There are some theoretical results available on the relations among the number of switching states, the number of wavelengths, and the traffic set, but the applicability is very limited as they change also depending on the network topology.

To demonstrate the efficiency of the WRNs in terms of wavelength reuse, we consider the four-station star-topology network shown in Figure 6.27. Figures 6.27(a–d), respectively, show a physical network topology, a logical connection graph, a bipartite graph, and the wavelength assignment table obtained assuming that a static wavelength router resides in the central node. From this table we find that the 12 directed connections in Figure 6.27(c) can be realized using only three wavelengths with a fourfold reuse. When compared with the broadcast-star topology that requires 12 wavelengths for the same connections, the static wavelength router network yields four times the efficiency in reuse factor.

6.3.2.2 Routing and Wavelength Assignment

The network design problem, in general, involves a trade-off between the optical layer equipment and upper-layer equipment. Within the optical layer, *routing and wavelength assignment* (RWA) is the most important problem. In essence, given requests for a set of lightpaths, or optical connections, it is important how to determine the routes and wavelengths to satisfy the requests using a minimum possible number of wavelengths. Here, wavelength assignment involves allocating an available wavelength to the connection and tuning the relevant transmitter and receiver to the assigned wavelength. Routing involves determining a suitable lightpath for the assigned wavelength channel and setting switches in the network nodes to establish that lightpath. A lightpath in a WRN is associated with a particular wavelength, so a lightpath can be set up only after a wavelength is assigned.

RWA in the WRNs is subject to two constraints: wavelength continuity and distinct channel assignment. *Wavelength continuity constraint* indicates

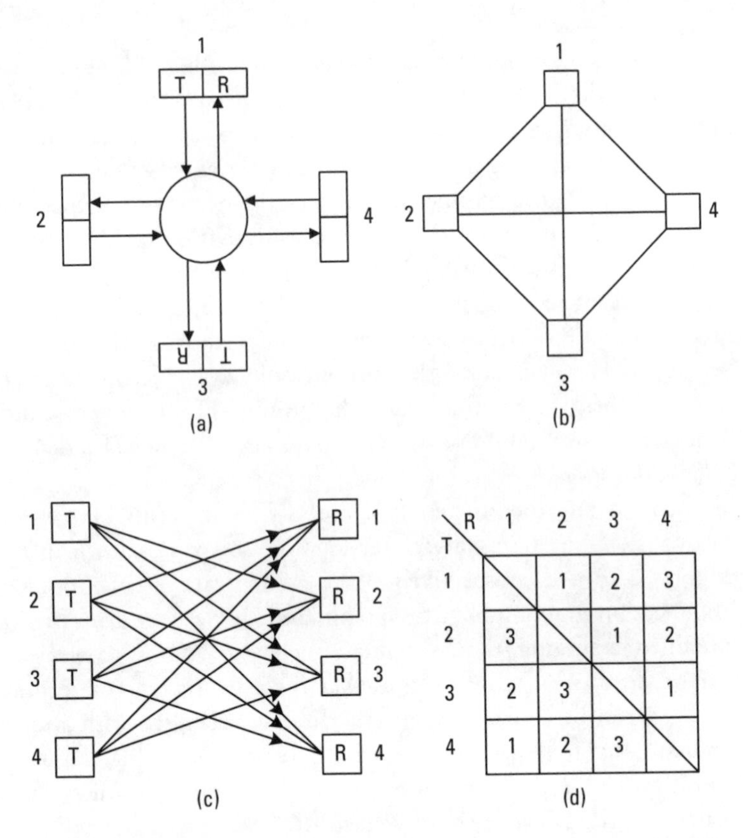

Figure 6.27 Example of a four-station star-topology network: (a) network topology, (b) logical connection graph, (c) bipartite graph, and (d) wavelength assignment table.

that the wavelength of each optical connection remains the same on all links it traverses from the source to the destination. This is a constraint imposed by physical phenomenon. The *distinct channel assignment constraint* implies that all connections sharing a common fiber must be assigned distinct channels. This is a design constraint required for proper network operation. Channels are said to be distinct if they can be distinguished at a receiver when superimposed on its access fiber. The distinct channel assignment condition is necessary to ensure distinguishability of signals on the same fiber, but it is also possible to reuse the same wavelength on fiber-disjoint paths. This includes lightpaths that may be carried on different fibers contained in the same (multifiber) link. Such channel assignment constraints exist in all types of communication networks. In contrast, however, the wavelength continuity condition is unique to transparent optical networks. This makes RWA a more challenging task than the related problems in conventional networks.

RWA becomes a different issue depending on whether the optical connections are dedicated or switched. Dedicated connections are normally maintained for a relatively long period of time, while switched connections are established and released on demand, with comparatively short holding times. Requests for dedicated connections occur in the form of a prescribed set, whereas requests for switched connections occur as a random sequence. Admission control of the connections is made considering the current network load, fairness, priorities, and other factors. Among them, the current state of the network, or the connection pattern that is currently active in the network, is most importantly considered.

The objective of RWA may be stated differently for dedicated connections and switched connections. In the case of the dedicated connections, the objective is to accommodate a prescribed set of optical connections using a minimum number of wavelengths or using the shortest optical paths for each connection. In contrast, in the case of the switched connections, the objective is to maximize the offered traffic for the given fixed number of wavelengths and specified limit on the blocking probability.

In general, the performances of a network depend not only on its physical resources but also on its control mechanisms. In fact, RWA is a fundamental control issue in large WDM optical networks; the objective of a routing and channel assignment algorithm is to achieve the best possible network performances within the limits of the physical constraints. The RWA problem for dedicated connections can be handled off-line so that computation-intensive optimization techniques are applicable. In contrast, RWA decisions for switched traffic must be made on the fly, and hence suboptimal heuristics are normally used.

6.3.2.3 Economic Trade-offs

In the WRNs, the optical layer provides an additional level of grooming for the upper-layer traffic. In some situations the optical layer grooming can help eliminate upper-layer grooming, thereby saving the equipment cost of upper layer. Instead, optical layer grooming cost will be added. For example, if the WRN carries the SDH/SONET traffic, installation of OADMs can waiver the installation of some SDH/SONET ADMs. In effect, it is desirable to have the pass-through traffic handled at the optical layer.

However, the optical layer processing is limited, as it cannot effectively perform any grooming at a rate lower than a wavelength capacity. In the case of a 16-wavelength WRN, with each wavelength carrying 16 of STM-1/OC-3 signals, for example, the optical layer can effectively drop off 16 STM-1/OC-3 signals multiplexed in one wavelength. However, it does not help at all if the 16 STM-1/OC-3 signals to be dropped off are evenly scattered in 16 wavelengths (e.g., one STM-1/OC-3 per wavelength). In this situation, it is helpful if a SDH/SONET DCS reconfigures the traffic in advance so that all those 16 STM-1/OC-3 signals get merged into one or a few wavelengths.

There are several key choices that effect the economic aspects of WRNs. First, wavelength is the core element to save in optical network. Use of more wavelengths can incur additional equipment cost in the optical layer. Line terminations in the upper layer, for example, the SDH/SONET line terminal equipment, is another important element to save, or to minimize. As a lightpath is established between two line terminations, this saving is directly connected to minimizing the lightpaths to set up to support the traffic. It is also desirable to design the network such that the number of hops taken by lightpaths be small because the network design becomes complicated and the resulting optical layer equipment becomes more costly as the number increases.

In WRN design, the economy factor changes depending on the choice of the virtual topology or the arrangement of lightpaths and wavelength assignment. In general, there is a trade-off among different elements. For example, the point-to-point ring arrangement shown in Figure 6.28(a) requires a smaller number of wavelengths than the full-mesh arrangement

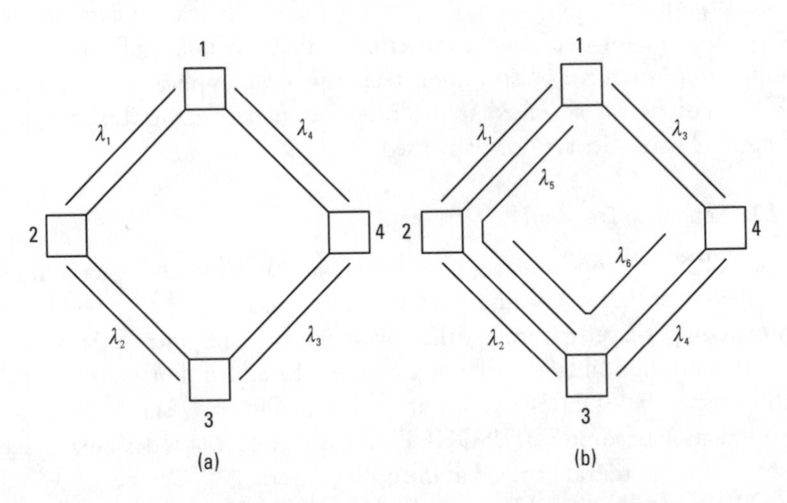

Figure 6.28 Virtual topologies for lightpaths and wavelength assignment: (a) point-to-point ring, and (b) full mesh.

shown in Figure 6.28(b) but requires more line terminations for upper-layer connections. When the carried traffic is light, the point-to-point connection could be a more economical choice than the full-mesh arrangement, but it would be reversed as the traffic becomes heavier.

6.3.3 Linear Lightwave Network

Static network and reconfigurable WRN represent two extreme cases as far as network node functionality is concerned: Static network node has neither wavelength selectivity nor wavelength control, whereas reconfigurable network has independent control of each individual wavelength. In the reconfigurable networks the nodes route light signals based on their wavelengths. A compromise of those two extreme architectures can be found when the nodes are arranged to route light signals but independently of the wavelengths. This type of optical network is called a *linear lightwave network* (LLN). LLN has advantages over those two networks in that the wavelength-independent routers are easier to build and less costly.

Figure 6.29(a) illustrates a four-node LLN. Each node in the network is equipped with a *linear splitter and combiner* (LSC). Figure 6.29(b) shows the

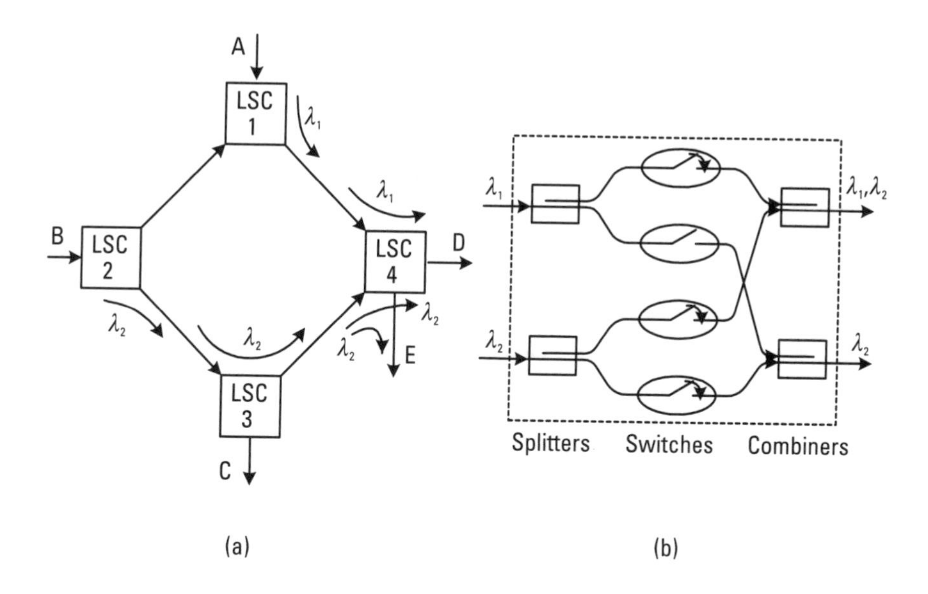

Figure 6.29 Illustration of an LLN: (a) four-node LLN, and (b) LSC in node 4.

block diagram of a simple LSC having two inputs and two outputs. As shown in the figure, an LSC is composed of splitters, switches, and combiners. A splitter does multicasting function on a single input fiber into multiple output fibers, and a combiner does the multiplexing function on multiple switched outputs to form a single output fiber. A switch determines the input-output connection setting based on the split-input signals. In the case of the LSC in Figure 6.29(b), for example, switches S_{11}, S_{21}, and S_{22} are connected and S_{12} disconnected, so input light signal λ_1 is transmitted to the upper output port and λ_2 to both output ports. Accordingly, at the upper output port two input optical signals come out together in mixed form. This example, in fact, illustrates the operation of the LSC in node 4.

If viewed from the standpoint of LLN, the static network is a special example of LLN that has no wavelength selectivity within the nodes (i.e., a single-waveband network) but still has combining and splitting functionality. In contrast, the WRN is another special case of LLN that has no combining and splitting functions but its spectral resolution is refined such that each waveband consists of only one wavelength. *Waveband,* here, refers to a bundle of wavelengths that are closely located. Conversely, LLN can be regarded as a WRN where the routing nodes are not wavelength-selective.

LLN retains wavelength reuse capability in limited degree. For example, in the case of the LLN in Figure 6.29(a), wavelength λ_1 is readily used for the connection of stations A and D. However, this wavelength may be reused for the connection of stations B and C as well, provided that the switches at nodes 2 and 3 are properly set.

Within LLN multicasting becomes easy. A multicast tree that contains all the links for the desired multicast can be easily established by setting the switch states in LSCs accordingly. Multiple multicast trees can be also set up as long as they are edge-disjoint—that is, they do not share any edges in common. This edge-disjoint condition ensures that no interference occurs among the multiple multicast signals.

As discussed in the previous subsection, there are two constraints applied on WRNs, which are wavelength continuity and distinct channel assignment. These two constraints equally apply to LLNs. In addition, there are two constraints that additionally apply to LLNs. They are the inseparability and distinct source combining constraints: To be more specific, the *inseparability constraint* means that wavelengths combined on a single fiber and positioned within the same waveband cannot be separated within the network. This constraint is imposed because LSCs operate on the aggregate power carried within each waveband, without distinguishing different wavelengths in the band. The *distinct source combining constraint* implies that only signals from distinct sources are to be combined on the same fiber. This constraint is readily illustrated in Figure 6.29(b). In fact, it is possible to specify the combining and dividing ratios of signals in the LSCs, which eventually determines the distribution of powers of the sources to the destinations.

6.3.4 Network Topology Design

The design of a WDM network topology is a comparatively complex matter since physical and virtual (or logical) topologies are both involved in the design. A *physical topology* represents the physical interconnection of the wavelength-routing nodes by means of optical fibers, whereas a *virtual* (or *logical*) *topology* represents the optical layer arrangement that results when the lightpaths are set up by suitably configuring the wavelength-routing nodes. In general, a lightpath-level network topology is different from the optical fiber-level network topology, as the requirements to take into account differ between the optical layer and the physical layer.

In a WDM network, the physical topology is the network topology seen by the optical layer, whereas the virtual topology is the network topology seen by the higher layer (e.g., SDH/SONET layer). In the physical topology, a node corresponds to an OXC and a link represents an optical fiber link connecting the corresponding nodes. In contrast, in the virtual topology, a node corresponds to a higher-layer node (e.g., SDH/SONET DCS), and a link represents a lightpath established between the corresponding nodes. A higher layer does routing and multiplexing of its connections over the links in the virtual topology. The virtual topology may take the same shape as the physical topology as in the case of Figure 6.28(a) but can also take a different shape as in the case of Figure 6.28(b).

The optical layer is the intermediate layer that lies between the physical layer of optical fibers and the higher layer of SDH/SONET systems, for example, adding flexibility and efficiency to the network operation. Without the optical layer, the higher-layer links would have to be nailed down on the physical links that cannot be changed easily. With the intervention of the optical layer, however, it becomes possible to change the lightpaths in response to the changes in the traffic at the higher layer. However, as taking down lightpaths can disrupt the traffic in the higher layer, reconfiguration of the lightpaths should be carefully coordinated, depending on the requirements of the higher layer.

Among the several elements that dictate the economy of WDM networks, wavelength is the core element to save as the use of more wavelengths can incur additional equipment cost in the optical layer. Thus, in designing WDM networks, the economy factor changes depending on the choice of the virtual topology, or the arrangement of lightpaths and wavelength assignment. In particular, it is important to design the virtual topology in such an efficient manner that would maximize the resource utilization while minimizing the congestion and the cost of the network.

The optical layer also provides an additional level of grooming for the upper-layer traffic in the WDM network. For example, if the WDM network carries the SDH/SONET traffic, the OADMs installed at the optical layer can share the grooming function at the lightpath level in supplement to the SDH/SONET level grooming or can waiver the installation of some SDH/SONET layer ADMs.

6.3.4.1 WDM Network Design Problem

A network design is normally done based on the information given on the location of the constituent nodes, on the projected traffic among the node pairs, and on the cost of implementing links between each node pair. Based on such information a network designer determines the topology of the network (or the set of links in the network) as well as the traffic arrangement over the topology. This is a complex but well-understood problem and several heuristic algorithms are readily available in the literature.

The network design of the WDM networks would follow a similar pattern to such ordinary network design if the optical link layer design were not taken into account. With the involvement of the optical link design, however, WDM network design turns into a two-level coupled design problem: the physical topology design at the optical-medium layer and the virtual topology design at the optical layer. The physical topology design should take into account the traffic, or the lightpaths, to carry in the physical layer, whereas the virtual topology design should reflect the constraints imposed by the physical topology. In practice, however, the two-level topology design is handled in two decomposed, independent designs, with the virtual topology design done first by reflecting the requirements of the higher layer as well as the constraints imposed by the physical layer. Once the virtual topology design is done, the physical topology design reduces to a conventional network topology design.

For example, in the case of the SDH/SONET services that are carried by the WDM network, the physical topology design is for the configuration of the optical fibers and the virtual topology design is for the configuration of the lightpaths to carry the SDH/SONET services. In this case, the SDH/SONET network in the higher layer would specify the node pairs and the number of lightpaths to transport the service traffic, then the specification would be incorporated in the virtual topology design in the optical layer. As a WDM network would normally support a number of different SDH/SONET networks that require different sets of lightpaths, the physical topology design must accommodate them all, while minimizing the network cost such as OXC node cost and optical link cost.

6.3.4.2 Network Design for SDH/SONET Over WDM

From the higher layer's point of view, the WDM network design means the determination of the set of lightpaths that would transport the higher-layer traffic—that is, the virtual topology design. In this context, the WDM network design issue reduces to a virtual topology design problem. The higher layer wants to see the virtual topology designed such that traffic requirements are met at a minimized cost. The cost in this case may differ depending on the configuration of the lightpaths and the QoS provided by the lightpaths. The cost needs to be minimized not only for the higher layer but for the overall network including both the optical and physical layers.

In the case of the SDH/SONET over WDM network, if we assume that the locations of the nodes and the traffic between node pairs are given, the network design problem then reduces to determining a set of interconnected SDH/SONET rings, each running at a fixed rate (e.g., at the STM-16/OC-48 rate). The SDH/SONET rings are composed of fiber links and SDH/SONET ADMs and are interconnected via SDH/SONET DCSs. The SDH/SONET links are realized by the lightpaths that are routed over the fiber links. The network design objective in this environment is to determine a set of interconnected SDH/SONET rings such that the total network cost is minimized. The total cost here accounts for the cost of the ADMs and DCSs at the SDH/SONET layer as well as the cost of OADMs, OXCs, and optical fibers at the optical layer.

The network design problem for the SDH/SONET over WDM networks consists of several steps that are interrelated. We first need to find a set of lightpaths to carry the traffic, then group the lightpaths into rings, next route the traffic onto the lightpaths, and finally assign wavelengths to the lightpaths. As these four steps are all interrelated, if we perform each step individually, one step at a time, it would result in a suboptimal solution. The problem could become even more complicated if each SDH/SONET ring is allowed to run at a different rate.

6.3.4.3 Virtual Topology Design Approaches

There are several different approaches that have been introduced to date in virtual topology design.

First, B. Mukherjee et al. [15] formulated the virtual topology design as a nonlinear optimization problem, with the objective set to either the delay minimization or the minimization of the maximum offered load. The problem was then divided into four subproblems: determining a logical topology, routing the logical links over physical links, assigning wavelengths to the routes, and routing packet traffic on the virtual topology. To solve the problem, simulated annealing and flow deviation methods were applied, respectively, for the first and the fourth subproblems. This approach is well systematized but the solution is incomplete.

D. Banerjee at al. [16] formulated the virtual topology design as a linear programming problem assuming that network nodes are equipped with wavelength converters. The objective of the formulation was set to minimizing the average hop length of logical links, with the hope of minimizing the number of wavelength converters used. As the approach did not include any traffic variables, it worked reasonably only when the traffic matrix was balanced. In addition, it performed well only when the physical topology was dense in the number of edges and the network size was small. For large networks, it is computationally expensive to solve the linear formulation.

R.M. Krishnaswamy et al. [17] took an exact linear formulation for designing a virtual topology assuming that network nodes are not equipped with wavelength converters. This formulation generalizes the formulations of the above two approaches in the sense that it includes a set of linear constraints that support wavelength continuity constraints. The objective of the linear formulation was set to the minimization of the *congestion*, or the maximum offered load on any logical link. The resulting logical topology reflects the traffic intensities in the constituent nodes. Due to the linearization approach employed, this virtual topology design becomes considerably simple for small networks. For large networks, it is possible to solve the linear formulation by relaxing the integer constraints, thereby obtaining a lower bound on congestion. Section 6.3.4.4 further examines this integer linear programming method.

6.3.4.4 Integer Linear Programming Method

In the integer linear programming method we may arrange a general linear formulation that incorporates routing/assigning wavelengths and routing traffic demands as a combined optimization problem. There are several constraints to take into account in this formulation such as the permitted number of hops, multiple links in the virtual/physical topology, and symmetry/asymmetry restrictions in designing the virtual topology.

We may set the objective function of the integer linear programming to the minimization of the *congestion*, or the maximum traffic at any link in the network, as discussed above. This objective function makes sense since throughput can be defined as the minimum value of the offered load for which the delay on any link becomes infinite, which happens when the maximum traffic at any link in the network is the same as the link capacity. Congestion here may be viewed as a function of various network parameters such as the traffic matrix, number of available wavelengths, resources at each node, hop lengths of the logical links, multiplicity restrictions on the virtual and physical topologies, symmetry/asymmetry restrictions, and propagation delays.

As indicated above, there are several parameters to take into account in the virtual topology design. The number of wavelengths that a fiber can support is an important parameter as wavelengths are scarce and thus have to optimally used. The hop length of a logical link, or the number of wavelength-routing nodes included in a logical link, should be taken into account, as a large hop length accompanies degradation of the optical signal. Symmetry restrictions are an important factor, as congestion is affected by the symmetry restrictions set on the virtual topology. Multiple edges are also an important issue in the virtual topology design as congestion varies if multiple edges are allowed.

There are several constraints to meet in solving the integer linear programming problem. Traffic flow should be conserved at each node in the network except for the source where net in-flow happens and the destination node where net out-flow takes place. Traffic flow at each link is limited by the link capacity. Each node should meet the degree constraints—that is, the numbers of incoming links and outgoing links should not exceed some specified numbers. Various numbers above should be nonnegative, with some of them (e.g., number of nodes) being integers.

In general, a feasible solution of a linear programming problem is a set of values of the variables that satisfy all the constraints, and an optimal solution is a feasible solution that optimizes (i.e., minimizes or maximizes) the objective function. In the case of the integer linear programming problem above, we can obtain the optimal solution by applying a readily established linear programming method under the above constraints such that the objective function, the congestion, gets minimized.

6.4 Add/Drop and Cross-Connect

As the switching and routing functions reside in the center of ATM and IP networks, ADM and cross-connect functions reside in the center of SDH/SONET and WDM/optics networks. In contrast to switches and routers that perform fine-fingered processing for connection among user signals, add/drop multiplexing and cross-connect do big-handed processing for the rearrangement of constituent fat-pipe signals.

While the ADMs and DCSs of the SDH/SONET networks deal with the PDH tributaries and ATM signals for internal processing, the OADMs and OXCs of the WDM/optics networks have STM-*n*/OC-*m* as the signals for internal processing. Consequently, the wavelength-routing capability of WRNs is incorporated in the OADMs and OXCs inside the networks. At the edge of such WDM networks OLTs are placed as the counterpart devices of the OADMs and OXCs in the center.

6.4.1 OLT

An OLT in WDM networks contains a pair of optical multiplexers and demultiplexers as the fundamental building blocks. There are several different ways to construct optical multiplexers and demultiplexers, as discussed in Section 6.2. In practice, in the lower-rate side, OLT interfaces with multiple optical signals that are generated out of SDH/SONET multiplexers or other equivalent devices. In the SDH/SONET case, each multiplexer combines multiple STM-*n*/OC-*m* signals into a high bit-rate optical signal at a particular wavelength. These light signals at different light wavelengths, which are independently generated from different SDH/SONET multiplexers or equivalent devices, should be rearranged into orderly allocated new wavelengths before being multiplexed by the optical multiplexer. The devices that rearrange the wavelengths before optical multiplexing are wavelength converters, which are also called *transponders.* A series of the reversed processings are done in the optical demultiplexer.

A wavelength converter plays an important role not only in constructing OLTs but also in constructing OADMs and OXCs. It helps to improve the utilization of wavelengths in the network and helps to set up lightpaths across different networks administered by different operators or across multivendor systems. It enables setting up a lightpath via a link even if the desired wavelength is preassigned to a third party, thereby enhancing the wavelength utilization.

In implementing wavelength converters, it is possible to use optical devices for the wavelength conversion but electrical implementation is equally possible. In the case of electrical wavelength converters, the incoming wavelength is converted back to an electrical signal by optical detector first and then converted back to another wavelength by laser. An optical multiplexer, in general, requires a number of transponders, which becomes the major source of cost in OLT systems. It is also possible to put SDH/SONET LTEs together with WDM OLTs in one system. In this case cost savings is possible by assigning the wavelengths of the LTE output streams in a orderly manner so that they can be connected to the optical multiplexer without wavelength converters.

On the higher-rate side of the optical multiplexer or demultiplexer, an optical amplifier is usually attached. This amplifier helps to increase the optical power for transmission in the multiplexer and helps to increase the received optical signal for demultiplexer. Figure 6.30 depicts the block diagram of an OLT.

6.4.2 OADM

OADM is the optical counterpart of the electronic ADM function carried out in the SDH/SONET devices. It refers to the function of dropping a small number of wavelengths from an input light signal composed of a multiplexed stream of multiple wavelengths, then adding that many new input wavelengths, finally rebuilding the output light signal that has the same constitution as the original input. The added/dropped wavelength signal may go through further processing in the SDH/SONET or upper layer before/after being added/dropped. If viewed from another viewpoint,

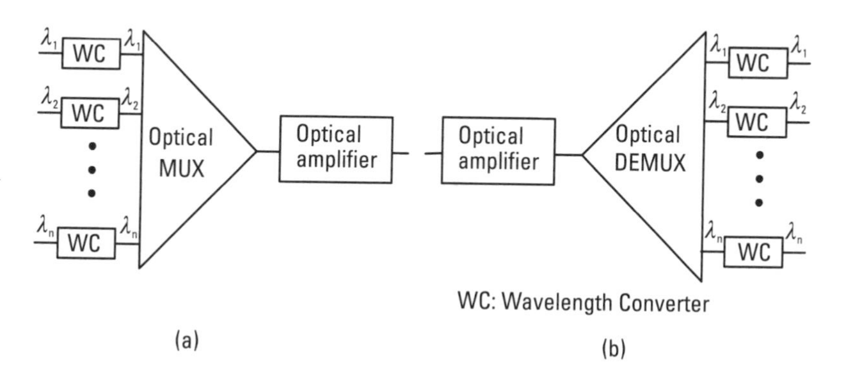

Figure 6.30 Block diagram of OLT: (a) multiplexer, and (b) demultiplexer.

OADM is a function that passes through most wavelengths in a cost-effective manner without applying further upper layer processing, which is applied only to a limited number of wavelengths. In support of ADM function, wavelength conversion processing is usually needed to realign the wavelengths before and/or after the add/drop processing.

OADM (or WADM) is the device that performs the optical add/drop multiplexing function in the wavelength level. As the name indicates, OADM is a special variation of optical multiplexer/demultiplexer. Simply put, an ADM can be constructed by joining an optical multiplexer and optical demultiplexer back to back. Figure 6.31(a) shows a block diagram of an OADM. The add/drop section in Figure 6.31(a) is the spot where add/drop takes place in the wavelength level. An OADM, in its basic functionality, will add/drop light signals as illustrated but can also integrate the SDH/SONET-

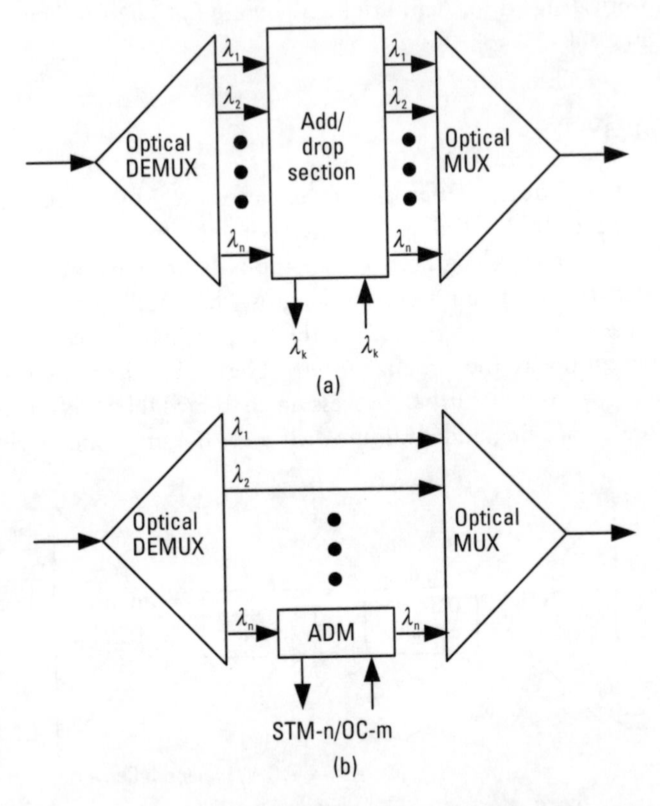

Figure 6.31 Block diagram of OADM: (a) generic form of configuration, and (b) example of single wavelength add/drop (with SDH/SONET ADM attached).

level ADM functions in addition, adding/dropping STM-*n*/OC-*m* signals directly from the OADM. The allowed number of wavelengths to add/drop varies depending on the design. Figure 6.31(b) shows a simple example of the OADM that adds/drops a single wavelength, which is then directly connected to SDH/SONET ADM

6.4.3 OXC

OXC is the function that reshuffles the constitution of the upper layer signals residing in multiple wavelength-multiplexed light signals. Apparently, it is an optical counterpart of the SDH/SONET cross-connect function. OXC is a central function of the wavelength-routing networks.

An OXC (or WXC) system realizes the optical cross-connect function in the optical wavelength level. In general, an OXC system takes interfaces with multiple fiber links, each of which carries multiple wavelengths, with each wavelength signal carrying multiplexed SDH/SONET or other upper layer signals. The reshuffling function, or switching, is then performed on those internal SDH/SONET or other upper-layer signals. Usually, some of such upper-layer signals can be accessed externally for test purposes, and can be added on or dropped off in the cross-connecting process. An OXC can also provide useful accesses for provisioning of lightpaths and protection/restoration of SDH/SONET or other upper layer signals. Figure 6.32(a) shows the generic structure of OXCs.

There are different ways of constructing the OXC systems, depending on the functionality and implementation means. Assuming an all-optical implementation, the configuration of OXCs differs depending whether or not the wavelength-routing functionality is incorporated. In case no wavelength-routing functionality is incorporated, the system structure becomes simple and the reconfiguring or reshuffling capability becomes that much limited as well. Figure 6.32(b) shows an example of such a case. We can observe that there take place w different sets of switching operations among the n sets of w wavelengths carried in n different fibers, where the w switching functions are all independent. Consequently, the reconfiguration capability becomes that much limited. In case wavelength-routing functionality is incorporated, the system structure changes to that shown in Figure 6.32(c). It requires a set of wavelength routers and the reconfiguration capability expands to encompass all the constituent signals.

An OXC can also be realized in a mixed form of optical and electronic processings. As shown in Figure 6.32(d), optically demultiplexed signals are converted to electronic signals and the cross-connect function is done in the

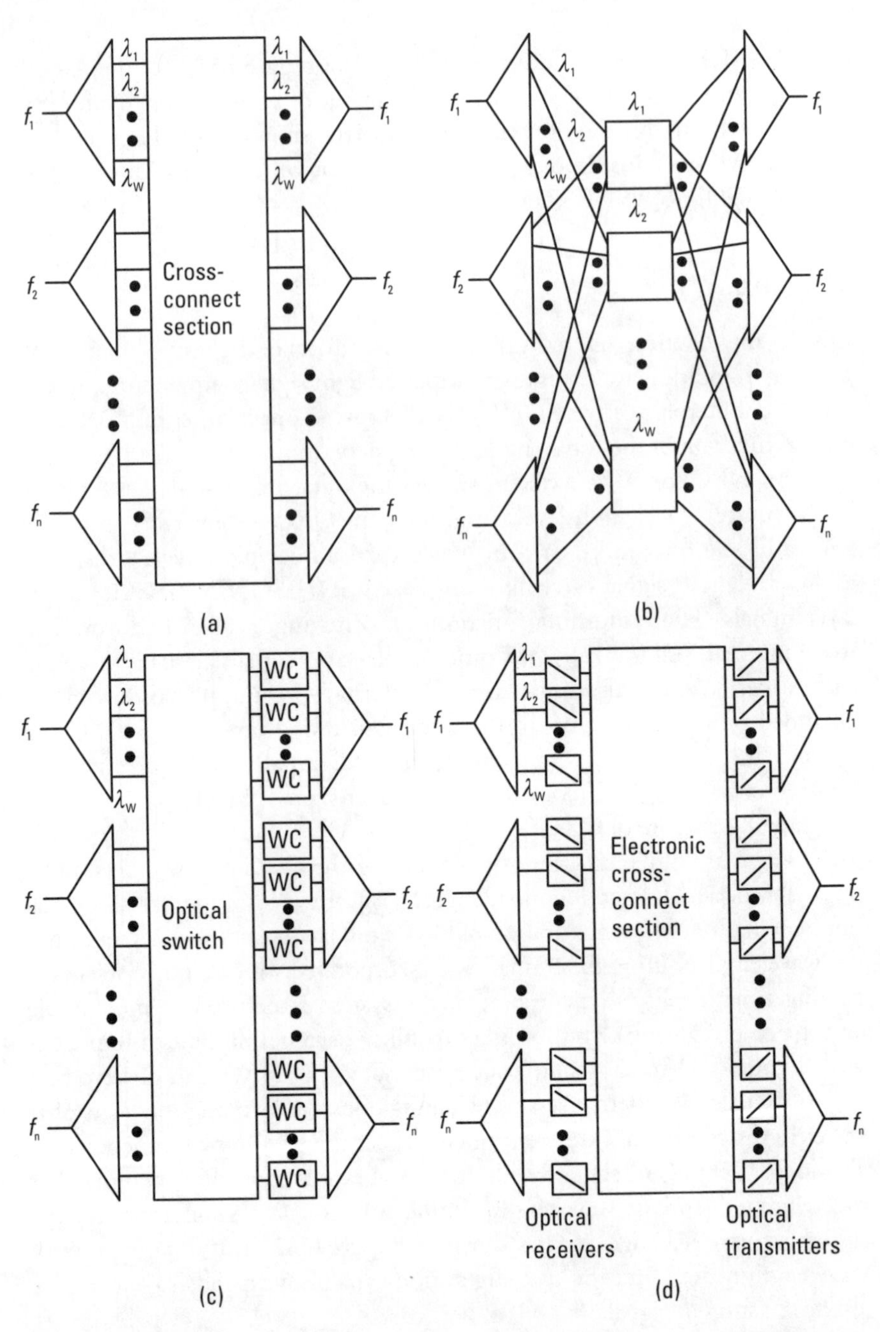

Figure 6.32 Block diagrams of OXC systems: (a) generic structure, (b) without wavelength-routing capability, (c) with full wavelength-routing capability, and (d) optoelectronic implementation.

electronic domain. In this case, wavelength converters are not needed, but optical receivers and optical transmitters are needed instead. Compared to the optical counterpart, an electronic cross-connect is easier and cheaper to implement, as well-proven electronic technologies replace the premature optical technologies. In addition, test access and internal signal monitoring can be more conveniently provided for the electronic cross-connect. However, OXC is advantageous in that it is fully transparent in terms of bit rate and format, and consequently can be applied to extremely high bit-rate signals.

Figure 6.33 shows an example of WDM networks that contains all three optical devices. OLTs are installed at every user interface, where a user meets SDH/SONET or other network service providers. OADMs can be installed at any node in the WDM network but are especially useful in the ring architecture. OADM rings match well with the ring-based protection schemes as will be discussed in Section 6.5. In the case of the OXCs, it is most appropriate to install them at the nodes in the mesh networks. Grooming as well as all wavelength-routing functions can be effectively performed in those OXC-equipped nodes.

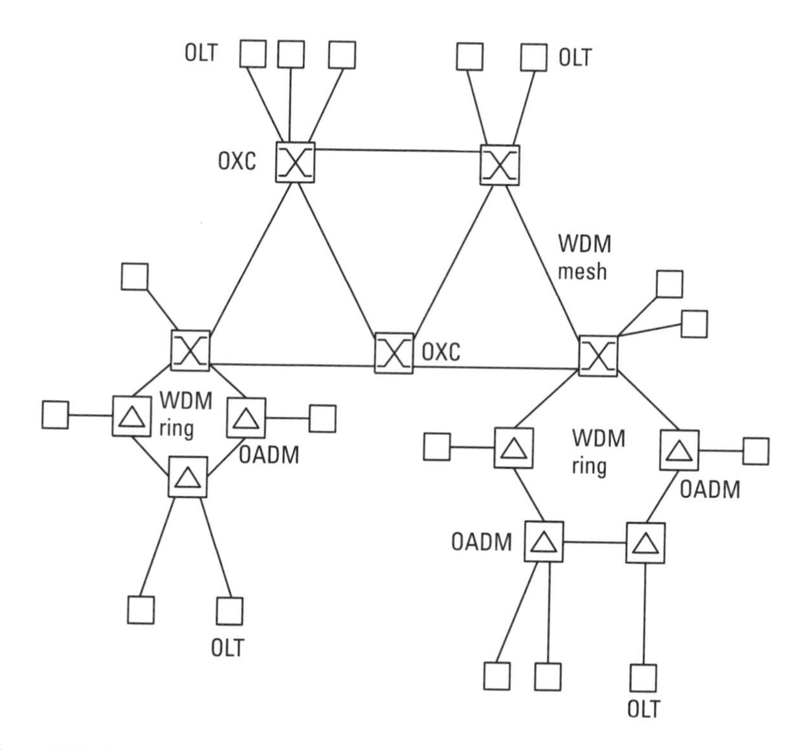

Figure 6.33 Example of WDM networks containing OLTs, OADMs, and OXCs.

6.5 Network Management

The WDM/optics network is an extremely large capacity system whose transmission rates range from gigabits to hundreds of gigabits per second. In such high bit rates, any small defect in the system could cause serious damages, as a car running at an extremely high speed does. Accordingly, it is fundamentally important to manage the WDM/optics network in healthy normal conditions so that services can be provided without interruption even when failures occur in the network. In this sense, network survivability is a crucially important function among all network management functions. WDM/optics networks should be protected from severe failures such as fiber cuts or node failures and should be restored within very short periods of time when such failures occur.

6.5.1 OAM

OAM function is essential to all networks that require reliable quality of services. The OAM function in most cases includes *performance monitoring, defect and failure detection, system protection and restoration, performance information transfer*, and *fault location.* More specifically, it first monitors the performance of the managed network entities continuously or periodically to verify that they are in normal operation. Second, it detects failure conditions through continuous or periodic inspections and generates alarm signals if failures occur. Third, it minimizes the effect of failures by replacing the failed entities and restoring the connections. Fourth, it delivers the performance information to the management entities so that appropriate actions can be taken. Fifth, it uses test equipment to determine the impaired entities, thereby allowing their replacement.

In addition to those OAM functions, a network requires the *network configuration management* and *accounting management.* The former is a network control function that involves installing and configuring network entities as well as establishing, maintaining, and releasing of network connections. The latter involves the billing function for the services provided. In recent years, *security management* has become an important function as network and data protection from unauthorized entities becomes a more and more serious issue. In the case of optical networks, in particular, *optical safety management* is an issue to pay attention to as optical signal, even if low in power, could cause serious damages to the eyes.

Each layer network (for example, ATM, SDH/SONET, and WDM/optics networks) has its own OAM scheme. For example, an ATM network has virtual channel-level and virtual path-level OAM functions, which are represented by the five OAM information flows F1 to F5. The OAM information is carried over the OAM cells. In the case of the SDH/SONET, it has full-fledged OAM features in the physical layer [namely, path, multiplex section (or line), and regenerator section (or section) layers], and has well-developed protection/restoration mechanisms. The OAM information is carried over the path and section overheads within the SDH/SONET networks and is connected to external dedicated networks such as TMN. Among the OAM channels in the SDH/SONET network are the performance monitoring channels, maintenance signal channels, orderwire channels, user channels, and data communication channels.

In the WDM/optics networks case, OAM functions are performed in the optical layer. Even though the aforementioned OAM functions, from performance monitoring to fault location, are all of interest to the optical layer as well, the most important one is the system protection and restoration function. This is critical to the optical layer due to the tremendously large capacity of information that WDM channels carry. While system protection and restoration can also be done in the SDH/SONET layer network, there are certain types of failures such as fiber cut that can be most effectively handled by the optical layer. In case the WDM network carries non-SDH/SONET traffic such as ATM or IP traffic directly, optical layer OAM functions become even more critically important.

6.5.2 Network Survivability

Network survivability refers to the capability of networks to continue providing services even in the presence of failures. Even if it is not possible to avoid failure completely, it is possible to detect the failure, identify the location, and restore the failure quickly enough to maintain the services without interruption. Network survivability may be pursued at each layer network depending on the type of failure, but is most critical in lower layers where high-speed transport services are provided. As a means for network survivability, networks may be designed to keep redundancy in links, capacity, or other facilities such that the traffic in the failed part can be moved to very quickly. Such an arrangement is called *protection*, and the process of shifting the service traffic to the protection facilities is called APS.

There is another way of achieving network survivability called *restoration*. Restoration refers to the arrangement of detouring the traffic based on the reconfigurable cross-connects and spare capacity in the network; if failure occurs, the upper-layer traffic is detoured to the spare capacity in an alternative path that is determined by the reconfigurable cross-connects. In contrast, in the case of protection, the network survivability relies completely on the redundant capacity and the APS is done on the physical layer traffic.

To prepare for network survivability, therefore, the network topology must be designed to have inherent survivability properties, making it capable of surviving single or more failures that occur in lines or equipment. Protection lines and equipment must be appropriately installed, redundant capacity must be secured, and rerouting capability must be implemented. In addition, rapid processing for failure detection, identification, and restoration must be incorporated. This is especially important to achieve fast protection and restoration, thereby minimizing the data loss and service interruption. To realize such fast protection and restoration, the control mechanism must be designed to operate locally in a distributed manner.

In the case of the WDM/optics networks, the failures may be categorized into link, node, and channel failures. The major source of link failures is fiber-cut. Nodes may fail due to power outages or equipment failures. Channels may fail due to the failures of the associated devices or components. In general, the protection and restoration mechanism may be designed against a single failure, assuming that the overall network is reliable enough not to have multiple failures occurring at the same instant. Likewise, it is assumed that no other failure event occurs while an earlier failure is being protected or restored.

WDM/optics network survivability can be pursued in three different ways, depending on the network architecture. In the case of the *point-to-point network* consisting of point-to-point links, a simple protection switching can be applied to the failed links. In the case of the ring network consisting of OADMs and protected fiber links, ring network protection mechanisms similar to that for self-healing SDH/SONET rings can be employed. In the case of the *mesh network* including reconfigurable OXCs, network restoration can be pursued, thereby establishing detour paths within the network. We will briefly consider the protection of the point-to-point network below, deferring the discussions of the other two cases to the following two subsections.[4]

In the point-to-point links, protection is normally done through 1+1, 1:1, or 1:*n* protection switching. In the case of 1+1 protection, the transmitter loads the traffic data on the working and the protection fibers

4. Refer to [7, 11] for more discussion of optical network survivability.

simultaneously. The receiver takes the data from the working fiber in normal operation, but switches to the protection line at the moment it detects any failure. This is the fastest possible protection scheme as the receiver makes its own decision without contacting any other nodes. Instead, it requires two fibers for the transmission of one-fiber capacity, and causes a 3-dB loss of power in case the optical splitter is used in the transmitter.

In the case of 1:1 protection, the traffic data uses only one of two fibers, with the protection fiber staying in standby. If a failure, or a fiber cut, occurs, both the transmitter and receiver switch to the protection fiber simultaneously. In the case of unidirectional fiber, the fiber cut is detected by the receiver only, so the failure information needs to be sent to the transmitter over the APS protocol that was devised for the SDH/SONET protection or something similar.

Apparently, 1:1 protection is not as fast as 1+1 protection in this case, as transmitter-receiver communication is needed. However, the 1:1 protection has the advantage that one of the two fibers is free to use for low priority traffic that may be interrupted whenever protection switching occurs, or for a shared protection with other working fibers. The latter case is the so-called 1:*n* protection. In the case of 1:*n* protection, *n* working fibers share one protection fiber, thereby increasing the utilization of fibers significantly. If a fiber cut happens, the switches at both ends simultaneously switch the failed fiber to the protection fiber, being helped by the APS protocol.

Figure 6.34 illustrates those three protection mechanisms.

6.5.3 Ring Network Protection

Ring networks have an inherent feature that any two points in the ring are connected by two disjoint paths: one is clockwise and the other counterclockwise. This feature makes ring networks resilient to failures. It enables them to incorporate protection mechanisms that detect failures and detour the traffic away from the failed links rapidly. In the SDH/SONET networks, rings having such capability were called SHRs. Such protection capability is equally applicable to WDM/optics networks.

We may classify ring networks depending on the directionality of traffic and the number of fibers, as was the case with the SDH/SONET networks. The three most commonly used ones among them are (two-fiber) UPSRs, *two-fiber bidirectional line-switched rings* (BLSR/2s), and *four-fiber bidirectional line-switched rings* (BLSR/4s). The first is unidirectional and the other two are bidirectional; and the first two are two-fiber-based and the last one is four-fiber-based.

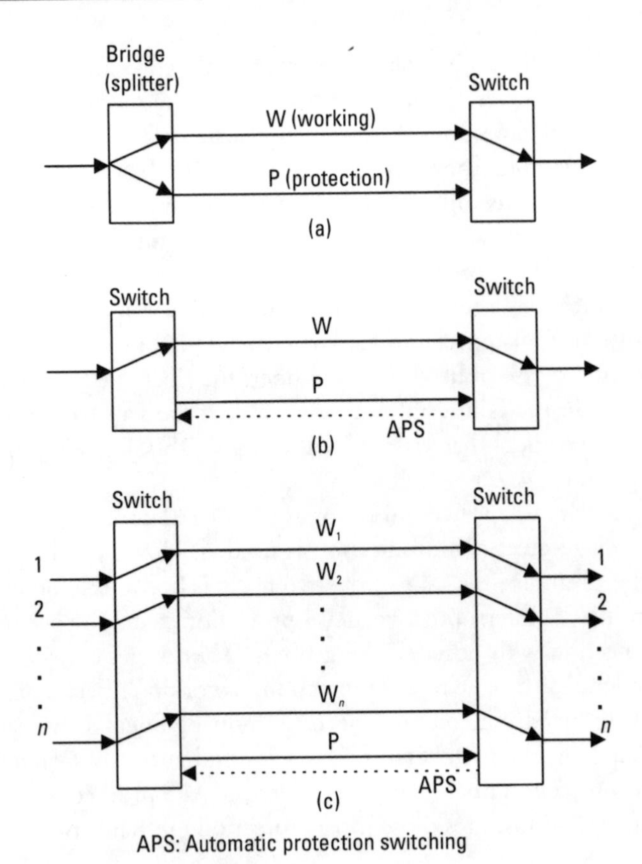

Figure 6.34 Illustration of protection switching in point-to-point networks: (a) 1+1 protection, (b) 1:1 protection, and (c) 1:*n* protection.

6.5.3.1 UPSR

A UPSR has two fibers, one for service and the other for protection. Figure 6.35(a) shows a four-node UPSR, where each node contains an OADM, a solid line indicates a working fiber, and a dashed line indicates a protection fiber. The traffic is sent on both working and protection fibers simultaneously, but to the opposite direction. For example, the traffic from node A to node B is sent counter-clockwise [see the outer solid line of Figure 6.35(b)] and its protection traffic is sent clockwise (see the inner dashed line). Likewise, traffic from node B to node A is sent counter-clockwise (see the outer solid line) with its protection traffic sent clockwise (see the inner dashed line). Node A, as well as node B, continuously monitors the traffic in both the

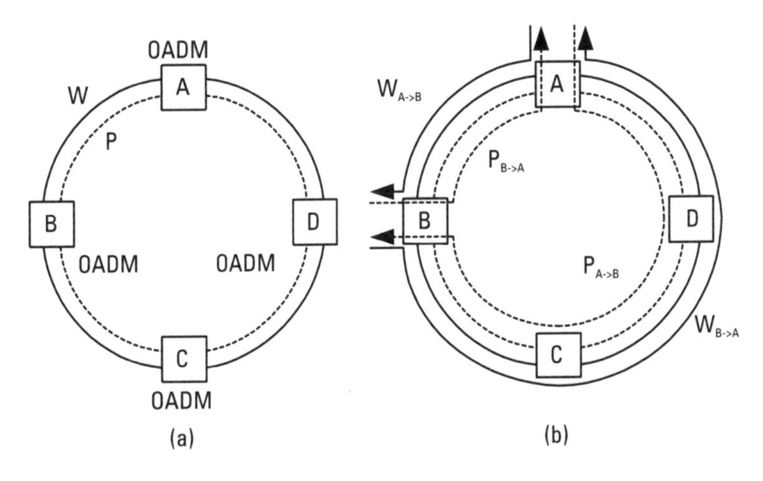

Figure 6.35 UPSR: (a) architecture, and (b) operation.

working and the protection fibers and selects the one with better quality. By doing so, it can perform very fast protection switching in case a failure occurs in the fiber link. In effect, this is essentially the same as the 1+1 protection for the point-to-point links. This protection scheme applies to the case of node failures as well.

UPSRs are easy to implement and operate very fast. Protection switching operation is taken by the receiver only, without any signaling protocols involved. So UPSRs are popular in lower-speed local exchanges and access networks. However, UPSRs have the drawback that half of the overall capacity is reserved for protection, and the fiber capacity is not spatially reusable. The length of the ring should be maintained within some limit such that the difference of delays in the clockwise and counter-clockwise paths does not affect the restoration time.

6.5.3.2 BLSR/4

A BLSR/4 has four bidirectional rings—two rings for service and the other two for protection—and the protection fibers do not carry the traffic in normal operation. This arrangement enables the BLSR/4 to retain the spatial reuse capability. Figure 6.36(a) shows a four-node BLSR/4. Two fibers are used as working fibers, and two for protection. Unlike the case of a UPSR, the working traffic in a BLSR/4 is carried on both directions over a pair of fibers. For example, the traffic from node A to node B is sent counter-clockwise on the working fiber, but the traffic from node B to node A is sent

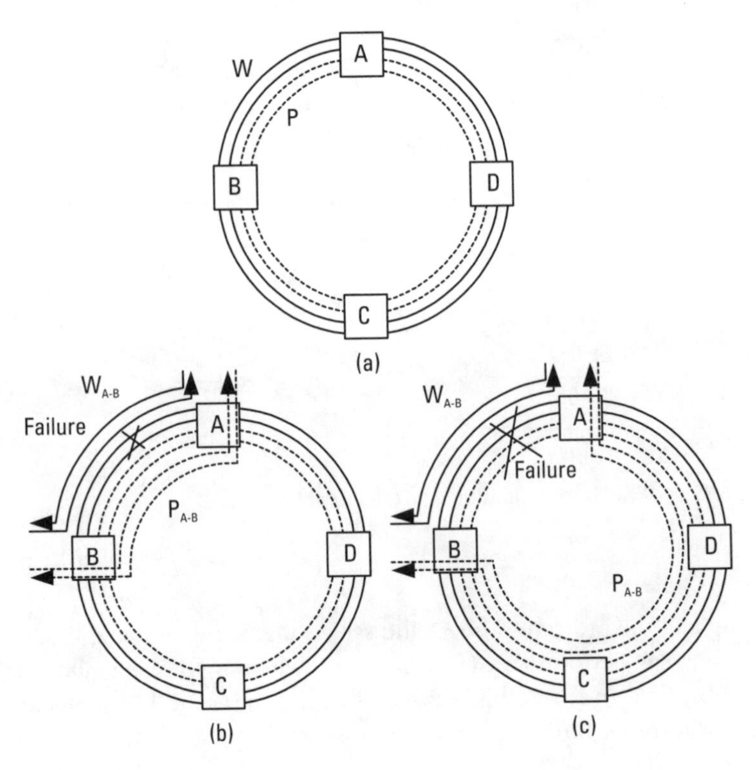

Figure 6.36 BLSR/4: (a) architecture, (b) span protection, and (c) line protection.

clockwise. Such pairs of fibers carrying the bidirectional traffic between a node pair are, in general, selected to be minimal in length, so that the spatial reuse of the fibers on the ring can be maximized.

A BLSR/4 employs two types of protection mechanisms: span protection and line protection. If the working fiber is cut or the transmitter and/or receiver in nodes A and/or B fails, for example, the traffic is switched onto the protection fiber in the same link, as illustrated in Figure 6.36(b). This is called *span protection.* If the working and protection fibers are both cut simultaneously, the traffic is routed through the protection line in the other part of the ring, as illustrated in Figure 6.36(c). This is called *line protection.* Line protection also applies to the case of node failure.

As UPSR is comparable to 1+1 protection, so is BLSR comparable to 1:1 protection on point-to-point links. Hence, the protection bandwidth, in the case of the BLSR, can be used to carry low-priority traffic during normal operation, which is to be preempted when protection switching takes place.

As the protection capacity in the ring is shared among all connections, BLSRs are more efficient than UPSRs.

6.5.3.3 BLSR/2

A BLSR/2 has two fiber rings, both of which carry the working traffic but up to half of the total capacity. The other half of the capacity of each fiber is reserved for protection use. Consequently, a BLSR/2 may be regarded as a BLSR/4, where the protection fibers are "embedded" within the working fibers. Figure 6.37(a) shows the BLSR/2 architecture. Line protection works in the same way as for the BLSR/4. In case a link failure occurs, the traffic on the failed link is rerouted to the protection channels in the two fibers in the other part of the ring, as illustrated in Figure 6.37(b). However, span protection is not possible, since the protection capacity shares the same fiber with the working capacity.

A BLSR/2 is not as effective as a BLSR/4. A BLSR/4 can handle more failures than a BLSR/2 can. A BLSR/4 is also easier to service than a BLSR/2 because multiple spans can be serviced independently. However, a BLSR/4 is more complicated than a BLSR/2 in ring management, because multiple protection mechanisms need to be coordinated.

6.5.3.4 Dual Homing

Ring architectures, in general, are vulnerable to node failures. This could be a serious problem if the failed node is a major hub node that handles a large amount of traffic. One of the most useful ways to resolve the node failure problem is to duplicate the connections between users and nodes in such a

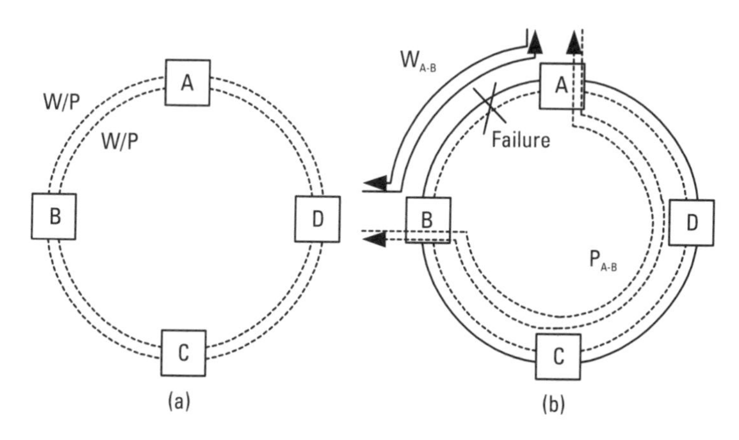

Figure 6.37 BLSR/2: (a) architecture, and (b) operation.

manner that each user is connected to two nodes in the ring. This is called *dual homing*. In this case, if one of the hub nodes fails, the other node can take it over such that the end user can receive the traffic without disruption. The OADMs in the nodes are supposed to be capable of dropping the traffic for the user and continue to carry the same traffic to the next node. In the dual homing arrangement, 1+1 path protection is employed such that the working traffic is sent in one direction of the ring, with the protection traffic sent simultaneously in the other direction.

6.5.4 Mesh Network Restoration

Whereas most nodes in ring networks are linked to two neighboring nodes, most nodes in mesh networks are connected to more than two nodes, in general. This difference in node connectivity is directly related to the node equipment; most nodes in ring networks are equipped with OADMs, whereas most nodes in mesh networks contain OXCs. Consequently, service restoration becomes more complicated in mesh networks than in ring networks.

It is simple and fast, as we have observed in the cases of point-to-point links and ring networks, to offer 1+1 path protection by establishing two edge-disjoint or node-disjoint paths for each connection. However, this approach is not efficient in bandwidth usage as the protection bandwidth is not shared among different connections. Accordingly, it is more efficient to use line protection, which enables the sharing of protection bandwidths among different connections. In this approach, if a link fails every connection carried on that link is rerouted to another path connecting the same node pair as the failed link. This approach is well fitted to mesh networks, as a larger number of alternative protection lines are available in mesh networks than in ring or point-to-point networks.

The most important issue in arranging line protection in mesh networks is how to do it simply and quickly without requiring coordination among multiple nodes on the path. It is most desirable to have the line protection scheme work through local protection switching at the both end nodes of the failed link, without requiring any coordination among multiple nodes. According to the graph theory, fortunately, it is possible to make such an arrangement only if each link has two or more edge-disjoint bidirectional paths. The issue is then how to implement such a line protection scheme without requiring the coordination/communication among multiple nodes in the path.

Figure 6.38 illustrates a line protection scheme based on predetermined protection paths. Figure 6.38(a) shows the architecture of a five-node mesh network in which directed protection links form closed protection loops. These loops render predetermined detouring paths when failures occur. Let us consider the lightpath that connects node A to node C via node B, which is indicated in solid line in Figure 6.38(b). If a failure occurs in the link connecting nodes B and C, then node B switches the connection of the original lightpath to the outbound protection link toward node D. This protection link belongs to the protection loop passing through nodes C and B. Now that the portion connecting nodes C and B in the protection loop is disconnected due to the link failure, the protection loop now goes up to node C only. If node C takes the optical signal out of this protection link, the detouring path is restored between nodes A and C, via nodes B and D. This is shown in Figure 6.38(c). Notice that the switching operation is done only at nodes B and C, which are the end nodes of the failed links. Consequently, this rerouting process can be done very quickly.

6.6 OTNs

As discussed in Section 6.3, OTN is the new network architecture that has been defined by ITU-T for the goal of fully utilizing the WDM technology that has been brought about by the optical networking revolution of recent years. This section concentrates on describing the OTN based on the ITU-T Recommendations G.872 [3] and G.709 [18].

6.6.1 Concept and Architecture

The OTN concept aims to use new technological advances, especially WDM, to support very large bandwidth networks, initially supporting 2.5-Gbps, 10-Gbps, and 40-Gbps links. OTNs will provide flexible point-to-point optical links called *optical channel trails* to users or clients.

Additionally the OTN aims to have a number of other characteristics. First, unlike existing SDH/SONET or PDH networks, the optical channels will provide a logical channel that can transport any digital signals in any format or data rate. The OTN will essentially provide a clear digital pipe for any use. Second, the OTN aims to have an OAM architecture like SDH/SONET that supports performance monitoring, protection, and other management functions. Third, the OTN channels will contain FEC fields to maintain a low BER.

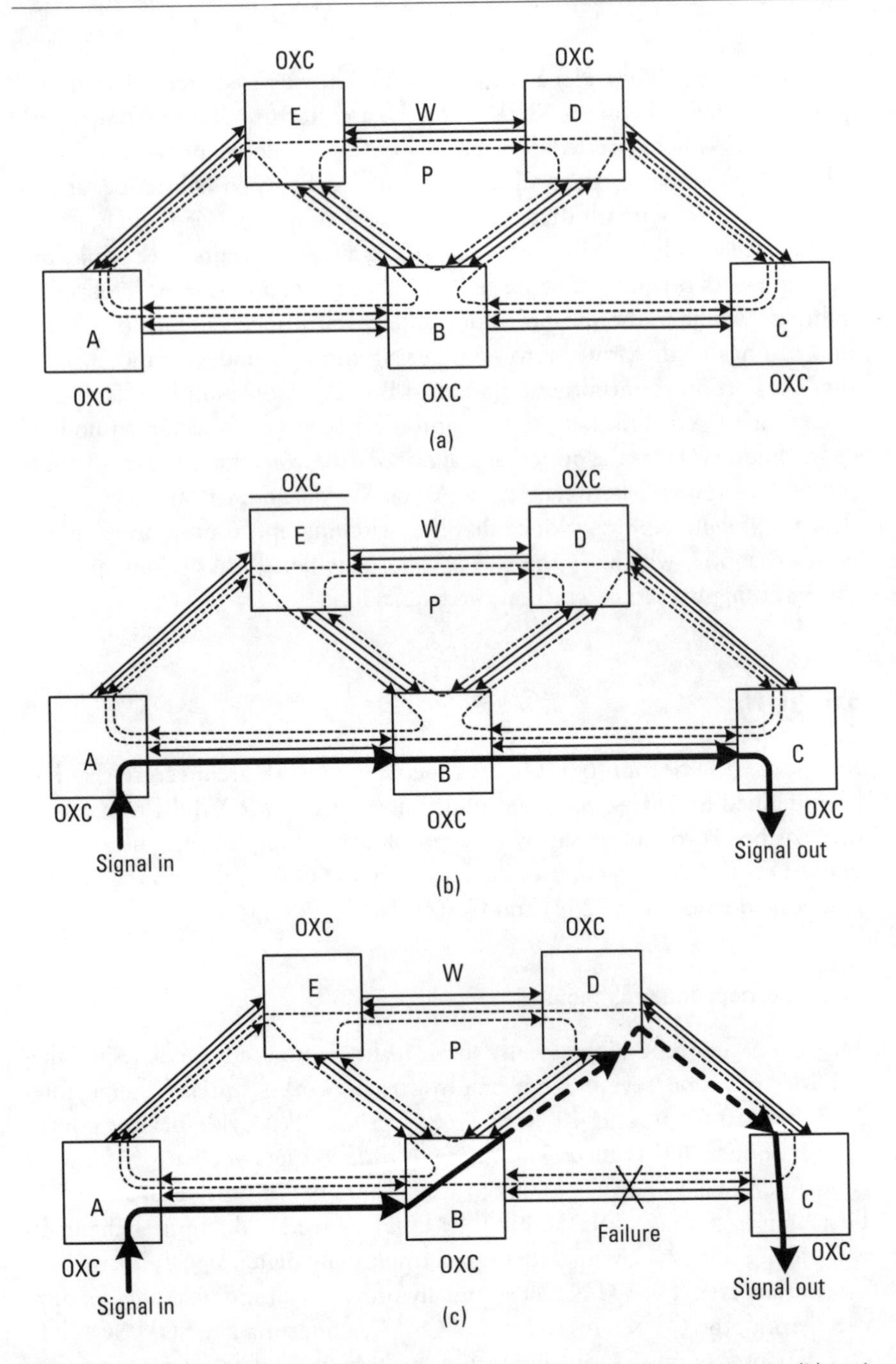

Figure 6.38 Illustration of line protection in mesh network: (a) architecture, (b) path A-B-C, and (c) detour path A-B-D-C.

Within an OTN, OXCs will provide the cross-connect function to the carried traffic data and additionally implement signaling and routing protocols necessary for realizing optical mesh subnetworks. An optical mesh subnetwork refers to a network of OXCs that supports end-to-end networking of optical channel trails.

Overall, an OTN is a group of networks under different administrative domains. Even within the administrative domain of a single network operator, there may be multivendor subnetworks. Under this assumption, G.872 and G.709 define two types of *optical NNI* (ONNI) interfaces, *interdomain interface* (IrDI) and *intradomain interface* (IaDI). The IrDI is the interface between the equipment of *different* vendors. In contrast, the IaDI is the NNI between equipment from the *same* vendor, which may be comprised of proprietary interfaces. Accordingly, the protocol stack for the IrDI is fully defined, but that for the IaDI is not, except for those necessary for network management. Figure 6.39 illustrates the locations of the IrDIs and IaDIs in an OTN.

The key interface structure used to support ONNI is the *optical transport module-n* (OTM-*n*). There are two types of OTMs—OTM-*n* ($n > 0$) and OTM0. OTM-*n* uses WDM to multiplex *n* optical channels, but OTM0 comprises a single optical channel and so does not use WDM. The OTM-*n* structure is composed of the three layers—OCh, OMS, and OTS. In contrast, the OTM0 structure consists only of the OCh and a simplified physical layer. Figure 6.40 shows the layer architecture of the ONNI. For the

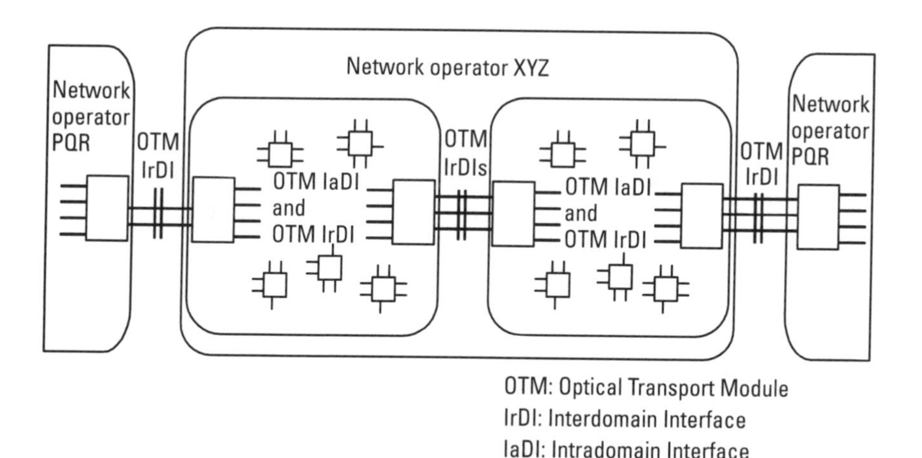

Figure 6.39 Locations of IrDI and IaDI in OTNs [18].

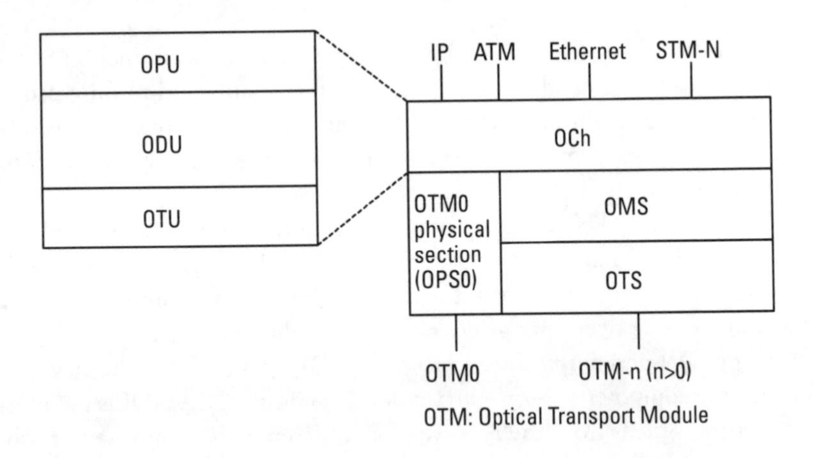

Figure 6.40 Layer architecture of the ONNI [18].

rest of this section, we concentrate on the OTM-n structure as it uses WDM and would be more widely used.

An optical channel trail (which is equivalent to a *lightpath* frequently used in Chapters 6 and 8) refers to a point-to-point optical layer connection between two access points in an optical network. The OCh layer basically provides an end-to-end network of optical channels for transparently conveying user payloads of different formats such as SDH, PDH, and ATM. The OCh layer itself is composed of the three sublayers—*optical-channel payload unit* (OPU), *optical-channel data unit* (ODU), and *optical-channel transport unit* (OTU). We will discuss these sublayers in more detail in Sections 6.6.3 and 6.6.4.

The OMS layer provides functionality for the networking of multiwavelength optical signals. It essentially maps the optical channels to individual wavelengths. The information contained in this layer is a data stream comprising a set of n optical channels, which have a defined aggregate bandwidth.

The OTS layer provides functionality for the transmission of the optical signal on optical media of different types.

6.6.2 Multiplexing Procedure

Figure 6.41 provides a pictorial view of the multiplexing procedure that takes place within the OTM-n structure. It shows how the client data unit is encapsulated in each layer, repeatedly, while being transmitted down to the OTS layer.[5]

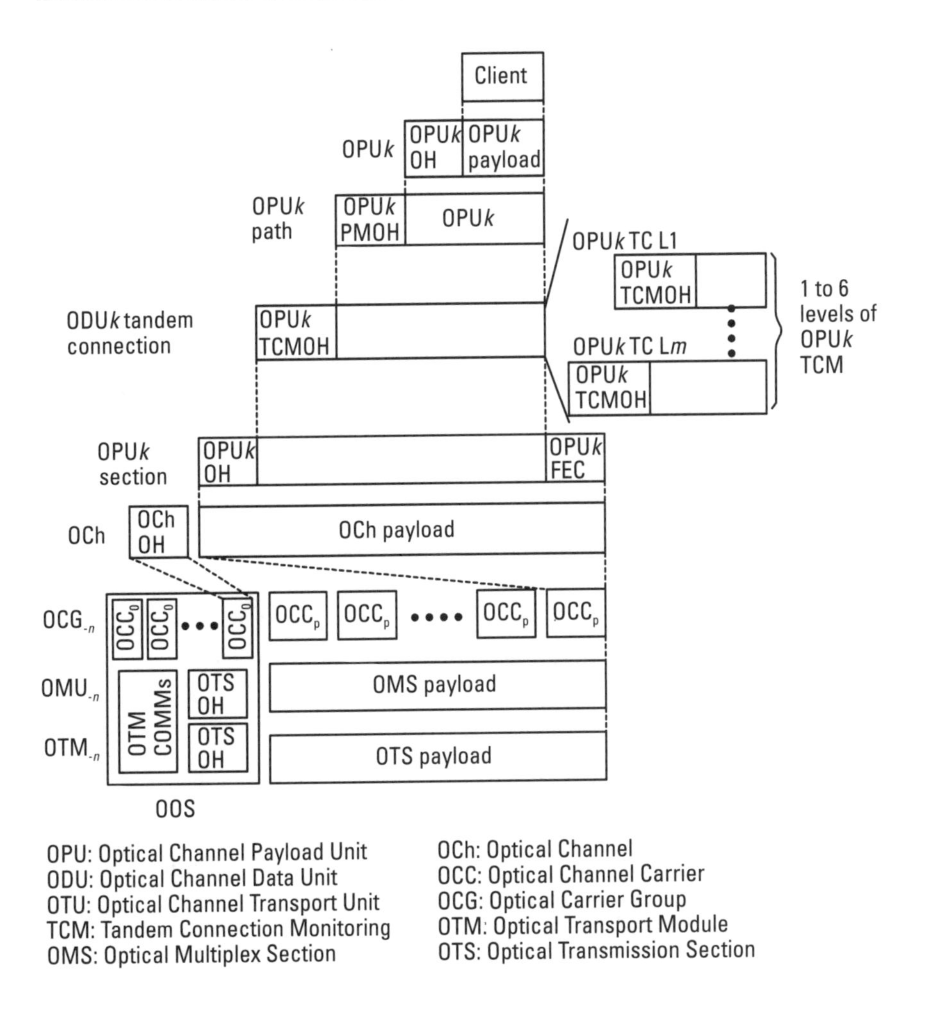

Figure 6.41 Multiplexing procedure for OTM-*n* [18].

The OPU sublayer performs the functionality to carry out rate adaptation between the client data rate and the OTN payload rate. The ODU sublayer supports end-to-end data connections and tandem connection

5. In Figure 6.41, the letter *k* is shown attached to the OPU*k*, ODU*k*, and OTU*k* layers. It can be a value of 1, 2, or 3 and indicates the channel data rates that are mapped to the 2.5-Gbps, 10-Gbps, and 40-Gbps user data rates, respectively. (The exact rates differ for each layer as the number of bits of the overhead and trailers differ at each layer.)

monitoring functionality. The OTU layer is to carry the ODU layer data over one or more optical channels. The OTU*k* is mapped into an OCh payload so that it can be transported over an optical channel.

The *optical channel carrier* (OCC) represents one wavelength channel. An OCh is modulated onto one OCC. An OCC itself consists of two parts, the *OCC payload* (OCCp) and the *OCC overhead* (OCCo). The OCCp carries the OTU information structure over its assigned wavelength, including the header and the trailer parts. Multiple OCCp's are handled as a single WDM group and carried in the OMS and OTS payload spaces. Likewise, multiple OCCo's are also grouped into an *OTM overhead signal* (OOS) and carried separately. Overhead channels for the OMS and the OTS layers are also carried in the OOS.

The *n* multiplexed OCCs are called an *optical carrier group n* (OCG-*n*), and the corresponding OMS and OTS layer information structures are called the *optical multiplex unit n* (OMU-*n*) and OTM-*n*, respectively. Note that these information structures include the physically separate but logically interrelated overhead channels that are being transported in the OOS as well. The OOS itself is carried over a separate optical supervisory channel.

6.6.3 Framing and Encapsulation

Figure 6.42(a, b) shows the frame structure of the OTU*k*, ODU*k*. The frames follow the same convention for transmission as for SDH/SONET—that is, left to right, top to bottom, and the MSB first. Before transmission, a frame-synchronous scrambling is applied on the OTU*k* signal after FEC computation and insertion.

The OTU*k* frame consists of *frame alignment* overhead, OTU*k* overhead, information field, and OTU*k* FEC field. The OTU*k* OH contains various fields for monitoring and reporting on the health of the OTU layer connection. In the OTU*k* OH there is a *general communication channel* (GCC) for proprietary uses between communicating equipment and a *reserved* (RES) field for future use. In addition, there is a *section monitor* (SM) field that contains a number of subfields such as *backward error indication* (BEI), *backward defect indication* (BDI), and *incoming alignment error* (IAE). The *trail trace identifier* (TTI) field in the SM contains the SAPI and the *destination access point identifier* (DAPI) to indicate the access points of the OTU section. The SAPI and DAPI fields are both *access point identifiers* (APIs). As such, they are assumed to be globally unique in its layer network.[6]

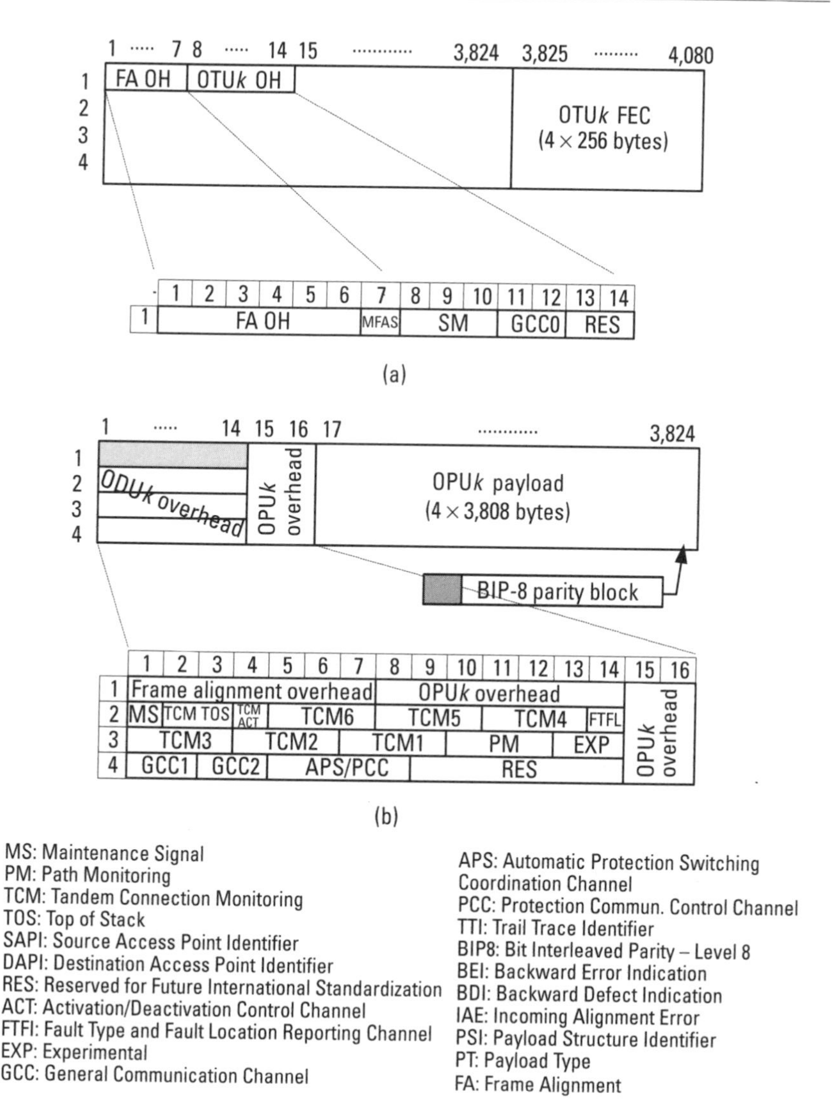

Figure 6.42 Frame structures: (a) OTU*k*, and (b) ODU*k* [18].

6. The access point structure contains two segments—the international segment and the national segment. The national segment is further divided into the ITU carrier code (ICC) followed by the unique access point code (UAPC). It is recommended that these APIs be structured in a tree-like format to aid routing and other control algorithms and mechanisms.

The ODU*k* OH also contains various fields for monitoring and reporting on the health of the ODU layer connection. The ODU*k* OH contains a large number of information fields for the maintenance and operational features needed to support optical channels. The ODU*k* contains fields for the end-to-end ODU*k* path as well as six levels of *tandem connection monitoring* (TCM). These are shown in Figure 6.42 as the TCM1-6 fields. The *path monitor* (PM) field is for the end-to-end path monitoring. In similar manner to the OTU*k* OH, the ODU*k* OH also contains GCC fields for general communication between adjacent equipment. The ODU*k* channel also contains overhead for protection switching in the form of the APS coordination channel and the *protection-communication control channel* (PCC). The *maintenance signal* (MS) field is used for the detection of maintenance signals such as those for indicating alarms or open connections [3, 18].

The OPU*k*/OTU*k* frames of the OTN render a general-purpose encapsulation means for transporting all possible digital transmission signals, thereby conducting the "digital wrapper" functions. Accordingly, the OPU*k* OH contains a large number of mapping-specific fields that depend on the type of the client payloads. In addition, it contains the *payload structure identifier* (PSI) field that includes the PT field. The PT field contains a large number of codes to differentiate the payloads to carry, for example, STM-*n*, ATM cells, and others. The PT code points of the carried payloads are listed in Table 6.1 [18].

Figure 6.43 shows two examples of payload mappings to the OPU*k* frame. Figure 6.43(a) shows how a 2.5G CBR signal (called CBR2G5) is mapped into the OPU1 payload. Negative and positive justifications are used to justify the bytes. The *negative justification opportunity* (NJO) and *positive justification opportunity* (PJO) bytes are for the justification actions and the *justification control* (JC) bytes are for their control. The basic functionality of the justification is similar to that used for justification in SDH/SONET frames.

Figure 6.43(b) shows how an ATM cell stream is mapped into the OPU*k* payload. In this case justification is not needed, so the NJO and PJO fields of the previous case are turned into RES bytes or data bytes. Unlike the SDH/SONET case where the H4 pointer is used to delineate the ATM cell boundaries, there is no such field in the OPU*k* OH. Consequently, the ATM cells must be delineated by using the ATM HEC-based delineation method.

Though not specifically mentioned in the table or recommendations, IP packets may also be mapped into the OPU*k* in a number of ways. Besides the indirect method of using ATM, IP packets can also be mapped into the OPU*k* directly by using the encapsulation methods such as gigabit Ethernet

Table 6.1
Payload Type Code Points

MSB	LSB	Hex code	Interpretation
1 2 3 4	5 6 7 8		
0 0 0 0	0 0 0 0	00	Not available
0 0 0 0	0 0 0 1	01	Experimental mapping
0 0 0 0	0 0 1 0	02	Asynchronous STM-*n* mapping
0 0 0 0	0 0 1 1	03	Bit synchronous STM-*n* mapping
0 0 0 0	0 1 0 0	04	ATM mapping
0 0 0 0	0 1 0 1	05	GFP mapping
0 0 0 1	0 0 0 0	10	Bit stream with octet timing mapping
0 0 0 1	0 0 0 1	11	Bit stream without octet timing mapping
1 0 0 0	x x x x	80–8F	Reserved codes for proprietary use
1 1 1 1	1 1 0 1	FD	NULL test signal mapping
1 1 1 1	1 1 1 0	FE	PRBS test signal mapping
1 1 1 1	1 1 1 1	FF	Not available

From: [18].

or SDL (refer to Section 8.3). All that the client would have to offer would be IP packets encapsulated in some continuous byte stream.

6.6.4 OAM and Survivability

One of the distinct characteristics of the ODU layer is its ability to support TCM. This means the function of monitoring multiple link layers. This ability uses the TCM fields and the TOS field defined in the ODU*k* OH. This is best shown by the example in Figure 6.44.

As shown in Figure 6.44, the six TCM fields are used to monitor in tandem separate overlapping links. The TCM3 field monitors the C1-C2 segment. The TCM1 field is used to monitor the A1-A2 segment. The TCM2 field is used in two different segments, the B1-B2 segment and the B3-B4 segment. The TOS field basically keeps track of the TCM to check which TCM field is the most "significant" in this link. In other words, the TOS field acts like a pointer, pointing to one of the TCM fields and thereby indicating that it is the current most significant TCM field.

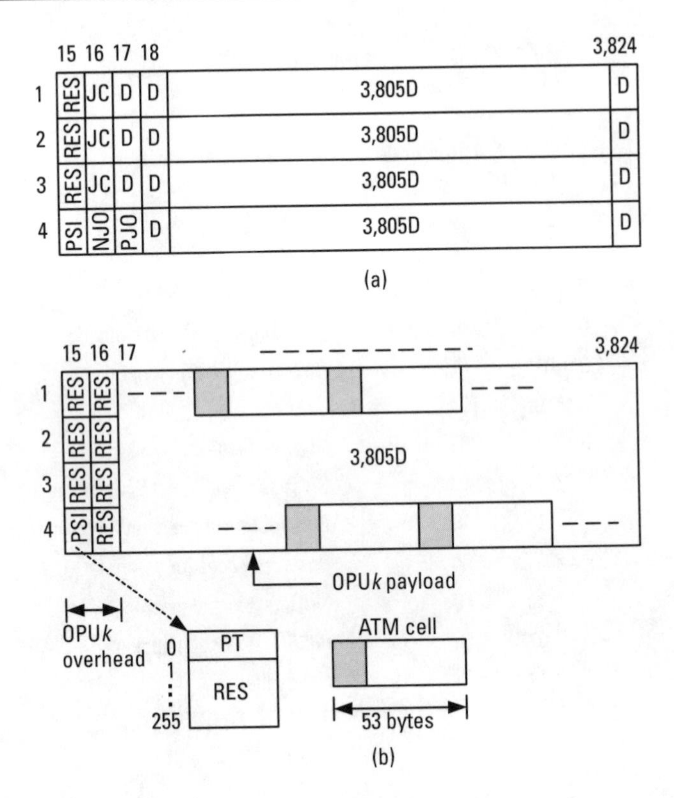

Figure 6.43 Mapping payloads to OPU*k*: (a) mapping 2.5-Gbps CBR signal to OPU1, and (b) mapping of ATM cells to OPU*k*.

The OTN architecture includes many other OAM and survivability features, for which the reader can refer to [3, 18].

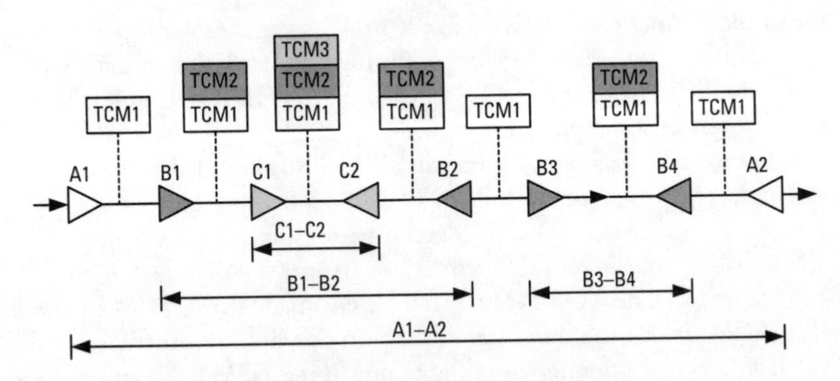

Figure 6.44 Illustration of monitoring of tandem ODU*k* connections.

References

[1] Lee, B. G., M. Kang, and J. Lee, *Broadband Telecommunications Technology,* 2nd ed., *Norwood, MA: Artech House, 1996.*

[2] Kartalopoulos, S. V., *Introduction to DWDM Technology,* New York: IEEE Press, 2000.

[3] ITU-T Rec. G.872 (02/99) "Architecture of Optical Transport Networks."

[4] Mynbaev, D. K. and L. Scheiner, *Fiber-Optic Communications Technology,* Englewood Cliffs, NJ: Prentice Hall, 2001.

[5] ITU-T Rec. G.651 (02/98) "Characteristics of a 50/125µm Multimode Graded Index Optical Fiber Cable."

[6] ITU-T Rec. G.652 (10/2000) "Characteristics of a Single-Mode Optical Fiber Cable."

[7] Ramaswami, R., and K. Sivarajan, *Optical Networks: A Practical Perspective,* San Francisco, CA: Morgan Kaufmann, 1998.

[8] Sudo, S., *Optical Fiber Amplifiers: Materials, Devices, and Applications,* Norwood, MA: Artech House, 1997.

[9] Becker, P. C., et al., *Erbium-Doped Fiber Amplifiers: Fundamentals and Technology,* New York: Academic Press, 1999.

[10] Mouftah, H. T., and J. M. H. Elmirghani, *Photonic Switching Technology: Systems and Networks,* New York: IEEE, 1998.

[11] Hunter, D. K., and I. Andronovic, "Approaches to Optical Internet Packet Switching," *IEEE Communications Magazine,* Vol. 38, No. 9, September 2000, pp. 116–122.

[12] Stern, T. E., and K. Bala, *Multiwavelength Optical Networks: A Layered Approach,* Reading, MA: Addison Wesley, 1999.

[13] Othonos, A., and K. Kalli, *Fiber Bragg Gratings: Fundamentals and Applications in Telecommunications and Sensing,* Norwood, MA: Artech House, 1999.

[14] Barry, R. A., and P. A. Humblet, "On the Number of Wavelengths and Switches in All-Optical Networks," *IEEE Transactions on Communications,* Vol. 42, Nos. 2, 3, 4, February/March/April 1994, pp. 583–591.

[15] Mukherjee, B., et al., "Some Principles for Designing a Wide-Area Optical Networks," *IEEE/ACM Transactions on Networking,* Vol. 4, No. 5, December 1996, pp. 684–696.

[16] Banerjee, D., and B. Mukherjee, "Wavelength-Routed Optical Network: Linear Formulation, Resource Budgeting Trade-offs, and a Reconfiguration Study," in *Proceedings of INFOCOM'97*, 1997, pp. 269–276.

[17] Krishnaswamy, R. M., and K. N. Sivarajan, "Design of Logical Topologies: A Linear Formulation for Wavelength-Routed Optical Networks with No Wavelength Changers," *IEEE/ACM Transactions on Networking,* Vol. 9, No. 2, April 2001, pp.186–198.

[18] ITU-T Rec. G.709 (02/2001) "Interface for the Optical Transport Networks (OTN)."

Selected Bibliography

Acampora, A. S., "The Scalable Lightwave Network," *IEEE Communications Magazine,* Vol. 32, No. 12, December 1994, pp. 36–43.

Ali, M. A., et al., "High-Speed Optical Time-Division Demultiplexer Using Semiconductor Optical Amplifiers," *IEEE/OSA Journal of Lightwave Technology,* Vol. 10, No. 11, November 1992, pp. 1735–1742.

Ayanoglu, E., and R. D. Gitlin, "Broadband Network Restoration," *IEEE Communications Magazine,* Vol. 34, No. 7, July 1996, pp. 110–119.

Bala, K., and T. E. Stern, "Routing in Linear Lightwave Networks," *IEEE/ACM Transactions on Networking,* Vol. 3., No. 4, October 1995, pp. 489–500.

Banerjee, S., and B. Mukherjee, "Fairnet: A WDM-Based Multiple Channel Lightwave Network with Adaptive and Fair Scheduling Policy," *IEEE Journal of Lightwave Technology,* May 1993.

Baroni, S., and P. Bayvel, "Wavelength Requirements in Arbitrary Connected Wavelength-Routed Optical Networks," *IEEE/OSA Journal of Lightwave Technology,* Vol. 15, No. 2, February 1997, pp. 242–252.

Barry, L. P., et al., "A High-Speed Optical Star Network Using TDMA and All-Optical Demultiplexing Techniques," *IEEE Journal on Selected Areas in Communications,* Vol. 14, No. 5, June 1996, pp. 1030–1038.

Barry, R. A., and P. A. Humblet, "Models of Blocking Probability in All-Optical Networks with and Without Wavelength Changers," *IEEE JSAC/JLT* (special issue on Optical Networks), Vol. 14, No. 5, June 1996, pp. 858–867.

Bigo, S., O. Leclerc, and E. Desurvire, "All-Optical Fiber Signal Processing and Regeneration for Soliton Communications," *IEEE Journal on Selected Topics in Quantum Electronics,* Vol. 3, No. 5, October 1997, pp. 1208–1223.

Chan, C., L. Chen, and K. -W Cheung, "A Fast Channel-Tunable Optical Transmitter for Ultrahigh-Speed All-Optical Time-Division Multiaccess Networks," *IEEE Journal on Selected Areas in Communications,* Vol. 14, No. 5, June 1996, pp. 1052–1056.

Chan, V. W. S., et al., "Architectures and Technologies for High-Speed Optical Data Networks," *IEEE/OSA Journal of Lightwave Technology,* Vol. 16, No. 12, December 1998, pp. 2146–2168.

Chew, Y. H., Tjeng Thiang Tjhung, and F. V. C. Mendis, "An Optical Filter of Adjustable Finesse Using an Amplified Fiber Ring Resonator," *IEEE/OSA Journal of Lightwave Technology,* Vol. 15, No. 2, February 1997, pp. 364–370.

Chraplyvy, A. R., "Limitations on Lightwave Communications Imposed By Optical Fiber Nonlinearities," *IEEE/OSA Journal of Lightwave Technology,* Vol. 8, No. 10, October 1990, pp. 1548–1557.

Derr, F., et al., "An Optical Infrastructure for Future Telecommunications Networks," *IEEE Communications Magazine,* Vol. 33, No. 11, November 1995, pp. 84–88.

Dutton, H. J. R., *Understanding Optical Communications,* Englewood Cliffs, NJ: Prentice Hall, 1999.

Ellinas, G., A. G. Hailemariam, and T. E. Stern, "Protection Cycles in Mesh WDM Networks," *IEEE Journal on Selected Areas in Communications,* Vol. 18, No. 10, October 2000, pp. 1924–1937.

Fujimoto, N., et al., "Photonic Highway: Broadband Ring Subscriber Loops Using Optical Signal Processing," *IEEE/OSA Journal of Lightwave Technology,* Vol. 7, No. 11, November 1989, pp. 1798–1805.

Ghafoor, A., M. Guizani, and S. Sheikh, "Architecture of an All-Optical Circuit-Switched Multistage Interconnection Network," *IEEE Journal on Selected Areas in Communications,* Vol. 8, No. 8, October 1990, pp. 1595–1607.

Gillner, L., C. P. Larsen, and M. Gustavsson, "Scalability of Optical Multiwavelength Switching Networks: Crosstalk Analysis," *IEEE/OSA Journal of Lightwave Technology,* Vol. 17, No. 1, January 1999, pp. 58–67.

Goto, N., and Y. Miyazaki, "Integrated Optical Multi-/Demultiplexer Using Acoustooptic Effect for Multiwavelength Optical Communications," *IEEE Journal on Selected Areas in Communications,* Vol. 8, No. 6, August 1990, pp. 1160–1168.

Iness, J., et al., "Elimination of All-Optical Cycles in Wavelength-Routed Optical Networks," *IEEE/OSA Journal of Lightwave Technology,* Vol. 14, No. 6, June 1996, pp. 1207–1217.

ITU-T Rec. G.650 (10/2000) "Definition and Test Methods for the Relevant Parameters of Single-Mode Fiber."

ITU-T Rec. G.653 (10/2000) "Characteristics of a Dispersion-Shifted Single-Mode Optical Fiber Cable."

ITU-T Rec. G.654 (10/2000) "Characteristics of a Cut-Off Shifted Single-Mode Optical Fiber Cable."

ITU-T Rec. G.655 (10/2000) "Characteristics of a Non-Zero Dispersion Shifted Single-Mode Optical Fiber Cable."

ITU-T Rec. G.661 (10/98) "Definition and Test Methods for the Relevant Generic Parameters of Optical Amplifier Devices and Subsystems."

ITU-T Rec. G.662 (10/98) "Generic Characteristics of Optical Fiber Amplifier Devices and Subsystems."

ITU-T Rec. G.663 (04/2000) "Application-Related Aspects of Optical Amplifier Devices and Subsystems."

ITU-T Rec. G.664 (06/99)" Optical Safety Procedures and Requirements for Optical Transport Systems."

ITU-T Rec. G.671 (02/2001) "Transmission Characteristics of Optical Components and Subsystems."

ITU-T Rec. G.681 (10/96) "Functional Characteristics of Interoffice and Long-Haul Line Systems Using Optical Amplifiers, Including Optical Multiplexing."

ITU-T Rec. G. 691 (10/2000) "Optical Interfaces for Single-Channel STM-64, STM-256 and Other SDH Systems With Optical Amplifiers."

ITU-T Rec. G.692 (10/98) "Optical Interfaces for Multichannel Systems With Optical Amplifiers."

ITU-T Rec. G.692 Corrigendum 1 (01/2000) "Optical Interfaces for Multichannel Systems With Optical Amplifiers."

ITU-T Rec. G.955 (11/96) "Digital Line Systems Based on the 1544 Kbit/S and the 2048 Kbit/S Hierarchy on Optical Fiber Cables."

ITU-T Rec. G.957 (07/99) "Optical Interfaces for Equipments And Systems Relating to the Synchronous Digital Hierarchy."

ITU-T Rec. G.958 (11/94) "Digital Line Systems Based on the Synchronous Digital Hierarchy for Use on Optical Fiber Cables."

ITU-T Rec. G.959.1 (02/2001) "Optical Transport Network Physical Layer Interfaces."

ITU-T Rec. G.982 (11/96) "Optical Access Networks To Support Services Up to the ISDN Primary Rate or Equivalent Bit Rates."

ITU-T Rec. G.983.1 (06/99) "Broadband Optical Access Systems Based on Passive Optical Networks (PON)."

ITU-T Rec. G.983.1 Corrigendum 1 (07/99) "Broadband Optical Access Systems Based on Passive Optical Networks (PON)."

ITU-T Rec. G.983.2 (04/2000) "ONT Management and Control Interface Specification for ATM PON."

ITU-T Rec. G.983.4 (03/2001) "A Broadband Optical Access System with Increased Service Capability by Wavelength Allocation."

Izumita, H., et al., "The Performance Limit of Coherent OTDR Enhanced with Optical Fiber Amplifiers Due to Optical Nonlinear Phenomena," *IEEE/OSA Journal of Lightwave Technology,* Vol. 12, No. 7, July 1994, pp. 1230–1238.

Jinno, M., and T. Matsumoto, "Nonlinear Operations of 1.55-Mu M Wavelength Multielectrode Distributed-Feedback Laser Diodes and Their Applications for Optical Signal Processing," *IEEE/OSA Journal of Lightwave Technology,* Vol. 10, No. 4, April 1992, pp. 448–457.

Kajiyama, Y., N. Tokura, and K. Kikuchi, "An ATM VP-Based Self-Healing Ring," *IEEE Journal on Selected Areas in Communications,* Vol. 12, No. 1, January 1994, pp. 120–127.

Kamamura, R., and I. Tokizawa., "Self-Healing Virtual Path Architecture in ATM Networks," *IEEE Communications Magazine,* Vol. 39, No. 9, September 1995, pp. 72–79.

Kaminow, I. P., et al., "A Wideband All-Optical WDM Network," *IEEE Journal on Selected Areas in Communications,* Vol. 14, No. 5, June 1996, pp. 780–799.

Kashima, N., *Passive Optical Components for Optical Fiber Transmission,* Norwood, MA: Artech House, 1995.

Koga, M., et al., "Design and Performance of an Optical Path Cross-Connect System Based on Wavelength Path Concept," *IEEE/OSA Journal of Lightwave Technology,* Vol. 14, No. 6, June 1996, pp. 1106–1119.

Kompella, V. P., J. C. Pasquale, and G. C. Polyzos, "Multicast Routing for Multimedia Communication," *IEEE/ACM Transactions on Networking,* Vol. 1, No. 3, June 1993, pp. 286–292.

Kovacevic, M., and M. Gerla, "A New Optical Signal Routing Scheme for Linear Lightwave Networks," *IEEE Transactions on Communications,* Vol. 43, No. 12, December 1995, pp. 3004–3014.

Labourdette, J. -F. P., and A. S. Acampora, "Logically Rearrangeable Multihop Lightwave Networks," *IEEE Transactions on Communications,* Vol. 39, No. 8, August 1991, pp. 1223–1230.

Lee, B. G., M. Kang, and J. Lee, *Broadband Telecommunication Technology,* 2nd ed., Norwood, MA: Artech House, 1996.

Madsen, C. K., and J. H. Zhao, *Optical Filter Design and Analysis: A Signal Processing Approach,* New York: John Wiley & Sons, 1999.

Maeda, M. W., "Management and Control of Transparent Optical Networks," *IEEE Journal on Selected Areas in Communications,* Vol. 16, No. 7, September 1998, pp. 1008–1023.

Masetti, F., et al., "High-Speed, High-Capacity ATM Optical Switches for Future Telecommunication Transport Networks," *IEEE Journal on Selected Areas in Communications,* Vol. 14, No. 5, June 1996, pp. 979–998.

Medard, M., S. G. Finn, and R. A. Barry, "WDM Loop-Back Recovery in Mesh Networks," in *Proceedings of INFOCOM'99,* 1999, pp. 752–759.

Members of technical staff, *Transmission Systems for Communications,* Madison, WI: Bell Laboratories, 1982.

Mokhtar, A., and A. Azizoglu, "Adaptive Wavelength Routing in All-Optical Networks," *IEEE/ACM Transactions on Networking,* Vol. 6, No. 2, April 1998, pp. 197–206.

Morioka, T., and M. Saruwatari, "Ultrafast All-Optical Switching Utilizing The Optical Kerr Effect in Polarization-Maintaining Single-Mode Fibers," *IEEE Journal on Selected Areas in Communications,* Vol. 6, No. 7, August 1988, pp. 1186–1198.

Mukherjee, B. "WDM-Based Local Lightwave Networks—Part I: Single-Hop Systems," *IEEE Network Magazine,* Vol. 6, No. 3, May 1992, pp. 12–27.

Mukherjee, B., "WDM-Based Local Lightwave Networks—Part II: Multihop Systems," *IEEE Network Magazine,* Vol. 6, No. 4, July 1992, pp. 20–32.

Mukherjee, B., *Optical Communication Networks,* New York: McGraw-Hill, 1997.

Nakagami, T., and T. Sakurai, "Optical and Optoelectronic Devices for Optical Fiber Transmission Systems," *IEEE Communications Magazine,* Vol. 26, No. 1, Jan. 1988.

Nederlof, L., et al., "End-to-End Survivable Broadband Networks," *IEEE Communications Magazine,* Vol. 33, No. 9, September 1995, pp. 63–71.

Ngo, N. Q., and L. N. Binh, "Novel Realization of Monotonic Butterworth-Type Lowpass, Highpass, and Bandpass Optical Filters Using Phase-Modulated Fiber-Optic Interferometers and Ring Resonators," *IEEE/OSA Journal of Lightwave Technology,* Vol. 12, No. 5, May 1994, pp. 827–841.

Nishikido, J., et al., "Multigigabit Multichannel Optical Interconnection Module for Broadband Switching System," *IEEE/OSA Journal of Lightwave Technology,* Vol. 13, No. 6, June 1995, pp. 1104–1110.

Offside, M. J., et al., "Optical Wavelength Converters," *Electronics & Communication Engineering Journal,* Vol. 7, No. 2, April 1995, pp. 59–71.

Okamoto, S., K. Oguchi, and K. Sato, "Network Architecture For Optical Path Transport Networks," *IEEE Transactions on Communications,* Vol. 45, No. 8, August 1997, pp. 968–977.

Pankaj, R. K., and R. G. Gallager, "Wavelength Requirements Of All-Optical Networks," *IEEE/ACM Transactions on Networking,* Vol. 3, No. 3, June 1995, pp. 269–280.

Potenza, M., "Optical Fiber Amplifiers for Telecommunication Systems," *IEEE Communications Magazine,* Vol. 34, No. 8, August 1996, pp. 96–102.

Ramaswami, R., and K. N. Sivarajan, "Routing and Wavelength Assignment in All-Optical Networks," *IEEE/ACM Transactions on Networking,* Vol. 3, June 1995, pp. 489–500.

Renaud, M., et al., "Network and System Concepts for Optical Packet Switching," *IEEE Communications Magazine,* Vol. 35, No. 4, April 1997, pp. 96–102.

Saruwatori, M., "All-Optical Signal Processing in Ultrahigh-Speed Optical Transmission," *IEEE Communications Magazine,* Vol. 32, No. 9, September 1994, pp. 98–105.

Sharma, M., H. Ibe, and T. Ozeki, "Optical Circuits for Equalizing Group Delay Dispersion Of Optical Fibers," *IEEE/OSA Journal of Lightwave Technology,* Vol. 12, No. 10, October 1994, pp. 1759–1765.

Sivalingam, K. M., and S. Subramaniam, *Optical WDM Networks—Principles and Practice,* Boston, MA: Kluwer Academic Publishing, 2000.

Smith, D. A., et al., "Integrated-Optic Acoustically Tunable Filters for WDM Networks," *IEEE Journal on Selected Areas in Communications,* Vol. 8, No. 6, August 1990, pp. 1151–1159.

SR-3928, "Optical Fiber and Fiber Optic Cable Certification," Telcordia, June 1996.

SR-4151, "Fiber Optic Passive Component Certification," Telcordia, June 1999.

SR-4226, "Fiber Optic Connector And Jumper Assembly Certification," Telcordia, January 2001.

SR-4263, "Single Mode Optical Fiber Splice and Splicing System Certification," Telcordia, April 1997.

SR-ARH-000015, "Switching Technology: Photonic Switch Application," Telcordia, September 1984.

SR-ARH-002744, "Single-Mode Fiber Connector Technology," Telcordia, August 1993.

SR-NOTES-SERIES-02, "Telcordia Notes on Dense Wavelength-Division Multiplexing (DWDM) and Optical Networks," Telcordia, May 2000.

Subramaniam, S., M. Azizoglu, and A. K. Somami, "All-Optical Networks with Sparse Wavelength Conversion," *IEEE/ACM Transactions on Networking,* Vol. 4, No. 4, August 1996, pp. 544–557.

Technical Personnel, Bellcore and Bell Operating Companies, *Telecommunications Transmission Engineering,* Third edition, 1980.

TR-NWT-001073, "Generic Requirements for Fiber Optic Switches," Telcordia, December 1993.

TA-NWT-001500, "Generic Requirements for Powering Optical Network Units in Fiber-in-The-Loop Systems," Telcordia, December 1993.

Tumolillo, T. A., Jr., M. Donckers, and W. H. G. Horsthuis, "Solid State Optical Space Switches for Network Cross-Connect and Protection Applications," *IEEE Communications Magazine,* Vol. 35, No. 2, February 1997, pp. 124–130.

Veitch, P., I. Hawker, and G. Smith, "Administration of Restorable Virtual Path Mesh Networks," *IEEE Communications Magazine,* Vol. 34, No. 12, December 1996, pp. 96–101.

Watanabe, S., et al., "Generation of Optical Phase-Conjugate Waves and Compensation for Pulse Shape Distortion in a Single-Mode Fiber," *IEEE/OSA Journal of Lightwave Technology,* Vol. 12, No. 12, December 1994, pp. 2139–2146.

Wei, J. Y., et al., "Connection Management for Multiwavelength Optical Networking," *IEEE Journal on Selected Areas in Communications,* Vol. 16, No. 7, September 1998, pp. 1097–1108.

Wong, Y.–M., et al., "Technology Development of a High-Density 32-Channel 16-Gbps Optical Data Link for Optical Interconnection Applications for the Optoelectronic Technology Consortium (OETC)," *IEEE/OSA Journal of Lightwave Technology,* Vol. 13, No. 6, June 1995, pp. 995–1016.

Yoo, S. J. B., "Wavelength Conversion Technologies for WDM Networks Applications," *IEEE/OSA Journal of Lightwave Technology,* Vol. 14, No. 6, June 1996, pp. 955–966.

7

Integration of IP and ATM

Before ATM, most data networks, such as Ethernet and token ring or FDDI networks, provided connectionless data services. They transmitted data using common MAC, LLC, or bridging protocols without setting up connections before sending data. Connectionless network technology such as TCP/IP also were connectionless service networks using the datagram routers. Connectionless networks are advantageous in that the end systems place no burden on connection management, thereby simplifying the design. Also, as discussed in Chapters 1 and 3, there are various other advantages in using connectionless networks.

In contrast, ATM networks are connection-oriented networks. In ATM networks virtual circuits are established between two end systems, and cells are switched according to their connection identifiers. Since resources are statistically allocated per connection basis, ATM networks can provide guaranteed QoS for connections. In addition, since routing is done only at the connection setup phase, ATM networks can support cell-sequence integrity.

Even though ATM networks were designed for connection-oriented services by nature, they are required to provide connectionless services as well, so that they can interoperate with other existing connectionless networks and services. Today many ISPs today use ATM network cores to transport TCP/IP traffic.

There are two major reasons for the importance of transporting IP traffic over ATM networks. Initially there was a cost-performance advantage as

ATM switches were widely available and cheap. While routers were also widely available, the link types traditionally supported by routers were either low in speed (e.g., T1 or T3) or not suitable for long-distance transmission (e.g., Ethernet). This first advantage has now disappeared in many cases due to the spread of cheap multifunctional routers with high-speed I/O ports. Second, ATM networks have the built-in ability to do bandwidth management and traffic engineering, mostly due to their connection-oriented attribute. This enables automatic distribution of traffic in contrast to the case of datagram routing, which does not utilize multiple parallel paths well. This second advantage is still valid as a key advantage that ATM technology has over IP technology. It can be viewed as derived from the fundamental differences in connection-oriented versus connectionless technologies.

There are also some other merits to using IP over ATM. While ATM supports high-speed links and multiplexing, it also supports multiplexing low-speed links into high-speed ATM links, thereby affording an efficient multiplexing solution for many edge-switching systems. As ATM supports various network management functions such as fast automatic rerouting and fault recovery, network management becomes easier. In addition, efficient traffic engineering and traffic control may be achieved by using various mechanisms defined in ATM technology.

This chapter discusses the integration of IP and ATM. We first describe various integrated/overlay architectures such as *classical IP over ATM* (IPoA), MPoA and MPLS over ATM, and then discuss the routing, multiplexing and switching, network control, traffic management, and QoS issues related to them.[1]

7.1 Concepts and Architecture

The problem of supporting TCP/IP protocols over ATM networks has been extensively studied during the last decade. The resulting architectures can be categorized into the *overlay model* and the *integrated model* depending how deeply their network control, signaling, and routing planes are entwined.

1. It should be noted that while both MPOA and MPLS contain the term "multiprotocol," they were both essentially designed for the support of IP. In the case of MPOA, support for other network layer protocols such as IPX was important initially but the interest has decreased considerably. In the case of the MPLS, the opposite has happened: While originally conceived as a way of improving IP technology, it is now being recognized as a technology that may be applied in a more generic manner, with optical networks being an obvious example (refer to Section 8.4).

In the overlay model it is assumed that the IP and ATM network nodes are unaware of each other. The ATM network appears as a cloud to the IP network, as is shown in Figure 7.1(a). The IP traffic is transported over ATM pipes between IP routers. The IP routers are unaware of the ATM switches that interconnect them. They only know of the ATM pipes (or virtual connections) that connect themselves to other IP routers. Basically, the routing protocols that run in the IP and the ATM networks are independent. The IP level routing protocols indicate how to go from one IP node to another, while the ATM routing protocols supply the same information for the ATM nodes. For example, ATM routing protocols may be used to find the correct route for the ATM virtual connection from an IP router *A* to another IP router *B*, but from the IP router's perspectives, routers *A* and *B* are only a single link apart. The two routers are unaware of the numerous ATM switches used to relay the virtual connection between the two routers.

The use of independent routing protocols also implies that there is no relation between the addresses from the ATM and the IP networks in the overlay model. There is also no algorithmic method for translating between the two types of addresses. From the IP network's perspective these facts mean that an ATM network will look like a network "cloud" offering connectivity to other points on the edges of the ATM network leading to other IP routers. However, the IP routers do not know how those connections are made through the ATM network. Such a structure implies the need for a protocol that is able to map the destination ATM address to the destination IP address. This is a basic problem that all the overlay architectures must solve. Later in this section we will examine various overlay architectures that differ primarily in how these problems are solved.

In the case of the integrated model, the IP and ATM nodes are peers; when ATM and IP networks interoperate, the ATM nodes and IP nodes are

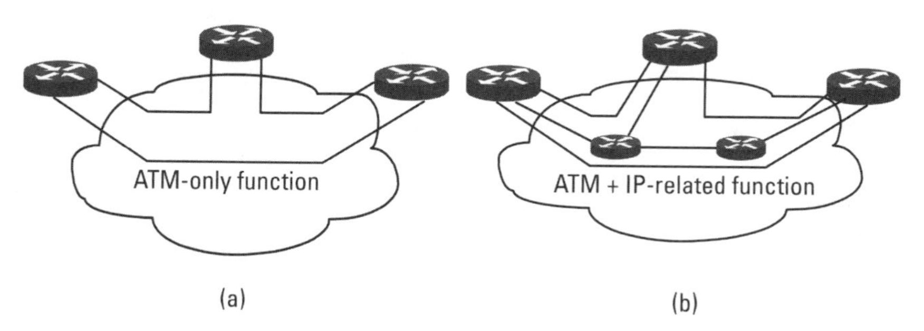

Figure 7.1 TCP/IP over ATM service models: (a) overlay model, and (b) integrated model.

specifically aware of each other. When IP traffic is transported over an ATM network, the IP router at the edge of the ATM network is fully aware of the ATM network's structure. In general, the IP-level routing protocols indicate how to go from one IP node to another, while the ATM routing protocols supply the same information to the ATM nodes. However, as shown in Figure 7.1(b), the IP router on the edge of the ATM network is aware of the structure of the ATM network. Consequently, it can route the IP packet to the next ATM switch along the path, instead of blindly putting it into an ATM virtual channel. In addition to the routing protocols, the signaling protocols used in the two networks must also interoperate. This may require new signaling schemes, such as the *label distribution protocol* (LDP) developed for use in MPLS-capable ATM networks. The integrated model implies that the ATM addresses and IP addresses can be translated from one to another by algorithmic methods, which is an obvious consequence of the interoperability of the routing protocols used in each network. As ATM addresses and IP addresses are translatable, additional address resolution protocols are not necessary to find out the destination ATM address when only the destination IP address is known. Examples of the integrated model include the integrated PNNI model from the ATM Forum and the MPLS protocol suite developed by the IETF.[2]

Figure 7.2 shows various methods for integrating IP networks and ATM networks. Basically, LANE, classical IPoA, MPOA, and NHRP methods belong to the overlay model, whereas the *integrated PNNI* (I-PNNI) and MPLS methods belong to the integrated model.

7.1.1 Classical IPoA

The basic classical IPoA method was defined by the IETF for transporting IP traffic over ATM networks. The basic method specifies how to operate an ATM network as a single IP subnet, or specifically, how to connect two nodes on the same IP subnet directly, by an ATM connection. However, connecting two nodes on different subnets requires routers. As such, the IPoA specified in RFC1577 does not change the fundamental nature of the IP protocol, so it relies on IP routers for interconnecting subnets consisting of LANs.[3]

---

2. The development of the integrated PNNI method, however, has not been done actively enough, so discussions will mainly be concentrated on the MPLS method in later sections.

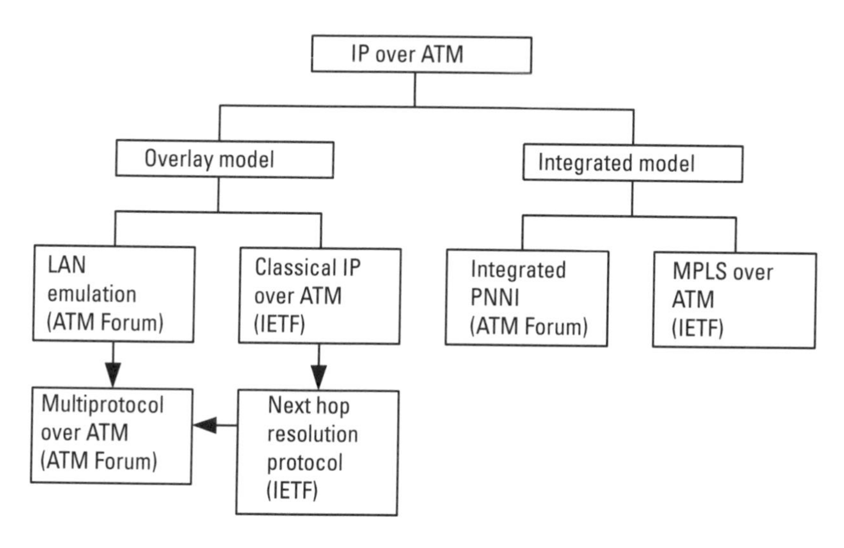

Figure 7.2 Taxonomy of IPoA models.

7.1.1.1 Elements of Classical IPoA

The classical IPoA approach is based on a special type of IP subnetwork called a *logical IP subnetwork* (LIS). An LIS corresponds to the concept of a subnetwork in traditional Ethernet LANs except that it consists of hosts and routers that are connected through an ATM network. An LIS consists of hosts and routers that have the same subnetwork mask and the same subnetwork address. Any two hosts in the same LIS communicate directly, but hosts in different LISs can communicate only through a router even if a direct ATM connection can be established between them.

As the LIS is an IP network constructed over the ATM network, an address resolution functionality is needed. In classical IPoA the *ATM address resolution protocol* (ATMARP) server provides this function. The ATMARP server basically allows a client to find the ATM addresses corresponding to a given IP address. Note that this ATMARP function only maps between ATM addresses and IP addresses. No other type of network level protocol,

3. The classical IPoA methods are defined largely in two main RFCs, RFC1577 and RFC1626. Recently these two RFCs have been updated and unified into a single specification RFC2455. Additionally, RFC2332 (formerly RFC1735 but currently upgraded to support UNI 4.0 features) defines how UNI 4.0 signaling methods and parameters may be used to set up and release SVCs for transporting classical IPoA IP traffic.

such as IPX or AppleTalk, is supported. This exhibits the IP-centric nature of the classical IPoA solution.

It is possible to construct multiple IP subnets over a single ATM network. Each of these subnets will constitute an LIS and, for communications between nodes on the same subnet, only the ATMARP function is required. However, for communications between hosts on different subnets, routers are needed to connect with other subnets. The current classical IPoA specifications also support IP multicasts through the *multicast address resolution server* (MARS) standards defined in RFC2012.

Classical IPoA operation over PVCs. The simplest mode of operation of a classical IPoA network is to use *permanent virtual channels* (PVCs) among all the routers connected by the ATM network [1]. This means that each router will have a PVC connection to all its neighbors. When the routers are powered on, all of them use the inverse ARP [2] protocol to find the IP address of the host on the other end of each of its default PVCs. Once all the IP addresses become available, the routers can use various routing protocols, such as RIP or OSPF, to find out routes to all the other points in the overall network. There are two major drawbacks to this PVC approach: It relies on the router being manually preconfigured to set up the PVCs, and it is not scalable to a large size. Manual configuration is not practical when the network grows to a large size.

Classical IPoA operation over SVCs. The use of *switched virtual channels* (SVCs) requires the setup of SVCs on demand [1]. To establish a VCC between two hosts in an LIS, a mapping is necessary between the ATM addresses and IP addresses of the source and the destination. Based on this mapping, IP addresses are resolved to ATM addresses through the ATMARP and the inverse process is done through the *inverse ATMARP* (InATMARP). For such address resolution, each host must register its IP address and ATM address to the ATMARP server in the same LIS. Then the ATMARP server can resolve the IP addresses in the same LIS.

Figure 7.3(a) shows an example of the classical IPoA model using an ATMARP server and Figure 7.3(b) shows the protocol stack of the classical IPoA. In Figure 7.3(a), Host 1 (IP address A, ATM address X) wants to send some connectionless data packets to Host 2 (IP address B, ATM address Y). So Host 1 sends an ARP request to the ATMARP Server 1 in the same LIS. Then ATMARP Server 1 resolves IP address B to ATM address Y and then sends an ARP response to Host 1. Next, Host 1 sets up a VCC to Host 2 using the ATM address Y. Packets are transmitted over this virtual connection

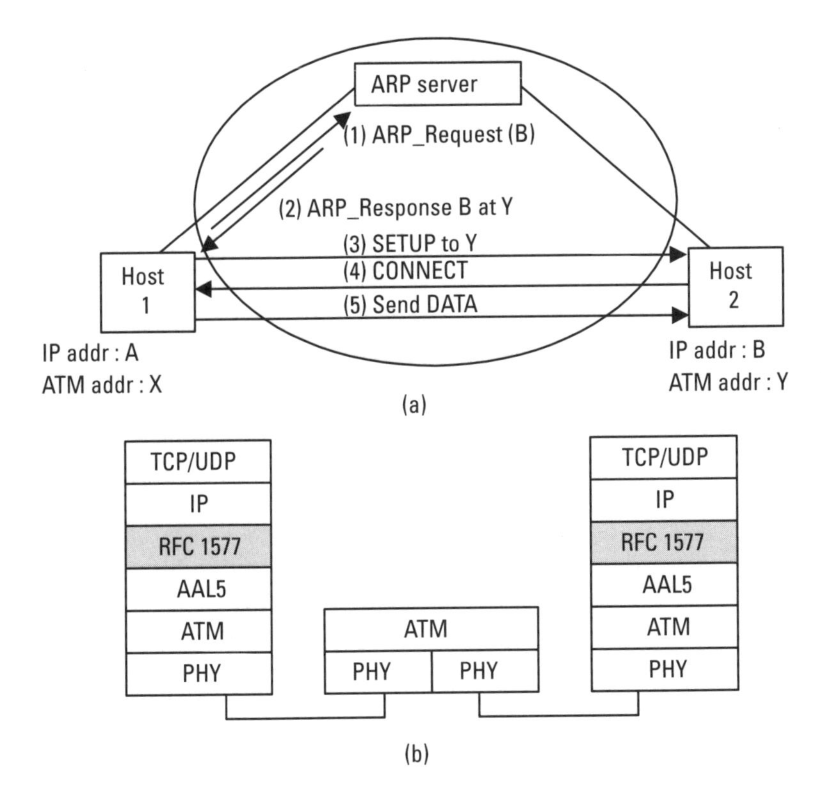

Figure 7.3 Classical IPoA: (a) operation example, and (b) protocol stack.

to reach Host 2. Host 1 preserves the mapping information indicating that IP address B is mapped to the ATM address Y for use in the next packet transmission. Because all these procedures are done below the actual IP layer, the IP layer need not care about the specifics of the ATMARP server interactions or the ATM connection setup procedures. Consequently from the IP layer point of view, there is no difference from transmitting data over traditional LAN or other data link protocols.

For this design to be robust, a number of timers are needed. First, the ATMARP server uses a timer to periodically test that a previously registered host is still alive. Whenever the timer goes off, the ATMARP server uses an InATMARP message to check that the host is still responding. If the host fails to respond the SVC will be torn down. Second, another timer is an inactivity timer used to test if data flow over an SVC setup between two hosts discontinues for a certain amount of time. If this happens, the connection is automatically released. Essentially, the above two timers are to release the unused VCs in a timely manner such that no VCs are wasted.

Note that the connection setup between Host 1 and Host 2 is an SVC. This is the key difference from the PVC method. Compared with the use of PVCs, the use of SVCs is more efficient as it would set up and use a virtual channel only when needed. This saves VPI/VCI label space and enables network resources to be utilized when needed. Saving label space is important in larger networks with many connections and in the networks where the hardware of practical ATM switches limits the number of available VCs. The use of timers is an important way of ensuring that these resources are not wasted.

Classical IPoA between different LIS networks. In classical IPoA networks, traditional hop-by-hop IP routing is used when routing between different LIS networks. Even when it is possible to use a direct ATM connection to connect two communicating hosts, if the two hosts are on different LISs, the packets must be transmitted through router(s) that connects the two LISs. This is a key concept in the classical IPoA model, which contributes to the term "classical" in the name. Accordingly, if two hosts are internal to an LIS (the *intra-LIS* case), it is possible to use direct ATM connections to communicate between two hosts. When those hosts lie on different LISs (the *inter-LIS* case), however, a hop-by-hop routing approach using routers is needed. Figure 7.4 demonstrates the classical IPoA operation when data flows

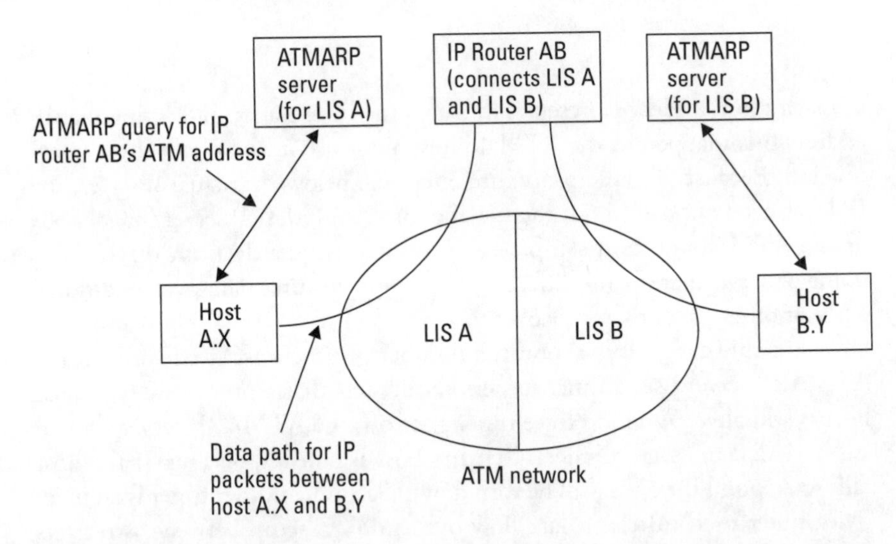

Figure 7.4 Operation of classical IPoA when data flows between different LIS networks.

between different LISs. For a host A.X on LIS A to send a packet to host B.Y on LIS B, the packet must be sent to router AB over an ATM connection between host A.X and the router AB. The router will then forward the packet to host B.Y again by using another ATM connection from itself to the host.

This method is inefficient because hop-by-hop routing over IP routers is needed for inter-LIS communication, even when a direct end-to-end shortcut is possible. For example in the network shown in Figure 7.4, the host A.X and host B.Y are on different LISs, but actually on the same ATM network. Thus, it is possible to set up a simple direct path between the two hosts instead of rerouting the packets every time they come. This inefficiency has triggered the development of the NHRP model, defined in the Section 7.1.2.

7.1.1.2 Classical IPoA and Encapsulation

Any method for transporting IPoA needs to address how to encapsulate IP frames in ATM cells. This is a basic problem for all IPoA models, and the basic solutions are outlined in RFC1483. According to this document, hosts can deploy two different methods to encapsulate different network layer protocols. The first method is to multiplex multiple protocols in a single ATM VCC. In this case the protocol of the carried PDU is identified by the LLC header and the *subnetwork attaching point* (SNAP) header. The other method is to set up a VC for each protocol. The first method is suitable for networks that use only PVCs, and the second method is suitable for the networks employing SVCs in which VCCs can be created or removed flexibly. These methods are further discussed in Section 7.2

The MTU must also be defined along with the encapsulation method. The actual MTU value may be negotiated while the connection is being set up by signaling. The size of the MTU that a host can transmit over the ATM virtual channel is 9,180 octets [3]. If all members in an LIS consent, this value can be changed. This is aligned with the default MTU defined for the *switched megabit data service* (SMDS) standards in RFC1209. Also all routers should use the IP path MTU discovery mechanism to find out the maximum MTU for a path [4].

Note that the address resolution–related messages, ATMARP request/ response and InATMARP request/response, which were defined for classical IPoA operation, also need to have their packet formats and encapsulation methods defined. As these messages are not layer 3 packets but ATM-level messages, they are directly transported over ATM connections in LLC/ SNAP-encapsulated format. The details and specific formats are defined in RFC1577.

7.1.1.3 Classical IPoA and ATM Signaling

Classical IPoA is based on the overlay model for IPoA transmission. As mentioned above, the IP routers on the edge of the ATM network view the ATM network as a cloud that offers the ability to connect with other IP routers connected to the ATM network. While the IP routers need not be aware of the ATM networks structure, the IP routers must be able to use the ATM signaling to set up connections to the destination IP routers. This means that the IP routers must be able to map IP parameters with the relevant ATM signaling parameters. RFC1755 specifies how UNI 3.0/3.1 signaling is used to support classical IPoA by defining how these parameters are mapped. (Refer to Section 7.4.2 for further discussions on signaling.)

7.1.2 NHRP

Although the classical IPoA method has the advantages of being conceptually simple and not requiring any change to existing systems, its performance is rather limited as communication among different subnetworks must be done through routers. This can cause a serious degradation of performance in an ATM network consisting of a large number of LISs. In ATM networks, hosts can communicate directly with each other without the involvement of IP layer switching in routers, and this fact can be exploited to enhance performance by removing unnecessary relay nodes.

As a means to set up direct connections in *nonbroadcasting multiaccess* (NBMA) networks such as ATM, the IETF has introduced NHRP, which relies on a new type of ARP server for ATM networks.[4] The aim of NHRP is to enable a source host to bypass all or some intermediate routers so as to establish a direct ATM connection to the destination host. When such a direct connection is set up, it is said that a "direct shortcut SVC" is used. In

4. The term NBMA comes from the fact that ATM networks allow multiple access but, unlike traditional LANs such as Ethernet, ATM networks do not allow broadcasting in a native manner. As pointed out earlier, this is one of the main differences between ATM networks and traditional LANs. While in our discussions we concentrate on the application of NHRP to IP and ATM networks, it must be noted that NHRP was defined not only for IP networks but also to support other protocols such as IPX and AppleTalk. As such, the NHRP specifications use the term internetworking address when referring to the address of the upper network layer's addressing scheme and use the term NBMA address to refer to the native addresses used in the NBMA itself. For the cases that we will consider, IP addresses correspond to internetworking addresses, while ATM addresses correspond to native NBMA addresses.

other words, NHRP is an interLIS address resolution protocol, a more complicated ARP that can be used in NBMA networks having multiple LISs.

Specifically, NHRP is used to determine the IP layer address and NBMA network addresses of the *NBMA next hop* toward a destination station. If the destination is connected to the NBMA network then the NBMA next hop is the destination host itself. Otherwise, the NBMA next hop is the egress router from the NBMA network that is "nearest" to the destination host. Usually this egress router would be the last router connected to the NBMA network that is on the routed path to the destination host.

7.1.2.1 Elements of NHRP

NHRP uses the concept of LIS introduced in classical IPoA. It is applied when a single NBMA network is divided into a number of disjoint LISs. In each LIS of the NBMA network, there is at least one *next hop server* (NHS) that resolves IP addresses to NBMA addresses. The NHS constructs an NHS address mapping table by utilizing information it gets through NHRP registration packets from clients on the same NBMA network or by applying dynamic address learning mechanisms. NBMAs as well as LISs are normally connected by routers, with the NHS usually coresiding in the inter-LIS routers.

NBMA stations are those stations that implement the NHRP protocol. They use the NHRP protocol to find the interworking layer address and the NBMA address of the NBMA next hop on the path to the destination host. NHRP stations can be divided into clients and servers depending on their operation, namely *next hop clients* (NHCs) and NHSs. All NHSs and NHCs maintain next hop resolution tables that map internetworking addresses to NBMA addresses.

Each NHS implements the NHRP protocol. Conceptually NHS may be considered as residing in a router. Normal IP routers can be both NHRP clients and servers. The NHRP specifications do not make any assumption that all routers implement the NHRP protocols. NHCs maintain an address cache that maps internetworking addresses to NBMA addresses. Examples of NHC are IP end hosts. For NHRP to operate correctly, all NHCs must know the address of at least one NHS. The NBMA address of this NHS may be obtained by various methods such as manual configuration or by using anycast addresses.[5]

Each NHC must register its NBMA address and IP address with its *serving NHS*. In the case of ATM networks this means that the ATM and IP address of the client is known by the serving NHS. The *last hop NHS* refers to the last NHS along the routed path to a client. This NHS is usually the serving NHS of the destination host or egress router.

7.1.2.2 Operation of NHRP

The NHRP protocol relies on various messages (see Table 7.1). Among them the main types are the NHRP resolution request/reply and registration messages. Figure 7.5(a) shows the basic message flows for client address registration. Each NHRP station registers its NBMA address and internetworking address with the NHS by using an *NHRP registration request.* This message contains the NHC's ATM address, the NHC's IP address, and the NHS's IP address. When the NHS receives this message, it may start to construct a cache based on this information. The NHRP clients use *NHRP resolution request* and reply messages to find the NBMA address of the next hop server

Table 7.1
NHRP Messages

Message Type	Direction	Description
Next-hop resolution request	Station → NHS	Sent to NHS to find the ATM address of the destination
Next-hop resolution reply	NHS → Station	Reply to the next-hop resolution request
Registration request	Station → NHS	Registration of next-hop information
Registration reply	NHS → Station	Reply to the registration request
Purge request	Station → Station	Requests a removal of next-hop information from the cache
Purge reply	Station → Station	Replies to the purge request
Error notification	Station → Station	Notifies sender of error indications and related problem descriptions

5. An anycast address is a special type of address that can be used to find the nearest host or server that is listening to that address. A client may use this address to automatically find the servers when first booting up. Anycast addresses have been defined for ATM networks and IPv6 networks, but not in IPv4 networks.

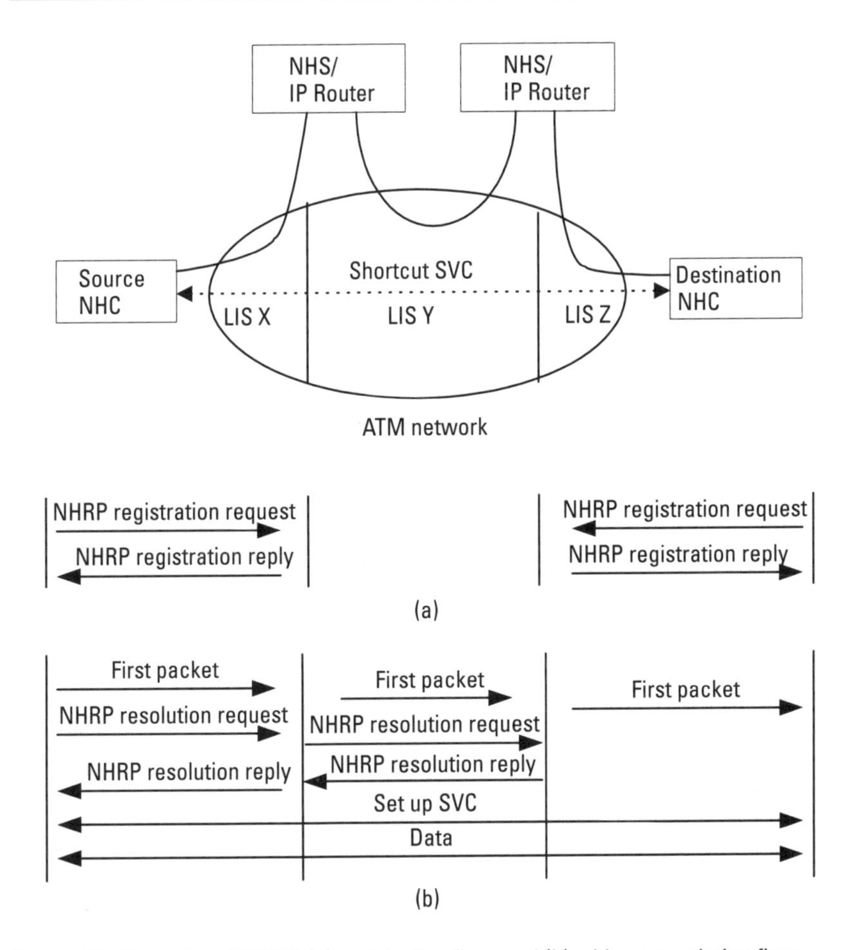

Figure 7.5 Operation of NHRP: (a) registration flow, and (b) address resolution flow.

on the routed path to the destination. The resolution request message contains [source NHC's ATM address, source NHC's IP address, destination's IP address].

Figure 7.5(b) shows the basic message flows for address resolution in NHRP. The basic procedure is as follows: When the source NHC has data to transmit it first checks its address cache to see if the destination host address has already been resolved. If not, the source host sends an NHRP resolution request to its NHS server. If the source host and the destination host are connected to the same subnetwork, the NHS will reply with the NBMA address of the destination host. If the destination host is not connected to the same subnetwork, the NHS looks up the next hop router's address of the destination host in its forwarding table and forwards the NHRP request to the next NHS.

The same procedure is carried out by the next-hop NHS. If the NHS contains a mapping for the destination IP address in its cache, it returns an NHRP registration reply with the destination's IP address and NBMA address. As mentioned above, this final NHS router that generates the initial NHRP registration reply is called the *last-hop NHS*. The NHRP registration reply is usually returned along the same path that it took in reaching this last hop NHS, thereby allowing all the NHS to update their cache entries with the information regarding the destination.

Later on, other hosts may request address resolution for the same destination host. The new requests will be forwarded through the NHS to the serving NHS for the destination host. Before arriving at the last hop NHS the request may first arrive at one of the NHSs that had previously cached the reply sent back to a previous request for the same destination host. In such a case the NHS may send back a reply based on the data in the cache but with the data marked as *nonauthorative*. By having the NHSs reply based on the cached data, the performance of the protocol may be improved and the scalability may be also increased for the cases where the destination host is a popular destination. It is important to mark the data as being nonauthorative as the requesting host should be able to choose between using cached data and always getting data from the last-hop NHS.

The NHRP request is always forwarded along the normal routed path to the destination host according to the network layer routing tables. It is possible that the NHRP request may finally reach the egress router of the NBMA network without reaching the final destination host. In such a case the NHRP registration reply will contain the address of this final egress router or the next hop router. An NHRP request never crosses the border of an NBMA network.

If neither the destination address nor the next-hop router to the destination is found within the NBMA network, the last NHS to receive the request will send back a negative NHRP resolution reply with a code indicating that no entry was found.

7.1.2.3 Encapsulation and Interaction with ATM Signaling

As stated above, NHRP can be viewed as an advanced version of the basic classical IPoA model. As we have shown above, most of the changes were within the address resolution procedures and architecture. In contrast, the actual methods for the encapsulating IP packets over ATM connections follow the basic classical IPoA methods. Also the basic interactions with the ATM signaling procedures are the same as the classical IPoA case. These are further discussed in Section 7.4.

As in the case of classical IPoA the NHRP specific messages are transported directly over ATM connections using AAL-5 framing with LLC/SNAP encapsulation. The details of the packet formats are described in [5].

7.1.2.4 Limitations of NHRP

As many researchers have pointed out, NHRP can suffer from poor scalability. At the NHRP client level, due to processing and memory limitations the NIC on the client may not be able to maintain the large number of mappings needed to support a direct ATM connection to each different destination. This problem would manifest itself not in the normal end user clients but in the nodes with a large amount of connections such as Web servers using an ATM interface.[6] At the NHRP server level, ATM-to-IP address mapping within large LISs would mean that the NHRP server or NHS would have to maintain a very large table of mappings. At the NHRP domain level, this leads to the connection scaling of the order of N^2 for the number of hosts on the network, N.

Another limitation of NHRP is that all the routers within an ATM network should be NHRP-aware. While the protocol does not require this, NHRP cannot resolve routing loops that can occur when NBMA networks and normal IP routers that do not understand NHRP are mixed

A third limitation is that when LISs have multiple NHSs, they need a mechanism to synchronize the cached information. While such a protocol [the *server synchronization cache protocol* (SSCP)] has been defined, any such mechanism is prone to failure and much harder to make robust. This may be due to the problems in the protocol design, but just as likely is from human errors in implementation or configuration.

A fourth limitation is that NHRP cannot set up multiple shortcut paths in ATM networks. This is because NHRP basically follows the destination-based routing paradigm used in traditional IP routing. It's therefore unable to set up multiple paths for different QoS and user requirements. This is especially disadvantageous when considering that ATM is capable of supporting this feature.

---

6. Note that though the Web server is a server with respect to its client, it is a client with respect to the NHRP protocol.

7.1.3 LANE

LANE was developed by the ATM Forum as a method of transporting LAN traffic over ATM networks. As the name suggests, the main idea was to emulate the behavior of popular LANs at the MAC layer, so that user applications could be run with minimal changes. This meant that LANE was designed to support the connectionless services offered by traditional LANs. Additionally, it supports broadcast and multicast services that all LANs offer. LANE currently emulates two LAN technologies, Ethernet (IEEE 802.3) and token ring (IEEE 802.5).

The two main types of LAN systems, Ethernet and token ring networks, have a number of common representative characteristics. First, messages may be characterized as connectionless, as opposed to the connection-oriented approach of ATM. Second, broadcast and multicast are easily accomplished through the shared medium of LANs. Third, LAN MAC addresses, which are basically the manufacturing serial numbers independent of the network topology, are a globally unique ID for whatever device that is used to connect to the LAN.

When the ATM Forum defined LANE across ATM networks the aim was to define an architecture and protocol suite that could offer the services based on the characteristics above. LANE was defined so that it could be implemented as a software layer in end systems, without affecting the layers above the MAC layer. Additionally, LANE supported the interconnection of ATM networks with traditional LANs by means of bridging methods. Consequently, LANE allows the interoperability between software applications residing in ATM-attached end systems and in traditional LAN end systems. In other words, LANE provides a simple and easy means for running existing LAN applications in the ATM environment. By offering different types of emulation at the MAC layer, LANE offers support for the maximum number of existing applications.

7.1.3.1 Characteristics of LANE

A significant characteristic of LANE is that it can support all network layer protocols. This is due to the fact that it operates below the MAC layer. In addition, LANE networks may be bridged with real (i.e., nonemulated) LANs, and may be interconnected with routers. LANE easily supports virtual networks over a single ATM network, while also offering the advantage of easy reconfiguration.

In LANE the point-to-point ATM switch provides the function of a virtual shared medium. From the protocol stack's point of view, the

ATM layer behaves like an IEEE 802 MAC protocol underlying the LLC. The key attribute of the shared medium connection is that communication is done as a broadcast. Every station in a LAN receives all the packets from all other stations, and filters out the packets destined to itself. This feature of broadcast can be emulated in ATM networks using broadcast servers even though ATM is originally connection-oriented.

LANE provides communication of user data frames among all its users, similar to a physical LAN. The communication channel between nodes on the same LANE consists of direct ATM connections between the nodes. Each LAN is an emulated entity on an ATM network based on the configuration data put into the LANE servers. There is no direct mapping between an emulated LAN and physical boundaries within a single ATM network. This means that there can be one or more *emulated LAN* (ELAN) running on the same ATM network. Each of the ELANs is logically independent, with the nodes connected to one ELAN being unable to directly communicate with the nodes connected to another ELAN. Any type of communication between ELANs requires some type of interconnection devices such as bridges and routers. This directly mirrors the characteristics of the real world.

The fact that a number of ELANs may run on the same ATM network is an important advantage of ATM networks. It enables the configuration and operation of virtual LANs. This was possible even before the *virtual LAN* (VLAN) specifications were defined by the IEEE [6, 7]. Emulating physical LANs also mean that LANE must have some other important characteristics. One is that it must be able to support connectionless services. That is, a sender must be able to send data without previously establishing a connection, which is a big problem for connection-oriented ATM networks. Additionally, LANE must support multicast, more specifically, the use of multicast MAC addresses (e.g., broadcast, group, or functional MAC addresses). This puts some constraints on the MAC driver interfaces in ATM stations. By supporting such characteristics, LANE enables existing applications to use an ATM network through existing protocol stacks and APIs such as IP, IPX, *advanced peer-to-peer networking* (APPN), NetBios, and AppleTalk.[7]

---

7. Since LANE emulates the IEEE 802 MAC layer below the LLC it can support not only IP but also various network layer protocols such as SNA/APPN, IPX, and NetBios. This contrasts the classical IPoA approach, which can support the IP suite only. While LAN emulation is capable of supporting many different network layer protocols, today the main protocols supported are IP and IPX, with IP becoming the de facto standard network protocol due to the popularity of the Internet.

ELANs enable configuration of multiple, separate domains within a single ATM network. The resulting configuration would be logically analogous to a group of LAN stations attached to an Ethernet/IEEE 802.3 or 802.5 LAN segment. Several ELANs could be configured within an ATM network, regardless of the physical location of each connected end system. An end system may belong to multiple ELANs, where each individual ELAN is logically independent.

However, the LANE has expansion limitations. While clients in an ELAN can communicate with each other directly, communications between ELANs are possible only through bridges or routers. Accordingly, LANE is suitable for small workgroup networks in a local area. ELANs interconnected by bridges and extended beyond a local area or small number of workgroups would be impractical. As the number of connected ELANs grows, the broadcast traffic passing over the bridge increases, and the bridge could become a bottleneck. To reduce the broadcast traffic, routers can be used instead of bridges, and to reduce the number of interconnection devices, multiple ELANs can be interconnected by direct ATM connections.

7.1.3.2 LANE Elements

LANE service architecture is based on a client-server model. Figure 7.6(a) shows the elements of a LANE-based network along with the connections that are used, and Figure 7.6(b) shows the protocol stack of LANE.[8]

As shown in Figure 7.6(a) the basic servers used in a LANE network are a *LANE server* (LES), a *LANE configuration server* (LECS), and a *broadcast and unknown server* (BUS). Broadly speaking, the LES is responsible for registering MAC addresses to ATM addresses and resolving the addresses. The LECS locates the LES and provides configuration information for each ELAN segment. The BUS delivers broadcast or multicast frames, and is responsible for delivering the unicast frames whose destination address is either unregistered or unresolved yet. Figure 7.6(a) also shows the multiple relationships and the types of connections between these servers and clients. The communication between the LECs and the communication between the LECs and the LE service servers are carried out over ATM VCCs. Each LEC communicates with the LE service servers over control VCCs.

8. Note that the architecture in Figure 7.6 is only functional, so LANE service configurations are not necessarily implemented in these three parts physically. For example, an LES and a BUS may be colocated on the same ATM switch, while LECS may be configured on a separate stand alone server so that it can function as a server for multiple LES/BUS ATM switches.

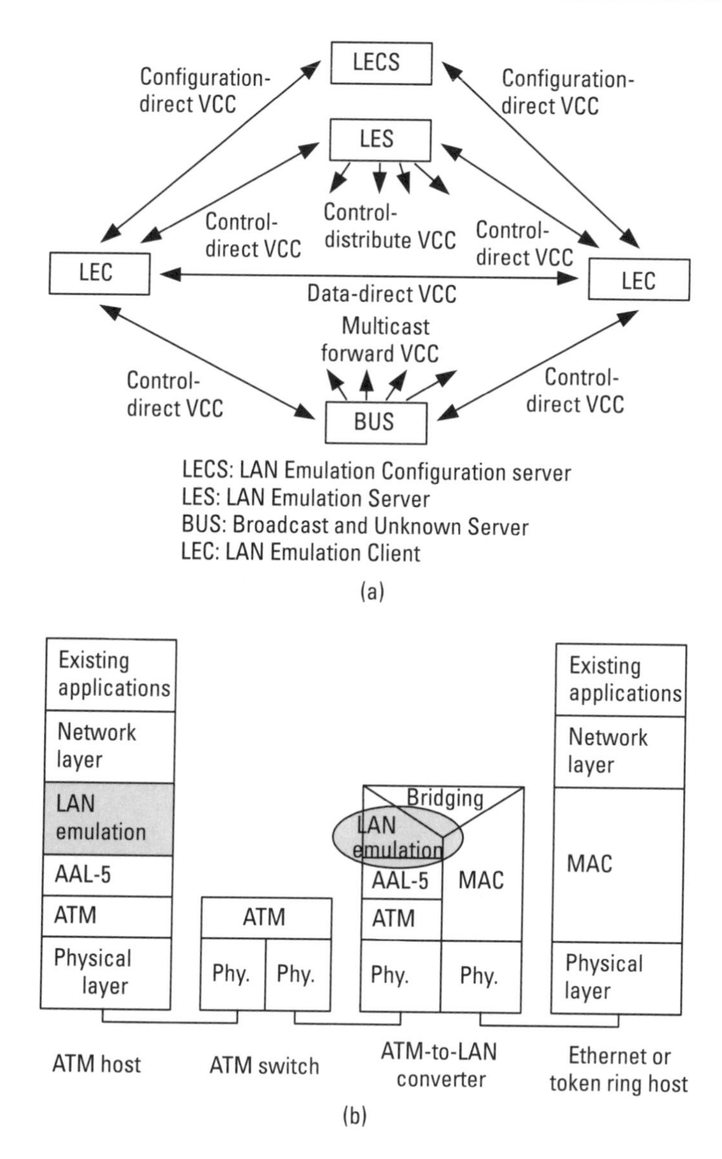

Figure 7.6 LANE: (a) elements and connections, and (b) protocol stack.

A single ELAN is a set of *LANE clients* (LE clients, or LECs) receiving *LANE service* (LE service) from a single group of servers. The LE service is offered by a group consisting of one or more LECS, one or more LES, and one or more BUS server. Conceptually this is shown in Figure 7.7. In the

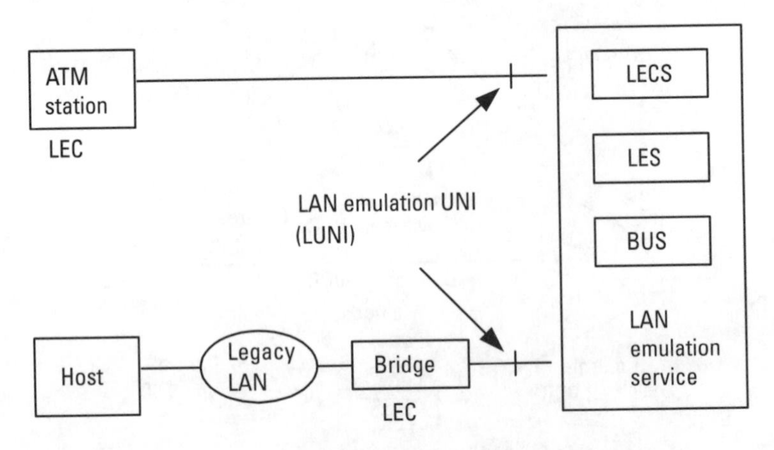

Figure 7.7 ELAN configuration and LUNI.

figure, *LANE UNI* (LUNI) defines the basic interface between an LEC and the servers providing the LE service. The LEC may be an ATM station, a LAN bridge, or a router with ATM interface. As shown in Figure 7.7, it is possible to connect a legacy LAN and an ELAN by using a layer 2 bridge. Note that in the same configuration it is also possible to use a layer 3 router and connect the two LANs at the layer 3 level.

LEC. An LEC is basically an ATM end station with an ATM address. As an LEC emulates a LAN node it is also assigned a MAC address, on the basis of which an LEC provides a MAC level emulated Ethernet/IEEE 802.3 or IEEE 802.5 service interface to the upper-layer applications. Any ATM end systems, for example, ATM workstations and ATM bridges, may be configured to be an LEC. It only needs to have the appropriate software and be configured with a method for finding the initial LECS to which it must connect. An LEC must implement the LUNI interface in order to communicate with other components within a single ELAN.

LES. An LES is the key control point of an ELAN, offering the control coordination function for the ELAN. It is basically a facility for registering and resolving unicast and multicast MAC addresses and/or route descriptors to ATM addresses. An LEC must be connected to only one LES. The LEC must register its MAC address with this LES. It is possible for an LEC to register multicast MAC addresses with the LES and thereby function as multiple LECs on a single ELAN.

An LEC will also query its LES to resolve a MAC address and/or route descriptor to an ATM address. That is, the LES takes on the responsibility of emulating the ARP functionality that is normally used in Ethernet or token ring networks to resolve MAC addresses. The LES will either respond directly to the LECs or forward the query to other servers. In such a case the LES is acting as a type of proxy server.

BUS. A BUS server provides a multicast server function to provide multicast connectionless data delivery. A BUS server basically handles the data sent by LEC to the *broadcast* MAC address. It is a server designed to emulate the broadcast ability that both Ethernet and token ring LANs have but that ATM networks do not. In addition, a BUS server is used to transmit initial unicast data sent by an LEC before the data-direct target ATM address has been resolved and, consequently, before a data-direct VCC has been established. Since the direct VCC has not been established, the data is sent on the multicast VCC to ensure that it is broadcast to all LECs on the same ELAN, as the packets are broadcast to all other hosts on the LAN. The BUS server also participates in the *LE address resolution protocol* (LE-ARP) to enable an LEC to locate its BUS.

A BUS sees all broadcast, multicast, and unknown traffic to and from an LEC and distributes data with multicast MAC addresses (e.g., group, broadcast, and functional addresses) to all the LECs it is connected to. An LEC is configured to always see only a single BUS server. If an LEC does not need to receive all multicast MAC addressed frames, the BUS server may then selectively forward multicast MAC-addressed frames to only those LECs that have requested them. To ensure that AAL-5 frames from different sources are not interleaved the BUS server must implement a serialization function in transmitting cells from different clients. Some LECs take advantage of the multiple interfaces of the BUS and send frames destined to a specific multicast MAC address to a different BUS interface.

LECS. An LECS is the main configuration server for an ELAN. An LECS assigns individual LECs to different ELANs. As such it must be configured beforehand along with the information regarding which LE clients are to be assigned to which ELAN. An LECS assigns any client to a particular ELAN service by giving that client the appropriate LES ATM address. The LEC then contacts that LES and thereafter operates as a member of the ELAN that is served by that particular LES. An LECS may assign a client to an ELAN based on either the physical location (ATM address) or the identity of a LAN destination (i.e., MAC address) that it is representing.

All LECs must be able to obtain information from an LECS using the configuration protocol. During the initial boot-up, an LEC must first contact its LECS. The LEC must either be preconfigured with the address of the LECS or may get it through other methods such as *interim local management interface* (ILMI).

Types of VCCs used in LANE. LANE uses a number of different types of VCCs to operate correctly. Broadly speaking, LANE uses two types of VCCs—point-to-point VCCs and point-to-multipoint VCCs. The *configuration-direct VCC*, *control-direct VCC*, and *data-direct VCC* belong to the point-to-point VCC category; and the *control-distribute VCC*, *multicast send VCC*, and *multicast forward VCC* belong to the point-to-multipoint VCC category.

Point-to-point VCCs. A *configuration-direct VCC* is a bidirectional point-to-point VCC that is used by the LEC to exchange configuration messages with the LECS. This VCC must be established during the initialization phase of operation. The LEC uses this VCC to receive configuration information from the LECS. The LEC and LECS may release the configuration direct VCC once the LEC is connected with the LES.

A *control-direct* VCC is a bidirectional point-to-point VCC that is used by the LEC to send control traffic to the LES. This VCC must be established during the initialization phase of operation. The LEC is required to accept control traffic from this flow. The LEC and LES must not release the control-direct VCC while participating in the ELAN.

A *data-direct VCC* connects the LECs with each other. It is established by the LEC once it knows the ATM address of the destination node by address resolution. The data-direct VCC is used to carry encapsulated Ethernet/IEEE 802.3 or IEEE 802.5 data frames. This VCC never carries control traffic.

Point-to-multipoint VCCs. A *control-distribute VCC* is a unidirectional, point-to-multipoint VCC that the LES may optionally establish for distributing control traffic to one or more LECs. This VCC may be set up by the LES as part of the initialization phase. If the control-distribute VCC is set up, the LE client is required to accept the control-distribute VCC. The LEC and LES must not release the control-distribute VCC while participating in the ELAN.

A *multicast-send VCC* is a bidirectional VCC that connects the LECs to the BUS. It is used to transmit multicast data frames from the LEC to the BUS. Additionally it is used to transmit data frames for the destinations whose ATM addresses the LEC does not know. This is frequently the case when the initial data frame for a destination is received and the LE_ARP reply from the LES has not been received. An LEC may also receive multicast frames over this VCC.

A *multicast-forward VCC* is a unidirectional point-to-point or point-to-multipoint VCC established between the BUS and the LEC. It is used by the BUS to forward multicast data frames to the connected LECs. The multicast-send VCC and multicast-forward VCC are used to carry encapsulated Ethernet/IEEE 802.3 or IEEE 802.5 data frames. This VCC never carries control traffic.

7.1.3.3 LANE Procedures

The LANE service function uses the procedure consisting of *initialization, registration, address resolution,* and *data transfer.* At first, a client contacts the LECS to locate the LES. Then an ATM connection is established to the LES, and the client registers its MAC address and ATM address to the LES. These functions are defined over a LUNI.

Initialization and configuration. The overall flow of the initialization and configuration flow in LANE is shown in Figure 7.8. The first process for joining an ELAN involves getting some basic initial parameters and fetching the LECS's address. Then the client determines which ELAN it is connected to and fetches configuration information and the LES's address. Next, the join phase takes place where the client sets up a VCC with the LES and joins the ELAN.

During the initial configuration phase, a connection with the LECS must be set up. Specifically, a configuration-direct VCC must be set up between the LEC and the LECS. The address of the LECS server can be found by a number of different ways. It can be found either by manual configuration of the information in the client, by using ILMI signaling data from the nearest ATM switch, or by using a well-known VCI (VPI = 0, VCI = 17) to connect with the LECS directly. By connecting with the LECS, the LEC obtains the address of the LES and other configuration parameters.

It is also possible for the LEC to skip this LECS connection phase and not set up a configuration-direct VCC but, instead, directly connect to an

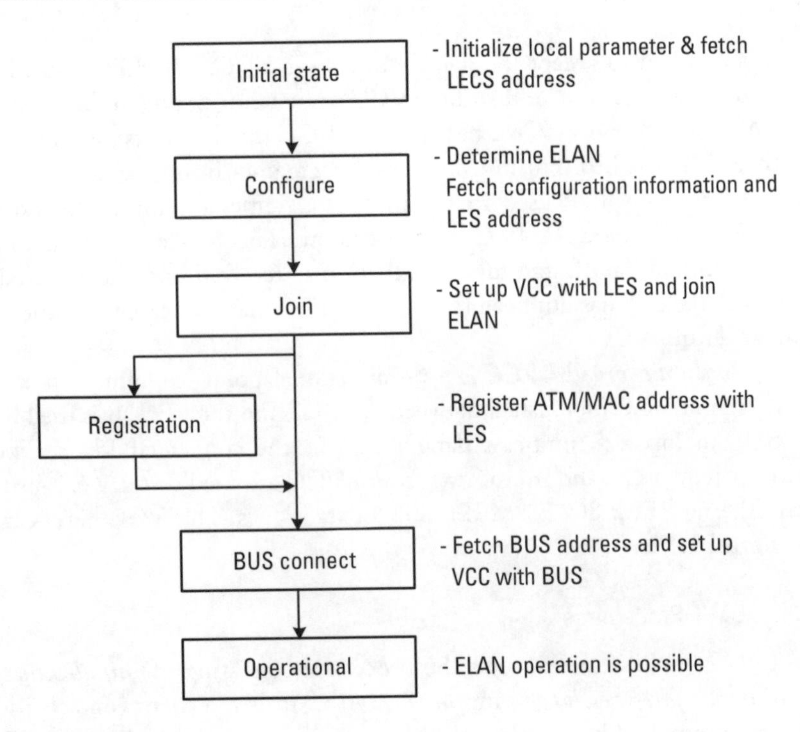

Figure 7.8 Initialization and configuration flow in LANE.

LES. For this to be possible the LEC must know the LES and other relevant information beforehand. How to find the address of the LES is not defined in this case. Manual configuration would be the most likely method.

The next phase following the configuration phase is the join phase, where the LEC establishes a control-direct VCC with the LES and attempts to join the ELAN with the name received during the configuration phase. Following this the client must register its MAC/ATM address with the LES. This is carried out by using the registration procedures to be described next.

Once the LEC is connected to the LES, the BUS connect phase begins. In this phase the LEC must connect with the BUS. This is done by first requesting the BUS's ATM address from the LES by using the LE-ARP function. The LEC sends an OK message to the LES and the LES responds with an OK message containing the BUS's ATM address. The LEC then establishes a multicast-send VCC with the BUS server. The BUS adds the LEC to a multicast forward VCC. The LEC is then connected to an ELAN and may start to transmit data.

Registration and deregistration. The registration procedure is used initially by the LEC to register its MAC/ATM addresses with the LES during the configuration and join phase. The procedures may be repeated to register extra MAC/ATM addresses. Deregistration is the opposite process by which previously registered MAC/ATM addresses are removed from the LES. As shown, clients may register multiple MAC addresses by using the registration procedure repeatedly. Bridges that want to register the MAC addresses of clients that are not LANE capable can also use such functionality. Such a case would apply to the bridge that hides multiple hosts on the legacy LAN segment as shown in Figure 7.7.

Address resolution. In LANE, each client maintains information on mapping MAC addresses to ATM addresses for connection setup. When a client does not have this information, it can obtain ATM addresses from MAC addresses using the LES. The LES makes a table for mapping MAC addresses to ATM addresses at the registration stage and updates it whenever change occurs.

The LEC obtains the ATM address by sending a request message to the LES. Then the LES normally replies with a response that contains the appropriate ATM address. The request message contains the source MAC address, the source ATM address, and the destination MAC address. The response message contains the destination ATM address in addition to the information in the request message. The request message is sent over the control-direct VCC, but the response message can be sent over either the control-direct VCC or the control-distribute VCC. The advantage of the sending the response message over the latter is that the response may be received by all the LECs connected to the control-distribute VCC. The other LECs may then cache the information for future uses.

There are three different operation scenarios depending on the type of MAC address and whether the destination MAC address has been registered with the LES. First, if the MAC address is a broadcast MAC address, the LES returns the address of the BUS. Second, if the MAC address is registered with the LES, then the LES will reply with the ATM address corresponding to the MAC address. Third, if the MAC address is not registered with the LES, then the LES forwards the request to all the LECs.

Data transfer. In LANE, data transfer may be done by either unicast or multicast frame. Once the MAC address to ATM address resolution is completed, the client begins the connection setup process. The connection setup at this time directly between the LECs is the data-direct VCC. After this process, unicast frames are transmitted directly to the destination client. The connection is automatically released after a fixed length of idle time. When transmitting multicast frames, the client takes a slightly different procedure: In this case the client transmits the multicast frame to the BUS using the multicast send VCC. The BUS then forwards this frame to all the other LECs by using the multicast forward VCC.

A client can also use the BUS to send frames before the address resolution process gets finished. By doing so the delay between transmitting a frame and its reception by a destination is minimized. In this case the BUS broadcasts the frames to all clients, because it is the only method available to send frames to destination clients with unresolved MAC addresses. In addition, for a delay-sensitive application, it is desirable to send frames using this method so that the service is unaffected by the delay that may occur during the address resolution and connection setup process. Once the address resolution procedure succeeds, a data-direct VCC is set up and data is no longer forwarded through the BUS server. To prevent the clients from abusing the broadcast channel, the number of broadcast frames that a client may send within a given time period is limited.

If the BUS were capable of delivering the unresolved frames only to their destinations without broadcasting, then the traffic in the network would be much reduced. However, it would result in complicated and costly BUS implementations.

7.1.3.4 LANE Implementation

The LANE specifications only define the functions and interfaces for LANE operation. From a practical point of view, the actual implementation details are left to the implementer. Typically, LECs are implemented in ATM end stations, either as a software driver or as an ATM adapter (i.e., ATM-specific hardware and software). ATM end systems can be either intermediate systems (e.g., bridges or routers) or end stations (e.g., hosts or PCs). The LE service might be implemented in any combination of ATM switches and ATM attached end stations (e.g., bridges, routers or workstations). An LE service component may be colocated with an LEC. Note that it is important that an LES can also be located in an ATM end station as the server communicates with the LECs and other servers by using LANE.

7.1.3.5 LANE Versions

Currently two versions of LANE have been defined. LANE version 1.0 defined only the LUNI specifications and operations as described above. There were a number of limitations in the initial standards: Only one LES could be defined for an ELAN making it a bottleneck and a potential single point of failure. Also it only supported UBR services, thereby rendering QoS-based services impossible.

LANE version 2.0 was later defined. It was comprised of LUNI version 2.0 and *LANE NNI* (LNNI) specifications. The former defines the interfaces between the LESs and LECs and focuses on the definition of LANE operation in a single emulated LAN, whereas the latter defines the interfaces between the LECS, LES, and BUS servers in the LE service and focuses on the definition of LANE operation when multiple LANE servers are involved. One of the main problems with LANE version 1.0 is the fact that as only a single LES can be used, there is a single point of failure. By defining interfaces between the multiple LESs and thereby enabling the use of multiple servers this problem can be eliminated in LANE version 2.0. The specifications allow for up to a maximum of 20 LESs and 20 BUSs. LANE version 2.0 also adds support for globally and locally administered QoS, enhanced multicast with selective broadcasting capability, and support for ABR rate-based flow control, and support for FDDI. Also added are support for LLC-multiplexed VCC and support for MPOA.

The enhanced multicast defined in LANE version 2.0 adds support for separating multicast traffic from the general broadcast path. It enables the possibility of determining which members of the emulated LAN are to receive multicast frames. The filtering function is performed through cooperation between the source and the LANE service.

7.1.4 MPOA

Multiprotocol over ATM (MPOA) was developed by the ATM Forum as a comprehensive solution for interconnecting various types of layer 3 networks by using ATM technologies. Where LANE solved the problem of using ATM to emulate LANs, MPOA aimed to efficiently use ATM to support the interconnection of LANs, both emulated and nonemulated.

The main goal of MPOA was to efficiently support internetwork traffic using ATM technology. The key idea was to use the direct ATM connections between end nodes where possible. This is similar to the concept of LANE, except that it is now extended to include the nodes not only in the same ELAN, but also in other LANs. This is illustrated in Figure 7.9. By using the protocols defined in MPOA along with other protocols such as NHRP, it is possible to set up a direct ATM connection to any other systems connected to the same ATM network.

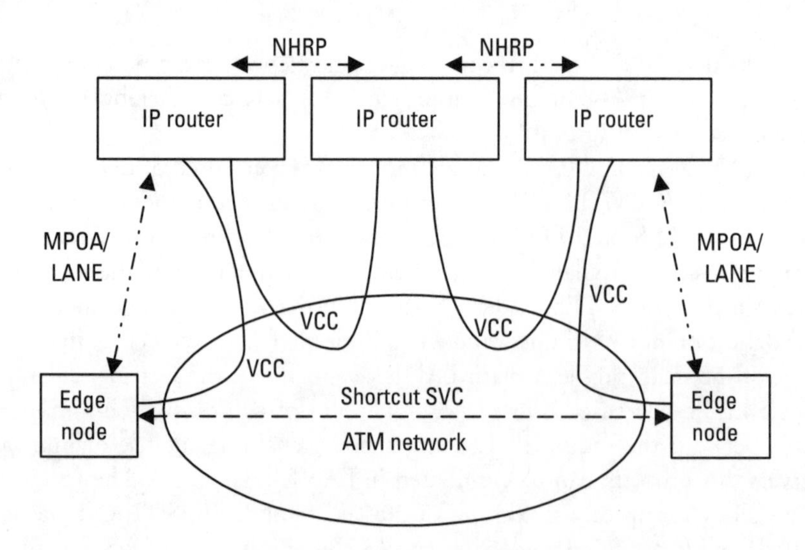

Figure 7.9 Use of MPOA to set up a shortcut SVC over an ATM network.

The method chosen for solving the above problem was to provide the functionality of a router over an ATM network using a distributed model and implementing it in a number of switches and servers in the network, along with the corresponding changes in the client software. It is a distributed model with the relevant functionalities distributed over the network in the form of server-client functions. At the same time it offers a single paradigm for overlaying internetwork layer protocols over ATM.

The characteristics of the solution include the efficient transfer of intersubnet unicast data in a LANE environment. It also provides a scaleable routing solution for IP networks by operating as a distributed virtual router in ATM networks. The routing functions are distributed over various physical boxes. While basic routing protocols such as OSPF and RIP are run on a server, IP data forwarding is carried out by ATM switches. For nodes in the same network, LANE is used to support data transfer in subnets, while the NHRP is used to support address translation between different subnets so as to enable the setup of connections between them.

7.1.4.1 Virtual Router Concept

One of the key ideas in MPOA is the concept of the virtual router. This concept is built on the idea of physically dividing the routing and forwarding functionalities in the internetwork layer; routing protocol and route computation are handled by separate router servers, while the actual packet forwarding is carried out by the ATM switches and edge devices. This can be construed as an example of separating the intelligent control function from relatively simple forwarding functions. This is a radical departure from the current generic router structure. The traditional single-box router contains both of the above functionalities in one box. It must run complicated routing protocols at the same time, also maintaining a fast path for forwarding user packets.

As the network speed and capacity increase, the routers must be speeded up at the same time. This may not be easily done in conventional structure, but in a virtual router structure, the servers and forwarding engines may be upgraded separately. Figure 7.10 helps to visualize what this means. It compares the structure of a single-box switch with that of an MPOA virtual router. A single-box switch may be roughly divided into a central processor, I/O cards, and a backplane. The central processor is the main controller of the router with the main responsibility for carrying out any route calculations. The I/O part classifies and forwards packets. The backplane interconnects the I/O ports and forms the path between the I/O ports (refer to Section 3.3.3 for more details). In a virtual router, these components can all be implemented in separate boxes. That is the central processor implemented as an MPOA server, the I/O ports as MPOA edge devices, and the backplane as the ATM-switched network that interconnects all the MPOA servers and edge devices.

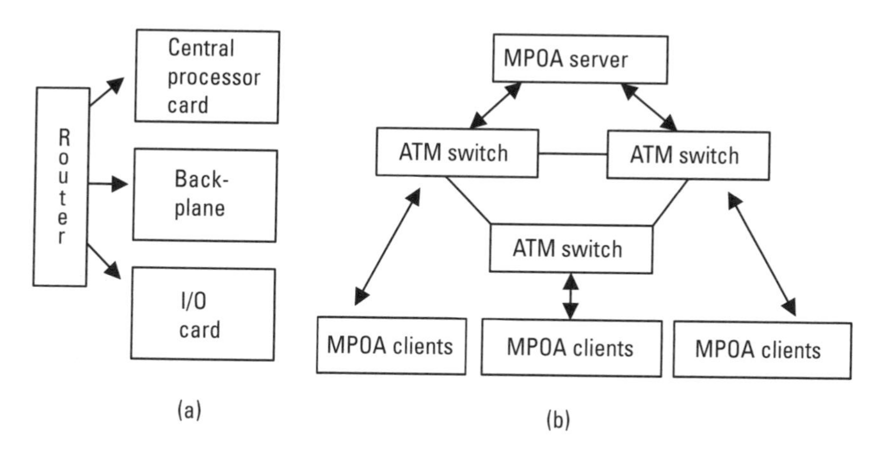

Figure 7.10 Comparison between (a) traditional router structure, and (b) MPOA-based virtual router structure.

Due to the specialized nature of each node, the virtual router will be easier to upgrade. For example, increasing the capacity of the ATM switch network may be done simply by adding more switches or by upgrading the switches themselves. Running more complicated control processing may be managed by increasing the number of servers. In contrast, increasing the speed or the processing capacity of a traditional router system requires increasing the routers themselves.

In addition, it is harder to introduce new routing functionality to a single-box router than to a virtual router. The idea of separating the routing functionality from the actual packet-forwarding component is also the main basis of the MPLS paradigm. (Refer to Section 3.3.5 for more detailed discussions on other benefits of such a separation.)

7.1.4.2 MPOA Elements

There are two main MPOA elements: one is *MPOA client* (MPC), and the other is *MPOA server* (MPS). An MPOA edge device contains an MPC and a LANE client. An MPS contains NHS servers and routing functions (and also a LANE client), as shown in Figure 7.11.

MPOA client (MPC) functions. The main function of the MPC is to support data transport by using a layer 3 short-cut. MPC only does layer 3 forwarding, not routing. The MPOA client uses an NHRP-based MPOA request/response protocol for short-cut transport. Once configured, the short-cut information is stored in an ingress cache table. Frames received over the short-cut path are passed to upper layers after appropriate data link layer encapsulation.

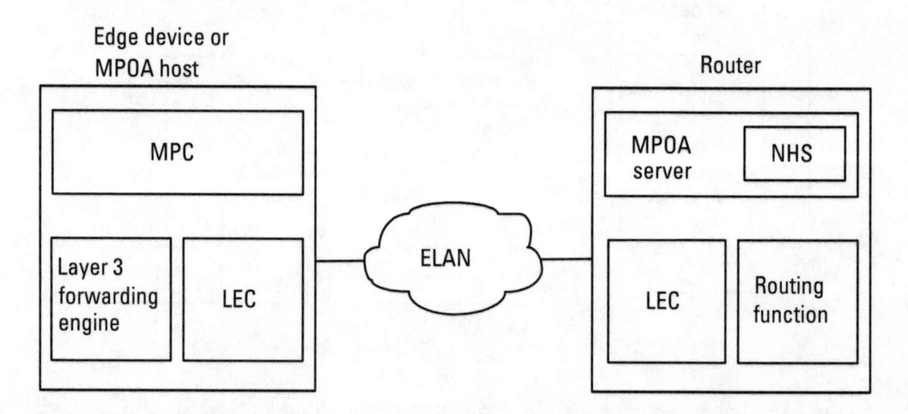

Figure 7.11 MPOA elements.

MPOA server (MPS) functions. The main function of the MPS is to implement the logical structure of a router. The MPS supplies to the MPC the layer 3 forwarding information learned by mapping the layer 3 address with the ATM address. It interacts with the local NHS routing functions to give information to the MPC. In addition, the MPS changes the MPOA request/response to an NHRP request/response for the MPC. Routers also run routing protocols such as OSPF, RIP, and IS-IS. An NHS is used to provide address resolution between ELANs.

Internetwork address subgroup (IASG). An important concept to understand in MPOA is the concept of IASG. An IASG is an address group that contains a wide range of internetwork addresses that can be summarized and used in an internetworking layer routing protocol. An example of an IASG is the IP network layer prefix.

7.1.4.3 MPOA Operation

MPOA consists of a number of basic operational procedures such as configuration of the various MPOA components including the MPOA clients and MPOA servers, discovery of the MPOA servers and their addresses in a subnet, resolution of the ATM address of the destination, connection management of the various control and data connections, and the actual data transfer. In general, one can view LANE services as providing the configuration, discovery, and legacy LAN support, while NHRP and LANE together provide support for destination address resolution.

Configuration and discovery. The configuration and discovery procedure aims to get configuration information to MPOA clients. The information is mainly on the MPOA servers to which the client may be connected. Note that as in all cases it is possible to configure all the relevant information manually on a client, but this is not a scalable method.

The configuration information consists of the ATM addresses of the servers and some information related to the IASG that the MPOA server supports. This information would include the IASG identifier and protocols, the name of the ELAN, and the IASG address and prefix. As MPOA is defined to support a number of protocols besides IP, the IASG configuration is more generic than would be if it had been designed for just to support only IP. The basic configuration and discovery procedures are based on the LANE configuration procedures and methods. The ATM address of the MPOA server can either be configured into the client or the client may use ILMI to retrieve it from the nearest ATM switch.

Address registration. The MPOA address resolution method is based on the NHRP resolution scheme. In the NHRP scheme the NHS is expected to have a database containing the mappings between the destination ATM addresses and the destination IP addresses. The NHS replies to the resolution requests by checking this database. In a similar manner, the MPS must also maintain such a database. Consequently, the MPC must register the IASG addresses that it supports with the MPS.

Address resolution. The address resolution procedures are the methods by which target addresses are mapped to ATM addresses. Depending on whether or not the destination is in the same internetwork layer subnet, the method used for destination address resolution differs. If the destination is on the same internetwork layer subnet then the destinations can communicate by using a LANE connection or by using a bridge with a LANE interface. In such a case the address resolution is carried out by using the mechanisms defined in LANE (refer to Section 7.1.3). If the destination is on a different internetwork subnet then the destinations can be normally reached only by going through a layer 3 router. In such a case the NHRP protocol must be used to resolve the ATM address of the destination node or its nearest exit router (refer to Section 7.1.2).

Data forwarding. MPOA data transport procedures can be divided into that for the default transport case and that for the short-cut transport case. The default transport method is to use LANE, so consequently all MPOA elements, MPCs, and MPSs must support LANE. The key edge device is the layer 2 bridge. However, in the short-cut transport method, a path is set up by using NHRP-based address resolution and cache management mechanisms. In this case the key edge device is the layer 3 router. All transport inside a single ELAN uses only the default transfer and all transport between ELANs can use either the default transfer or short-cut transfer.

The data forwarding operation by an MPOA client follows the following steps: When a frame arrives at the MPC it is first forwarded along the default routed path. This means that it is forwarded by using LANE either to the destination or to the default router if the destination is not reachable by LANE. As more packets arrive, the MPC must decide whether the flow of packets warrants a short-cut path setup. If this is the case, MPOA address resolution is carried out to find the ATM address of the destination to which a short-cut path must be set up. The method short-cut path is then set up and data transfer may start.

Cache management functions. One of the functions in MPOA is the cache management function used to support short-cut connections. All MPCs must keep an *ingress* cache and an *egress* cache. A simplified example of the ingress and egress cache tables is given in Figure 7.12.[9] The ingress cache maps a combination of the MPS's MAC address and the destination IP address to the correct ATM connection. Additionally, it maintains a count entry that is used to decide whether the connection has enough traffic to warrant a short-cut path setup.

When the first packet of a flow first arrives at an MPC, the ingress cache table is checked for an entry for this packet's flow. If there is a cache miss, a new entry is made with the information available. At this stage ATM connection entry is empty. The count entry is set to 1 and incremented with each packet transmitted. If more than a predefined number of packets are transmitted within a certain period of time then a short-cut path is setup for the flow. The VPI/VCI of the short-cut path is put in the ATM connection entry. Thereafter, any packet of this flow is automatically transmitted by that ATM connection. The information for updating the ingress cache (i.e., VPI/VCI of the ATM connection) is updated when the MPC receives the MPOA resolution response. If the packet flow is idle for a certain amount of time the cache entry is removed.

The egress cache maps a combination of the incoming ATM connection and the destination IP address to the correct MAC address and port

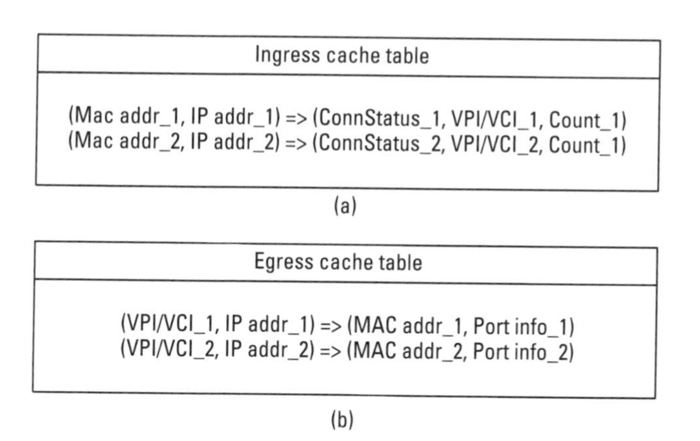

Figure 7.12 Example of the caches in MPOA: (a) ingress cache, and (b) egress cache.

9. Note that the exact cache entries and forms may differ from implementation to implementation.

information. When the egress MPC receives a packet, if a corresponding entry can be found in the egress cache table, the MAC address is appended to the packet and the packet is sent out on the appropriate port. The egress cache entries are updated when an MPOA egress cache imposition request is received from the egress MPS.

7.1.4.4 MPOA Message Types

As explained in the previous section, MPOA is based on the LANE and NHRP protocols. As LANE version 2.0 was designed from the beginning to support MPOA, no additional messages or information types need to be added. However, to support some extra functionalities of MPOA, a number of extra messages and information elements have to be added to the base NHRP protocol. All such messages follow the basic NHRP format. The *initial resolution request* is sent by the (ingress) MPC to the (ingress) MPS to request that the ATM address for an IASG address be resolved. The *resolution reply* is sent by the (ingress) MPS to the (ingress) MPC and contains the resolved ATM address. The *cache imposition request* is sent by the (egress) MPS to the (egress) MPC, instructing it to set up an egress cache entry for the connection. The messages needed for correct operation of the protocol include egress cache purge, keep alive, trigger, and data plane purge.

7.1.4.5 Example of MPOA Operation

The steps shown above give a rough idea how MPOA operates. A more detailed explanation on MPOA operation can be found in the sequential steps given below. Figure 7.13 illustrates the basic operation of MPOA.

- Step 1: The source transmits a packet to the destination host. The first packet arrives at the MPOA edge device (C1) on its legacy LAN interface. On checking the ingress cache for an entry, there occurs a cache miss indicating that this packet is a new flow. A new ingress cache entry is made and the packet is forwarded to the next hop router by way of the ELAN—that is, the first packets are forwarded along the default router path.
- Step 2: As further packets arrive, the ingress cache entries are updated. After a while, the MPOA client decides that the flow is sufficiently long-lived to warrant a new connection to be set up.
- Step 3: The MPOA client sends an MPOA resolution request to the nearest MPOA server (S1). (The MPOA client is configured with the address of the nearest MPOA server beforehand.)

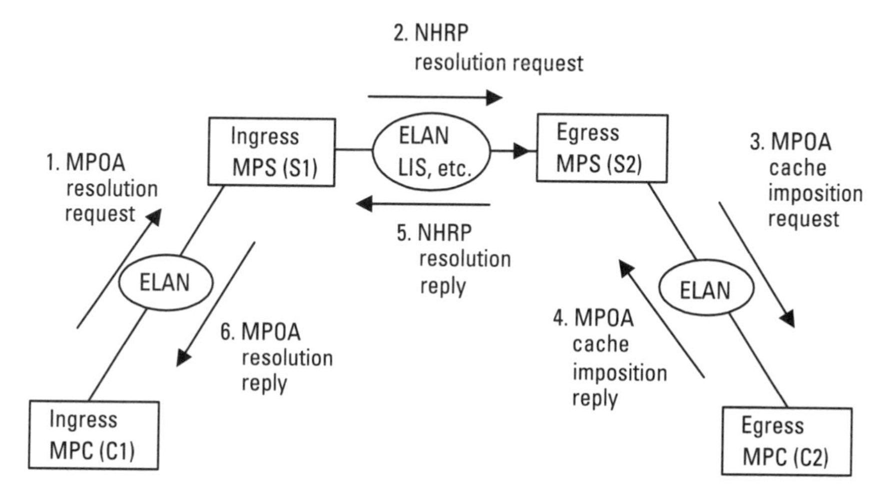

Figure 7.13 Basic operation of MPOA.

- Step 4: The MPOA server transforms the MPOA resolution request message into an NHRP resolution request and forwards it to the appropriate next hop router according to the internetwork destination.

- Step 5: The NHRP resolution request eventually arrives at an MPOA server (S2) that knows how to directly reach the destination. Usually this means that the MPOA server is on the same emulated LAN as the MPOA client.

- Step 6: This last MPOA server locates the MPOA client (C2) to which the destination is directly connected and sends a cache imposition request to the MPOA client. The client creates an egress cache entry for this flow.

- Step 7: The MPOA client (C2) generates a cache imposition reply that is sent back to the MPOA server.

- Step 8: The MPOA server (S2) receives the cache imposition reply, transforms it into an NHRP resolution reply, and sends it back to the MPOA server that originally generated the request. The reply follows the path that the original NHS resolution request took when coming to the destination (refer to Section 7.1.2).

- Step 9: The first MPOA server (S1) receives the NHRP registration reply and relays this back to the first MPOA client (C1) as an MPOA resolution reply.
- Step 10: The MPOA client (C1) now has enough information to set up a direct ATM SVC connection with the final MPOA client (C2). Once this connection is set up data is forwarded along this connection directly to the final MPOA client, instead of going through the complicated default router path.

7.1.5 I-PNNI

I-PNNI is an attempt to extend PNNI to support the integrated model for IP and ATM interoperability. As explained in the introduction to this chapter, the previous methods (classical IPoA, LANE, NHRP, MPOA) are all overlay network model–based architectures. The IP nodes at the edges are completely unaware of the underlying ATM network structure. In contrast, I-PNNI results in the IP nodes becoming aware of the ATM network's topology. The ATM nodes and the IP nodes are peers communicating control messages and signals to each other.

By using I-PNNI both ATM and IP nodes become fully aware of each other's network topology. This is achieved by using common (or interoperable) routing protocols and addressing schemes. More specifically, I-PNNI achieves these aims of supporting both ATM and IP by using a single routing protocol based on link state routing protocol. I-PNNI supplies basic information needed for SVC setup to both ATM and IP networks. Originally it was planned to be an extension of PNNI-based ATM switch functionality. I-PNNI is applicable to ATM and non-ATM (or datagram) media. It can be used between ATM switches, between routers, and between ATM switches and routers.

The basic flow of operation that was originally envisioned by its proponents is as follows: I-PNNI nodes exchange hello messages with neighbors to determine local topology. Nodes then announce their topology, with the announcements being flooded throughout the area. Based on this information, nodes are able to understand complete topology of the area and calculate routes.

As yet the protocol is not completely developed and no further active work has been carried out lately. It has now been mostly supplanted by MPLS as the means to integrate ATM and IP networks with a single routing technology.

7.1.6 MPLS over ATM

As explained previously in Chapter 3, MPLS is conceptually a combination of layer 3 routing and layer 2 switching. It employs level 3 routing protocol functions, level 3 forwarding at the edge nodes, and level 2 forwarding in the core nodes. By combining the IP control paradigm with the label-switching–based traffic forwarding, it is possible to get higher bandwidth, enable traffic engineering through explicit routing mechanisms, and garner various other advantages. (For a more detailed explanation on the principles and practical aspects of MPLS, refer to Section 3.3.5.)

As noted above MPLS has many advantages, including that MPLS enables the conversion of ATM switches into IP routers. This means that cheap ATM switches may be converted into high-performance IP routers by applying minimal changes. This can be easily understood if one recalls the relationship between MPLS and ATM.

Currently, MPLS in general is being designed to explicitly allow *label switch routers* (LSRs) to be based on traditional router platforms and on ATM switch platforms as well. The reason for the emphasis on ATM platforms is a natural progression from the fact that ATM networks pioneered many of the basic underlying building blocks that constitute MPLS. These include the concepts of simplified forwarding, traffic engineering based on explicit routing, and QoS routing. Any MPLS devices must inevitably incorporate these building blocks and ATM devices, which already dealt with these features within ATM networks, can support them very easily.

A simplified view on how an ATM switch functions as an LSR is as follows. The labels of an MPLS packet are carried in the VPI/VCI fields of the ATM cell. Conceptually, an ATM VPI/VCI header is in fact equivalent in semantics to an MPLS label. The operation of an MPLS LSR on the MPLS label mirrors the action that the ATM switch carries out on the VPI/VCI header of the ATM cell. The VPI/VCI header fields of an ATM cell may be used as a single label or it may be divided up into two different labels. In other words, the VPI/VCI fields may be viewed as a label stack, with the VPI used as the top label in the stack and the VCI used as the second label.

While the mapping of MPLS functionality to ATM network hardware is conceptually simple, it is not a completely cut and dry affair. There are a number of issues that must be solved before it becomes possible to use MPLS control architecture over ATM network hardware. We examine a number of these issues in the following sections.

An ATM switch may have multiple interfaces, with some working to the original ATM specifications and some operating as MPLS interfaces. To distinguish the two types, we may call an ATM interface that operates under the "MPLS label switch controlled" paradigm an LC-ATM. An LC-ATM operates in a different mode from the normal "conventional" ATM.

7.1.6.1 Basic Operation of an MPLS-ATM Switch

The MPLS-ATM switch preserves the ATM user plane. That is, user data is switched and transmitted in the ATM switch as in any other normal ATM switches. When an ATM cell is received its VPI/VCI is used for table lookup, based on which a new VPI/VCI label is assigned. Additionally traffic parameters and policing may be also carried out on the connection according to the data lookup.

The main difference between the MPLS-ATM switch and the ITU-T/ATM Forum-defined ATM switch is that the control plane of the MPLS-ATM switch mainly runs IP routing protocols such as OSPF, RIP, BGP, and PIM rather than UNI and PNNI protocols. The forwarding operation is determined by the underlying ATM switch fabric, whereas the control functionality is similar to that of a router. This is because cell forwarding is dictated by the switch fabric, while switch functionality is largely defined by the control component.

7.1.6.2 Encapsulation of Labeled Packets on ATM Links

As mentioned previously, the MPLS label is carried in the VPI/VCI field of the ATM cell. As the size of the VCI field is 16-bits-long, there are up to 2^{16} labels available. If the VPI field is used as well, additional 8 bits (or 12 bits) are available for the labeling. As the tag stack can have two layers, it is possible to make the VCI field define one layer and the VPI field the other.

There are two points to note in using MPLS over ATM networks. First, the label stack field header that is normally used by MPLS packets is also included in the ATM AAL-5 PDU in front of the actual network layer packet. The top level is carried in the ATM VPI/VCI while the lower layers are in the stack. The whole stack is carried, but the top level is ignored as it is already in the ATM VPI/VCI. This approach enables label stacks of arbitrary depth. As mentioned previously the use of the VPI/VCI fields in the ATM cell header essentially limits the label stack to only two layers. By including the whole stack in the payload, it is possible to support arbitrary depth label stacks. This approach also helps to solve the problem of transporting TTL and *experimental* (EXP) bits. ATM VPI/VCI fields do not have any fields for carrying the TTL and EXP bits that are in the labels defined in MPLS.

Second, remote binding is always acquired on demand. This means that an ATM LSR will not advertise its local binding to another ATM LSR before the other ATM LSR specifically requests the binding. In order to acquire remote bindings the ATM LSR must specifically request it from the other ATM LSR. By requiring that bindings are only acquired on demand, the number of bindings actually needed is minimized. This is helpful in practical situations because many commercial ATM switches have a built-in hardware limit in the VCs that they can support. Though the use of fixed labels (VPI/VCI fields) is helpful in simplifying hardware design, it does not necessarily mean that lookup tables based on 24 bits are feasible. Most ATM switches are designed to use a smaller portion of the VPI/VCI fields. On-demand remote binding of labels also helps solve the cell interleave problem to be discussed below.

7.1.6.3 Cell Interleave Problems

As discussed in Section 3.2.1, in IP routers, destination-based forwarding is normally used. This can lead to an interesting problem called the *cell interleave problem.* This problem occurs when multiple MPLS streams meet at an ATM switch and are all routed toward the same destination, as shown in Figure 7.14(a). As all the packets are to be sent to the same destination, they are all given the same MPLS label. In the case of ATM networks, this means that the cells from each stream are given the same VPI/VCI values. Consequently, the receiving host would have no way to discriminate the cells sent by source *A* from the cells sent by source *B* as they would all have the same VPI/VCI value. In such a case the receiver will not be able to correctly reassemble the packets from those cells as they would be mixed up.

One proposed solution to this problem is the so-called *VC-merge* method shown in Figure 7.14(b). In this scheme, VC-merge occurs at the ATM switch where the two streams meet. While the cells of the first stream are being transmitted, the ATM switch buffers the cells coming from the other stream. This is done until a full packet has been sent. The switch recognizes that a full packet has been sent by examining the end-of-frame bit in the ATM headers of the cells in the stream. This function is usually implemented in the switch to support frame-based discard methods for improving TCP/IP traffic performance. Once all the cells from one stream have been sent, then the cells from another packet are sent. This second packet may or may not be from the same stream. The key point is that it is ensured that cells belonging to different end user packets are not interleaved.

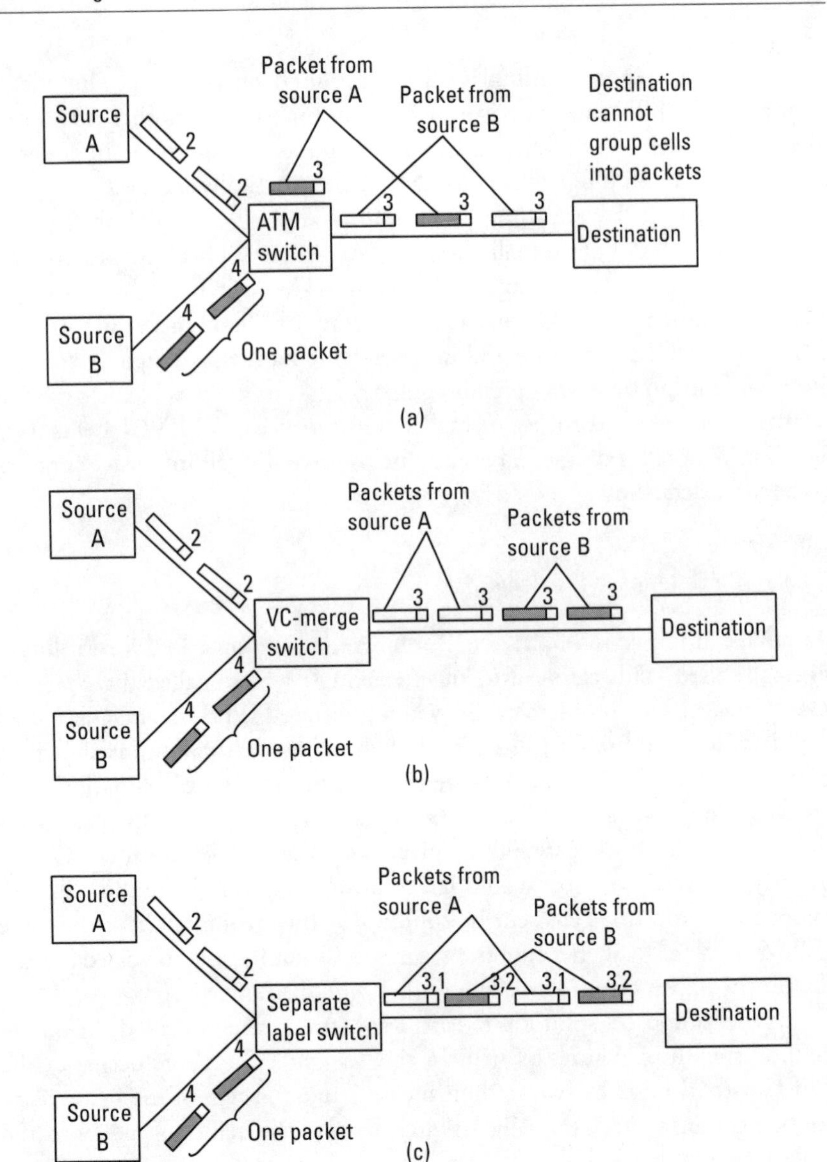

Figure 7.14 Illustration of the cell interleave problem in MPLS/ATM: (a) basic cell interleave problem, (b) VC-merge solution, and (c) separate label solution.

Another proposed solution is to use separate labels for different streams, as illustrated in Figure 7.14(c). As each stream is allocated a separate label, there is no problem at the receiving node in differentiating the different streams. The price, however, is that a large strain is placed on the number of labels (i.e., VPI/VCI values) needed. Consequently, the on-demand method must be used for assigning the labels.

7.1.6.4 Looping and TTL Adjustment

Another problem with LC-ATM links is that when the packet is sent over an ATM link there is no TTL field in the ATM cell. As such, the problem of protecting against routing loops is an important issue. A solution may be found in applying the loop-mitigation method used in the MPLS to MPLS-ATM networks. This can be accomplished by having the TTL values decremented by the number of ATM LSRs that the path traverses. As mentioned above, ATM cells do not contain any TTL field and ATM LSRs themselves cannot look inside an ATM cell stream and decrement the TTL field. So, instead, the non-ATM LER at the starting point of the MPLS path decrements the TTL field by the number of the ATM LSRs that the path traverses. This number is learned during the path setup time from the label mappings. If the TTL value happens to become a negative number due to the decrement, the packet is handled in an appropriate manner.[10] Otherwise, the TTL value is decremented and put into the packet. When the packet comes out of the path at the end, the TTL value is then corrected.

7.1.7 CLSF-Based Data Services

The ITU-T recommends two different configurations to support connectionless data services in BISDN: the *indirect* and *direct* methods. In the indirect method, *connectionless service functions* (CLSFs) and the associated adaptation layer entities are located outside the BISDN, whereas in the direct method, CLSFs are located inside the BISDN. Figure 7.15 shows the connectionless service reference models for these two methods.

7.1.7.1 Indirect Method

In the indirect method, connectionless services are provided through the virtual connections that connect ATM *interworking unit* (IWU) pairs. Each IWU provides an interface between a connectionless LAN and the ATM network. Virtual connections connecting IWUs form a dense mesh of connections in the network. Such connections may be established all together by

10. For example, the packet may be dropped and an ICMP error message may be generated and sent without being mapped to a label path.

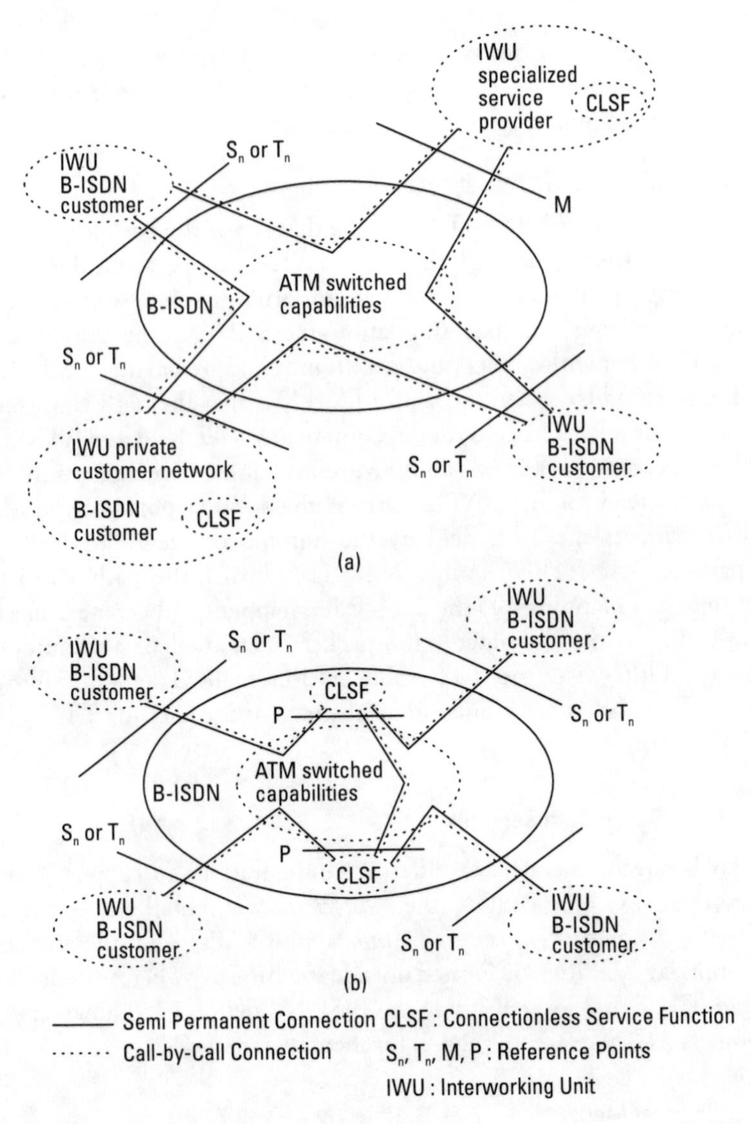

Figure 7.15 ITU-T connectionless server model: (a) indirect method, and (b) direct method.

using PVCs or may be established only when needed by using SVCs. The choice of PVC versus SVC depends on the network size and the service requirements.

SVCs can support large-size networks because it is not necessary to maintain connections when there is no data to transmit. So PVCs are used only when the size of the network is small enough to fully interconnect all IWUs with one mesh network. On the other side, SVCs require some connection setup overhead; each IWU must buffer packets until the connection is established, which causes long transmission delays. However, PVCs help to eliminate such delays.

The indirect method has the drawback (whether it uses PVCs or SVCs) that it cannot efficiently utilize the network resources, especially the bandwidth. When establishing a connection for a connectionless datagram, if the allocated bandwidth is too large, it means that bandwidth is being wasted. In contrast, if the allocated bandwidth is too small, it causes excessive delays in data transmission. Further, the indirect method makes it difficult to scale up the network since an increase of the number of end systems accompanies a rapid increase of the number of connections to support them all.

7.1.7.2 Direct Method

The direct approach can resolve the problems of scalability and bandwidth utilization. The direct method implements the CLSF using *connectionless* (CL) servers and IWUs. An IWU interconnects connectionless networks and ATM networks, and segments and reassembles the connectionless data. Each CL server may be integrated as a part of an ATM switch or may be attached to an ATM switch within the BISDN. In general, CL servers are interconnected through PVCs so that connection setup delays can be reduced. Each CL server makes routing decisions to have each connectionless datagram packet delivered to the next-hop CL server or the destination IWU.

Figure 7.16 depicts the protocol architecture of the connectionless service using CL servers. The *connectionless network access protocol* (CLNAP) in the source IWU encapsulates connectionless datagrams before delivering them to the ATM adaptation layer. The CLNAP frame is encapsulated in an AAL-3/4 convergence sublayer PDU, and is then segmented into many ATM cells. The ATM cells are delivered to the CL servers through ATM switches, and are forwarded, with or without the reassembly/processing/segmentation treatment, to the next CL server or to the destination IWU. At the NNI each CLNAP frame is encapsulated with an additional four-octet header by the *connectionless network interface protocol* (CLNIP). The *mapping entity* (ME) is responsible for the necessary encapsulation and decapsulation processes. The destination IWU reassembles the received cells into CLNAP frames and then decapsulates them into the connectionless datagrams.

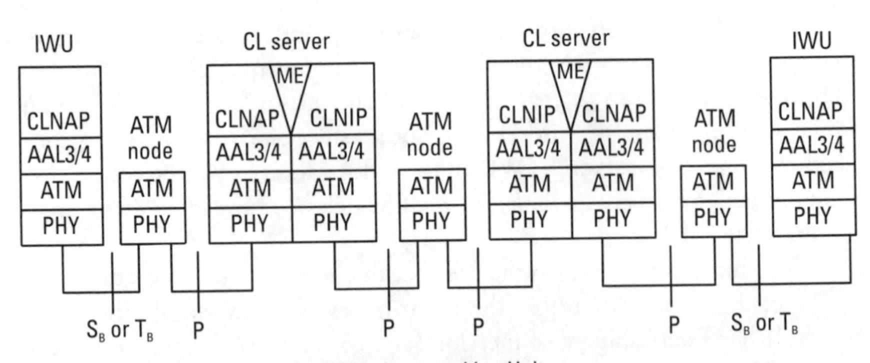

Figure 7.16 Protocol architecture of the connectionless service using CL servers.

The connectionless datagrams are finally delivered to the appropriate end systems.

The direct method has advantages over the indirect method in delivering connectionless data over public networks. First, each IWU in the direct method requires only one connection to deliver connectionless data to ATM networks, so the IWUs are not required to make the routing decisions. Second, all connectionless data traffic is aggregated in some connections between CL servers, which can increase the statistical multiplexing gain and can make the network management simple. Third, the number of required connections is much smaller in the direct method than in the indirect method because only the CL servers are interconnected, so the direct method is scalable.

On the other side, the direct method has some potential weak points: CL servers and connections between CL servers may become bottlenecks. The direct method has not resolved the complicated routing job but, instead, has shifted the job to the ATM network. There may be an interoperability problem when interoperating with LANE or IP over ATM, which has decided to use AAL-5, because the direct approach is likely to utilize AAL-3/4.

The direct method is likely to use AAL-3/4 because cell-based forwarding is much simpler than frame-based forwarding, among the two forwarding schemes that the CL servers support. In the case of the cell-based scheme, a CL server forwards cells to the next CL server or IWU as soon as it receives them, but in the case of frame-based scheme, cells are reassembled into frames at each CL server. The reason that the cell-based forwarding scheme is required to use AAL-3/4 is as follows: When a CL server receives the first cell of a frame, it finds out the output VPI/VCI and MID in the routing table using the input VCI/VPI and MID information stored in the cell. Then the mapping information is preserved and utilized so that the other cells of the same frame, which have the same VPI/VCI and MID, can be forwarded as soon as they arrive. To keep pace with the flow of the traffic, a cell-based CL-server must be able to perform the three-phase process—receive a cell, look up the routing table, and forward the cell—within one cell transmission time. This implies that for an STM-1 155-Mbps interface all processes must be finished within 2.7 ms, and for STM-4 622-Mbps, within 680 ns.

On the other hand, in the case of the frame-based forwarding scheme, a CL server buffers all cells to be accommodated in a frame, and then reassembles them into a frame. Then it makes the routing decision or carries out other processes, and finally resegments the frame before forwarding. Since the frame is processed in one time, the MID field is not necessary, so AAL-5 can be used in the frame-based CL servers. This is why the frame-based forwarding scheme is considered to be an appropriate means for interconnecting ATM LANs. The frame-based scheme is less restricted in processing time. It can avoid useless transmissions because if a cell in the frame is lost the CL servers can detect the loss and can drop the whole frame. However, the frame-based scheme has the drawback that it requires a large reassembly buffer and causes processing delays.

7.2 Routing

The basic problem of routing can be viewed as the driving concept behind the various architectures that we have examined in the preceding section. There is a major conceptual and architectural problem in trying to use a connection-oriented network architecture to support the transport of connectionless network traffic. These methods were all advanced as a way of solving this intrinsic problem. In contrast, the problem of encapsulation (which we examine in Section 7.3) is solved in a similar manner by all the architectures. In fact they all reference the same specification, RFC1483, defined by the IETF.

There are basically two models for supporting the transport of TCP/IP traffic over ATM networks—the overlay model and the integrated (or peer-to-peer) model. LANE, MPOA, NHRP, and classical IPoA are overlay model types, and I-PNNI and MPLS are the integrated model types. The next section examines the routing issue for both models and considers how the basic problem of routing is solved in both models.

7.2.1 Overlay Model and Routing

From an architectural point of view LANE and classical IPoA are both completely aligned with the overlay models. Both use the RFC1483 encapsulation method and allow both LLC-type and VC-type multiplexing (refer to Section 7.3 for more details). This section reviews their characteristics from the perspective of how routing is performed. In addition, we examine in detail two new mechanisms recently defined to aid the autoconfiguration and operation of IPoA networks using the overlay model. One is the newly defined PAR and *proxy PAR* (PPAR) functions that offer a way for the PNNI topology database to be updated with the information on the IP routers and clients that are connected to the ATM cloud, thereby simplifying the autoconfigurability and scalability of IPoA networks. The other is the ILMI-based server discovery methods that have been defined by the IETF, which offer a way of automating the discovery of server addresses, thereby simplifying the configuration of IP clients connected to ATM networks.

7.2.1.1 Communications Within the Same Subnet

Both LANE and classical IPoA allow only ATM connections for communications between nodes in the same subnet. The nodes defining a subnet in an ATM network are named differently for each of these methods. In LANE it is called an ELAN, while in classical IPoA it is called an LIS. In both methods setting up ATM connections in the subnet requires that ATM signaling is used and that some sort of IP address-to-ATM address resolution function is available. In classical IPoA this problem was solved by the ATMARP function. As explained before, this acts as an IP address-to-ATM address resolution function. It is conceptually equivalent to the ARP function used in Ethernet subnetworks.

In LANE, the approach is slightly different as the LANE model operates at a slightly different layer than the classical IPoA model. As described before, the LANE protocol aims to appear like a LAN to the upper layers. Consequently, a node connected to a LANE will operate, at least with respect to the upper layers, as if it were connected to a LAN (either Ethernet or token ring, depending on the configuration). This means that the node will use ARP as defined for Ethernet to find the correct MAC address to which to send any IP layer traffic. The main function in LANE is to map the MAC address of the destination node to the ATM address of the destination node.

Consequently, when a node is operating in a LANE environment the resolution of the destination address would go through a two-step process. First, the IP address of the destination would be used to find the MAC address of the destination, and then the MAC address of the destination would be used to find the ATM address. In the first step, normal ARP as defined for Ethernet/token ring would be used, and in the second step the LANE functions would be used.

For both LANE and classical IPoA, which nodes in an ATM network belong to which subnet is defined by the configuration data in the centralized servers. In LANE the servers are LES/LECS, and in classical IPoA the servers are the ATMARP servers. As the definition of subnet boundaries is dependent on the configuration data, this means that the logical boundary of a subnet does not have to be equivalent to the boundaries of the ATM network itself. This aspect brings up the concept of VLANs. In fact ELANs and LIS can both be regarded as some types of VLANs, differing only in the methods used for implementation.

7.2.1.2 Communications Between Subnets

As mentioned above, both LANE and classical IPoA only define the use of ATM connections for communicating between nodes on the same subnet (ELAN or LIS). Therefore, communicating with any node connected to some external subnet requires that a router be used as the gateway to external networks. This model essentially follows the normal routing model of TCP/IP networks. All communications inside subnets is done by methods specific to that subnet, while intersubnet routing is done through routers.

The essential problem with such a model happens when the ATM network becomes very large. In such a case it can be presumed that there are many routers connected to this network. The number of routers must be increased as the number of ATM nodes on a subnet is increased, since the router capacity is a finite resource.

If all the ATM nodes are members of a single ATM subnet, then all the routers are *adjacent* to each other. That is, they all exchange routing information with each other. The routing information will be in the form of routing protocols, such as OSPF or RIP (see Section 3.2) but this leads to a scalability problem. OSPF and RIP essentially do not work with a very large number of adjacent neighbor routers. Having too many neighbors increases the number of protocol messages that must be sent and also increases the time for the routes to get stabilized.

To get around this problem the ATM network may be cut into reasonably small subnet groups with all communications to/from outside the subnet going through a router. However, this method brings up the possibility that even though two ATM nodes are on the same ATM network, they must communicate through a router. Note that the NHRP and MPOA protocols were defined to solve this problem, essentially by using the short-cut routing method.

7.2.1.3 PAR and PPAR

As explained in Section 4.2.2, PAR is an extension of PNNI that facilitates the distribution of information about non-ATM services in the ATM network. One example of the use for such service is overlay networks such as the ones used to support TCP/IP over ATM. Information regarding the routers, such as their IP addresses, subnet masks, and the routing protocols supported (such as RIP or OSPF) can be distributed through PAR. By using this information, it is possible for the IP routers on the edge of the ATM network to discover the other routers on the edge of the ATM network and thereby construct an IP network level topological map. This provides an efficient and dynamic way of supporting IP networks over ATM networks.

The information used in PAR is carried in a new PTSE type defined for carrying non-ATM related information. This PTSE can carry various IGs that are the actual containers of the non-ATM information. Currently IPv4-specific IGs including OSPF, BGP-4, and DNS are defined. There is also a system-capabilities IG that may be used by vendors to carry experimental or proprietary information. Table 7.2 lists the defined information groups.

PPAR. A potential problem with using PAR is that for PAR to be used in the above suggested manner the edge router must implement PNNI as well as PAR functionality. Due to the high cost of implementing PNNI, it may be better if PAR could be supported by some sort of proxy method. PPAR provides such a method. In PPAR a PAR server exists that is PAR- and PNNI- capable. Any PPAR client may then connect to this proxy PAR server and get information from it regarding the non-ATM services offered over the ATM network. The client is also able to register its own services such as its own IP address on the server. PPAR has deliberately been designed to use a separate protocol from ILMI, as the information that must be exchanged for

Table 7.2
Information Group Summary

Type	IG Name	Nested In
768	PAR service IG	PTSE (64)
776	PAR VPN ID IG	PAR service IG (768)
784	PAR IPv4 service definition IG	PAR VPN ID IG (776) / PAR service IG (768)
800	PAR IPv4 OSPF service definition IG	PAR IPv4 service definition IG (784)
801	PAR IPv4 MOSPF service definition IG	PAR IPv4 service definition IG (784)
802	PAR IPv4 BGP-4 service definition IG	PAR IPv4 service definition IG (784)
803	PAR IPv4 DNS service definition IG	PAR IPv4 service definition IG (784)
804	PAR IPv4 PIM-SM service definition IG	PAR IPv4 service definition IG (784)

PPAR operation would be much more than that which could be transferred through the use of ILMI. Of course a PPAR client may use ILMI to retrieve other configuration information.

The client/server interaction in PPAR consists of discovery, query, and registration functions. The discovery function is used by clients and servers to discover adjacent neighbors. The registration function is used by the client to register the services that it offers at the server, and the query function is used by the client to retrieve the services registered by other clients. Figure 7.17 shows an example of the operation of the interaction between a PPAR client and a PAR server. The client and server initially use the hello protocol to establish connectivity. Following this, the client registers the services that it offers (which may be the information such as its IP address) and then queries the server for information on other clients. Either side keeps the state on what information the other side may or may not have. Accordingly, the burden of maintaining correct operation falls on the client side software. The main responsibility of a PPAR server is to ensure that the information registered by the clients are delivered to all other PAR-enabled devices in the ATM cloud.

Interaction with other protocols. Neither PAR nor the PPAR specification itself defines how the service information retrieved by the client is to be used. Instead, it is expected that for each protocol, the relevant standardization organizations will define the interaction between the protocol and PAR/PPAR elements. For example, for the case of OSPF, the IETF has

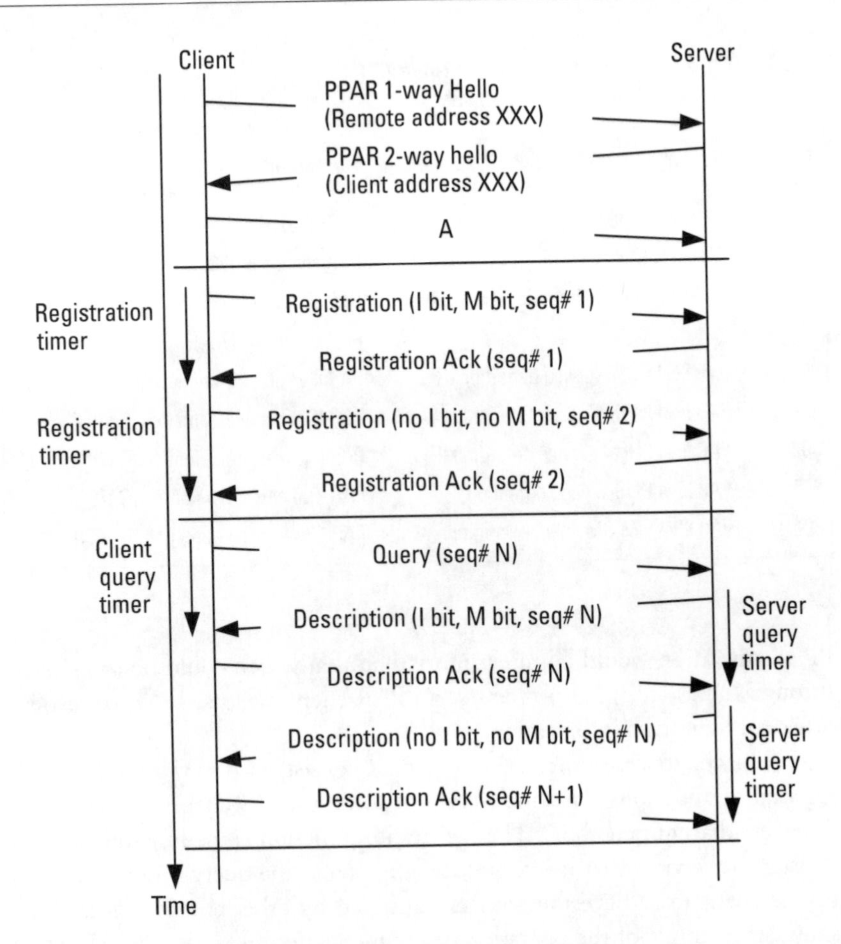

Figure 7.17 Example of the PPAR protocol operation.

defined an RFC2844 [8] that explains how OSPF is to be run over PPAR. A good example of how this may be used is the case when PPAR is used to support OSPF routing. We assume that the OSFP router is the PAR client and that a PAR server is the ATM switch to which it is connected. The OSPF IG provides information about the routers on the edge of the PNNI ATM network that supports OSPF. The information included is the OSPF area, the router priority ID, and the interface types of that OSPF router.

The interaction operation can be implemented in the following manner. For each OSPF area in which the router participates over an ATM interface, the OSPF router will transmit to the PAR server (on the ATM switch) a

PAR OSPF IPv4 service definition IG that has the OSPF and router priority information. It is assumed that the OSPF routers will use an interface type of NBMA when connected through an ATM network.[11] From the PAR server the OSPF router will get information on the other OSPF routers connected to the PNNI network. As each OSPF will have indicated its OSPF area, the OSPF routers in the same area will be informed of the presence of other routers. Based on this the OSPF routers will elect a designated router and set up VCs appropriately so as to be able to communicate with this designated router and themselves. The connection setup will be of UBR type by default unless differently configured.

Relationship with overlay models. PPAR can be viewed as a way of discovering other IP routers connected to the same ATM cloud. As explained in Section 3.2.2, a fundamental operation in IP routing protocols such as OSPF and RIP is to discover and maintain contact with neighboring routers. Usually this is achieved through the use of the hello protocol. When the neighboring routers are connected to this router by a broadcast/multicast capable networks, it is easy to design and use a hello protocol based on this broadcast/multicast ability for discovering other routers. As explained above, LANE and classical IPoA have the ability to emulate the broadcast/multicast ability by the use of servers (BUS and MARS). Consequently for both models, it is relatively straightforward to adapt the neighbor discovery functions used in the original OSPF definitions for broadcast networks.

Besides the NBMA model, OSPF also supports other network models such as the point-to-multipoint model and the simple point-to-point connections. By using PPAR it is possible to support these other network models, which usually require a large amount of configuration to be done at the OSPF enabled routers. An advantage of using PPAR in such situations is that this configuration information may be automated. For example, to use OSPF over NBMA networks the routers may be configured with the ATM addresses of all neighboring routers and with information related to the routers' OSPF settings. In the case of point-to-multipoint networks the OSPF routers are configured only with addresses of directly reachable routers. Another advantage is that the servers used in LANE on classical IPoA often represent single point of failure. By relying on PAR and PPAR the problem of single point of

11. This means that the routers will need to define a designated router to act as the representative for this network and generate LSAs to other OSPF areas.

failure disappears as the information is distributed throughout the ATM network in the PNNI database.

PAR and PPAR can also be considered to be an alternative to LANE or classical IPoA as a method for ATM address resolution. This may be an advantage as in some cases this would decrease the use of separate broadcast emulation functions defined in LANE and classical IPoA. This would result in many of the steps used to retrieve information from the ATMARP or LANE servers being skipped. To support multicast the broadcast/multicast servers are still necessary. Also, while the broadcast ability offered by LANE or classical IPoA may not be used, the information used by PAR and PPAR is always broadcast throughout the ATM cloud by being piggy-backed on top of the normal PNNI messages. Accordingly, a broadcast function must be used in some manner at some level of the protocol stack.

7.2.1.4 ILMI-Based Server Discovery

The basic classical IPoA and NHRP models both assume either that when a client first boots up the client node will have various configuration and initialization information or that it will be able to locate and connect with an initial server that can supply the needed information. The former approach is frequently subject to errors. Moreover, it is hard to reconfigure or change information dynamically as the network develops. The latter server-based approach is a more desirable method from that point of view. The ILMI-based server discovery methods defined by the IETF is such a server-based approach. The basic aim is to use ILMI as a method of automatically configuring the clients with the appropriate server addresses. Three different methods have been defined respectively for discovering ATMARP servers, NHRP servers, and MARS servers [9–11]. All these methods are similar in basic functionality, so we consider only the ATMARP server discovery method below.

ILMI offers a way for ATM attached devices to retrieve information from ATM switches and other ATM devices [9–11]. It is based on SNMPv1 and uses the *get*, *get-next*, and *trap* operations. For IP over ATM operation the ATM switch (or network side) must support service registry *management information bases* (MIBs). This is the MIB that is queried by the user (client) to get the information on the servers. The information included in the MIB includes the service identifier, full ATM address of the service, and a service parameter string. The service identifier indicates whether ATMARP, NHRP, or MARS is supported by this entry. The service parameter is service specific. For the ATMARP case, it contains information on the protocol type (IPv4 or IPv6), length of the protocol address, the network address, and the network mask. The ATMARP server must have its service registry MIBs correctly configured with this information. It may be possible to use information supplied by the PAR protocol to keep this information updated if the ATM switch is also PAR-enabled. The ATMARP client uses the SNMP get-next operation to do a search through the service registry MIB table to find the server address needed.

7.2.2 Integrated Model and Routing

From an architectural point of view, I-PNNI and the MPLS methods are examples of the integrated model for routing TCP/IP packets over ATM networks. The integrated model is also called the peer-to-peer model as it essentially regards all the ATM nodes and IP routers as peers. They are expected to have equal and the same knowledge of the topology of the whole network, including both the ATM and IP network parts. Based on this knowledge both types of nodes are expected to set up connections or routes that will essentially take into account the whole topology. For example I-PNNI tries to be a unified routing protocol for both ATM and IP networks.

There are problems with such an approach. First of all, it is not obvious that such a method is feasible due to the differences between connectionless and connection-oriented networks. More specifically, the IP packets are routed based on the routing table. When the network topology changes, the routing table is modified automatically, changing the path between the packet. This is done by propagating the routing information and recalculating the routing tables in all the relevant routers along the path. In contrast, the paths in an ATM network will not be changed that easily since paths are set up only at connection setup time.

While both I-PNNI and MPLS may be said to be examples of the integrated model, this is not entirely correct. To be exact, I-PNNI can be viewed as an almost exact example of the integrated model, as it aims to share the topological data among separate ATM networks and IP networks, while each network essentially operates according to its own protocols. The basic operation of the ATM networks, which is based on UNI/NNI signaling, VPI/VCI table lookup, and detailed OAM, does not change. In contrast, MPLS over ATM should be viewed as a way of using ATM hardware for TCP/IP control software. It is thereby a weaker form of merger of TCP/IP and ATM. This point will become clear by recalling the operation of MPLS described in Section 7.1.6. In MPLS, the ATM protocols are essentially removed from ATM switches. ATM switches no longer use ATM UNI/NNI signaling or PNNI routing functionality in any manner, either for the setup or release of traffic flows. Instead, the ATM switch sets up its VPI/VCI tables based on the MPLS control protocols and IP routing protocols.

As we mentioned above, I-PNNI is currently not being actively pursued by the ATM Forum. In contrast, MPLS is becoming a basic solution to many problems in networking. MPLS is also being applied to solve similar problems in the optical domain (refer to Section 8.6).

7.3 Multiplexing and Switching

For IPoA networks one of the most important problems that must be addressed is the method of encapsulation: how to transport IP packets in AAL-5 payloads and how efficient is the transportation of IP traffic over ATM cells. The action of encapsulation involves adding headers and trailers to an IP packet and essentially defines how the multiplexing of the IP packets will occur in the ATM network. In the following we consider these two issues one by one.

7.3.1 Encapsulation and Multiplexing

Encapsulation of IP packets in ATM networks is defined in RFC1483. This standard defines the specific formats for using AAL-5 to transport routing protocols or the bridging of common protocols. When used to support bridging of common protocols a MAC address must be included in the encapsulation. For the cases where the aim is to only transport routing protocols no such MAC address is needed.

Basically there are two methods of encapsulation: LLC/SNAP encapsulation and VC multiplexing. In LLC/SNAP encapsulation, an LLC/SNAP header is used to multiplex different protocols over a single ATM connection. The LLC/SNAP header provides a protocol ID field, which means that even though a single ATM connection is being used, a number of different protocol connections may be multiplexed. In VC multiplexing, each protocol is carried in a separate ATM connection. In this method, it is implicitly assumed that each VC connection will carry only one packet flow. Which of the two encapsulation methods to use is decided by the configuration in the case of PVCs, and by the signaling protocols in the case of SVCs.

More specifically, LLC/SNAP encapsulation adds an LLC/SNAP header to the PDU. This was designed for the environments where VCs are scarce as in public ATM networks. It is useful in such networks because the VCs can be used more efficiently. For example, by using this method, multiple LECs can share a single VCC in the LANE architecture.

Figure 7.18(a) shows the LLC/SNAP encapsulation methods for both routed and bridged protocols. For routed protocols, the *organizational unique identifier* (OUI) in the SNAP header is set to 0x000000 to indicate

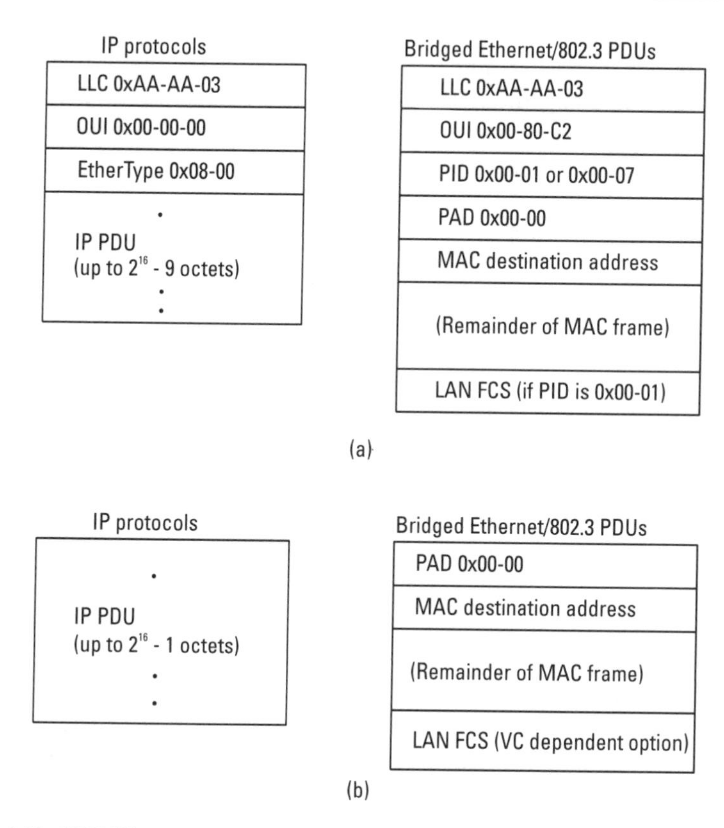

Figure 7.18 RFC1483 encapsulation methods: (a) LLC/SNAP encapsulation, and (b) VC multiplexing encapsulation.

that the PID field uses the same semantics as in the normal Ethernet [12]. For bridged protocols, the OUI is set to 0x0080C2 to indicate that the PID field follows the semantics defined in RFC1483. Two different PIDs are allowed for each case, one indicating that the LAN FCS is included, the other indicating that it is not.

As noted before, VC-based multiplexing essentially does not add any overhead to the packet itself. All users are differentiated on the basis of the VC values used. The encapsulation for such cases is shown in Figure 7.18(b), for both routed and bridged protocols. Essentially this results in supporting only a single protocol over a single ATM VC.[12] The payload efficiency is

12. Note that this does not necessarily mean that each protocol will be supported over only one platform.

high as there is no need to add an LLC/SNAP header. A disadvantage is that it requires using a large number of VCs, which will lead to scalability problems. For routed protocols, the encapsulation just consists of using the whole AAL-5 payload to transport the protocol to be routed. As noted in this section, this should be the maximally efficient method for transporting TCP/IP packets over ATM networks as there is no overhead needed. For bridged protocols, the encapsulation is similar to the LLC bridged case but without the LLC, OUI, and PID fields.

7.3.2 ATM Cell Tax Problem

The use of TCP/IP over ATM with AAL-5 encapsulation results in a large waste of bandwidth, because ATM has a fixed cell size while TCP/IP has variable length packets. The resulting wasted bandwidth is usually called the *ATM cell tax.* In general, ATM cell tax depends on the size of the original IP packet: Smaller IP packets will lose more bandwidth to ATM overhead, while larger packets will be less effected.

Normally TCP ACK packets are 40-bytes long, which consist of a 20-byte TCP header and a 20-byte IP header. In case LLC/SNAP encapsulation is used, a total of 8 bytes must be added to the TCP ACK packet. In contrast, if VC multiplexing is used, then no extra bytes need to be added. Now when this packet is encapsulated with AAL-5, an extra 8-byte overhead is incurred due to the trailers in the AAL-5 payload format. Accordingly, if LLC/SNAP encapsulation is used then the whole payload becomes 56 bytes, which is longer than what can be fit into a single ATM cell. When transported over ATM, such a payload will be divided into two ATM cells, with the second cell mostly filled with padding. In contrast, if VC-based multiplexing is used, since no extra encapsulation bytes need be added, the ACK packet may be transported in a single ATM cell. As can be seen in this example, ATM cell tax can be high when packet size is small. However, for most common IP packet sizes, for example, 576 bytes, ATM cell tax drops sufficiently low, making transport efficiency reach as high as 90% plus.

It is interesting to compare the effect of various different packet sizes. This is illustrated in Table 7.3. The table shows that while the choice of physical layer has very little effect on the efficiency, the size of the packet has a very large effect. Due to this dependency on packet size, the distribution of IP packet sizes has a large effect on the overall efficiency observed. Under the rule of thumb that 80% of the IP packets are small (e.g., 44 bytes) and 20% are large (e.g., 500 bytes), the overall efficiency becomes about 80%.

Table 7.3
Transmission Efficiency for IP Over ATM Networks

IP Datagram	Physical Layer	Efficiency	Raw Link Bandwidth	Max. Effective Bandwidth
44-byte	SONET OC-3c	39.97%	155.52 Mbps	62.16 Mbps
	SONET OC-12c	40.08%	622.08 Mbps	249.33 Mbps
	TAXI	40.00%	100 Mbps	40 Mbps
	DS-3	37.77%	44.736 Mbps	16.90 Mbps
576-byte	SONET OC-3c	80.51%	155.52 Mbps	125.21 Mbps
	SONET OC-12c	80.74%	622.08 Mbps	502.27 Mbps
	TAXI	80.56%	100 Mbps	80.56 Mbps
	DS-3	76.07%	44.736 Mbps	34.03 Mbps
1,500-byte	SONET OC-3c	85.18%	155.52 Mbps	531.38 Mbps
	SONET OC-12c	85.42%	622.08 Mbps	85.23 Mbps
	TAXI	85.23%	100 Mbps	85.23 Mbps
	DS-3	80.48%	44.736 Mbps	36.00 Mbps
9,180-byte	SONET OC-3c	86.88%	155.52 Mbps	135.12 Mbps
	SONET OC-12c	87.12%	622.08 Mbps	541.96 Mbps
	TAXI	86.93%	100 Mbps	86.93 Mbps
	DS-3	82.09%	44.736 Mbps	36.72 Mbps

The use of TCP/IP *header compression* [13] techniques is one way of mitigating the effects of the ATM cell tax. TCP/IP header compression is a method used in most TCP/IP connections running over low-speed links. It decreases the size of the TCP/IP header significantly from 40 bytes to about 8 bytes, so could be ideally suited for TCP/IP over ATM links. However, this does not necessarily justify the use of the header compression technique in the ATM links. Such a software-based header compression may not be much of a load when used over low-speed links, but becomes a big problem over high-speed links. In addition, this header compression technique assumes a different point-to-point connection for each TCP connection, such as those used over PPP or SLIP links in low-speed modem lines, but for ATM networks, this would mean that a separate VC would have to be opened up for each TCP connection. But this would cause a big problem in ATM networks as it would deplete the VCs and cause a large increase in signaling.

As far as the cell tax or link efficiency is concerned, in general, IP over SONET may be more favorable than IPoA as it also provides statistical multiplexing gain while cell tax is avoided by bypassing the ATM processing. However, it should be noted that the flexibility of the ATM, with respect to its ability to support QoS levels, a small bandwidth granularity and traffic engineering requirements through the use of VCs, is then sacrificed when using SONET links. This topic will be dealt with in more detail in Section 8.3.2.

7.4 Network Control

There are various signaling capabilities that are required for IPoA. Among them, the most notable are the capability to provide point-to-point communication with shorter on-demand setup delay, the capability to provide point-to-point communication with/without QoS guarantee, the ability to provide point-to-point communication with appropriate transfer capability and symmetric/asymmetric bandwidth, the capability to provide point- to-multipoint communication, and the ability to provide multipoint-to-multipoint communication. There are a number of problems that must be examined when investigating the interaction of ATM signaling functions with TCP/IP traffic. In this section we first examine some of these problems and then we examine the IETF specifications on using ATM signaling to set up connections to support TCP/IP traffic.

7.4.1 Basic Problems in Using VCs with TCP/IP

The basic problem in using connection-oriented VCs to support connectionless TCP/IP traffic is setting up VCs in an efficient manner. It must satisfy the network operator's need to minimize the number of active VCs at any given time. At the same time the connection setup delay should be minimized so that the delay experienced by the TCP/IP traffic and, consequently, by the user can be minimized. A number of options affect these problems with the most basic one being the question of whether to use PVCs or SVCs to support the IP traffic.

7.4.1.1 Connection Setup Delay

One of the most important requirements is the need for short ATM connection setup time. This is because ATM connection setup delay has a large effect on the performance of TCP/IP over ATM networks. It is due to the large difference between the basic conceptual architecture of the two protocol suites. TCP/IP is based on the connectionless packet network paradigm, while ATM is a connection-oriented architecture. Connectionless packet networks do not maintain state in the network and only need to transmit packets with addresses in the headers. The network will route the packets to the correct destinations based on these addresses. However, since ATM networks always require the setup of a connection before the transmission of traffic, whenever a TCP/IP packet is to be sent over an ATM network an ATM connection must be set up beforehand.

Figure 7.19 demonstrates the effect of ATM VC setup delay for TCP/IP traffic. To send some IP data over an ATM network, the sender first sets up an ATM connection. This results in a large delay as the signaling message sent by the source has to be processed by each switch in the path to the destination. Once the destination receives the signal it then returns the signaling message. As a result, a full roundtrip time is needed to set up the ATM connection. After this connection setup data transport begins. If the user is using TCP, then another roundtrip delay is incurred for the initial TCP connection setup between the source and the destination. Consequently, the setup time of an ATM connection can substantially affect the performance of TCP/IP over ATM connections.

As an example, we consider a Web browser application. A typical Web page is made up of many different text files and image files. In HTTP1.0 every time a Web page is opened up, a TCP connection has to be set up to retrieve each text object and image object. If the connection was run over an

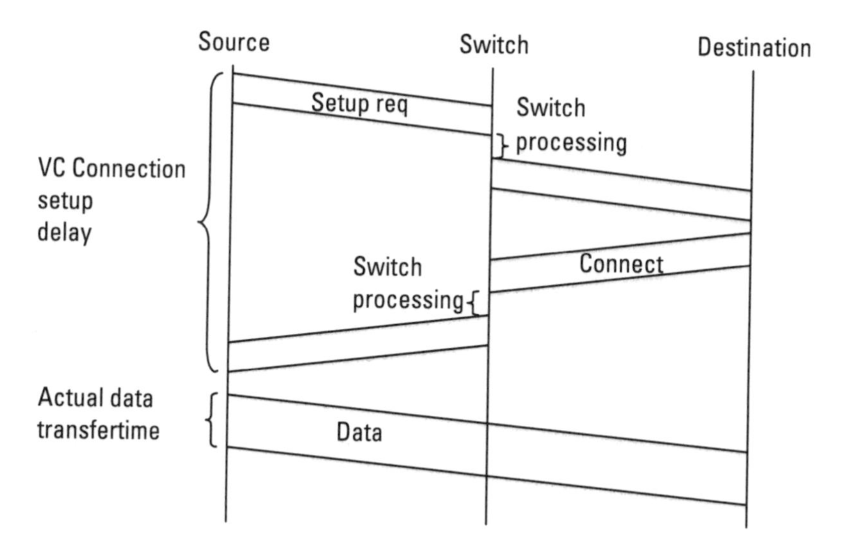

Figure 7.19 Effect of ATM VC setup delay for TCP/IP traffic.

ATM link, then in the worst case an ATM connection would have to be opened up and released for each and every text object and image object. This would increase the delay in retrieving the Web page significantly.

7.4.1.2 PVCs and SVCs

There are basically two main types of ATM virtual connections—PVCs and SVCs. PVCs are the connections that are set up and that stay up continuously. In most cases, they are set up manually by the operators. SVCs are set up on demand by using signaling mechanisms. Both types of connections have their respective advantages and disadvantages depending on a number of factors, such as the burstiness of the connection and the type of network topology in which the connections are deployed.

For bursty traffic sources, SVC may be the best fit. For such connections, the VC will probably not be needed all the time, so a PVC would be extremely inefficient. In most cases we may expect that the connection carries live data only when traffic arrives. The very definition of bursty traffic implies that for an extended period of time there may be no such traffic. In contrast, for sources such as gateways that generate steady streams of traffic, PVC may be acceptable. Though the individual traffic coming into the gateway from the various connected tributaries may be bursty, there will be a steady stream of packets that use the PVC.[13]

Using PVCs. If PVCs are used, the ATM connection setup time normally has no effect as the connection is a permanent one and will have no discernible effect on the total time observed by the user. This will be especially true for cases in which manually configured static PVCs are used. If soft PVCs are used, then some signaling will occur, but it will only occur when the network is in an unstable state or a link is down. That is, these events will be unsynchronized with the normal request for connection sent out by normal calls. As such, they can be expected to have no effect on the overall performance perceived by the user.

The use of PVCs in either of the above forms is suitable for small networks. Once the network becomes large the use of PVCs to interconnect the sites becomes very hard to manage. Basically, such a solution has problems in scaling to very large networks. The operator of the network would have to

---

13. This does not necessarily imply that the stream will be steady in CBR. Mixing bursty traffic may often result in more bursty traffic. This phenomenon is studied in the field of "self-similar" traffic research.

configure the connections on each switch and router. Whenever a new router or edge device is added the PVCs to all the other devices would also have to be reset up to ensure routability. This leads to the classical N^2 complexity problem.

Using SVCs. If SVCs are used, it brings in a signaling load every time an SVC is set up. Accordingly, the frequency of SVC setup will be an important determining factor of the overall performance; how often SVCs will be set up and how fast SVCs can be set up are the major factors. How fast the network sets up SVC connection is an implementation issue that could vary depending on switch manufacturers. As such, we cannot pursue this aspect of the problem any further. If the switches are capable of handling a large amount of signaling load, it is not necessary to attempt to minimize the signaling load.

The important question here is how to decide when to set up an SVC. This determines the frequency of SVC setup and consequently the performance perceived by the end user. This decision may be done in a number of ways. The simplest way is to allow the first packet to always trigger a connection setup. Another way is to use the amount of packets transmitted as a trigger. Also available are more complicated methods that depend on the current state of the network and router.

Another related issue is when the connection should be torn down. This is really the other side of the previous problem. In most cases, this will also rely on a timer, or an inactivity timer. The inactivity timer is used in such a way that if no traffic is observed over the connection for a certain amount of time the connection is released. Any time the connection carries real data the connection is reset. So the timer value has a critical effect on the performance and the efficiency of the connection. If the timer is too long, the connection will be kept alive even though there is no traffic to transmit, resulting in a waste of resources in the ATM switches along the path. If the timer is too short, too much signaling processing will take place for the setup and release of the ATM connections.

There are two basic approaches that can be obtained from the above methods. One is to set up SVC when there are packets to transmit but to delay the closing of the connection until the inactivity timer goes off. The other is to delay opening the connection until a certain amount of packets has arrived and been buffered but to close the connection immediately after all the currently buffered packets have been sent. The former minimizes the delay that the first packet will experience, but at the cost of network efficiency. The latter ensures that the opened connection is used efficiently but sacrifices delay. There are various other approaches that have been made by using queuing theory and simulation tools, but the results vary depending on the applied assumptions.

7.4.2 Signaling Support for Classical IPoA and NHRP

The signaling support defined by the IETF in RFC1755 defines the signaling methods that must be used to set up connections to transport best-effort traffic, while the signaling support defined in RFC2381 (and the related specifications RFC2379, RFC2380, and RFC2382) defines the same information for controlled-load service and guaranteed service. Note that these specifications essentially define how the IntServ model of IP QoS is to be supported when TCP/IP is used over an ATM network.

RFC1755 and RFC2381 describe the ATM call control signaling exchanges needed to support classical IPoA implementations as described in RFC 1577 [1]. ATM endpoints are assumed to incorporate the ATM signaling services as specified in the ATM Forum's *UNI Specification Version 3.1 and 4.0* [14, 15]. Clients such as classical IPoA, LANE, and MPOA implementations utilize the services of local ATM signaling entities to establish and release ATM connections to support IP traffic.

7.4.2.1 VC Establishment

The owner of an existing VCC is defined to be the entity within the ATM end system that establishes the connection. An ATM end system may establish an ATM call when it has a datagram to send but there is no existing VCC that it can use for this purpose or the VCC owner does not allow sharing.

To reduce the latency of the address resolution procedure at the called station, the following procedure may be used: If a VCC is established using the LLC/SNAP encapsulation, the calling end station of the VCC may send an InARP_REQUEST to the called end station after the connection is established (i.e., received a CONNECT message) but before the calling end station sends the first data packet. In addition, the calling end station may send its data packets without waiting for the InARP_REPLY. An end station may respond, generate, and manage its ATMARP table according to the procedures specified in RFC1293 [2], during the lifetime of the VCC.

To avoid establishing multiple VCCs to the same end station, a called end station may associate the calling party number in the SETUP message with the established VCC. This VCC may be used to transmit data packets destined to an end station whose ATMARP resolution results in an ATM address that is the same as the associated calling party number.

Support for multiple VCs. An ATMARP server or client may establish an ATM call when it has a datagram to send but there is no existing VCC that it can use for this purpose, it chooses not to use an existing VCC, or the owner of the VCC does not allow sharing. Note that there might be VCCs to the destination that are used for IP, but an ARP server might prefer to use a separate VCC for ARP only. The ATMARP server or client may maintain or release the call as specified in RFC 1577. However, if the VCC is shared among several protocol entities, the ATMARP client or server does not disconnect the call as suggested in RFC 1577.

While allowing multiple connections is specifically desired and allowed, implementations may choose (by configuration) to permit only a single connection to some destinations. In such a case, if a colliding incoming call is received while a call request is pending, the incoming call is rejected. Note that this may result in a failure to establish a connection. In such a case, it is recommended that each system wait at least a configurable collision retry time in the range of 1 to 10 seconds before retrying.

7.4.2.2 VC Teardown

Either end system may close the ATM connection. Systems configure a minimum holding time (i.e., the time the connection has been open) for connections to remain open as long as the endpoints are up. A suggested default value for the minimum holding time is 60 seconds.

Some public networks may charge for connection holding time, and connections may be a scarce resource in some networks or end systems. Accordingly, each system implementing a public ATM UNI interface should support the use of a configurable inactivity timer to clear connections that are idle for some period of time. The timer's range includes a range from a small number of minutes to "infinite." A default value of 20 minutes is recommended in RFC1755.

7.4.2.3 Call Establishment Message Content

Signaling messages contain mandatory and optional variable-length *information elements* (IEs). The IEs are further subdivided into octet groups, which in turn are divided into fields. IEs contain information related to the call, which is relevant to the network, the peer end point, or both. The called end station and the type of communication channel opened over the ATM connection are determined by the IEs that are contained in the call establishment message. For example, the call establishment messages will differ between a call that sets up an AAL-1 connection for CBR video and a call that sets up an AAL-5 connection for IP.

A SETUP message that establishes an ATM connection to be used for IP and multiprotocol interconnection calls must contain the following IEs: AAL parameters, an ATM traffic descriptor, broadband bearer capability, broadband low-layer information, a QoS parameter, the called party number, and the calling party number.

There are IEs in a SETUP message which are important only to the endpoints of an ATM call supporting IP. These are the AAL parameter IE and the broadband low-layer information IE; the AAL parameter IE carries information about the AAL to be used on the connection. RFC 1483 specifies encapsulation of IP over AAL-5. Selection of an encapsulation to support IP over an ATM VCC is done by using the *broadband low layer information* (B-LLI) IE.

7.4.2.4 ATM Traffic Descriptor

The ATM traffic descriptor characterizes the ATM virtual connection in terms of PCR, SCR, and maximum burst size. This information is used to allocate resources (e.g., bandwidth and buffer) in the network. In general, the ATM traffic descriptor for supporting multiprotocol interconnection over ATM will be defined based on factors such as the capacity of the network, conformance definition supported by the network, performance of the ATM end system, and (for public networks) cost of services.

The default model of IP behavior corresponds to the best-effort capability. If this capability is offered by the ATM networks, it may be requested by including the best-effort indicator, the PCR-forward, and PCR-backward fields in the ATM traffic descriptor IE. When the best-effort capability is used, the network does not provide any guarantees, and in fact, throughput may be zero at any time. This type of behavior is also described by RFC 1633 [16].

If the user (or network) desires to use a more predictable ATM service for IP traffic, it must be possible to use more specific traffic parameters. In such cases the basic traffic descriptor IE is used along with a broadband bearer capability IE and the QoS parameter IE. These elements essentially define the signaling aspects of ATM traffic management.[14]

---

14. The specific elements are defined in a number of RFCs including RFC1755, RFC2331, and RFC2381. RFC1755 also defines a set of combinations of traffic parameters that the ATM signaling modules in all IPoA end systems must support. These include the best-effort traffic, a type of traffic description that is intended for ATM "pipes" between two routers or IP systems, and a type of traffic description that allows use of token-bucket style characterizations of the RFC1363 and RFC1633.

7.5 Traffic Management

When IP is used over an ATM network various traffic problems appear that degrade the network efficiency. They basically stem from the inherent differences between IP and ATM—connectionless against connection-oriented, fixed-length against variable-length, and so on. To resolve such traffic problems and thereby enhance the network efficiency, it is important to examine the effects of ATM cell loss on TCP/IP performance and then devise methods to handle the resulting limitations in proper ways.

7.5.1 Effects of ATM Cell Loss on TCP/IP Performance

In a seminal study, it was shown that if ATM networks are used to transport TCP traffic without any special provisions, then the throughput seen by the TCP/IP user could become very low, in some cases dropping to 34% of the available link capacity [17]. This effect is basically due to the fact that the ATM layer has no knowledge of the packets carried by the upper layers. This can be analyzed in more detail as follows.

A single TCP segment is usually transported in one IP packet, whose normal size depends on the path MTU but is usually 1,500 bytes or 576 bytes. As this single TCP segment does not fit into a single ATM cell in general, it is segmented into plural ATM cells after going through an appropriate encapsulation process. (Refer to Section 7.3 for more details on encapsulating TCP packets in ATM cells.) The cells are then transported across the network and, when all the cells for a packet are received by the destination, they are reassembled into the original packet. The destination recognizes the last cell of a packet by examining the packet indicator in the ATM cell header. Once the last cell is determined, the reassembler engine assumes that all the cells have arrived, checks the 32-bit CRC field at the end and, if it is correct, delivers the packet to the upper layer application.

When the ATM cells happen to confront network congestion while traversing the ATM network, some cells may be dropped. If this happens, the reassembler in the receiver will find that the CRC value does not match, and the packet will then be dropped. This is the only mechanism that can take place at the node as ATM does not offer any retransmission or upper-layer error recovery mechanism. Consequently, even a single ATM cell loss can result in dropping of the whole packet.

Such a large multiplicative effect decreases the efficiency of the network operation. First, it means that all the ATM cells belonging to the dropped packet are discarded even if they were successfully delivered.

Second, it means that the packet must be retransmitted, with a high probability of confronting congestion again and thereby suffering cell loss. More importantly, it triggers TCP's congestion control mechanism, pushing TCP connections into slow-start mode. (Refer to Section 3.5 for the basic TCP congestion control mechanism.)

Various solutions to this problem have been suggested. Simplest among them may be to increase the buffer sizes in the network to minimize the cell loss due to congestion. However, this solution is not very efficient and possibly leads to a large network delay, which is not conducive to effective network operation. Basically, there are two approaches that have been proposed to deal with this problem. One is to use a feedback mechanism to ensure that the cell loss due to congestion is minimized, which is the basis of the ABR rate control mechanism examined in Section 4.5.2. We examine this approach in Section 7.5.2. The other approach is to use a frame discard mechanism to ensure that unnecessary cell transmissions do not take place. We consider this approach in the following.[15]

7.5.1.1 Frame Discard Methods

There are two different types of frame discard, one is *partial packet discard* (PPD) and the other is *early packet discard* (EPD). Both PPD and EPD basically try to emulate the loss behavior observed in normal packet networks and thereby minimize the bandwidth used by the traffic that is useless at destination. The main difference is in the decision principle on which cells to drop once congestion has been detected.

The PPD method is to immediately drop the cell and all the following cells with the same VPI/VCI until a cell with an end-of-packet indication is found. This method will ensure that no useless cells are transmitted through the congested switch beyond those already sent before the congestion was detected. Note that to ensure successful operation the cells with the end-of-packet indication bit must not be dropped and reach the destination. Otherwise, the next packet would also be dropped due to CRC check failure.

The EPD method is to drop a whole packet, or, to be more precise, all the cells that carry the segmented data of the next whole packet if the queue size grows larger than the EPD threshold. This mechanism is illustrated in

15. Note that all the cells carrying the data belonging to a single packet must have the same VPI/VCI, as they must all be transported over the same virtual circuit.

Figure 7.20. This operation can be easily accomplished by looking at the incoming cells for the next end-of-packet indication and then dropping the cells incoming from that point on until the next end-of-packet indication.

Table 7.4 compares the performance of the PPD and EPD methods along with the normal packet network TCP [18]. The normal UBR case is denoted as UBR-PLAIN, the implementation using PPD as UBR-PPD, and the implementation using EPD as UBR-EPD. The last is the case when a normal packet network TCP is used. We observe that the throughput increases steadily from the UBR-PLAIN to the UBR-PPD, to the UBR-EPD, and then to the packet TCP. In the end, the UBR-EPD and the packet TCP cases basically exhibit the same throughput.

Two of the critical issues in the frame discard methods are the criteria to decide when to drop the cell of a frame, and which VPI/VCI cells to drop. The first issue is usually solved in a straightforward manner by using a threshold value and comparing the current queue length in the switch.

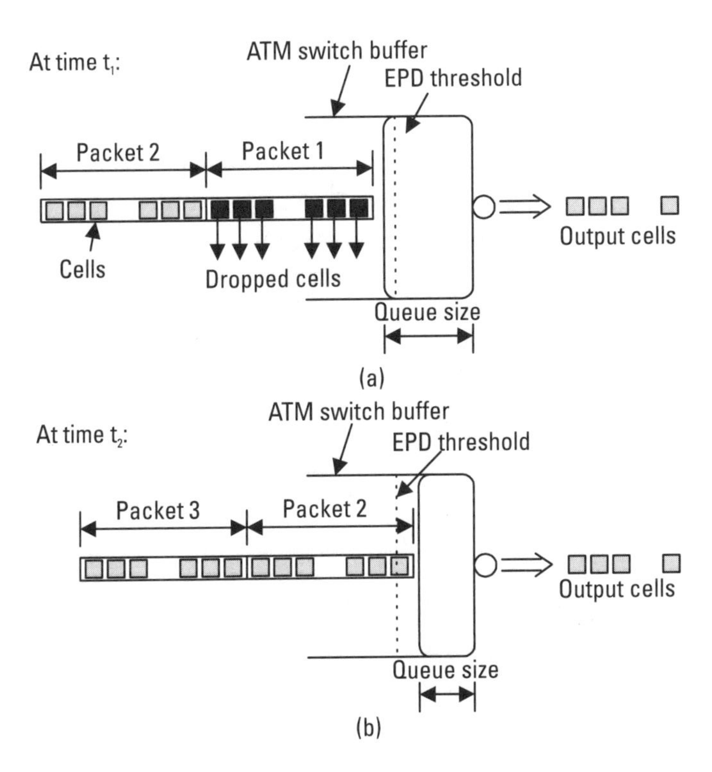

Figure 7.20 Operation of EPD algorithm: (a) when the queue size is larger than the EPD threshold, and (b) when the queue size is smaller than the EPD threshold.

Table 7.4
TCP Throughput over Various Flavors of UBR Service

Buffer Size (in Kbytes)	UBR-PLAIN	UBR-PPD	UBR-EPD	Packet TCP
100	0.62	0.8	0.98	0.98
200	0.7	0.88	0.98	0.98
300	0.8	0.92	0.98	0.98
400	0.84	0.95	0.98	0.98

From: [18].

However, the second issue is rather tricky, as it is related to the fairness in packet dropping. As noted in Section 3.5.2, the queue management policies will have a large effect on the end throughput seen by each user and consequently the fairness of the whole system.

7.5.1.2 GFR-Based Methods

GFR is a new traffic management category defined by the ATM Forum to more efficiently support TCP/IP traffic-based on the initial research on frame discard methods and other optimizations [19]. GFR originates from the idea of making the frame discard mandatory and giving to TCP/IP traffic a guarantee of the minimum rate as well. By making frame discard mandatory for all switches, GFR ensures user level performance to be better than pure UBR switches. Moreover, GFR allows the user to reserve some bandwidth in the network. This value usually equals the minimum rate that the relevant application would need.

The GFR traffic contract contains PCR, MCR, MBS, and MFS attributes.[16] An important characteristic of GFR is that the end system is not required to shape the traffic. Consequently a packet will usually be transmitted as a burst of cells at the PCR. For policing such a traffic pattern the normal GCRA must be modified. *The frame-based GCRA* (F-GCRA) was defined for this purpose [19]. While the ideal F-GCRA algorithm is exact, it is not very practical, so the simple F-GCRA(T, L) algorithm was also

16. The MFS is the size of the AAL-5 that is used. The MCR is usually negotiated to be equal to the long-term average of the connection.

introduced in the specifications. Figure 7.21 describes this algorithm. While it is a simplified form of the ideal F-GCRA algorithm, it is equivalent to the ideal F-GCRA(T, L) algorithm for the connections that only contain conforming frames or cells.

GFR has a number of advantages over other service classes in supporting TCP/IP traffic. Compared with UBR, GFR is more efficient as the network supports frame-discard strategies. Compared to VBR, GFR has advantages as QoS is guaranteed at the frame level, which is a more meaningful metric for the end user. In contrast to ABR, GFR imposes no complex scheduling or queuing mechanisms on the end points. Nor does GFR need to use RM cells as ABR does.

7.5.2 Using ABR for TCP/IP Traffic

The feedback mechanism renders a viable solution to traffic control when transporting TCP/IP traffic over ATM network. TCP/IP traffic is bursty and not ideal for rate-guaranteed traffic contracts and efficient transport over virtual circuits. In contrast, ATM is optimized for transporting fixed rated traffic over configured virtual circuits. The feedback-based traffic control mechanisms used in this situation help to enhance network efficiency by preventing network congestion by controlling the data rate and thereby avoiding ATM cell dropping.

The ATM Forum has defined a new parameter, *minimum desired cell rate* (MDCR), to use during call setup to indicate to the network that a TCP/IP application would like to get a certain minimum bandwidth [20].

Cell arrival at time t_a: First cell of an AAL-5 frame:	Middle or last cell of an AAL-5 frame:

if(($t_a < TAT-L$) OR (IsCLP(cell)) {
/* non-eligible cell */
eligible = FALSE;
} else {
/* eligible cell */
eligible = TRUE;
$TAT = max(t_a, TAT)+T$;
}
	if(eligible) {
	/* eligible cell */
	$TAT = max(t_a, TAT)+T$;
	} else {
	/* non-eligible cell */
	}

Figure 7.21 Simple F-GCRA (T, L) algorithm.

The MDCR differs from other ATM traffic parameters in that it does not define a service commitment to guarantee the minimal requested rate. For example, if a user requests a connection with a certain MDCR value, the network internally commits a minimum bandwidth of MDCR to the connection, and polices the traffic according to the PCR with a GCRA (1/MDCR, T) test.

In contrast to the MDCR-based rate control, ABR-based rate control is an active form of feedback-based traffic control mechanism. It defines a structure by which data traffic can fairly and efficiently share all the resources of a network, by controlling the data rate based on the feedback information. Accordingly, it is natural to expect much enhanced network efficiency by ABR rate-controlled VCs to support TCP/IP traffic. However, the use of ABR rate control brings in a number of problems yet to be solved as well. In the following section we will discuss the ABR-based traffic control in detail.

7.5.2.1 Effects of ABR Rate Control

When ABR rate control was first proposed, extensive simulations were carried out to show that the use of ABR rate control results in an efficient transport of TCP/IP traffic over ATM networks. Many results indeed exhibited the network utilization sustaining nearly 100% without any cell loss or abnormal delay. However, a key point to note on the simulations is that an end-to-end ATM network was assumed. In reality, however, ATM networks do not appear at the desktop but are mostly used to interconnect various LAN-based IP edge networks; this is therefore called the *subnet ATM model.* This means that the ATM connection employing ABR flow control is normally used between LAN ATM edge devices. This introduces an interesting interaction phenomenon between TCP's congestion control and ABR rate control: Even if, as noted before, the use of ABR in an all-ATM network may ensure a maximum network utilization and fair network resources sharing, neither efficiency nor fairness can be guaranteed when the ATM network, or the consequent ABR feedback loop, covers only part the network. In the following we shall call such a basic interconnection scheme for transporting TCP/IP traffic over ABR connections the *ABR+ scheme.*

Figure 7.22 shows an ATM network implementing ABR rate control to connect to IP LAN networks. If congestion occurs in the ATM network, then the ABR rate control mechanism signals the edge LAN ATM devices that congestion has developed in the network so the rate of transmission should be reduced. This should effectively cause the congestion in the ATM

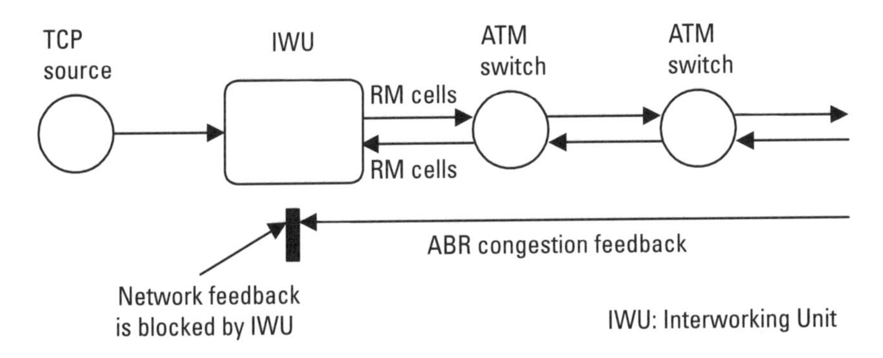

Figure 7.22 Effect of using ABR connections to support TCP/IP traffic.

network to disappear. However, what really happens is that congestion then starts building up in the LAN ATM device. In other words, the use of ABR simply shifts the point of congestion from inside the ATM network to the edge of the ATM network.

When all nodes are ATM capable, then "congestion" occurs inside the source node. This is not a problem, as it will only result in the source sending out cells/packets more slowly, with no cell or packet loss occurring at the source. The real problem takes place when the congestion occurs at the LAN ATM device, with the IP source being a completely separate device. In this case, the IP source does not receive any information that congestion has occurred, so does not slow down the transmission rate. In fact, if it is a TCP implementation, it will slowly increase the rate, only aggravating the situation and resulting in more packet losses at the LAN ATM device. In the end, TCP implementation will know congestion has occurred, but only after the congestion point has moved from the inside of the network to the LAN ATM device. As the buffers at the LAN ATM device can be expected to be large, this will cause a very large feedback delay even before the source TCP realizes congestion has occurred.

There is another facet to this problem that has to do with the fairness of the bandwidth usage. Since one of the basic aims of the ABR mechanism was to share the available bandwidth fairly among the connections, fair allocation of bandwidth should be easily attained. Also, as ABR aims to maintain a minimal or zero cell loss rate, the fragmentation effect should not be a problem. In the end-to-end connection case, it has been shown that almost perfect fairness and high throughput can be achieved. However, the same result does not automatically carry over into the subnet ATM case. This is because the gateway at the edge of the ATM cloud acts as a barrier to the feedback coming from the network, consequently nullifying the ABR mechanism by hiding the network congestion signals coming from the end host.

ABR expects the traffic source to control its output rate according to the feedback information. However, in the subnet ATM model, though the gateway will immediately drop its output rate according to the ABR feedback messages, this will only cause a queue buildup at the gateway. The TCP source will lower its output rate only after packet loss occurs in the gateway due to the overflow of this queue. Unless some special buffer management schemes are employed at the gateway, the connection whose packets are dropped will not necessarily be the connection whose rate was lowered by the ABR mechanism. This leads to the wrong connection decreasing its window size and thereby its output rate. The effects of this behavior become predominant when large and small delay connections share a single output buffer. In this case the ABR+ scheme behaves unfairly and with low goodput.

Basically, this problem is due to *hogging* of the available buffer space by the short delay connections and the dynamic allocation of service rates of the ABR mechanism. The short delay connections will fill up the available buffer space, causing the long delay connections to be squeezed out of the buffer and leaving an uneven distribution of packets in the buffers. However, as the link bandwidth is fairly divided among the connections by the ABR mechanism, each connection will be served at its allocated rate. Consequently, there is a shortage of packets when the long delay connections queues are served, and hence the ABR+ scheme becomes underutilized and unfair. This leads to the allocated bandwidth not being fully used, since the long delay connections see less buffer space than they need to fully utilize their allocated bandwidth. This effect is especially predominant with relatively small buffers since the short delay connections can easily fill up the small amount of available buffer space.

In summary, though ABR rate control may *allocate* the available bandwidth fairly among the connections, it does not guarantee that the connections will be able to effectively *use* the allocated bandwidth. This is basically due to a mismatch between the rate control-based ABR mechanism, which relies on explicit rate feedback, and the window-based TCP congestion control mechanism, which relies on implicit feedback in the form of packet loss.

The fact that ABR rate control pushes the queues to the edges of the network may also be helpful in solving the problem it creates. Since the gateway now becomes the single point of congestion for that connection, we can concentrate on designing an intelligent gateway congestion control scheme that utilizes the ABR rate feedback information. The basic idea is to design a fair buffer allocation scheme so that the bandwidth allocated by the ABR mechanism may be fully utilized.

7.5.2.2 ABR with Fair Buffer Allocation (ABR+FB)

A number of solutions have been proposed to solve this problem. Two of the main solutions are based on the observation that to improve efficiency some sort of intelligent queue control algorithm must be implemented in the LAN ATM device.

Jagannath and Yin proposed such an intelligent approach [21]. In their approach they do not wait until the queue is full, but drop a packet immediately after the ABR rate control signals a rate decrease. Additionally, this drop is implemented by using a drop from front policy, thereby minimizing the delay in notification to the TCP source. It has been shown that this method is effective.

Kim and Lee proposed a more complicated but more efficient scheme by noting that for an effective ABR rate control some method must be devised to reflect the rate information in the congestion control scheme at the gateway on the edge of the ATM cloud [22]. They proposed a *fair buffer* (FB) allocation scheme that aims to use this feedback information and obtain better utilization of the allocated bandwidth. Along with the fair bandwidth allocation due to ABR, this resulted in fair and high throughput for the TCP connections. This method is called the *ABR+FB scheme.*

The fair buffering mechanism is based on the following two observations. First, the total achieved throughput of a TCP connection with a fixed service rate depends only on the normalized buffer size, and is maximized when its buffer space exactly equals the connection's bandwidth-delay product.[17] Consequently, for fair utilization of the allocated bandwidth by each connection, buffer space must be allocated in proportion to its *bandwidth-delay product.* Second, the TCP congestion control mechanism works most effectively when its loss is spread out and does not occur in bursts. The question of how much the loss should be spread out can be decided by observing that TCP Reno operates optimally when there is only one loss per congestion

17. The ABR+FB algorithm relies on the correct calculation of the bandwidth-delay product. In traditional best-effort networks, calculating the bandwidth-delay product for a connection was not possible as neither the bandwidth nor the delay of the connection was known. In contrast, when ABR connections are used, though bandwidth is variable over time, it is explicitly and fairly allocated for each connection during discrete intervals. Also, as ATM is connection-oriented, the fixed propagation delay of the path is also known.

avoidance cycle. Ideally, if one packet per congestion avoidance cycle is dropped, this should be optimal.

The ABR+FB scheme is based on the following observations. First, to filter out noise due to rate fluctuations, it is desirable to use the average of the bandwidth in all calculations. Second, it is necessary to allocate buffer space in proportion to the bandwidth-delay product. Third, it is important to minimize the problems that occur during the slow-start congestion avoidance cycle when the connection is started. However, the ABR+FB scheme relies on the characteristics of TCP congestion control, so may not effectively control UDP-based traffic.

7.6 QoS

As discussed in Section 3.6, there are two different models defined by the IETF for QoS support in TCP/IP networks: IntServ and DiffServ. The IntServ QoS model is logically very similar to the ATM model in that it relies on specific resource allocation, resource reservation, and a signaling protocol to guarantee services to users. As such, the model offers a relatively obvious logical model of integration, whereby the resource allocation and reservation information must be mapped to appropriate ATM functions, most importantly, the signaling aspects. An important condition for supporting the IntServ model is that the RSVP path messages must follow the same path as the path used by the data traffic towards the destination. The DiffServ QoS model is slightly different in that it relies on marking individual packets to be classified into aggregate classes. It is a more incremental approach that is more compatible with the current state of TCP/IP networks and technology.

Additionally, there are two basic models for supporting IP traffic over ATM networks: the overlay models and the integrated model. While there are a number of methods classified as being based on the overlay model, there is only one existing method based on the integrated model, that of MPLS over ATM. As the integrated model aims to integrate the ATM switches and IP routers on a peer-to-peer basis, this means that the same control function must be used on all nodes. This means that the QoS functionality must be supported in a similar manner on both ATM and IP nodes. In other words, all nodes must be aware of the MPLS control traffic, some of the IP control traffic, and the QoS classes and services supported in the network. In contrast, in the overlay model, while the IP routers on the edge of the ATM network must be aware of such information, the ATM switches internal to the network may not know of the IP packet's QoS class. Instead, such information must be passed in an indirect manner, which usually consists of setting up appropriate connections based on a mapping of IntServ parameters to ATM QoS parameters.

Various mixes of models for supporting QoS and IPoA are shown in Figure 7.23. The mapping of the two QoS models with the two IPoA models results in four different basic models. Each model uses a slightly different solution to the basic problem. The overlay models map the QoS of IP connections to ATM connection type, while the integrated model relies on the ATM switches understanding and processing RSVP messages or carrying IP level information (e.g., DSCP values) in the ATM cells themselves. We discuss Figure 7.23 in more detail in the following sections, where we examine the specific solutions.

7.6.1 Overlay Model and QoS Support

As mentioned above, from a QoS point of view the overlay model basically uses the ATM protocol suite as currently defined, employing ATM connections as layer 2 links over which IP packets are transmitted. In such a model the routers on the edge of the ATM cloud, or the edge routers, must carry out the processing needed to support QoS interaction between the ATM network and the IP network.

QoS models / IPoA models	IntServ	DiffServ
Overlay	Mapping between ATM connection & RSVP info.	Differentiated UBR
Integrated	Carry RSVP information	Map DSCP to ATM cell header

Figure 7.23 Mapping IPoA models to QoS models.

Figure 7.24 shows the edge router functions in an abstract manner. The edge router must be able to maintain multiple VCs per IP flow and maintain mappings of QoS classes to the appropriate VCs, and implement sophisticated queuing and scheduling algorithms. In addition, if IntServ is supported, it must be able to translate QoS parameters between those used by RSVP and those used in ATM. The edge router must be able to use the information contained in the RSVP signaling message to guide the establishment of ATM VCs of the appropriate QoS class toward the next-hop router connected to the same ATM cloud. If DiffServ is to be supported, the edge router must be able to map the DSCP and PHB of the flow to the appropriate output queues and scheduling classes.

Figure 7.25 shows on a larger scale how the edger router functions and the various QoS techniques developed for ATM can be used in a large IP over ATM network to achieve the aims of supporting QoS for IP traffic. The key is to have most of the complicated QoS functionality in the edge routers with minimal functionality (mostly congestion control functions) kept in the core of the ATM network itself.

A number of problems must be addressed in this model. One problem is the management and use of VCs, which was dealt with in Section 7.4. A key question that must be answered is how the mapping between the IP QoS service classes and the relevant ATM QoS classes should be done. Table 7.5 shows how the mapping between ATM QoS classes and the various classes in the IntServ and the DiffServ models can be done. It shows that ATM QoS classes defined in UNI 4.0 and TM 4.1 can be used to support both types of IP QoS models. In addition, the new differentiated UBR service may be used

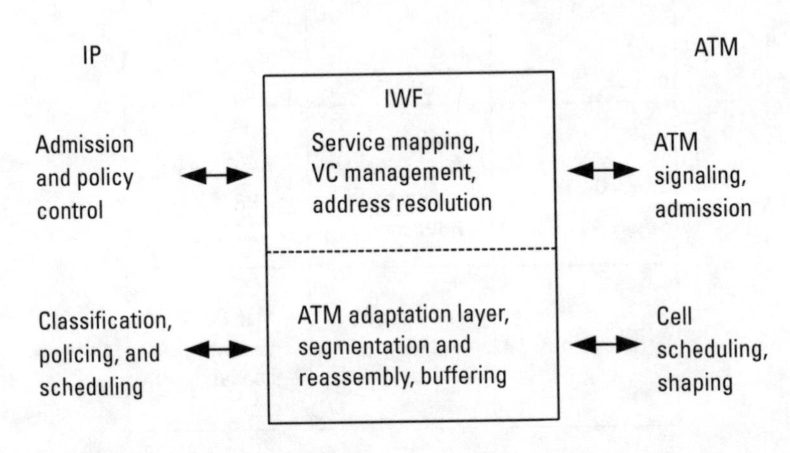

Figure 7.24 Edge router functions [23].

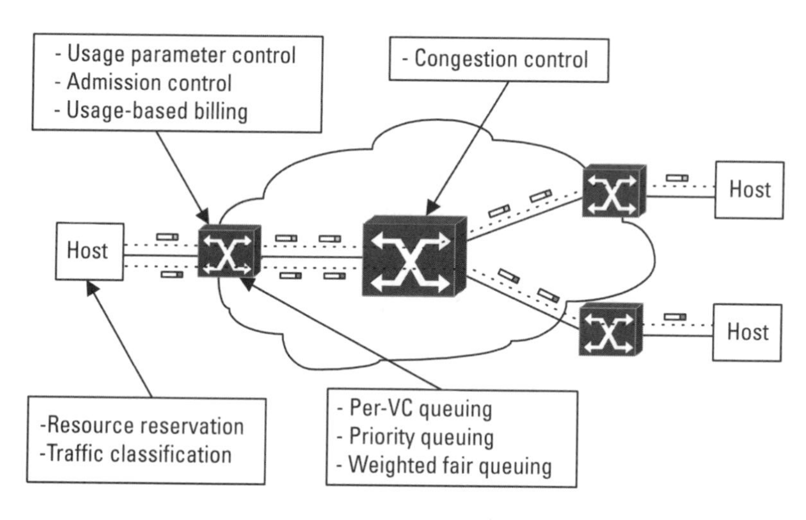

Figure 7.25 Example of different techniques used to support QoS in ATM networks.

to support DiffServ QoS in a more natural manner [24]. However, Table 7.5 does not necessarily provide an absolute guide, as there may be a number of ways of mapping the ATM QoS classes to the IP QoS classes.

7.6.1.1 Overlay Model and IntServ over ATM (UNI 4.0)

To use the IntServ QoS model over networks using the overlay IP over ATM model a number of problems must be resolved, including VC management and QoS mapping. The critical condition of the RSVP path messages that follow the same path as the forward data path is satisfied by ensuring that the ingress and egress points of the path over the ATM network is the same for the RSVP messages as well as the data packets. Note that the actual path

Table 7.5
Different Ways of Mapping IP QoS with ATM QoS in the Overlay Model

QoS Model	Service Class	ATM QoS Class
IntServ	Guaranteed service	CBR or rt-VBR
	Controlled-load service	nrt-VBR or ABR (MCR)
	Best-effort service	UBR or ABR
DiffServ	Expedited forwarding (EF)	CBR or rt-VBR or UBR (diff)
	Assured forwarding (AF)	nrt-VBR or ABR or UBR (diff)

taken inside the ATM path need not be the same for the two types of packets, as the information needed to set up QoS paths for the data packets is only needed at the ingress and egress points. The default best-effort VC will always be established before any QoS-specific VCs as it is needed to transport the initial RSVP PATH messages to the receivers. The initial RSVP RESV messages are also received along this VC.

One of the main differences between the ATM QoS and IntServ is the signaling model. RSVP is a receiver-oriented protocol where the receiver signals to the routers along the path what amount of resources must be reserved, while in ATM it is the sender that decides the amount of resources to reserve. Actually, this is not a big difference because the RSVP receiver bases its reservation decision on the sender's TSPEC that was included in the PATH message. Consequently, the reservation may also be viewed as being initiated by the sender in an indirect manner.

Another important problem is how to manage the VCs that must be set up to support different QoS levels and connection topologies. As discussed in Section 7.4, two types of connections, PVCs and SVCs, may be used. Using PVCs simplifies the support of the IntServ model as the connections can then be considered as being simple point-to-point circuits and there is no difference from when RSVP is used over other leased line configurations. In contrast, if SVCs are used the problems of complexity and efficiency of setting up SVCs must be considered again as the use of RSVP complicates the situation. Basically, the problem is in deciding how many ATM connections must be set up to support the RSVP flows. Various options exist, ranging from using a single connection for all flows (which would be similar to the leased line solutions) to using different flow for each RSVP flow.[18]

Depending on the solution chosen, it decides where to concentrate the queuing and scheduling function of the edge router. If a single VC is established for all RSVP flows, the scheduling and queuing functionality must be done on the IP layer at the edge router so as to satisfy the QoS requirements. In contrast, if multiple VCs are established, the problem is simplified for the IP layer as it only needs to map the packets to the appropriate VCs. From this point on the ATM layering, queuing, and scheduling mechanisms would have to ensure the QoS guarantees.

Note that these problems of VC to QoS class mapping are complicated when the use of multicast connections is also considered. This is because RSVP basically supports heterogeneous receivers—receivers that differ in the QoS

18. While this same problem exists for the PVC case as well, complexity is more for the SVC case due to the increased flexibility.

requirements. This again means that there are a number of options in setting up the VCs for the relevant QoS classes. One method is to simply set up a different VC for each receiver. While this solution is simple to consider, it would be prohibitively expensive to implement. Another method is to set up a single multicast VC of a single QoS for all receivers. In this case the single QoS must be equal to the best QoS requested. Yet a third method would be to set up multiple multicast VCs for each different level of QoS such that each receiver could attach to the appropriate VC tree depending on its QoS level.

As mentioned in Section 7.1.1, in classical IPoA, the connection is normally released if the VC is idle for a certain amount of time. While this behavior is suitable for best-effort traffic, when the connection is established by RSVP, it must be assumed that the connection has been set up to offer a certain level of services. Consequently, the connection should not be released by the idle timeout function. The connection must be under the control of the RSVP state machine only.

Another major problem is how to map the IntServ and RSVP QoS parameters to the ATM connections. The IntServ QoS parameters are defined in the sender TSPEC, receiver TSPEC, and the receiver's RSPEC (Refer to Section 3.4). These parameters must be mapped to the ATM QoS parameters (PCR, SCR, and MBS). Details on how these mappings should be done for the three IntServ QoS classes are given in [23].

There are of course many more ATM traffic parameters than those mentioned above. Several ATM QoS parameters have no equivalent in the IntServ model's parameters, for example, the CLR, CDV, and CTD parameters. These must be set at the edge routers at the ingress point of the ATM network based on the configured parameters or other network operator–specific methods.

Guaranteed service and ATM QoS classes. The CBR and rt-VBR classes may both be used to support the guaranteed service class. While CBR is an obvious choice, CBR will lead to an inefficient use of bandwidth, as CBR will use up a certain amount of bandwidth whether or not traffic is present. When CBR is used to support guaranteed service, the PCR of the connection must be set to the peak rate of the guaranteed service.

In contrast to the CBR class, rt-VBR is a better match to supporting the bursty nature of Internet traffic. When rt-VBR is used to support guaranteed service, it is suggested that the SCR, PCR, and MBS of the ATM connection be related to the QoS parameters of the IntServ model [23, 25].

Controlled-load service and ATM QoS classes. The nrt-VBR and ABR may both be used to support the controlled-load service class. The UBR class by itself is not appropriate for supporting the controlled-load service model, as it is very susceptible to network congestion. The CBR or rt-VBR classes are also not appropriate, as they will result in an inefficient use of network resources when used to support the bursty data traffic for which controlled-load service was defined. When nrt-VBR is used the PCR, SCR, and MBS are also related to the QoS parameters of the Intserv model [23].

When ABR is used the problem is simplified as the basic ABR service class offers similar service to that of the controlled-load service. The only parameter that needs to be set is the MCR of the ABR connection, which should be set to the minimum rate needed by the data source.

While UBR may be unacceptable to use as a basis for controlled load service, it may be possible to support controlled-load service under the newly defined differentiated UBR service type. As the service offered and guaranteed to a differentiated UBR connection is completely network-defined, this possibility may be a viable solution in the future.

Best-effort service and ATM QoS classes. Both UBR and ABR may be used to support the best-effort QoS class. The UBR class is an obvious choice for mapping best-effort IP traffic to ATM traffic classes. However, as explained in Section 7.5, naively mapping best-effort traffic to UBR connections can lead to a serious drop in performance as perceived by the user and aggravate any network congestion. Therefore, the network should implement some sort of frame discard method.

Using the ABR class is another way to support the best-effort QoS class. This would result in an efficient use of the ATM network resources while at the same time ensuring minimal cell loss for the best-effort packets. However, as was pointed out in Section 7.5, when ABR is used to support best-effort traffic in the overlay model, the performance of TCP/IP connections can drop seriously unless some intelligent queuing schemes are used in the edge routers.

7.6.1.2 Overlay Model and DiffServ over ATM (UNI 4.0)

It is also possible to support DiffServ QoS models over ATM connections. As with the IntServ case a number of mappings are possible.[19] The most obvious mapping is the use of CBR or rt-VBR connections to support the EF PHB. This is because the EF PHB tries to offer a "virtual circuit" like service

19. Note that unlike the IntServ model, there is no working group in the IETF working on the problem of mapping DiffServ QoS models and parameters to specific link layers. Consequently, this section is based on the authors' opinions.

with guaranteed bandwidth and minimal queuing delay for the packets that use this PHB. This is an obvious choice of mapping to the CBR QoS class. The use of the rt-VBR would also satisfy this criterion. The AF PHB may be supported in a number of different ways. Both nrt-VBR and ABR connections would be able to support the basic semantics of AF PHB. A crucial item in supporting the AF PHB is to ensure that the packet drop priorities are mapped to the appropriate CLP values. While the AF PHB defines three different loss priorities, the ATM cell is only able to carry a single CLP bit that can only indicate two cell loss priorities. Consequently, when mapping the AF PHB to the CLP bit either the top two or bottom two loss priorities must be tied together and marked with the same CLP bit. Note that it would be inappropriate to try to use different VCs for the different loss priorities in a single AF class as this may lead to out-of-sequence delivery at the receiver. This does not rule out different VCs for each AF class.

7.6.1.3 Overlay Model and DiffServ over ATM Using Differentiated UBR

As explained in Section 4.6.1, differentiated UBR is one of the new methods developed recently by the ATM Forum to more efficiently support TCP/IP traffic.[20] This is done by defining a new attribute to be associated with a UBR connection. Normally a UBR connection does not offer any service guarantees, basically offering a best-effort service. However, when differentiated UBR is used, a new attribute, the *behavior class*, may be associated with the UBR connection. The network may offer differentiated services to the user based on the behavior class associated with the user's connection. What the specific behavior classes are and whether the resources are specifically allocated are not specified. These are all up to the network operator's discretion to define and are employed as needed.

A BCS parameter is used to indicate the behavior class for the connection when the connection is initially set up. The values that the BCS parameter may take and the mapping with the actual behavior class in the network are not defined and may differ from network to network. The capability is applicable to all types of ATM connections including VPs, VCs, point-to-point, and point-to-multipoint connections.

It is easy to see that the definition and the specified functionality of differentiated UBR are extremely similar to that of the differentiated service

20. The actual specifications show two possible uses for the differentiated UBR mechanism. One is the support of differentiated services QoS model of the IETF. The other is the support of IEEE 802.1D user priorities that are used to provide service differentiation at the MAC layer.

model defined by the IETF. We can illustrate how this may be employed in an overlay network based on Figure 7.25. To support DiffServ in this case, multiple differentiated UBR ATM VC connections must be set up between each router. The edge routers (and also the interLIS routers) must also maintain a DSCP to VPI/VCI mapping table for each port. When a packet arrives at the router, the packet is classified based on its DSCP value to a certain PHB. Along with the "next-hop router" information from the router's routing table, this information is used to select the appropriate output port and the corresponding differentiated UBR ATM connection. Cell loss priority information is also passed to the ATM layer and encoded into the CLP bit of the ATM cell. The IP level QoS is guaranteed at the edge routers by the use of various queuing techniques. These QoS levels are maintained in the ATM level by using queuing and other traffic management techniques such as UPC and per-VC queuing.

7.6.2 Integrated Model and QoS Support

The integrated model currently consists of the MPLS-based solution for supporting IP traffic over ATM. As shown before, this model basically removes the whole ATM control stack and replaces it with an IP/MPLS control stack. As such, none of the predefined ATM methods for supporting QoS by using signaling and resource reservation are available. Instead, the MPLS control stack and the basic functions of the underlying ATM hardware must be used to offer this functionality.

7.6.2.1 Integrated Model and IntServ over MPLS/LC-ATM Switches

The key idea in supporting IntServ with MPLS/*label switch-controlled ATM* (LC-ATM) switches is to use RSVP for both label distribution/*label switch path* (LSP) setup and also for resource reservation. As resource reservation is the original function of RSVP only a method for label distribution with RSVP need to be added. This can be easily accomplished by defining a new object, the LABEL object, in all RESV messages [26]. Figure 7.26 shows

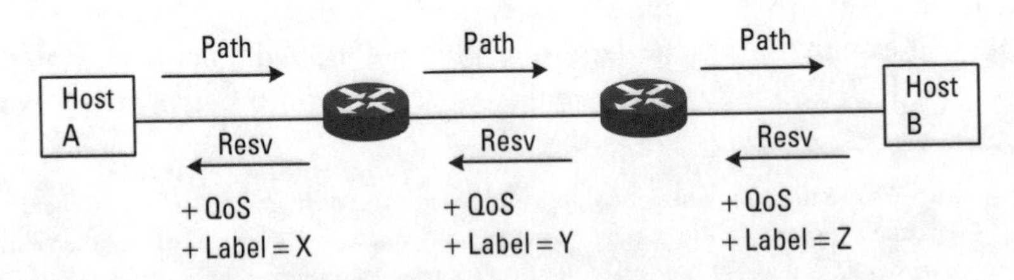

Figure 7.26 Label distribution and QoS reservation by RSVP messages [26].

how the labels may be distributed in the RESV messages. An LSR receiving an RESV message containing a LABEL object would update its *label-forwarding information base* (LFIB) with this label used as the outgoing label. The LSR should then allocate a new label to be used by the upstream node and include it in the RESV message and transmit it upstream. At the same time it should update the LFIB with this label. RSVP is shown here to support IntServ and label path setup. By the same mechanisms RSVP can also be used to solve traffic engineering problems when it is used to set up explicit paths.

7.6.2.2 Integrated Model and DiffServ over MPLS/LC-ATM Switches

To support DiffServ over MPLS there must be a method to map DiffServ DSCP values to the appropriate labels. This can be supported in two ways: one is so-called the *EXP bit-inferred PSC LSP* (E-LSP) method and the other is the *label-inferred PSC LSP* (L-LSP) method.[21] Figure 7.27 shows the difference between these two methods. Basically, E-LSP uses only one LSP and uses a field in the MPLS header to carry the DSCP code information, whereas L-LSP simply sets up a different LSP for each DSCP defined class.

In E-LSP, the three-bit EXP field in the MPLS label header is used. The packet is routed based on its label value, but at the output queue the queuing and scheduling behavior is decided by the EXP value. The EXP field

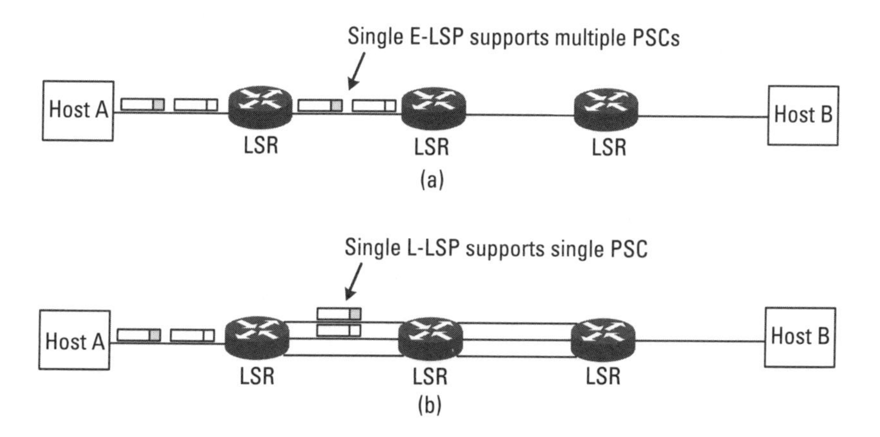

Figure 7.27 Ways of supporting DiffServ in MPLS networks: (a) E-LSP, and (b) L-LSP.

21. PSC is defined as the PHB scheduling class.

is only a 3-bit field, so the mapping of the 6-bit of DSCP to this field may result in loss of some information. The MPLS switches must be configured to map the EXP bits to the appropriate PHB behavior. However, for the case where MPLS-ATM switches are used, E-LSPs cannot be supported over LC-ATM interfaces because the ATM cells cannot carry the EXP bits in the label headers. Thus, only the L-LSP method may be used.

The L-LSP method maps each PHB to a different label or *forwarding equivalence class* (FEC). When DiffServ is run over MPLS-ATM, any number of L-LSPs per FEC may be allowed within a single MPLS ATM DiffServ domain. The basic compliance requirement is that the forwarding behavior experienced by a behavior aggregate forwarded over an L-LSP by the ATM LSR must be compliant with the DiffServ PHB specifications.

As only one CLP bit is available for encoding the drop priority, three PHB drop preference levels must be mapped to two levels. All ATM-LSRs must implement a frame discard mechanism such as EPD or PPD for performance improvements.

References

[1] Laubach, M., "Classical IP and ARP Over ATM," *RFC 1577*, January 1994.

[2] Bradley, T., and C. Brown, "Inverse Address Resolution Protocol," *RFC 1293*, Wellfleet Communications, Inc., January 1992.

[3] Atkinson, R., "Default IP MTU Over ATM AAL5," *RFC 1626*, Naval Research Laboratory, May 1994.

[4] Mogul, J. C., and S. E. Deering, "Path MTU Discovery," *RFC 1191*, November 1990.

[5] Luciani, J., et al., "NBMA Next Hop Resolution Protocol (NHRP)," *RFC 2332*, April 1998.

[6] IEEE, "IEEE Standards for Local and Metropolitan Area Networks: Media Access Control (MAC) Bridges," *IEEE 802.1D/p*, 1990.

[7] IEEE, "IEEE Standard for Local and Metropolitan Area Networks: Virtual Bridge Local Area Networks," *IEEE 802.1Q*, 1998.

[8] Przygienda, T., D. P. Siara, and R. Haas, "OSPF Over ATM and Proxy-PAR," *RFC 2844*, May 2000.

[9] Davidson, M., "ILMI-Based Server Discovery for ATMARP," *RFC2601*, June 1999.

[10] Davidson, M., "ILMI-Based Server Discovery for MARS," *RFC2602*, June 1999.

[11] Davidson, M., "ILMI-Based Server Discovery for NHRP," *RFC2603*, June 1999.

[12] Postel, J., and J. K. Reynolds, "Standard for the Transmission of IP Datagrams Over IEEE 802 Networks," *RFC 1042*, February 1988.

[13] Jacobson, V., "Compressing TCP/IP Headers for Low-Speed Serial Links," *RFC 1144*, February 1990.

[14] ATM Forum, "ATM User-Network Interface Specification, Version 3.1," Upper Saddle River, NJ: Prentice Hall, 1995.

[15] ATM Forum, "ATM User-Network Interface (UNI) Signaling Specification, Version 4.0," July 1996. Available at ftp://ftp.atmforum.com/pub/approved-specs/af-sig-0061.000.ps.

[16] Braden, R., D. Clark, and S. Shenker, "Integrated Services in the Internet Architecture: An Overview," *RFC 1633*, June 1994.

[17] Romanow, A., and S. Floyd, "Dynamics of TCP Traffic Over ATM Networks," *IEEE Journal on Selected Areas in Communications,* Vol. 13, No. 4, May 1995, pp. 633–641.

[18] Hassan, M., and M. Atiquzzaman, *Performance of TCP/IP over ATM Networks,* Norwood, MA: Artech House, 2000.

[19] ATM Forum Technical Committee, "Traffic Management Specification v4.1," af-tm-0056.000, April 1996.

[20] ATM Forum, "Addendum to Traffic Management Version 4.1 for an Optional Minimum Desired Cell Rate Indication for UBR," at-tm-0150.000, July 2000.

[21] Jagannath, S., and N. Yin, "End-to-End TCP Performance in IP/ATM Internetworks," ATM Forum Contribution 96-1711, December 1996.

[22] Kim, W. J., and B. G. Lee, "On Supporting TCP Traffic Over ABR Connections," *Proceedings of ICC' 98,* June 1998.

[23] Borden, M., and M. Garrett, "Interoperation of Controlled-Load and Guaranteed Service With ATM," *RFC 2381*, August 1998.

[24] ATM Forum, "Addendum to TM4.1: Differentiated UBR," at-tm-00149.000, July 2000.

[25] Ferguson, P., and G. Huston, *Quality of Service,* New York: John Wiley & Sons, 1998.

[26] Awduche, D., et al., "RSVP-TE: Extensions to RSVP for LSP Tunnels," *RFC 3209*, December 2001

Selected Bibliography

Andersen, N. E., et al., "Applying QoS Control Through Integration of IP and ATM," *IEEE Communications Magazine,* Vol. 38, No. 7, July 2000, pp. 130–136.

Armitage, G., "Support for Multicast Over UNI 3.0/3.1–Based ATM Networks," *RFC 2022*, November 1996.

ATM Forum, "ATM User-Network Interface Specification, Version 3.0," Englewood Cliffs, NJ: Prentice Hall, 1993.

ATM Forum, "ATM Traffic Management Specification, Version 4.0," April 1996. Available at ftp://ftp.atmforum.com/pub/approved-specs/af-tm-0056.000.ps.

ATM Forum, "MPOA Baseline Version 1," May 1997.

ATM Forum, "PNNI Augmented Routing (PAR) Version 1.0," AF-RA-0104.000, January 1999.

ATM Forum Technical Committee, "LAN Emulation Over ATM, Version 1.0 Specification, af-lane-0021.000," January 1995.

ATM Forum Technical Committee, "LAN Emulation Over ATM Version 2—LUNI Specification," December 1996.

ATM Forum Technical Committee, "Traffic Management Specification v4.1," af-tm-0121.000, March 1999.

ATM Forum Technical Committee, "Private Network-Network Interface Specification v1.0 (PNNI)", March 1996.

Azcorra, A., et al., "IP/ATM Integrated Services Over Broadband Access Copper Technologies," *IEEE Communications Magazine,* Vol. 37, No. 5, May 1999, pp. 90–97.

Berger, L., "RSVP Over ATM Implementation Guidelines," *RFC 2379,* August 1998.

Berger, L., "RSVP Over ATM Implementation Requirements," *RFC 2380,* August 1998.

Borden, M., et al., "Integration of Real-Time Services in an IP-ATM Network Architecture," *RFC 1821,* August 1995.

Braden, R., et al., "Resource ReSerVation Protocol (RSVP)—Version 1 Functional Specification," *RFC 2205,* September 1997.

Broadband Integrated Service Digital Network (B-ISDN), "Digital Subscriber Signaling System No.2 (DSS2) User Network Interface Layer 3 Specification for Basic Call/Connection Control," ITU-T Recommendation Q.2931, (International Telecommunication Union: Geneva, 1994).

Cocca, R., M. Listanti, and S. Salsano, "Interaction of RSVP with ATM for the Support of Shortcut QoS Virtual Channels," *Proceedings of 2nd International Conference on ATM,* June 1999.

Crawley, E., et al., "A Framework for Integrated Services and RSVP Over ATM," *RFC 2382,* August 1998.

Eichler, G., et al., "Implementing Integrated and Differentiated Services for the Internet with ATM Networks: A Practical Approach," *IEEE Communications Magazine,* Vol. 38, No. 1, January 2000, pp. 132–141.

Floyd, S., and V. Jacobson, "Link-Sharing and Resource Management Models for Packet Networks," *IEEE/ACM Transactions on Networking,* Vol. 3, No. 4, August 1995, pp. 365–386.

Garrett, M. W., "A Service Architecture for ATM: From Applications to Scheduling," *IEEE Network Magazine,* Vol. 10, No. 3, May 1996, pp. 6–14.

Georgatsos, P., et al., "Technology Interoperation in ATM Networks: The REFORM System," *IEEE Communications Magazine,* Vol. 37, No. 5, May 1999, pp. 112–118.

Heinanen, J., "Multiprotocol Encapsulation Over ATM Adaptation Layer 5," *RFC 1483,* July 1993.

Hong, D. P., and T. Suda, "Performance of ATM Available Bit Rate for Bursty TCP Sources and Interfering Traffic," *Computer Networks,* January 1999.

ITU-T Rec. I.311 (08/96) "B-ISDN General Network Aspects."

ITU-T Rec. I.311 Amendment 1 (03/2000) "B-ISDN General Network Aspects."

ITU-T Rec. I.321 (04/91) "B-ISDN Protocol Reference Model and Its Application."

ITU-T Rec. I.356 (10/96) "B-ISDN ATM Layer Cell Transfer Performance."

ITU-T Rec. I.361 (02/99) "B-ISDN ATM Layer Specification."

ITU-T Rec. I.362 (03/93) "B-ISDN ATM Adaptation Layer (AAL) Functional Description."

ITU-T Rec. I.363 (03/93) "B-ISDN ATM Adaptation Layer (AAL) Specification."

ITU-T Rec. I.363.1 (08/96) "Type 1 AAL."

ITU-T Rec. I.363.5 (08/96) "Type 5 AAL."

ITU-T Rec. I.371 (03/2000) "Traffic Control and Congestion Control in B-ISDN."

ITU-T Rec. I.371.1 (11/2000) "Guaranteed Frame Rate ATM Transfer Capability."

ITU-T Rec. I.381 (03/2001) "ATM Adaptation Layer (AAL) Performance."

Laubach, M., "Classical IP and ARP Over ATM," *RFC 2225,* April 1998.

Maher, M., "ATM Signaling Support for IP Over ATM—UNI Signaling 4.0 Update," *RFC 2331,* April 1998.

Mir, N. F., "An Efficient Multicast Approach in an ATM Switching Network for Multimedia Applications," *Journal of Network and Computer Applications,* Vol. 21, January 1998, pp. 31–39.

Mountzouris, I., et al., "Evaluation of the TCP Traffic Over the ABR Service Targeted To Support Mass Storage Applications," *Proceedings of ICATM,* 1999, pp. 85–90.

Orphanos, G., et al., "Compensating for Moderate Effective Throughput at the Desktop," *IEEE Communications Magazine,* Vol. 38, No. 4, April 2000, pp. 128–135.

Parekh, A. K., and R. G. Gallager, "A Generalized Processor Sharing Approach to Flow Control in Integrated Services Networks: The Multiple Node Case," *IEEE/ACM Transactions on Networking,* Vol. 2, No. 2, April 1994, pp. 137–150.

Partridge, C., "A Proposed Flow Specification," *RFC 1363,* BBN, September 1992.

Pazos, C. M., M. R. Kotelba, and A. G. Malis, "Real-Time Multimedia Over ATM: RMOA," *IEEE Communications Magazine,* Vol. 38, No. 4, April 2000 pp. 82–87.

Perez, M., et al., "ATM Signaling Support for IP Over ATM," *RFC 1755,* February 1995.

Rajagopalan, B., et al., "A Framework for QoS-Based Routing in the Internet," *RFC 2386,* August 1998.

Shenker, S., C. Partridge, and R. Guerin, "Specification of Guaranteed Quality of Service," *RFC 2212,* September 1997.

Shenker, S., and J. Wroclawski, "General Characterization Parameters for Integrated Service Network Elements," *RFC 2215,* September 1997.

Shiomoto, K., et al., "Scalable Multi-QoS IP+ATM Swith Router Architecture," *IEEE Communications Magazine,* Vol. 38, No. 12, December 2000, pp. 86–92.

Siara, D. P., and T. Przygienda, "OSPF Over ATM and Proxy-PAR," *RFC 2843,* May 2000.

Wroclawski, J., "Specification of the Controlled-Load Network Element Service," *RFC 2211,* September 1997.

Yashiro, Z., T. Tanaka, and Y. Doi, "Flexible ATM Switching Architecture for Multimedia Communications," *Proceedings of IEEE BSS'97,* 1997, pp. 58–64.

8

Integration of IP and Optics

Today it is widely acknowledged that IP traffic will become the dominant traffic in all telecommunications networks. The growth and wide usage of the TCP/IP networks mandate that IP networks must now not only meet the expected growth in volume but also deliver other aspects of traditional communication networks such as survivability and control methods for facilitating network operation or performance [1, 2]. These requirements have been analyzed by the IETF and a number of solutions have been devised.

On the other hand, optical networks are being widely introduced to meet an exponentially increasing demand for bandwidth. By integrating IP network technology and optical network technology more tightly it should be possible to increase the efficiency and ease of running such optical IP networks. To realize such demands, the optical transport network would have to become versatile, reconfigurable, and capable of supporting various protection and restoration schemes [3].

There are many ways of integrating IP networks with optical networks. The first item that must be clarified is what an optical network exactly is. We define an optical network to be any transport system that uses optical components and technology as the basic transport entity. This includes the widely deployed SDH/SONET networks, the optical WDM transport networks, and other optical link technologies such as the newly defined optical gigabit Ethernet and 10-gigabit Ethernet. The trend is that the optical transport network is moving away from the traditional point-to-point transport, from simple linear and/or ring topologies and toward mesh-oriented topologies. In support of this, the optical network protocols needed to facilitate such a transition are also being studied [3–7].

As discussed in Section 6.6, the ITU-T G.872 specifications define the OCh trail, OMS, and OTS as the basis for an optical transport network. In the optical transport network, OXCs provide the cross-connect function of the carried traffic data and implement signaling and routing protocols necessary for realizing an optical mesh network. An optical mesh network refers to a network of OXCs that supports end-to-end networking of optical channel trails. Based on this understanding we examine three possible options for combining IP networks with optical networks—IPoA, IPoS, and IP-over-optical lightpaths (or WDM) directly [5, 8].

This chapter is organized as follows: We first examine the basic architectural choices that are available from a bird's-eye view. Following this basic overview, we examine each of the basic architectures in more detail, considering their advantages and disadvantages. The overall structure is similar to that of the previous chapters; following the conceptual and architectural overview, we discuss layering and encapsulation, routing, switching and multiplexing, signaling and control, and traffic management and QoS issues.[1]

8.1 Concepts

Overall, the IP optical networking model assumes that there are many high-speed IP routers interconnected by optical core networks. The IP routers are assumed to exist mostly on the edge of the optical networks. Figure 8.1 shows this model, where an optical network is connected to various other client networks through UNIs. Client networks would include IP networks and other types of networks such as ATM networks. The optical network itself could also be assumed to be comprised of a number of optical subnetworks that are interconnected by NNIs. The optical network itself is considered to be a transport network that carries the user data from various networks. Ideally, this OTN would be transparent to the rate and encoding of the client traffic [6, 7].

1. Much of this field is in still in the state of research and development. Many of the protocols and architectures that we discuss in this chapter are in the process of being defined by various standards organizations. As such, many references are in draft form. The reader is advised to follow the pointers to the latest standards and specifications for the latest information.

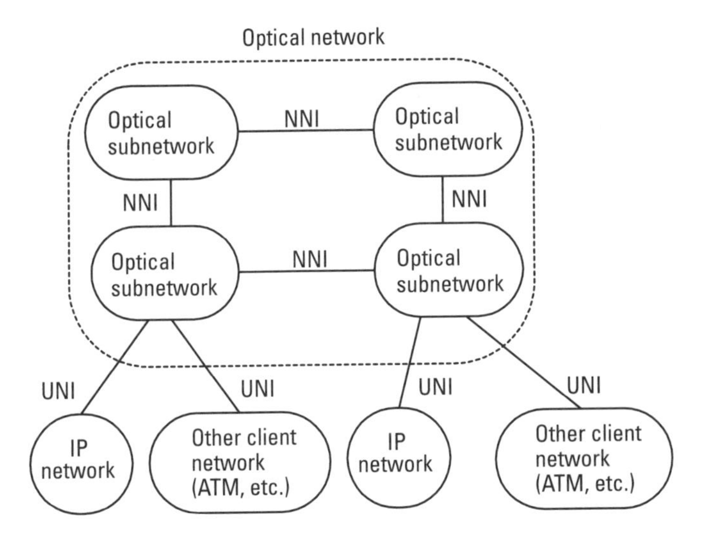

Figure 8.1 An optical network model.

The optical subnetworks are assumed to be mesh networks built out of optical fibers and OXCs. An OXC essentially implements a switching function that is dictated by the current state of the cross-connect fabric. The controller of an OXC changes the interconnection state of the cross-connect fabric. In its simplest form the OXC will just map the incoming lightpath from one port to the outgoing lightpath on another port. Depending on the functionality available, the port may just physically map a lightpath from the ingress port to the egress port. More advanced OXCs may remap the wavelength of the lightpath if the wavelength used by the incoming signal is unavailable on the egress port.

There is a growing consensus for using IP-based protocols as the control plane of the optical networks. By doing this, the dynamic provisioning and restoration of lightpaths becomes possible within and among optical subnetworks. Note that the control plane of OTNs have traditionally relied on network management systems, but it has a number of problems. The system suffered from slow convergence after network failures and ineffective interoperability among different vendors' equipment. This caused problems in internetworking between different operators as they each had a different network management system. Consequently, an important facet of the new optical network is the design and implementation of a new control plane with the relevant signaling functionality to increase the responsiveness and the interoperability of OTNs [6].

There are two fundamental issues that must be addressed. The first is how to adapt and use IP control plane protocols in the optical network control plane. This involves designing new signaling and control protocols or retrofitting existing IP-based signaling and control protocols for use in the optical core network in order to manage the coherent end-to-end provisioning and restoration of lightpaths across multiple optical subnetworks. Note that this is irrelevant to what is actually transported over the optical network itself, as it only deals with the control plane aspect. The second issue is how to transport IP traffic over such an optical network. This involves various problems related to establishing paths from one IP endpoint to another over an optical core network, including the traditional problems of determining IP reachability. Note that this deals specifically with the issue of how to support the transport of IP traffic over the data plane of the optical core network. These topics will both be covered in this chapter.

8.1.1 OXCs and LSRs

As mentioned above OXCs are considered to be dynamic switchable nodes that are able to set up paths.[2] The OXCs are programmable and may support wavelength conversion/translation. Basically OXC can be viewed as a programmable and reconfigurable cross-connect system.

There is an isomorphic relation between OXCs and MPLS LSRs in that both separate the control plane from the data plane. The signaling and setup/release of OXC lightpaths or MPLS LSRs are done through a control plane that is separate from the data plane. MPLS LSR uses labels in the header of the packet to map a packet to a predefined path, whereas OXC will use a mapping of the incoming port and optical channel/wavelength to identify the correct output port. The control functions in both cases must maintain relevant state information to set up and maintain the connections. However, there is a difference between them in that LSR carries out packet-level processing but OXC does not. Also, OXC cannot support hierarchical label encoding that can be done with an LSR.

Likewise, there is similarity between an explicit LSP and an optical channel. They both offer a path between an ingress point and an egress point in the network. The payload carried by an LSP or the optical channel is transparent to the underlying network. Constraint-based routing algorithms are used often to choose the path for LSPs and optical channels. This is related to

2. This section considers the logical model of an OXC. A physical description of it is given in Section 8.5 as well as in Section 6.4.

the fact that both LSPs and optical channels may be chosen to satisfy certain restrictions such as performance, survivability, and other requirements.

8.1.2 Layering and Encapsulation Models

Depending on how many layers are allowed to be involved in switching or routing packets there are a number of different realizations. An additional point of difference is the encapsulation method to be used for transporting IP packets over optical networks. To transport IP packets over any media there must first be a method for delineating the IP packets. As the IP packet itself has no means for delineation, the possibility of running IP over anything should be supported by appropriate encapsulation means.

8.1.2.1 Layering

The main protocol layering architectures for transporting *IP over WDM* (IPoW) networks are currently via the pre-existing ATM and SONET networks, while direct transportation over WDM networks is being pursued for future systems [3]. Figure 8.2 shows four different layering models for transporting IP traffic over optical transport networks. As shown, there are four basic layering models, depending on whether ATM and/or SDH/SONET is used between the IP layers and the WDM optical transport layers.

In Figure 8.2(a–c), the optical adaptation layer may be assumed to follow the OTN model of ITU-T's G.872 Recommendations, which contains the OCh, OMS, and OTS layers. The optical adaptation layer (or *optical layer*) will manage the WDM channel setup/release and provide limited protection and survivability. The physical layer performs such functions as optical amplification, wavelength switching, wavelength conversion, wavelength add/drop, and O/E, E/O conversion.

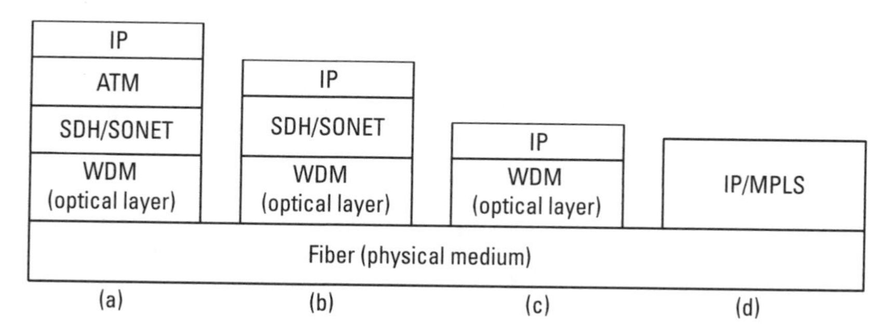

Figure 8.2 Protocol stack development for four IPoW solutions: (a), (b), (c), and (d).

IP/ATM/SDH/WDM. In this case the packets may be switched at the IP layer, ATM layer, SDH/SONET layer, or WDM layer. This layer stack has the highest amount of framing overhead and results in four management layers.

IP/SDH/WDM. In this case the packets may be switched at the IP layer, the SDH/SONET layer, or the WDM layer. This requires PPP or SDL framing methods to put IP packets on SDH/SONET frames. It results in three management layers.

IP/WDM. In this two-layer case, only IP layer routing/switching functions along with WDM-based OADM or OXC are used to switch or route packets. Encapsulation may be done using SDH/SONET, but SDH/SONET cross-connects or ADM will not be used.

There are a number of approaches to IP/WDM networking. One is the current approach in Figure 8.2(c), which is based on transporting IP packets over optical lightpaths such as those defined in the G.782 Recommendations. This would require some sort of framing and fault recovery methods to take the place of those functions offered by SDH/SONET, and it still has two layers of management. Another is to take the approach shown in Figure 8.2(d). In this view, MPLS is used over the optical network and the IP network as well, which results in a single management and control plane. This model takes the extreme view that the optical network will be directly controlled by the same methods as those used to control the IP network. As such, framing/monitoring, fault detection, provision/survivability and other functions will be carried out by the MPLS control plane. Therefore, out of the two IP/WDM models, the former may be viewed as a model for the *overlay routing* and the latter as a model for the *peer routing* (refer to Section 8.2).

Note that in Figure 8.2, we only explicitly mention MPLS as a control protocol for Figure 8.2(d) but not for Figure 8.2(a–c). This is not because the MPLS control plane cannot be used in the other cases but because other control methods may be used in the cases shown in Figure 8.2(a–c). In contrast, the case shown in Figure 8.2(d) is explicitly designed on the premise of using MPLS as the control plane for the IP level and the optical network.

8.1.2.2 Encapsulation Models

In support of each layering architecture above, a specific method of encapsulation is needed. An encapsulation method must define a clear way to decode the IP packets out of the received bit stream.

As discussed in Section 7.3.1, IPoA networks use the RFC1483 encapsulation model, which basically defines how to use AAL-5 to transport IP packets. It uses a combination of LLC/SNAP and AAL-5 frames to provide a framing method for the transport of IP packets over ATM networks.

A *point-to-point protocol* (PPP) may be used when transporting IP packets directly over SDH/SONET networks. For the IPoS case the current widely used solution is to use PPP in *high-level data link control* (HDLC)-like framing [9–11]. However, it is also possible to use other framing methods such as SDL for encapsulating IP packets in SONET payloads. For IPoW there are many other possibilities including SDL and gigabit Ethernet frames. We discuss these issues in more detail in Section 8.3.

In the future, IPoW networks that do not use ATM or SDH/SONET technology need new encapsulation methods. Such new encapsulation methods are demanded for other reasons as well—for example, to improve the efficiency of the transporting IP packets and to reduce the complexity of implementation.

There are a number of options being discussed as candidates for the encapsulation or framing layer for IP packets in the IPoW networks. SDL is a new framing method that has recently been proposed [12]. Methods for transporting IP packets over various transport media, such as Ethernet framing in the form of the gigabit Ethernet standards and the new 10-gigabit Ethernet standards, are also being discussed as candidates for framing [13, 14]. A new framing method similar to SDH/SONET but simpler (namely, SDH-lite) is also being advocated as another possible solution. The ITU-T has gone a step further and started developing a digital wrapper, which would act as a framing method that could be used to transport all types of digital data, without regard to the data rate (refer to Section 8.3.4 for details).

8.1.3 Data Plane and Control Plane

It is important to understand that the architectural models for the data plane and the control plane can be different in optical IP networks.

In most of the current architectures, it is assumed that the data plane will follow an overlay model, in which the optical network offers pipes through which IP packets may be transported and where the optical switches are invisible from the IP packets' point of view. The key characteristic of this overlay model is that the optical switches are unaware of the individual IP packets. If the optical switches were able to process the IP packets individually then a true peer-to-peer model would be possible on the data plane. In such a case the optical switches would be able to process each IP packet separately based on its IP header. Currently, however, this type of solution is not deemed to be practical.

As a result, the dynamically controllable OXCs that are being developed today are viewed as the solution of choice. Such controllable OXCs lead to a model in which the optical path is set up through an optical network first before data is actually transmitted. Consequently, the IP traffic can be viewed as being put into an optical tunnel at the ingress point of the optical network and pulled out of that optical tunnel at the egress point of the optical network.[3]

In contrast to the data plane, the control plane may take a number of different models. Like the data plane, one can assume an overlay model for the control plane but other models are also possible. One possibility is the peer-to-peer model and another possibility is the augmented model (see Section 8.2). It is not easy to predict which of these control plane models will become dominant. Nevertheless, many envision evolutionary scenarios that lead to more tightly coupled IP and optical layers as technology advances. These will be discussed in the following sections.

8.1.4 Signaling and Routing Models for the Control Plane

In the network model shown in Figure 8.1 we observe that there are two important boundaries for signaling and control. One is the network interface between the node outside the optical network (i.e., the client), and the other is the network interface between the nodes inside the optical network. They are important in two aspects—one is the signaling that must be carried out between the two entities and the other is the routing protocol that must be run between the two.

8.1.4.1 Signaling Aspects

As shown in Figure 8.1, UNI is the interface between the client and the optical network, whereas NNI is the interface between two optical subnetworks. We concentrate on the UNI interface model below, as the NNI interface has not yet been extensively studied.

There are several possibilities for the UNI, such as direct and indirect interfaces. A direct interface uses an IP control channel that is directly set up

3. OXCs and their uses are discussed again in Section 8.5. Also discussed in Section 8.5 are some other experimental methods that are more closely related to a true optical packet switch, but based on using electrical processing and O-E/E-O translation functions. These include the OTPN and optical burst switching (OBS) methods. These methods introduce per-packet processing functionality into some variants of OXC.

between the edge router and OXC. Both in-band and out-of-band methods may be used. The control channel may be used to exchange both routing and signaling information. For exchanging routing information, link state protocols such as OSPF or *intermediate system to intermediate system* (IS-IS) and also BGP are being studied. These protocols would have to be extended along the lines of the currently suggested mechanisms for supporting TE extensions. Signaling protocols would also include *RSVP with tunneling extension* (RSVP-TE) or *constrained route LDP* (CR-LDP) variants. Both of these are being developed within the G-MPLS framework.

An indirect interface uses an out-of-band IP control channel between the client and a server in the optical network. Based on the signaling messages from the client the server sets up appropriate optical paths through the network. For example, the server may be the network management system for the optical network or it may be a specialized server. Note that this model would also include the case where direct communication between network management systems in the client IP network and the optical network are used to set up optical paths.

The provisioned interface is a trivial case where the optical paths are set up manually.

8.1.4.2 Routing Protocol Aspects

Routing protocols must be changed to allow the exchange of information beyond simple connectivity. Unlike the case of IP networks, providing just enabling connectivity between nodes is not sufficient for optical networks. This means that the traditional SPF routing algorithms are not the preferred method. Instead, *constrained route SPF* (CR-SPF) is preferred in most cases.

This leads to various changes in the routing protocol. Information regarding the types and characteristics of the switchable channels must be exchanged. This may include such items as the number of wavelengths available and the type of protection available along the lightpath. The routing aspects will be dealt with in more detail in later sections.

8.1.5 Optical Domain Service Interconnect, Optical Internetworking Forum, and IETF

Currently a number of organizations are pursuing standardization actions on the architectures, layering stacks, signaling protocols, and control protocols for the optical IP network. Overall, there are two main groups—the *Optical Domain Service Interconnect* (ODSI) coalition and the *Optical Internetworking Forum* (OIF)—in addition to the IETF [15–20].

Both ODSI and OIF agree basically on using IP-based protocols as the basis for the control plane of the optical core network. In addition, both agree that the IP core network will most probably use the G-MPLS specifications as the basis for the control plane signaling and control protocols in the core network itself. As such, the two organizations are actively supporting the development of G-MPLS specifications driven by the IETF [21, 22].

The main differences between the two organizations are in the control architecture and the actual protocols to be used. The ODSI is more restrictive in that it will support only the overlay model and will concentrate only on defining protocols that operate over the UNI between the client network and the IP core network. It is in the process of defining a service discovery and address registration protocols. The main signaling protocol to be used between the client network and some optical network entity is also being defined. The OIF is a bit less restrictive than the ODSI in that it aims to progress to support the integrated model as well as the overlay model, even if it initially started by defining a UNI signaling and control protocol for the overlay model.

The IETF is defining the G-MPLS framework and the related extensions to the signaling and routing protocols [21, 23–27]. In contrast to the ODSI and OIF, the IETF is neutral with respect to the various models from the beginning and is trying to define protocols that will work well for all different types of models.

8.1.5.1 Combining MPLS TE with OXCs

One of the original proposals for integrating IP technology with optical networks was to combine the MPLS TE control plane defined in [28] with the newly developed, dynamically controllable OXCs. The main idea was to use the control plane defined for MPLS TE as the control plane for OXCs. The use of OXCs would be strictly limited to using optical networks as the transport plane for Internet traffic [29, 30].

To satisfy this functionality the OXC control plane must satisfy a number of basic requirements. This emphasizes the routing and path setup capability of the OXC. An OXC must be able to set up OCh lightpaths on demand. This means that the time scale of path setup must be within the subsecond range. In addition, it must be able to support TE functionality and various other mechanisms for enabling path protection and restoration.

The OXC control plane would use MPLS messages with TE extensions for signaling, and either OSPF or IS-IS with TE extensions for routing. An OXC is considered to be a path switching element in an optical transport network that establishes routed paths for optical channels by locally connecting an optical channel from an input port to an output port on the switch element.

Figure 8.3 shows a basic OXC system architecture that satisfies the above requirements; the OXC switch fabric or data plane is controlled by an MPLS control plane through the OXC switch controller. Note that the structure does not mandate that the control traffic must flow through the same paths or connections as the data traffic.

The OXC control plane is described in more detail in Section 8.6, while G-MPLS and other TE extensions to the routing protocols are explained in their relevant subsections.

8.2 Architecture and Routing Models

As stated in Section 8.1, from an architectural point of view, the various models differ in signaling and control architectures, but basically assume the same data transport model.

The IP data plane over optical networks is realized over an overlay network of optical paths. In contrast, IP routers and OXCs can have a peer relationship or a client–server type relationship on the control plane, as shown in Figure 8.4. Figure 8.4(a) shows the case where the user equipment (or client) accesses network services through a well-defined UNI signaling protocol, with the request/release of point-to-point connections done through the optical network. This is called the *overlay model* as the user equipment is unaware of the internal structure of the optical network. Figure 8.4(b) shows

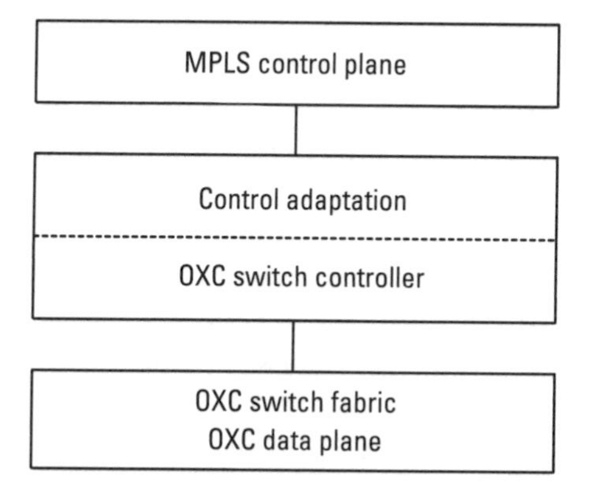

Figure 8.3 A basic OXC system architecture with MPLS control plane [30].

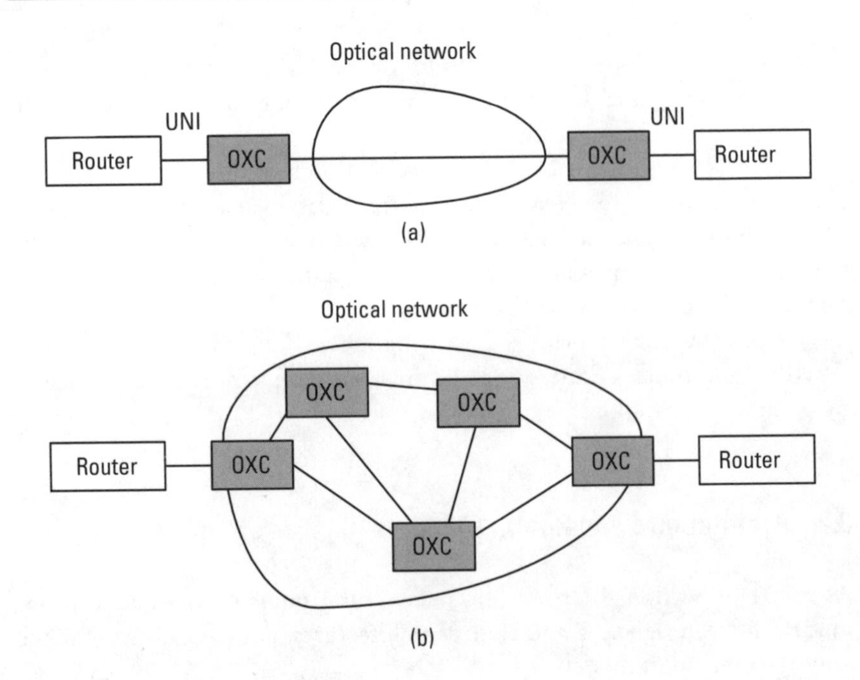

Figure 8.4 Relationship of IP routers and OXCs on the control plane: (a) overlay model, and (b) peer model.

the *peer model* in which the user equipment is fully aware of the optical network structure, which results in a peer relationship between the OXCs of the optical network and the client equipment. As such, the IP-over-optical network architecture is defined essentially by the organization of the control plane.

8.2.1 Service Models

There are two main types of interfaces in the Figure 8.1 optical network model—the UNI interface between a client network and the optical network, and the NNI interface between two optical subnetworks. For both cases, one side of the interface (i.e., the optical network side in the case of the UNI, and one of the optical subnetworks in the case of the NNI) can be viewed as offering services to the other side. Note that both models assume that there will be a neighbor discovery and service discovery function at the interface. This function would be used at boot-up time to automatically find the nodes that are connected to this node and to discover the services that are offered.

Depending on the types of services and methods by which these services are invoked there are two main models. The first is the *domain service model* and the second is the *unified service model* [6]. Figure 8.5 depicts these two service models. The two models contrast in that the former has completely separated addressing and control planes while the latter has a common address space.

8.2.1.1 Domain Service Model

In this case the interface offers a clean, well-defined set of services to the client. Completely separate addressing and control planes are assumed to exist in the optical network and the client network. That is, each domain is assumed to use its own signaling and addressing model. This is shown in Figure 8.5 by the separate dotted lines enclosing the routers and the OXCs. The routers run traditional IP layer control protocols and/or MPLS control protocols. The OXCs use proprietary or G-MPLS signaling/control protocols. This model easily maps to a control plane that uses the overlay model concept. As such, it assumes that the client network nodes (i.e., IP routers) will use the optical core network as a cloud that will offer data paths and interfaces for setting up and maintaining those paths.

This interface enables the client network to easily set up and maintain a lightpath. Basically four types of functions are offered, namely, lightpath

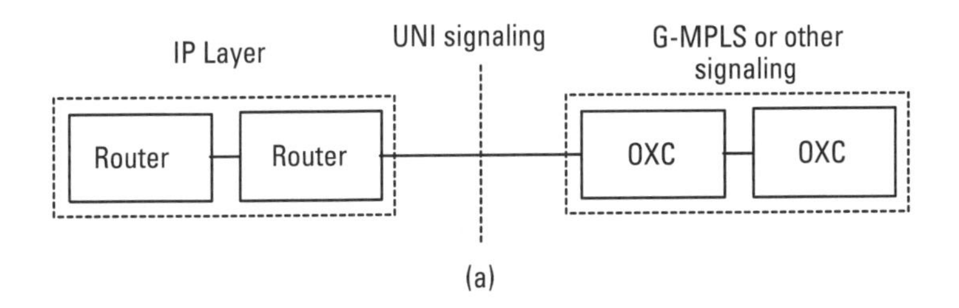

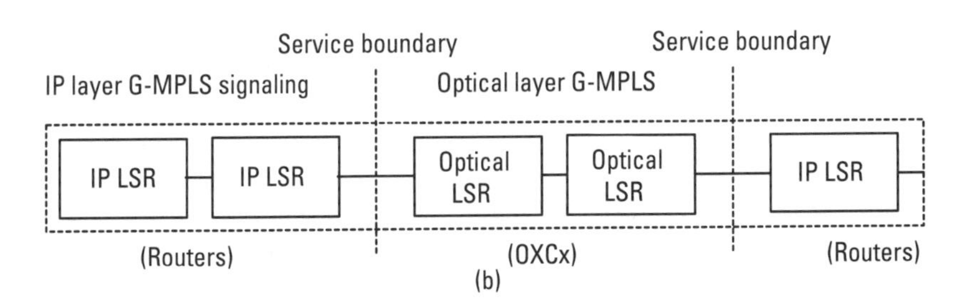

Figure 8.5 Service models: (a) domain service model, and (b) unified service model [6].

creation, lightpath deletion, lightpath modification and lightpath status enquiry.[4]

Specifically, lightpath creation is the function that enables the client to create lightpaths connecting the specified end points with the desired properties such as BER and path survivability characteristics. The lightpath deletion function allows the client to delete existing lightpaths whenever desired. The lightpath modification function enables the client to modify the properties of the lightpath when desired. Finally, the lightpath status enquiry function enables the end systems to query the current state of the lightpath so as to determine its connectivity, BER, and other relevant data.

A second group of functions would also have to be defined that would enable address resolutions and registration by the client network nodes. These functions could be defined as client registration, client deregistration, and address query.

The client registration function is used by the client to register its address and identification with the optical domain. This information may be considered to be kept in a central database or repository in the optical domain.[5] The client deregistration function is used by the client to remove or deregister its address and identification from the optical domain. The address query function is used by a client to find out the optical address of another client that is connected to the optical domain and with which it wishes to set up a connection.

8.2.1.2 Unified Service Model

In this case the interface would be seamless and the IP and optical networks would not be differentiated. The OXCs and the IP routers would consider each other as peers and a single unified control plane would be used for both the OXCs and the IP routers.

As shown in Figure 8.5(b), this is a natural extension of the G-MPLS signaling. Figure 8.5(b) shows how IP routers and OXCs can both function as G-MPLS LSRs and communicate by using G-MPLS signaling. From a signaling or routing point of view the IP routers and the OXC are the same. With such a service model a common address space is used to identify all

---

4. In addition to the above functions, there may be a notification function that may be used by the optical network to notify the end systems about the problems in the lightpath. This function, while useful, is not necessarily needed by all systems as there are other ways of getting similar functionality (e.g., lightpath status enquiry function may also be used).
5. This does not necessarily mean that the implementation is in any means similar to this.

routers and OXCs. Optical network services are obtained implicitly by the IP routers through the use of G-MPLS signaling methods. Therefore, in contrast to Figure 8.5(a), the dotted line box in Figure 8.5(b) encompasses both the IP routers and the OXCs.

While the IP routers and the OXCs may use the same signaling methods based on G-MPLS, the specific semantics of the attributes and messages may differ between the IP routers and the OXCs. For example, the protection services offered by the optical networks may be different from those of the IP routers, or the bandwidth granularity available may differ. This necessitates that there may be some sort of service translation at the network boundaries. In Figure 8.5(b) this is shown by the specific demarcation of the service boundary between the IP routers and the OXCs.

8.2.1.3 ODSI and OIF Service Models

These services are triggered by the use of a signaling protocol. The signaling protocols for the different cases may use different types of protocols. For example both the ODSI and the OIF initially concentrated purely on the domain service model. In contrast, the IETF tried to define a single G-MPLS that is flexible enough with the possible extensions to support both types of service models. Using a single control protocol in both networks and thereby enabling a unified service model has a number of advantages for the operators as mentioned above. These include simplification of management and operation.

However, there are also various advantages for operators using the domain service model. It enables operators to introduce optical networks even though the protocols that are to be used between optical equipment is not fully specified. As long as the interface between the edge of the optical network and the IP routers is clearly defined there is no problem in operation. Another aspect is that many operators may not want to show their internal network topologies to outside clients, and by using the domain service model this problem can be easily sidestepped.

8.2.2 Interconnection and Routing Models

There are two different basic service models for IP-over-optical networks. On this basis, we may consider three different interconnection and routing models depending on the tightness of the coupling between the control planes of the IP and the optical network. The coupling of control planes in this case can be viewed from the following perspectives: first, the amount and level of details of the routing and topological information exchanged over the UNI;

second, the amount of control the IP router has in selecting a path over the optical network; and third, how the problems of access control, accounting, and security are resolved in dynamic provisioning of optical paths connecting routers. The resulting three models are the *peer model*, the *overlay model*, and the *augmented model* [6, 31].

8.2.2.1 Peer Model and Integrated Routing

In the peer model the IP router's control plane and the OXC control plane are peers and use the same protocols. Basically there is a single unified control plane, and a single signaling protocol is used that runs seamlessly over the IP and optical networks. This was shown previously in Figure 8.4(b). From the service perspective, the peer model fits naturally with the unified service model discussed in Section 8.2.1. A single routing protocol is running that manages topological information of both networks and allows all routing entities to be aware of the topology of both IP and optical networks. The routing protocols would be common IGPs, such as OSPF or IS-IS, but would have to be modified with the appropriate optical extensions.

The peer model assumes that there will be a single addressing scheme used over both IP and optical networks. This scheme would presumably be based on the current IP addressing schemes. This means that the optical network entities would have to use IP addresses.

8.2.2.2 Overlay Model and Overlay Routing

In the overlay model the IP router's control plane and the OXC control plane are independent. Each plane is assumed to be running a separate signaling/control protocol. A UNI is defined over which a signaling protocol is used to convey messages. This was shown previously in Figure 8.4(a). From the service perspective, the overlay model fits naturally with the domain service model discussed in Section 8.2.1. Either static or dynamic optical link provisioning is possible. Static optical link provisioning refers to the case where the end user or IP router cannot set up and release optical links for the IP traffic in a dynamic fashion. It relies on the optical network operator to set up and release connections on a preprovisioning agreement. If dynamic link provisioning is used the IP router may use a signaling protocol to request that a dynamic link be set up for IP traffic on need basis.

For the overlay model the routing protocols of the IP and optical networks are separate and do not exchange routing and topological information as in the peer model. Note that this does not mean that changes in optical network topology will not affect the IP routing topology. If an optical link that connects two IP routers disappears, the IP routers will have lost connectivity, and this will be propagated throughout the IP routing domain. However, when optical links are set up for IP traffic, the IP routers at the end points must add this information to the IGP routing protocol routes. For some cases it may be appropriate to add optical network characteristics. The overlay model does not assume that there will be a single addressing scheme used over both IP and optical networks.

8.2.2.3 Augmented Model and Domain-Specific Routing

In the augmented model the IP router's control plane and the OXC control plane are peers but do not run the same control protocols. This contrasts to the peer model where the two control planes are unified by a single control protocol. In the augmented model the two control planes are separate but exchange information through a predefined interface. A signaling protocol is used on the interface of the two networks to exchange routing information. The two networks each run separate routing protocols but exchange information regarding the routing changes with each other. For example, external IP addresses reachable from routers on the edges of the optical network could be passed to the other IP routers, thereby allowing reachability information to be passed to other IP clients. Note that this bears many similarities to the functional model of the PNNI PAR function (see Section 4.2.2) and its use in IPoA network models (see Section 7.2.1). For both cases the two domains run separate routing protocols and maintain separate topology maps, but the IP routers interconnected by an ATM network running PNNI PAR are able to exchange IP reachability information dynamically. Note that these routing protocols running in the IP domain and the optical domain could be the same protocol, such as OSPF or IS-IS, but each would be unaware of the others. In the augmented model the IP and optical networks can use different addressing schemes.

8.2.3 End-to-End Protection Issues

One of the most important aspects of using optical lightpaths is the support of fast protection and restoration methods. In SDH/SONET networks the availability of fast-APS and restoration abilities is considered to be a major advantage. Any IP-over-optical network solution must also supply such a solution. Assuming that an LSP is set up across an ingress IP network, an ingress optical network, an egress optical network, and an egress IP network, there are basically two methods to protect the LSP—1+1 protection and 1:n protection (or shared protection).

In the case of segment-wise or local protection, the end-to-end LSP is assumed to be a concatenation of three segments—segment *A* covering the ingress IP network segment, segment *B* covering the optical network segment, and segment *C* covering the egress IP network segment. If a certain protection model is offered end-to-end, each individual segment is responsible for mapping the protection that it offers with the overall protection model. For example, if the whole LSP offers IP endpoints with 1+1 protection, each individual segment must also offer 1+1 protection. This includes not only the LSP segments in the IP networks but also the optical network and the links between the edge routers and the OXCs. This basically assumes that the IP layer can utilize the optical layer protection services.

In the case of the simple layer or end-to-end protection, the end-to-end IP layer does not rely on any protection services offered by the optical network. Instead, two separate SRLG-disjoint LSPs must be established between the two endpoints. The corresponding lightpaths in the optical network must be separate.

The main difference between these two methods is that in the former the optical network is actively involved in providing protection to the path, whereas in the latter method the lightpath only supplies an optical path. There is no distinction in this point between the active lightpath and the protection lightpath.

Another difference is with respect to scalability and sharing of protection paths. In the first method, the optical protection path in the optical network may be shared with other end-to-end LSPs that have different IP endpoints but use the same optical end points. For the latter case, this is impossible, so every single LSP that is protected must use a separate optical path.

A third difference is that in the former it is possible to use different restoration methods in the optical network from the IP network. This can lead to cases where the optical restoration speed can be faster. The fact that in the former case restoration only needs to occur in the failed segment itself can also lead to much faster optical restoration speed.

8.2.4 Miscellaneous Control Plane Issues

There are a number of miscellaneous control plane issues that must also be clarified to derive an IP optical network. These are related to the main issues of services, interconnection, routing, and protection models discussed above but are viewed as being relatively minor issues compared with them [6, 7].

8.2.4.1 Addressing

A fundamental issue in networking is addressing. Addressing for networks defines how to identify the entities from a signaling and routing point of view.

In optical networks a number of identifiable entities exist such as OXCs, optical links, ports, optical channels and subchannels, and *shared risk link groups* (SRLGs), as illustrated in Figure 8.6(a). The main issue is what the granularity of identification should be. The identification must be done in such a way that the termination point of the lightpath can be identified in a clear manner.

For example, in the case of Figure 8.6(a), an OXC can have a number of ports. Each port may terminate many optical channels, and each channel may contain many subchannels. The connections between two OXCs can be viewed from various granularities. OXCs are connected by optical links, with each optical link connected to a single port on the OXC. Multiple optical links are combined into an optical fiber through WDM technology. Optical fibers themselves are also usually grouped into link bundles. In the case of SDH/SONET, each optical link can be expected to support various channels and subchannels. Depending on the position of an OXC along a lightpath, the OXC may not need to know the full granularity (e.g., up to the subchannel level) but a knowledge of a less detailed level (e.g., the channel level) may suffice. This naturally leads to an identification where the endpoint is identified by an IP address and a "selector" that contains fine-grain information that can be used by the OXC up to the granularity that it needs. This selector may be structured such that it can identify the port, channel, and/or subchannel as needed.

Another important item that must be identified is the SRLG. An SRLG identifies links that use a common resource. For example, all the optical channels that share a single optical fiber can be identified by a single SRLG. Consequently, a single link may belong to multiple SRLGs; two links belonging to a single SRLG may also belong to different SRLGs individually. This is illustrated in Figure 8.6(b); two optical channels *A* and *B* have one fiber link in common, which is marked as an SRLG with the value 2112. At the same time, channels *A* and *B* also pass other fiber links that they do not share. In Figure 8.6(b), these are the fiber links with the different SRLG values, 1701 and 1313, respectively. Consequently, the SRLG identifier for channel *A* would be (2112,1701) while that for channel *B* would be (2112, 1313). From this we can recognize that the two paths share a common path and therefore are both susceptible to failure if the shared path fails.

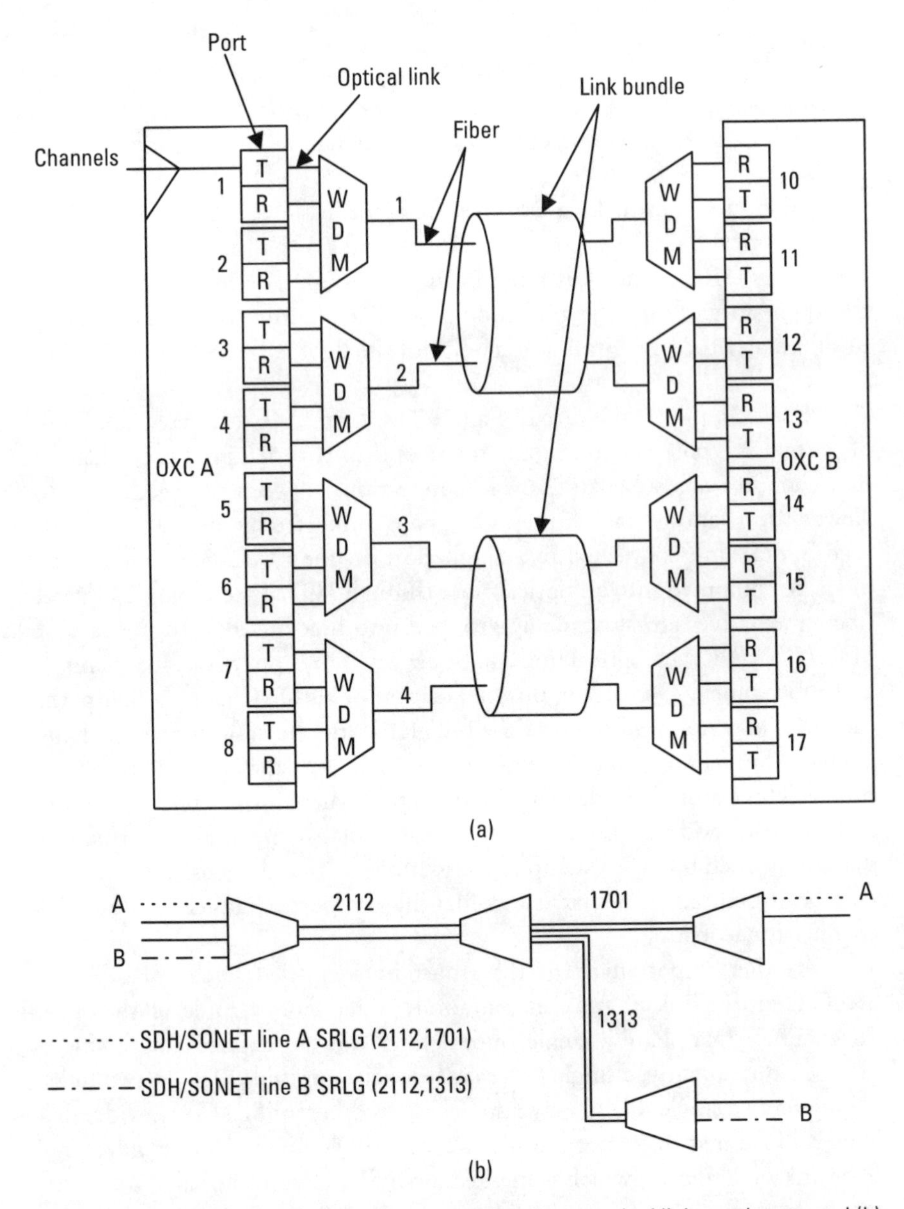

Figure 8.6 Addressable elements: (a) link bundles, fibers, optical links, and ports, and (b) SRLGs [7].

SRLGs are used to identify SRLG-disjoint paths through a network. As a result, the SRLGs must be unique within a network. The SRLG identifier may be a flat identifier. While the resources are classified into SLRGs manually, the uniqueness of the identifiers must be guaranteed.

An SRLG may be used for protection purposes in the following manner: When routing an alternate protection path, the alternate path optical links must not share an SRLG identifier with any optical links of the original path. This arrangement assures that the failure of any single optical link in the original path may be safely passed by when the protection path is used.

In many cases the optical links between adjacent OXCs are bundled when they are advertised in link state routing protocols such as OSPF. It is important in such cases to arrange it so that the component links of an optical bundle are identifiable. By doing so it becomes possible to reduce the level of the routing adjacency, which is critical to the scalability of OSPF networks [6].

8.2.4.2 Neighbor Discovery

In all the networks that rely on distributed control the first step of operation is to check the status of the links that connect the relevant elements. In the case of optical networks, this means that each OXC must check the status of the links and the OXCs on the opposite end of the links. Specifically, the checkpoint includes the up/down status of the links, the parameters of the links including bandwidth, and the identity of the OXC on the opposite side.

This function is usually achieved by manual configuration data and the use of some protocols that automatically discover or negotiate the relevant parameters. The protocol that carries out such functionality is known generically as a *neighbor discovery protocol* (NDP).

An example of the NDP lies in the overhead bytes in an SDH/SONET connection, which deliver the status information to the opposite side of the link. Another example is the recently defined *link management protocol* (LMP), which supports basic NDP functionality as well as the link management and fault management functions. The ODSI is defining an NDP protocol that is based on the PPP protocol.

8.2.4.3 Topology Discovery

Topology discovery is the procedure by which the elements in the network discover the topology of the network. By understanding the network topology the constituent elements can make intelligent decisions on how to route packets or circuits.

It is possible to use a management protocol from a centralized server to discover and build topology tables. Such a method is simple to accomplish but it suffers from the need for a centralized server, which may become the single point of failure or point of congestion.

Another method for finding the topology of the network is to use dynamic routing protocols such as OSPF or IS-IS. They distribute the link state information of each individual node globally in a reliable manner. Based on this globally distributed information each node is able to build and maintain its own global topological map.

When link state routing protocols are used in optical networks, the link state information must support the concept of link bundles and may have to include restoration-related parameters as well. In addition, the link state protocol may need to be capable of setting the thresholds dynamically to trigger the link state advertisements so that it can signal the need for a new link state advertisement if the amount of available bandwidth falls below a threshold.

8.2.4.4 Route Computation

Route computation in optical networks is essentially a constraint-based routing problem. The key initial constraint is the availability of bandwidth along the optical path. Another constraint is that when determining the protection paths the links that compose the protection paths must not share the same SRLG with any of the links on the protected path. This second constraint complicates the problem of route computation. A third constraint may be given in relation to QoS support but, at the optical path level, this constraint may not be visible at all.

The above issue has some implications for the LSAs. Specifically, the LSAs must carry enough information to compute routes under the above constraints. For example, the LSAs must indicate the resources available at each link and OXC as well as the relationships between links and SRLGs.

8.2.4.5 Signaling Issues

Signaling is concerned with the issues regarding the methods and messages to exchange to establish and provision a lightpath. As noted above, route computation calculates the correct route for each lightpath and the OXCs through which the lightpath is to be established. Then signaling must ensure that each OXC along a lightpath is notified to set up its connections appropriately to support that lightpath. This can be done by RSVP-TE or CR-LDP when MPLS is used.

While signaling is fairly clear-cut there are several issues that must be solved. First, as lightpaths are inherently bidirectional, it can happen that two adjacent OXCs may simultaneously choose the same input/output combinations for different lightpaths, resulting in a collision. Second, while individual node failure may occur in the optical network, it should not cause the lightpaths that pass through that node to be torn down. Third, signaling must support the setup and activation of restoration or protection paths, subject to the constraints discussed in the previous section.

8.2.5 Evolution Path and Other Issues

The architectural models described above naturally fit into an evolution path for IP-over-optical network development. This is based on the complexity and feasibility of the various functions needed for the architectural models.

8.2.5.1 Evolution Path

First, it is expected that the overlay model using a domain service model will be the easiest to deploy. At this stage the routing of the IP domain and the optical domain would not exchange any routing information at all. This would require minimal intelligence in any of the optical domain components, even leaving the possibility of using manual configuration methods in the optical domain.

At the UNI, it is expected that using a signaling interface based on G-MPLS at the first stage will facilitate the introduction of the next stage, which would be the augmented routing model. At this stage, while routing information would not be exchanged directly between the two domains, the optical domain would relay routing information in the form of the reachability information of external addresses.

Finally, the architecture would be developed to the final integrated routing model based on the unified service model. At this stage the routing of the IP domain and the optical domain would be fully integrated.

Using G-MPLS as a common signaling protocol from the beginning for both IP and optical domains will also facilitate the evolution from the domain service model to the unified service model. In this case it should be possible to implement those increasingly sophisticated models by changing the semantics of the signaling protocols, instead of developing completely new signaling protocol.

8.2.5.2 Other Issues

There are a number of other issues that are related to the use of IP-over-optical networks. Among them, we discuss the following three major issues below—the use of TDM technology in conjunction with WDM, the use and availability of wavelength conversion, and the rate of lightpath setup and its architectural consequences.

Coexistence of WDM and TDM

It is possible to use SDH/SONET as a method of partitioning the bandwidth that is offered by a single lightpath. For the purpose of optical networking based on G-MPLS, an LSP may be considered to be the path defined by a timeslot in the SDH/SONET channel. Any signaling or control method must be able to set up and control such paths. This can lead to a hierarchy of paths, where the lowest level path is a WDM lightpath carrying SDH/SONET frames, which in turn carry various TDM channels. While WDM and TDM are physically different methods of multiplexing, they are fundamentally controllable by the same single mechanism as both are fundamentally circuit-switched networks. For the case of G-MPLS this would be similar to the case where label stacking or LSP hierarchies are used.

Wavelength Conversion

In an optical network wavelength converters may not be available at all optical switching points (i.e., OXCs). This is an important point as it imposes a constraint on the possible paths that may be set up in a WDM-based optical network. Due to the limited number of wavelengths available in current systems, wavelength converters are an important means for enabling wavelength "reuse" and routing. For example, in case all the wavelengths between node A and node B are in use, except for the color "blue," if a new lightpath setup request comes to node A where the incoming lightpath's color is "red," a wavelength converter is definitely needed to switch the color from "red" to "blue" for the lightpath setup.

As such, any IP-based control model must be able to deal with the fact that the optical network may be constrained in the possible routing paths that the network can support by the limited number of wavelengths and wavelength converters available in the network. The G-MPLS must be able to support this sort of model. This can be carried out by the use of the "label sets" definitions in G-MPLS [22].

Rate of Lightpath Setup

Rate of lightpath setup will be an important issue in IP optical networks. The available architectural choices differ depending on the assumptions made on this setup rate. If this setup delay is minimal it is possible to use dynamic data-driven connection setup models along the lines of many IPoA proposals. However, if the setup delay is large this would not be possible. As yet, in IPoA networks, the former method is not widely used due to various concerns regarding the CPU overhead, complexity of solutions, and stability. For similar reasons, it is expected that dynamic data-driven connection setup methods will not be widely used in the initial stage of the optical networks.[6]

8.3 Layering and Encapsulation

As discussed in the introduction, there are various layering models for transporting IP traffic over WDM networks. The protocol layering dictates where the multiplexing and switching will occur in the network. The encapsulation method used for transporting IP packets also depends on the protocol layering used.

Encapsulation is carried out by one of the layers in between the IP layer and the WDM transport layer. The encapsulating entity must group the bits transmitted by the WDM layer into the frames that the IP layer can deal with. Different protocol layering models differ fundamentally regarding how this framing is carried out. Depending on the method chosen for this, many other aspects of the network architecture are affected.

Figure 8.7(a) shows the variety of architectures that realizes IPoW. Starting from the left, the first two architectures, Figure 8.7(a, b), use ATM as the basic framing technology, forming an IPoA layering model. More specifically, they use AAL-5 as the method for transporting IP packets over an ATM network. The architecture in Figure 8.7(a) assumes that ATM runs directly over a WDM/optics networks, whereas the architecture in Figure 8.7(b) assumes that a SDH/SONET layer exists in between. Both architectures suffer from the overhead of transporting IP packets over ATM networks, so-called the ATM "cell-tax." However, they also have the advantage of being able to use many of the features of ATM such as the built-in OAM functionality and the ability to do traffic engineering through VCs.

The next two architectures, in Figures 8.7(c, d), use SDH/SONET technology, forming IPoS layering. The architecture in Figure 8.7(c) can be viewed as the classical *packet over SONET* (POS) method, where the IP packet is sent over PPP and HDLC-like frames over SDH/SONET links.

---

6. Instead, long-term lightpaths based on traffic engineering considerations and service agreements are expected to be prevalent initially. (Again this is similar to the path followed in IPoA development.)

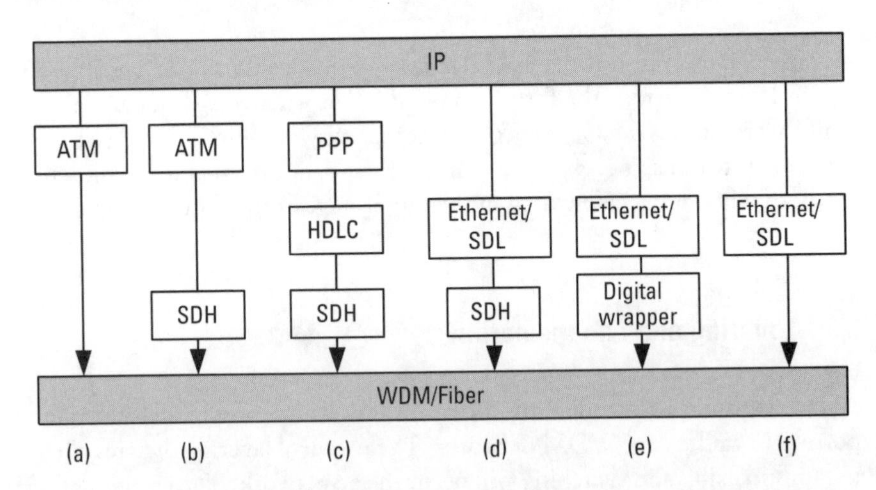

Figure 8.7 Architectures to realize IPoW: (a) IPoA, (b) IPoA over SDH/SONET, (c) IPoS via PPP/HDLC, (d) IPoS via Ethernet/SDL, (e) IPoW via digital wrapper , and (f) IPoW via Ethernet/SDL.

The architecture in Figure 8.7(d) may be viewed as a newer model for POS with the framing method replaced by either Ethernet or SDL.

Ethernet framing basically aims to use gigabit Ethernet frames and signaling as the method for transporting high-speed data over fiber, which is advantageous as Ethernet renders a widely spread user interface. SDL is a new framing method for variable-length packets that has been recently invented. It applies the ATM cell delineation technique of using a CRC check over a fixed length of bytes to find packet boundaries in a bit stream of packets. It is a simple and efficient method for framing with minimal overhead.

The next architecture, Figure 8.7(e), replaces the SDH/SONET layer of architecture, Figure 8.7(d), with the newly defined ITU-T digital wrapper or optical channel to form a pure IPoW layered architecture. Similar to the architecture in Figure 8.7(d), it employs Ethernet or SDL framing. It is expected that the ITU-T's digital wrapper–based OTN will provide similar capabilities to the SDH/SONET network but for WDM networks in a bit stream-format–independent manner.

The last architecture, Figure 8.7(f), runs IP packets directly over WDM without relying on any underlying optical transport protocol such as ATM, SDH/SONET, or the ITU-T digital wrapper, thereby forming a pure IPoW layered architecture. This architecture employs a framing technology such as Ethernet or SDL to directly derive the bytes and packets from the physical layer.

There are four different layers—IP, ATM, SDH/SONET, and WDM—where multiplexing function can be incorporated. Figure 8.8 illustrates the related four different types of multiplexers. Multiplexing at the IP layer has a number of advantages. In particular, assuming an efficient framing method is used, multiplexing at the IP layer will yield an efficient link utilization. One possible disadvantage is that non-IP traffic must be encapsulated over IP before transportation. Consequently it is a good choice if IP is the dominant traffic to multiplex.

Multiplexing at the ATM layer has the advantage of being able to efficiently aggregate both circuit-type and packet-type traffic. However, this also means that neither circuit-type nor packet-type traffic will naturally fit into ATM networks. Therefore, all types of traffic must be adapted to run over ATM. In the case of IP traffic this takes the form of using AAL-5 as the framing method and the choice of one of the various models explored in Section 7.3.1. In supporting IP traffic the cell overhead can go up as high as 20%, which decreases channel efficiency substantially.

Multiplexing at the SDH/SONET layer helps to take advantage of the reliable transport capability of SDH/SONET which is basically a transport technology that can transparently aggregate various types of services, both in circuit and packet modes. Additionally, the built-in OAM capabilities of SDH/SONET provide link failure detection and fast protection switching capabilities.

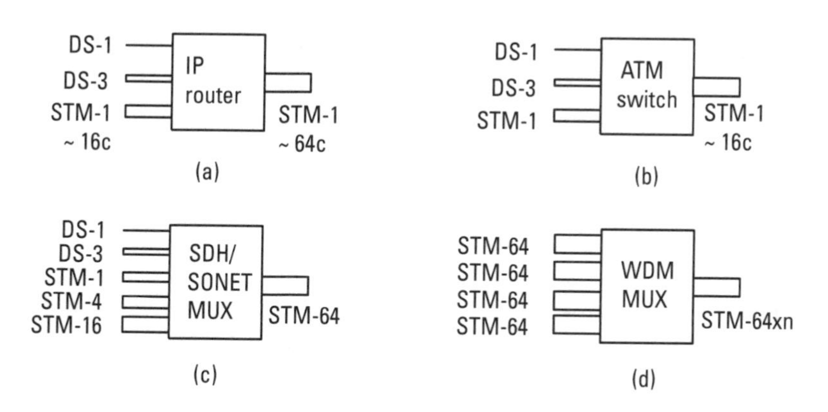

Figure 8.8 Comparison of multiplexing at different layers: (a) IP, (b) ATM, (c) SDH/SONET, and (d) WDM.

A fourth point of multiplexing is the WDM layer. The protocols for this multiplexing have not been standardized but are rapidly evolving due to the advantage of getting high capacity bandwidth. In addition to bandwidth increase, WDM also has the advantage that it works for all types of traffic, that is, it is blind to the type of traffic. A disadvantage is that it cannot yet offer the full-fledged protection switching and OAM functionalities that SDH/SONET has, though this is expected to change with the development of new standards such as the ITU-T OTN architecture [5].

8.3.1 IPoA

One of the most obvious and widely deployed methods of transporting IP-over-optical networks is the use of the IPoA architecture. As ATM was designed from the beginning to be transported over optical networks, directly or via SDH/SONET, the use of IPoA automatically translates into a method for transporting IP packets over optical fiber. The case of running ATM over SDH/SONET optical transport links (i.e., SDH/SONET-based ATM) has become much more widely used, while the direct mapping case (i.e., cell-based ATM) has seen only limited usage. Both methods can have their optical capacity increased by using WDM technology at the optical link level.

As a complete description of IPoA networks and their operation are already provided in Section 7.1, we will only briefly discuss the advantages and disadvantages of using IPoA in this section. Specifically, we will concentrate on the IP over ATM over SDH/SONET network architecture, as the cell-based ATM transport technology is not widely used. In this architecture the IP packets are adapted to be transported over ATM cells through AAL-5 encapsulation, and the ATM cells are then carried over the STM-1/STS-3c, STM-4/STS-12c, or STM-16/STS-48c frame of SDH/SONET.[7]

8.3.1.1 Architecture and Protocol Layering

Currently the IP over ATM over SDH/SONET architecture is widely deployed in Internet backbones for data communications. An interesting point of this architecture is that it has a number of points where services and users are multiplexed. The most obvious layer is the ATM layer which allows service providers to transport both circuit and packet data traffic. So an operator can use ATM's bandwidth management abilities to divide up the

7. Note that STS-m is the electrical domain counterpart of OC-m.

network between ATM cells transporting IP packets and the ATM cells transporting circuit traffic. However, another point of multiplexing is the SDH/SONET layer. The SDH/SONET layer can be used to transport ATM cells, but at the same time it can also carry other traffic in different VC-4/SPEs (refer to Section 5.1 for terminology).

As the architecture offers multiple points of multiplexing, different legacy services may be easily added together and transported over a single network. This is because both ATM and SONET can offer integrated legacy services. Therefore, using this architecture, it is easy to offer universal services, including IP, easy to construct VPNs, and easy to offer QoS. In addition, the architecture retains the granularity and flexibility of the ATM network on one side and the reliability and survivability of the SDH/SONET network on the other.

Figure 8.9 shows how the IP packets can be carried in the IP over ATM over SDH/SONET architecture. An IP packet is first encapsulated in the AAL-5 frame, which is then segmented into ATM cells of 53 bytes, and the cells are finally mapped into the STM-1/STS-3c payload space.

Each layer in the protocol stack adds a specific functionality and advantages. The IP layer support ensures that interoperability be maximized and that the network be optimized for data transport. The use of ATM underneath the IP network ensures that a single network supporting unified formats and control mechanisms can be used for data and voice traffic. It additionally allows the offering of guaranteed QoS and bandwidth management abilities.

Using the SDH/SONET optical network layer ensures that the upper layers are provided with a high-reliability transport mechanism that will offer ease of operation and management along with an efficient optical transport mechanism. Additionally SDH/SONET offers a simple way to support circuit bandwidth management at a lower level. The last layer, WDM, offers the possibility of efficient high-bandwidth transport that is independent of the bit rate and format.

In the future IPoA over SDH/SONET over WDM networks it is expected that IPoA will be used mostly at the edges of the network, SDH/SONET transport near the edges, and WDM in the core network.

8.3.1.2 Advantages

One of the main advantages that ATM networks have over pure IP networks is the QoS support; this is due to small, fixed-size ATM cells and VC/VP-based connection setup on the ATM side and also due to the "serialization

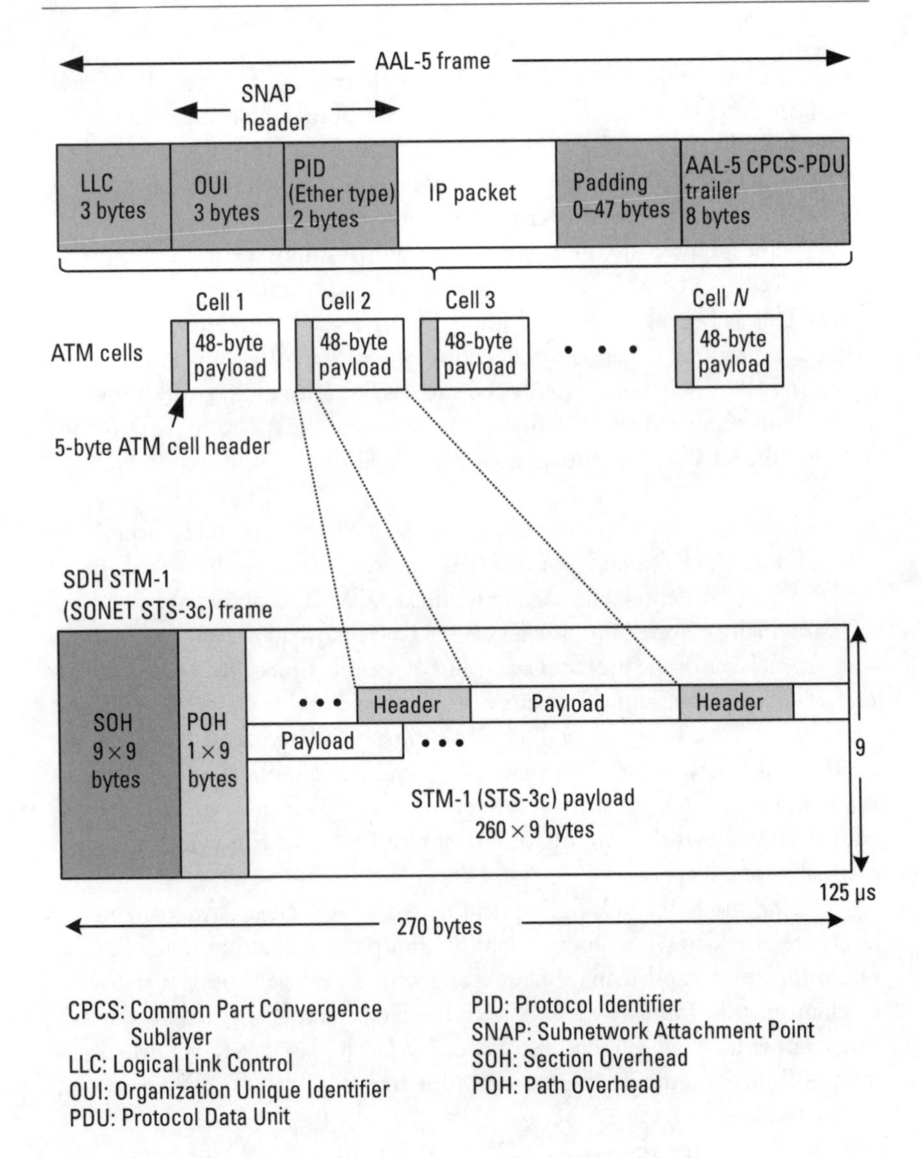

Figure 8.9 Framing of IP over ATM over SDH/SONET [23].

delay" that large IP packets can have when multiple packets are contending for transmission over a single link.

Another advantage of using ATM and SDH/SONET networks for supporting IP traffic is that both networks offer an easy management platform because they were designed to support OAM efficiently. As a consequence, ATM monitors performance of the links continuously and SDH/SONET also carries out performance, fault, and alarm tests on the links continuously. In addition, SDH/SONET enables the use of fast protection and restoration.

ATM usage also offers the advantage of what may be viewed as "virtual connectivity," as it offers the possibilities of easily constructing service or customer specific VPNs and multi-QoS networks. Such networks can be easily built using ATM or SDH/SONET connections and different user's IP traffic can run over them completely segregated without noticing such mechanisms.

As mentioned above, the use of an ATM over SDH/SONET network architecture results in the possibility of coexistence with other non-IP services in the same unified network. This would be a major cost saver for network operators.

8.3.1.3 Disadvantages

The main disadvantages of the IP over ATM over SDH/SONET architecture may be grouped into three main areas—complexity, efficiency, and scalability.

One of the main disadvantages of using this four-layer architecture is the complexity problem. As IP link speed is continuously increasing, it becomes very hard to get the SAR engines that divide up IP packets into ATM cells and work correctly at high speeds, for example, above 2.4 Gbps. Specifically, ATM SAR engines running at STM-16c/OC-48c or higher speeds do not exist today and are viewed as being very hard to implement. In contrast, network interface chips that support STM-16c/OC-48c packet over SDH/SONET interfaces already exist.

Another facet of the complexity issue is the complexity of the architecture, which means that multiplexing, traffic engineering and bandwidth management functions are duplicated partially or totally in the IP, ATM, and SDH/SONET layers. This may possibly enable dynamic and efficient operation of the network if very well managed but depending on the network operator's proficiency, it can also lead to confusion and thus very inefficient operations.

Another major problem with using IPoA networks is that the network links are not used very efficiently. This is due to the "cell-tax" which goes up as high as 20% and the duplication of functions in the network architecture. In addition, the related fragmentation issues have to be dealt with together (see Section 7.3.2).

A third problem is that IPoA networks are not very scalable. Basically for any IPoA network to be scalable, SVCs must be employed. However, problems occur because SVCs must be set up and torn down for each IP flow. ATM link needs to buffer the data while setting up each SVC, the setup time normally is long, and frequent setup/tear-down causes a heavy load on the network control nodes.

8.3.2 IPoS

Another main candidate for using IP over fiber optic networks is the idea of transporting IP packets directly over SDH/SONET networks, which is also known as POS by many experts in the field.[8] POS basically relies on the PPP for framing IP packets directly over the byte streamed transport offered by SDH/SONET. PPP with HDLC-like framing basically offers a method for transporting IP frames over the byte stream offered by SDH/SONET.[9]

IPoS has several advantageous features, most of which are inherited from SDH/SONET. The basic bandwidth management and OAM functionalities are carried over to IPoS by virtue of the underlying SDH/SONET equipment. By taking advantage of the scalability of SDH/SONET, it is possible to aggregate IP/PPP traffic up to STM-1/OC-3c, STM-4c/OC-12c, and STM-16c/OC48c rates using the current technology.

This architecture can provide survivable and reliable network services that SDH/SONET networks support. Owing to the ubiquity of SDH/SONET networks today, this architecture stands on an evolutionary development path from the existing networks. This is clearly advantageous to operators.

8.3.2.1 Architecture and Protocol Layering

PPP was originally developed and designed by the IETF as a standard method of communicating over point-to-point links.[10] It was designed as a successor to the SLIP protocol which had been developed as a simple way of transporting IP packets over serial communication lines. The initial

8. This method was detailed originally in RFC 1619 and updated in RFC 2615 [9, 11].
9. Note that PPP is used mostly as a framing method, that is, as a method for retrieving an IP packet frame from the byte stream service that SDH/SONET offers to upper layers. This is done by the use of flag bytes and FCS. See the subsequent sections for details.
10. PPP is described in a number of RFCs defined by the IETF. The main specification for the packet-over-SDH/SONET architecture is RFC 2615. Originally it was RFC 1619 but this was recently updated to include scrambling functions.

deployment of PPP was over short local serial lines, leased lines, and POTS using modems. However, the general efficacy and applicability of the PPP design allowed the protocol to be easily deployed in other environments as well. This has been especially true recently as new packet services and high-speed lines have been rapidly introduced. In many cases PPP has been used as the layer 2 framing method of choice, of which a typical example is SDH/SONET. Since SDH/SONET is a point-to-point circuit, PPP is well suited to use over these links.

Figure 8.10 shows how to map IP packets into SDH/SONET frames by using PPP in HDLC-like framing: the IP packet is first encapsulated in a PPP frame, which is then flagged with HDLC-like frame flags. The resulting frame is mapped into the STM-1/STS-3c frame of SDH/SONET. This framing for octet-synchronous links is termed "PPP in HDLC-like framing" in the RFC 1661.

PPP treats SDH/SONET transport as octet-oriented synchronous links. SDH/SONET links are full-duplex by definition. PPP in HDLC-like framing presents an octet interface to the physical layer. There is no provision for suboctets to be supplied or accepted. The octet stream is mapped into the VC-4/SPE within the STM-1/STS-3c frame, with the octet

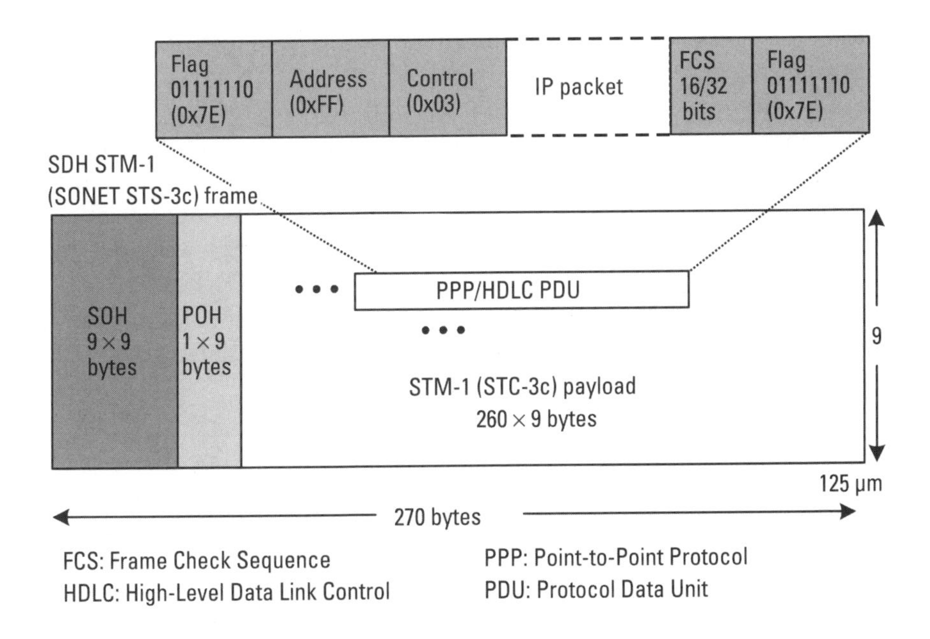

Figure 8.10 Framing of IPoS using PPP in HDLC-like framing [23].

boundaries aligned with the VC-4/SPE octet boundaries. Because frames are variable in length, the frames can cross the VC-4/SPE boundaries.

Scrambling is performed during insertion into the VC-4/SPE to provide adequate transparency and to protect the data against potential security threats.[11] The entire SDH/SONET payload (i.e., VC-4/SPE except for the path overhead and any fixed stuff bytes) is scrambled using a SSS of polynomial $x^{43} + 1$, which is the same scrambler used when ATM cells are mapped into SDH/SONET payloads (refer to Section 4.5 of [32]). The scrambler runs continuously and is not reset per frame. The initial seed is randomly chosen by transmitter to improve operational security.[12]

The proper order of operation of the IPoS system is as follows: When transmitting, the packets are handled in the order: IP packet, PPP header, FCS generation, byte-stuffing, scrambling, and SDH/SONET framing; and when receiving, the packets are handled in the order: SDH/SONET deframing, descrambling, byte destuffing, FCS detection, PPP header removal, and IP layer processing.

8.3.2.2 Advantages

Basically, the main advantage that POS has in comparison with the IPoA methods is that it is more efficient in transporting IP packets. This is due to the lower overhead of the POS architecture with respect to the IPoA network architectures.

As shown in Figure 8.10, with the POS framing, the size of an IP packet increases as much as the PPP header, FCS trailer, and the HDLC flags. This adds only a small number of bytes, especially when compared with the normal IP packet size of 1,500 bytes. In this case the overhead takes up only up to 5% of the available bandwidth, which is much lower than the ATM "cell-tax" required by the IP over ATM over SDH/SONET.

In addition, POS also takes advantages of the well defined and widely implemented SDH/SONET networks. Some critical ones among them are OAM, performance monitoring and protection switching.

---

11. For backward compatibility with RFC 1619 (VC-4/STS-3c-SPE only), the scrambler may have an on/off capability where the scrambler is bypassed entirely when it is in the off mode. If this capability is provided, the default must be set to scrambling-enabled.
12. In the SDH/SONET framing process, the POH C2 (i.e., path signal label) byte in the VC-4/SPE is encoded to digit 22 (or 0x16) when PPP is carried with $x^{43} + 1$ scrambling and to digit 207 (or 0xCF) when PPP is carried without scrambling. This label is decoded during the SDH/SONET deframing process and is led to proper follow-up actions. Refer to [32] for more discussion of scrambling in SDH/SONET.

8.3.2.3 Disadvantages

The POS architecture also suffers from several problems. Some of the problems are due to the use of PPP while others are due to the use of SDH/SONET.

While PPP is advantageous in various respects as previously noted, it is disadvantageous in that it does not have the flexibility of ATM; PPP-based links cannot offer the flexible traffic engineering capabilities that an ATM VCs can provide. For example, ATM VCs can offer various degrees of QoS and guaranteed bandwidth but PPP links as yet do not have the ability to offer this.

There are several problems that arise from the use of the SDH/SONET framing. First, even if scrambling is used, the POS scheme is still subject to sync attacks. Secondly, as tributary services require SDH/SONET multiplexing, the edges of the network may require complex SDH/SONET devices. However, the cost of SDH/SONET ADM devices is high even in the core of the network. Third, the PPP frame processing is becoming more complex as there can be multiple packets per frame, or packets can cross frame boundaries. This causes a heavy load on the equipment as each byte stream must be searched byte for byte. Fourth, there are problems in using symmetrical SDH/SONET transport paths to transport asymmetrical IP data traffic patterns. However, the traffic engineering issue related to mapping bidirectional links to asymmetric IP flows is a problem not limited to SDH/SONET networks but to all optical networks supporting IP traffic.

8.3.2.4 Other Issues of POS

POS is not a perfect solution to the problem of transporting IP packets over optical networks. While POS has numerous advantages as noted above, it has some disadvantages at the same time that become more obvious as the link speed increases. They can be divided into two types as before—one related to the use of PPP as the framing protocol and the other related to the use of SDH/SONET as the base transport link.

In the high-speed environment, the byte stuffing/destuffing operations required by the PPP in HDLC-like framing becomes expensive as they must be done by examining each byte one by one. This operation becomes more expensive as the link speed increases. The stuffed bytes also interfere with the bandwidth management strategies implemented for QoS support, because byte stuffing affects the length of the packet that is actually transmitted and consequently the amount of bandwidth used by that packet as well. Bandwidth management and scheduling are based on the assumption that a packet uses up only a fixed amount of bandwidth but this assumption does not hold in this situation.[13]

The use of SDH/SONET itself accompanies the problem of the inflexibility in bandwidth granularity. Due to the bandwidth structure of the SDH/SONET hierarchy all bandwidth must be a fixed multiple of OC-1 (52 Mbps), STM-1/OC-3 (155 Mbps), STM-4/OC-12 (622 Mbps), etc. In the aspect of efficiency, SDH/SONET has comparatively high overhead, which occupies at least 3% of the overall overhead bandwidth. The fact that SDH/SONET networks are synchronous networks that require the use of high precision clocks and clock distribution networks may become a burden to many network nodes.

As noted so far, there may be many reasons for using POS architectures, and at the same time there are many reasons not to use POS. The main reason for the latter is the cost of SDH/SONET equipment and the associated operation cost. The SDH/SONET products take up a large amount of space and power, and are expensive. However, synchronous multiplexing, which is a core functionality of SDH/SONET products, is much less demanding when transporting IP packets is of main interest. In other words, multiplexing is available at the SDH/SONET level but most of the traffic multiplexing is being carried out at the IP level and SDH/SONET is used only as a traffic link. Consequently, the high cost of SDH/SONET products may not justify their value in IP traffic dominant networks.

8.3.3 IPoW

The transmission capacity of optical fibers has been increasing dramatically, going over 160 Gbps in the case of a DWDM fiber. This huge increase of capacity is expected to trigger many changes in future communication networks, not only in transmission systems but also in network architectures. Transporting IP packets directly over WDM may be one of the major changes that we can naturally expect on the network architecture. The resulting vast link capacity might give freedom for IP technology to fully exercise its potentials, thereby resolving the QoS and other problems.

There exist no firm standards for the IPoW architecture yet but there are a number of proposals that are commonly perceived as valid approaches. In this section we consider some architectures that have been perceived as

13. These problems point to the need for a new type of framing method. In later sections we deal with two new options—SDL framing and gigabit Ethernet framing.

the most promising to gain an overall view of this field. Discussions on specifics and details of the related framing techniques will be dealt with separately in Section 8.3.4.

8.3.3.1 Architecture and Protocol Layering

There are several functional requirements that must be met to directly transport IP packets over WDM. The first issue is how to frame the IP packets that are transported. The method of framing is directly connected to the methods used for frame recognition, protection switching, scrambling, and scaling issues. The second issue is to develop efficient management and control functionalities along the lines of SDH/SONET equipment, including fault detection, recovery, and provisioning. Network transparency is another important issue, as old and new services will have to be commonly supported. Support for legacy synchronous networks and their clock distribution may be among the most important items to support. In addition, QoS is a key issue, as it is supposed to be gradually introduced into all future networks and thus must be basically supported.

Currently the main IPoW architectures mainly offer the use of point-to-point physical connections to offer large bandwidth. They usually use a simplified interface between the IP network layer and the physical layer that uses some sort of framing mechanisms. The popular framing mechanism is PPP in HDCL-like framing, but due to its deficiencies other methods are being explored. The most important candidates are the SDL technology, the use of gigabit Ethernet framing, and the recently proposed ITU-T digital wrapper functionality.

We may view the IPoW problems from two different applications—one for using IPoW in long-distance networks and the other for using IPoW in metropolitan areas. While both aim at the same objective of supporting IP directly over WDM, the needs and consequent architectures are different.

In the case of long distance IPoW networks the cost of fiber and wavelengths is very high, due to very high infrastructure cost of laying out long-distance fiber links. In such cases the majority of network cost is in fiber plants. So electronics is used for multiplexing signals to minimize fiber requirements. As a consequence, SDH/SONET multiplexers may still be used in many WDM links to raise the data rate even with IPoW. In fact some experts expect that IP/WDM may need to employ SDH/SONET as the layer 2 protocol to remain compatible with SDH/SONET and enable multiplexing to higher rates.

In the case of metropolitan IPoW networks, link distances are short, on the order of 10 to 20 km, and the rates between switches are on the order of DS-3 or STM-1/OC-3. In such a case SDH/SONET may be commonly used for electrical multiplexing. WDM direct interface appears economical at the STM-4/OC-12 level today, and at the STM-1/OC-3 level in the future. IPoW may still use SDH/SONET to remain compatible with existing equipment

8.3.3.2 Advantages

The main advantage of an IPoW architecture would be its highly simplified architecture. SDH/SONET or ATM equipment will not be needed any longer, and thus equipment cost is expected to drop significantly. The resulting network will be well optimized for IP. Nevertheless, such WDM-based infrastructure evolution will maximize the reuse and minimize the life-cycle cost of existing fiber-optic facilities.

It is noteworthy that introducing IPoW equipment will allow flexible incremental capacity growth. This is due to the inherent ability to increase WDM link capacity gradually by adding another wavelength. This is very useful for today's operators that are unable to accurately predict the growth of network traffic. In comparison, traditional SDH/SONET-based equipment had a highly structured capacity expansion plan (STM-1/OC-3, STM-4/OC-12, STM-16/OC-48, etc.) that would increase the capacity rather abruptly.[14]

WDM equipment allows multiple interface types to coexist on a single fiber. Accordingly, in addition to IPoW, there can also be IP over ATM over SDH/SONET and IP over PPP/HDLC over SDH/SONET links. It also eases the introduction of higher speed SDH/SONET or other TDM equipment into the network. This allows the use of current legacy systems while allowing flexibility to accommodate newly developed equipment.

8.3.3.3 Disadvantages

A main disadvantage of the IPoW architecture is that it is not easily connected to existing networks. Also, it has the disadvantage of not supporting

---

14. For example, in the SDH/SONET cases when an operator using a 155-Mbps STM-1/OC-3 link wants to expand the capacity to support 200 Mbps of traffic, the operator has no choice other than to use an STM-4/OC-12 (at 622 Mbps) equipment. In contrast, WDM can offer the choice of just adding another wavelength, increasing the bandwidth in units of STM-1/OC-3.

fine-granularity bandwidth management. QoS support is another major challenge.

Basically, the disadvantages of the IPoW architecture originate from the fact that the intermediate ATM and SDH/SONET functionalities are missing. From the ATM aspect, granularity and QoS become challenging issues as stated previously. Likewise, diverse OAM functionality as well as network reliability and survivability could be the issues from the SDH/SONET aspect. In addition, technological immaturity could be another big disadvantage of the IPoW architecture. Besides the optical storage problem aforementioned, there are a number of technological barriers to overcome in the optical domain (see Section 6.2).

IPoW may become a widely applied technology in the future. Currently, however, the main use of IPoW is limited to increasing link capacity and simplifying network systems. The near-term outlook is that IPoW will be used mostly as a link multiplier technology. It can be used mostly over long-distance fiber as a way to squeeze in more traffic into fiber, another way of multiplexing data on an expensive optical-fiber link.

As a medium-term outlook, we can expect that the technology will be expanded to include the development of virtual WANs and the notion of backbone peering. This can be done by operating several SDH/SONET rings on a single physical mesh, leading to multiple WANs operating over a single WDM network.

The long-term outlook is that label-switching-based systems will become widely deployed, leading to all-optical networks that are closely interleaved with the upper layer IP network. However, it must be noticed that the shortage of wavelengths also means that optical label switching (or lambda switching) may not be quite as straightforward as hoped.

8.3.4 New Framing Methods for IPoW

There are advantages to using the SDH/SONET framing method, with one of the main advantages being complete compatibility with the existing SDH/SONET networks. However, as has been pointed out above, the use of SDH/SONET framing also has several problems including the loss of throughput/performance, and the high cost of equipment. To resolve these problems, there have been developed several light-weight framing technologies, of which gigabit Ethernet and SDL are well-known. In addition, the digital wrapper framing function has been developed by the ITU-T as part of the OTN architecture; it presents a general method for encapsulating not only IP but also other digital signals such as Ethernet, STM-1 over an optical WDM network.

8.3.4.1 Gigabit Ethernet and 10-Gigabit Ethernet

Use of standard LAN frames such as gigabit Ethernet frames for encapsulating IP packets is advantageous for many reasons. First of all, they would maintain compatibility with existing LANs. This is useful as there would be no need for separate WAN interfaces from the LAN interfaces used by computers. WAN interfaces frequently need special processing such as SAR engines for ATM and byte-stuffing mechanisms for PPP in HDLC-framing. With the frame size being the same as the packet size, it is easier to implement packet switching. In addition, as the data format will be the same as the LAN format, no format translation will be needed, again making implementation simpler. Also, unlike SDH/SONET, Ethernet is a natural low-cost tributary service that does not need special tributary termination equipment such as SDH/SONET DCS equipment. Use of Ethernet can also be expected to enable the use of standard SNMP MIBs, which should make it easy to extend current network management tools to the new equipment as well.

Even today, when compared with SDH/SONET systems, gigabit Ethernet interfaces are cheaper than SDH/SONET interfaces. In addition, customers are more familiar with the simple Ethernet technology than the rather sophisticated SDH/SONET and ATM technologies. As most network management and signaling are done at the IP layer, anyone with LAN experience can build even a long-haul WAN on a given fiber.

A new gigabit Ethernet frame now being defined features reduced overhead and the use of SDH/SONET-like synchronous coding, which increases the effective distance over which it can operate. Additionally, the new 10-gigabit Ethernet, currently being defined, will be approximately equivalent to STM-64/OC-192 in capacity. This standard may possibly move into the WAN links even ahead of STM-64/OC-192 SDH/SONET systems.

Figure 8.11 shows the encoding of the gigabit Ethernet frame in comparison with those of the 100-Mbps fast Ethernet and 10-Mbps original Ethernet frames. As can be seen, gigabit Ethernet basically keeps the same encoding as the original Ethernet frame. All three cases follow the basic Ethernet frame structure of using a *start of frame delimiter* (SFD), *destination address* (DA), *source address* (SA), data, and *frame check sequence* (FCS), with a minimum frame size of 64 bytes. The EPD mark the start and the end of a packet by using control characters. The carrier extension appears only in gigabit Ethernet as explained below.

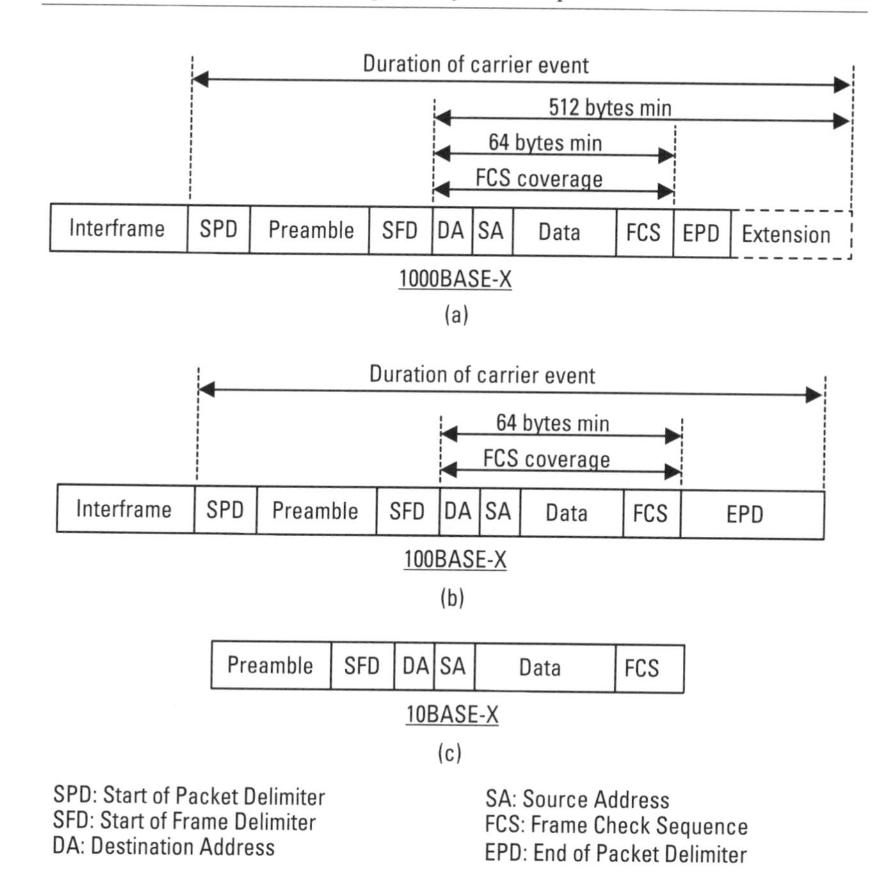

Figure 8.11 Ethernet framing: (a) gigabit Ethernet, (b) fast Ethernet (100 Mbps), and (c) Ethernet (10 Mbps).

Figure 8.12 shows the gigabit Ethernet layer diagram. Currently, the gigabit Ethernet standard supports both optical fiber and copper wire transmission technologies. The optical fiber standard uses the 1000 Base-X *physical coding sublayer* (PCS) protocol stack based on the *fiber channel* (FC) FC-1 and FC-0 standards, while the copper wire transmission uses the 1000 Base-T protocol stack, which was defined for use over category 5 UTP wires.

The fiber channel–based PCS uses an 8-bit/10-bit encoding scheme whereby 8 bits of user data are represented by a 10-bit "code group." The unused code groups are used by the PCS to support such features as autonegotiation, flow control, start/end of packet delimiters, and carrier extension characters. Additionally, the gigabit Ethernet standard uses the serializer/

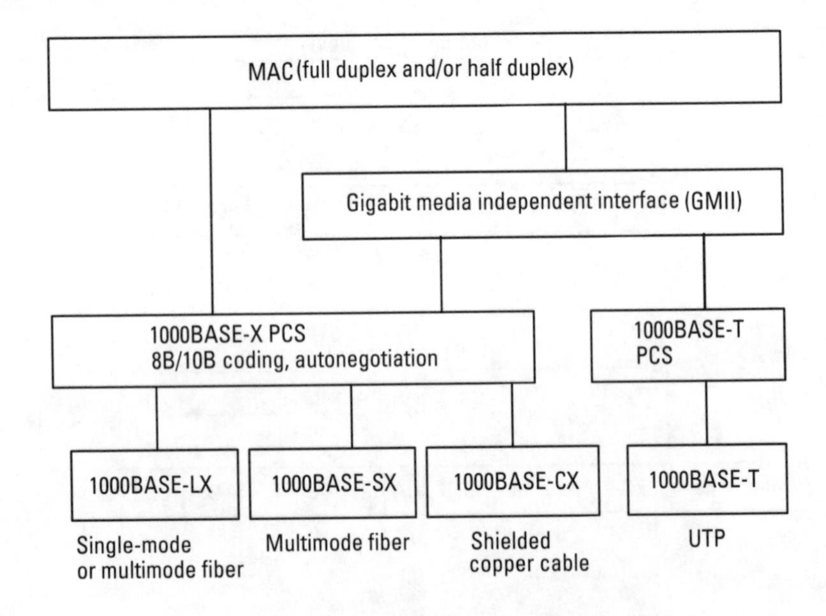

Figure 8.12 Gigabit Ethernet layer diagram.

deserializer defined in the fiber channel standard. As shown in Figure 8.12, the gigabit Ethernet standard also supports the three main types of physical medium supported by the fiber channel standard—short wavelength (SX), long wavelength (LX), and coaxial (CX).

The copper-wire stack was designed to enable the use of the wide base of category 5 UTP wiring with 4 twisted pairs that have been installed due to the popularity of fast Ethernet. Due to various physical limitations the copper wire stack will only operate up to a maximum of 100m.

Gigabit Ethernet has various other characteristics and aims [13]. First, full-duplex operation is supported. This is a natural extension of the full-duplex operation of fast Ethernet and the consequent switch-based network topologies that were developed. Gigabit Ethernet also uses flow-control mechanisms as in fast Ethernet but with the important characteristic that they are asymmetrical. Accordingly, only the switches can control the flow of traffic coming in from the attached end station, while the end station cannot control traffic being received from the network. By doing so all packets that would need to be buffered would be buffered only by the end systems and not by the switches.[15] The specification also supports half-duplex operation for shared connections in the CSMA/CD operation mode.[16] This was enabled by various changes put on the MAC layer to compensate for the inefficiencies of operating CSMA/CD over gigabit speeds.

Basically, if full duplex is supported, CSMA/CD functionality is not used by the MAC layer. However, if CSMA/CD is used the basic algorithm must be modified when the link speed extends up to gigabits. For this purpose *carrier extension* and *frame bursting* were developed. When the network speed increases to gigabits, the network diameter decreases. That is, there is a dependency between the minimum frame size and the maximum distance between the two farthest end stations. If the 10-Mbps/100-Mbps Ethernet frame sizes of 64 bytes were used at 1,000 Mbps the maximum network size would be limited to less than 20m. Consequently, it was decided to extend the size of the physical transmission frame (but not the logical frame) to 512 bytes by using the concept of carrier extension. This is shown in Figure 8.11 where the dotted line shows the carrier extension that is added to packets that are not larger than 512 bytes. This effectively results in an increase of the frame size by 8 over the 64-byte case and thereby extends the operation of CSMA/CD up to 200m.

While carrier extension solves many of the problems involved in basic CSMA/CD operation it is obvious that in the case where there are many small packets to be transmitted the frequent use of carrier extension could result in an inefficient use of bandwidth. Frame bursting was developed as a way to solve this problem. Frame bursting basically enables the gigabit Ethernet transmitter to group a number of small frames together and transmit them at the same time. This is accomplished by having the transmitter not relinquish the link after it has transmitted a packet (and the carrier extension) but, instead, transmit a control frame (thereby blocking other nodes from trying to transmit) and then transmit the next packet. In this manner the node may transmit up to 8,000 bytes of data.

There are some disadvantages to using gigabit Ethernet framing at higher speeds. Among them is the use of 8-bits/10-bits block coding, which is not an efficient coding method. Also, as with the original Ethernet standards, there is no standard for out-of-band management or monitoring. This is a feature that is provided in SDH/SONET networks and is very much

15. It was realized that by allowing the end system to control the traffic coming to itself from the network, it could happen that the packets have to be buffered in the network, leading to HOL blocking or other problems.

16. However, to the knowledge of the authors, this is not widely used in current systems.

appreciated by operators. Currently several companies have announced long-haul gigabit Ethernets and *coarse WDM* (CWDM) with transceivers at 50-km spacing. It is expected that the new 10-Gbps Ethernet standards will extend these numbers.

The IEEE 802.3ae group is in the process of defining a new 10-gigabit Ethernet standard to address these very issues [14]; 10-gigabit Ethernet is expected to operate at STM-64/OC-192 speeds. The new standard aims to maintain and use the current IEEE 802.3 Ethernet MAC protocol, Ethernet frame format, minimum and maximum frame sizes. Additionally the group has decided that 10-gigabit Ethernet will only support full-duplex operation and not support shared media, thereby officially removing the need for continued support of the CSMA/CD protocol. Essentially this enables Ethernet to be used as a pure framing format over any distance.

Currently, two families of PHY are being designed, one for LANs and one for WANs. Note that the PHY includes the PCS (which, as in the gigabit Ethernet, encompasses the coding and the serializer/deserializer functions) and the PMD layer. The LAN PHY and WAN PHY operate over the same PMDs but differ in the encoding sublayers used. The LAN PHY is expected to support existing gigabit Ethernet applications. In the future, this is expected to include pure optical switching case. The WAN PHY uses a simplified SDH/SONET framer. This should enable the direct connection of 10-gigabit Ethernet LAN modules with SDH/SONET WAN interfaces. Additionally, to enable the development of cheap modules, the standards explicitly states that the SDH/SONET jitter, stratum clock, and certain optical specifications need not be met. Essentially this leads to an asynchronous network using "cheap" SDH/SONET components. The WAN PHY will provide a subset of the SDH/SONET management information and will not support complex functions such as protection switching.

8.3.4.2 SDL

SDL is a new framing protocol to be used for encapsulating packet data over general-purpose point-to-point links [1, 12]. SDL was originally proposed as a framing means of realizing IPoS. It is now used for the implementations that use PPP over high-speed point-to-point circuits and over public telecommunications networks, both with the so-called "dark fiber."[17] SDL can

---

17. The SDL protocol was originally proposed by Lucent Technologies to various standards organizations including the IETF. It was defined over other media types and for other data link protocols, but the current RFC 2863 specification covers the use of PPP over SDL on SDH/SONET only.

be viewed as a generalization of the HEC-based cell-delineation method defined originally for ATM networks to variable-length packets. Also, like ATM, SDL uses scrambling techniques to ensure uniform transition density on the link. SDL, as currently defined, has very low implementation overhead and good hardware scaling characteristics. Accordingly, it is anticipated that significantly higher throughput can be attained, when compared to other possible SDH/SONET payload mappings, at a significantly lower cost for line termination equipment.

There are many problems with the traditional flag-based delineation methods, such as those used in frame relay and PPP/HDLC data link, and they drove the development of SDL. Flag-based packet delineation methods require byte/bit stuffing to be carried out on the packets. This results in high-complexity circuits or software to carry out real-time pattern matchings and data manipulations. Also the use of byte/bit stuffing results in the packet length increase by variable amounts, which makes it hard to implement modern QoS queuing schemes.

Figure 8.13(a) shows the basic SDL frame format, which consists of *PDU length indicator* (PLI), HEC, offset field, information field, and FCS. As shown, SDL framing is in general accomplished by putting a four-octet header and an FCS trailer on the packet. This fixed-length header allows the use of a simple framer to detect synchronization.

The PLI is a 16-bit number indicating the size of the PDU in the information field in octets. The PDU size is 4 octets at minimum and 65,535

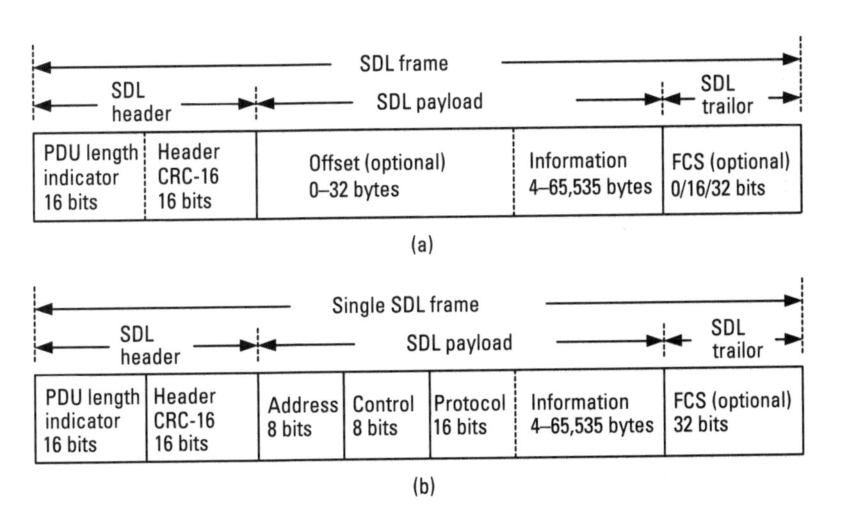

Figure 8.13 SDL frame formats: (a) basic SDL frame format, and (b) simple encapsulation of HDLC/PPP over SDL.

octets at maximum. The PLI values 0 through 3 are reserved to indicate special SDL protocol messages.

The HEC field is generated over the PLI field using the ISO CRC-16 polynomial with an initialization value of 0.

The information field consists of an optional offset field and the actual information PDU. The offset field supports some OAM functions such as message exchange or information indication regarding the PDU, such as protocol type. The size of this field must be fixed either by preconfiguration or by negotiation. The information field contains the user PDU such as PPP encapsulated IP packets. This is shown in Figure 8.13(b). The information field may support other types such as Ethernet frames or IPX packets.

The FCS field is also an optional field. When used it may be used either with an ISO CRC-16 or ISO CRC-32 polynomial to generate 16-bit or 32-bit check fields. As for the offset field, the use and the size of the FCS field must either be preconfigured or negotiated.

As mentioned above, the PLI field values of 0 through 3 indicate special SDL packets. Figure 8.14 shows those special SDL PDU messages. A PLI value of 0 indicates an idle cell to be transmitted when there is no data to be transmitted. A PLI value of 1 indicates that the payload field contains the value of the scrambler's current state. PLI values of 2 and 3 are to be used for OAM messages.

As mentioned above, a key feature of SDL is the use of the HEC-based self-delineation procedure, which was first defined for ATM. In fact the two are practically the same except that the SDL messages are of variable length (see Section 4.1.3). Figure 8.15 shows the basic state machine for the HEC-based delineation.[18]

Another important facet of SDL is that it employs scramblers to ensure adequate bit density and transparency. This need for payload scrambling is similar to the problems faced by ATM and is shown below with similar solutions. The first solution is to use an SSS. An independent, self-synchronous $x^{43} + 1$ scrambler is used on the data portion of the message including the 32 bit CRC. This is done in exactly the same manner as with the ATM $x^{43} + 1$ scrambler on an ATM channel.[19] Another solution is to use an FSS, taking preconfiguration between the two end sites. In this case the SDL endpoints both use an independent scrambler function based on a 48-bit polynomial. The scramblers in both ends seek for synchronization by periodically

---

18. The SDL delineation state machine starts in the HUNT state, where it carries out a bit-by-bit or byte-by-byte search of the data stream until a four-octet sequence with a valid HEC field value is found. Such a valid sequence is an indication that the previous four

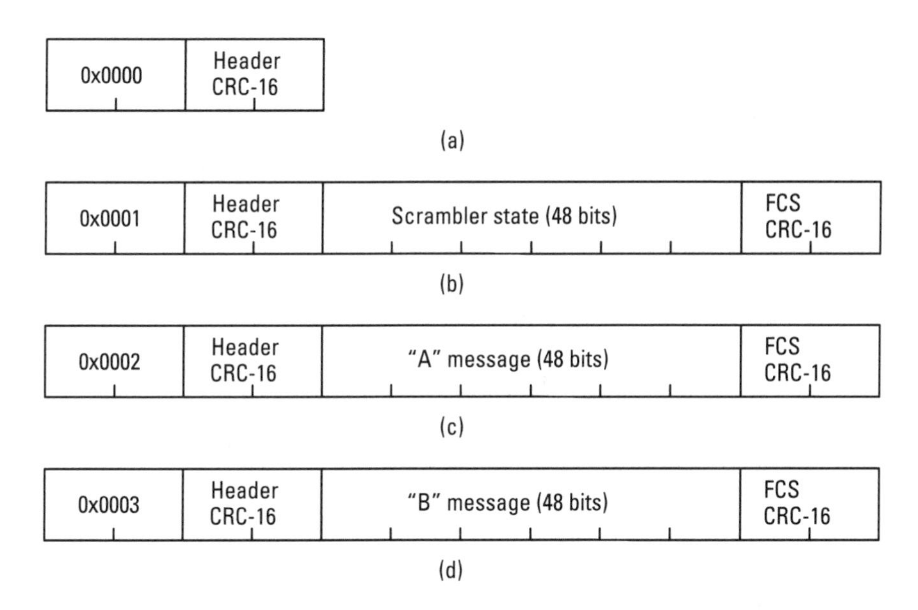

Figure 8.14 Special SDL PDU messages [12]: (a) PLI = 0 (no payload, no FCS), (b) PLI = 1 (scrambler synchronization status), (c) PLI = 2 (OAM "A" message), and (d) PLI = 3 (OAM "B" message).

exchanging the scrambler states in special SDL messages with the PLI value set to 1 (see above).

The designers have shown that SDL is scalable up to STM-64/OC-192c rates and above, ensuring that it can be a good candidate for framing in future optical networks. Additionally its inherent support of OAM and ability to support QoS implementation are advantageous.

---

bytes may be a valid SDL header. Once a valid HEC field value is calculated the state machine moves to the PRESYNC state. If N consecutive valid SDL headers are seen after entering the PRESYNC state, then the link enters the SYNC state and data transmission is enabled. If an invalid SDL header is detected, then the link is returned to HUNT state without enabling data transmission. Once the link enters SYNC state, the SDL header single bit error correction logic may be enabled. In such a case any unrecoverable header CRC error returns the link to the HUNT state, disables data transmission, and disables the error correction. Refer to [32] for the details of HEC processing in ATM.

19. The scrambler is not clocked when SDL header bits are transmitted. Thus, the data scrambling may be implemented in an entirely independent manner from the SDL framing, and the data stream may be prescrambled before insertion of SDL framing marks. Refer to [32] for more discussions on scrambling in ATM and to [33] for an in-depth description of scrambling techniques in general.

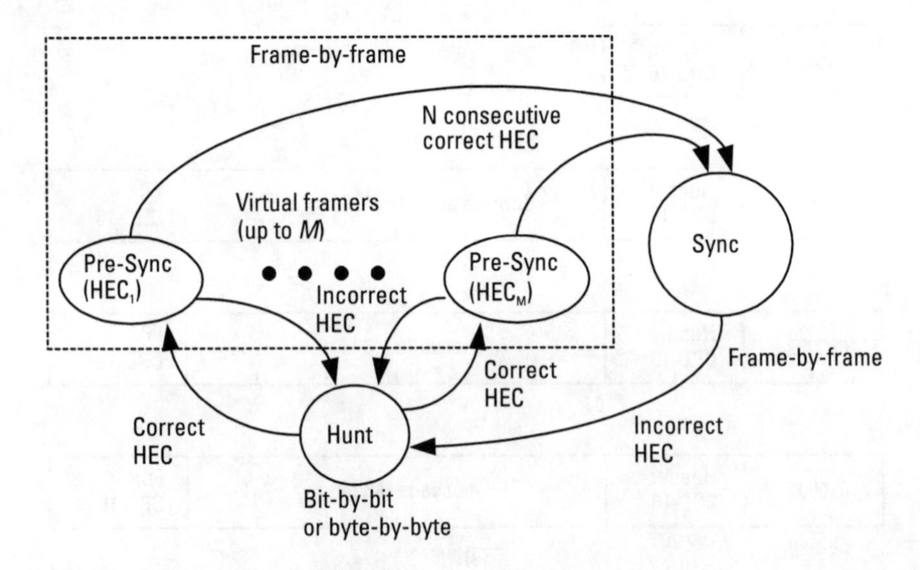

Figure 8.15 Generic SDL delineation state machine [12].

8.3.4.3 ITU-T Digital Wrappers

As explained in Section 6.6, the ITU-T has been studying new architectures for optical transport networks. One of the key ideas of this architecture was to develop methods for the transport of any type of digital signal over an optical channel. This basic digital wrapper concept and the optical transport architecture to support are defined in the ITU-T Recommendations G.872/G.709 [5, 8]. This specification aims to define a universal digital wrapper that could be used to transport all different types of digital data, while offering the advantages of SDH/SONET-like protection and restoration abilities and OAM. As most of the specific aspects of the technology are readily explained in Section 6.6, we concentrate in this section on highlighting the differences between the digital wrapper concept and the other framing technologies discussed above.

Figure 8.16 shows what the basic concept of the digital wrapper would aim to achieve. It aims to be a digital signal transport frame that can carry all forms of digital communications channels. This would include all traditional digital channels such as PDH, SDH/SONET, FDDI, and ATM, and also includes the new framing methods such as SDL and gigabit Ethernet. This contrasts to SDL and gigabit Ethernet, which are specifically designed for encapsulating packet data.

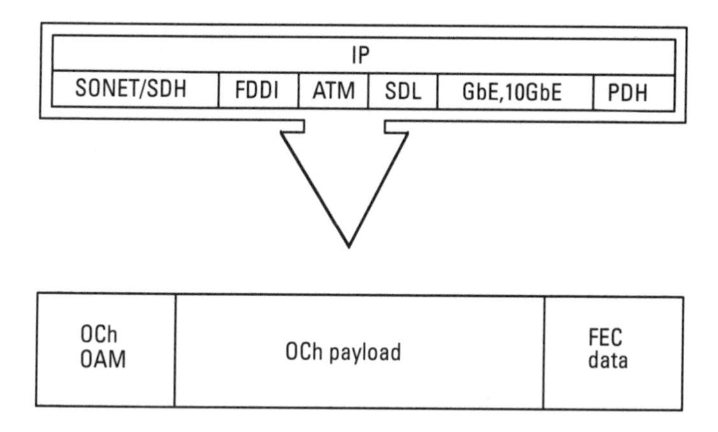

Figure 8.16 An OCh digital wrapper [3].

The basic format of the digital wrapper contains various overhead and headers for OAM in a manner similar to that of SDH/SONET. This reflects the importance in high bit-rate transport networks of built-in management capability.[20] The OTN aims to use these overhead bytes to provide a suite of standard OAM abilities that are at least as good as those provided by SDH/SONET [5, 8]. In many cases the OAM abilities of a network are intimately related to the amount of OAM overhead that the encapsulation or framing mechanism provides. We observed previously that SDL reflects this concern by allowing certain special PLI values (i.e., 2 and 3) to indicate OAM messages. While gigabit Ethernet itself does not offer much in the way of OAM in the framing itself (and as a consequence it relies on traditional IP network management tools such as SNMP), 10-gigabit Ethernet is expected to support OAM more adequately by using the SDH/SONET OAM overhead as appropriate.

The digital wrapper includes an FEC field to enhance the BER capabilities of the channel. Such a field is nonexistent in SDL or gigabit Ethernet. As such it reflects the different targets of the approaches. SDL and gigabit Ethernet were aimed at a generic network, with the basic assumption that the network's BER would be satisfactory. The ITU-T G.872/G.709 Recommendations specifically aim at the rates of 2.5 to 40 Gbps. As a result, they are willing to sacrifice some extra overhead bits to add error detection

20. Note that OAM ability is important in all networks at any speed and over any media.

and correction capabilities, as retransmission of packets at such a high speed would be inefficient.

The digital wrapper would form the OCh layer of the OTN protocol stack. The OMS and OTN layers would then use the OCh digital wrapper and other fields in the OCh header. As such it is part of a much more comprehensive network architecture than either gigabit Ethernet framing or SDL, neither of which has a comprehensive network-wide architecture plan. However, it must also be pointed out that gigabit Ethernet and SDL can both point to IP networks as being their comprehensive network architecture.

This last point brings up the final difference, which is that these technologies basically operate at different levels. As mentioned in Section 6.6, the OTN is an architecture for the future underlying optical WDM network. From the IP optical network point of view, it is the underlying transport network offering basic point-to-point link services. Both SDL and gigabit Ethernet are for framing IP packets when such a point-to-point digital link is available.

In fact, while some new method may be defined, the digital wrapper as currently defined does not have a universal standard method for delineating the frames or packets in its payload. As such it must rely on some other methods that are embedded in the payload itself. For example, SDL encapsulated packets may be transferred in the OCh payload, so that the receiving node may retrieve the bitstream from the OCh payload, and then delineate the packets by using SDL.

8.4 Routing

For routing in IP networks, a normal link state routing protocol that only supports basic connectivity information may work fine. However, this information would not be sufficient for routing in optical networks. This is because the optical path in question may be constrained by other factors depending on the technology that is being used. For example, in the case of an optical WDM network, the routing environment changes substantially depending on whether or not the OXCs in the network are capable of converting wavelengths. Accordingly, in this case routing is constrained by the wavelength conversion capability of the constituent OXCs.

As such, in optical IP networks, routing protocols must be changed to allow the exchange of information beyond simple connectivity. As connectivity information is insufficient for optical networks, the routing protocols devised for IP networks should be changed to incorporate other constraint information as well. This means that the traditional SPF routing algorithms are not suitable in current forms but must be modified to CR-SPF algorithms that accommodate optics-specific constraints.

This calls for various changes in the existing routing protocols. Information regarding the types and characteristics of the switchable channels must be exchanged. This information may include such items as the number of wavelengths available and the type of protection available along the lightpath. In response to such demands, the IETF has been extending the link state protocols to include the information needed by the G-MPLS signaling entities to determine correct paths through the network that can support a given optical path. Both the OSPF and IS-IS protocols are being extended to include this information.

8.4.1 Issues of IP Routing in Optical Networks

In traditional packet networks one of the most important aspects of routing protocols was to ensure that all the nodes in the network maintained and used the same routing table. Especially, each node along the path from the source to the destination must have the same routing table, since otherwise the packets could end up with looping and cause other routing problems as well. In fact, many of the problems observed in packet networks occur when the routing tables in the routers are not synchronized.

In optical IP networks there may be less emphasis on maintaining the same routing table than on getting the correct topology and resource availability information to each node. This is because, in the optical IP networks discussed in this chapter, we assume that the edge nodes will set up connections in a circuit-switch fashion by using G-MPLS and other methods. This means that each node in the path does not need to have the same view of the topology as the route that would have been defined by the originating source and the path information included in the original setup packet.[21]

Traditional link state protocols such as OSPF and IS-IS will supply enough information to the nodes to ensure connectivity, the ability to find a path from the source to the destination. For optical IP networks, the optical aspects, resources, and other factors can be added by using extensions to the

21. Note that this discussion assumes that a full optical packet network is not available. If the optical processing advances to the point where optical packets may be processed in the same manner as the packets are processed today electronically, then the salient features may become more similar to today's packet networks.

routing protocols—for example, the OSPF TE and IS-IS TE extensions. BGP may also be used to pass routing information and extended with extensions to add optical specific information. Based on this information the constrained path route computation can be done.

Based on these observations, we may consider various models of routing information exchange in optical networks [31]. We consider the optical network model in Figure 8.1. This model assumes that optical paths can be dynamically set up by the client edge routers (i.e., the routers connected to the edge of the optical network) using UNI signaling. To make this possible, the signaling request must specify the destination endpoint in the optical networks (as mentioned in Section 8.2). In addition, the characteristics of the link, which may include various optical aspects, may have to be specified as well. Then the problem is how the client will become aware of such information.

There are basically three possible ways for the client to be aware of this type of routing information. First, the information may be configured in the client. Second, it may be obtained by running a limited-function routing protocol across the UNI that only conveys the reachability information. The border routers would then have to run some sort of routing protocol among themselves based on this information. Third, it may be obtained by the client edge router and the border OXCs (i.e., the OXC at the edge of the optical network) by running a full functional routing protocol across the UNI that conveys all topological and routing related information. The basic possible routing models that we discuss below are based on these possibilities.

8.4.1.1 Aspects of the Integrated Routing Model

As mentioned in Section 8.1, the integrated routing model assumes that a single IP routing protocol instance is being run on the optical network and the IP network. As such, all the IP routers and OXCs will have the same information with regard to the topology of the whole network. Hence the "integrated" nature of this routing model is that the IP routers and the OXCs essentially use a single routing protocol.

To ensure the integrated nature, the IP routing protocol is assumed to have been extended in such a way that it can capture the optical specific parameters and constraints in addition to the normal IP-related information. By using such an extended routing protocol, the resulting topological and link state information of all the IP routers and the OXCs will become identical. Note that in contrast to traditional IP networks where this identical topological information base would have been critical for blocking routing loops, this information now forms the basis for the edge routers to set up optical lightpaths across the optical network.

While such an integrated architecture enables utilization of an expanded amount of routing information for efficient route calculation, it also results in a "flat routing organization." That is, it results in the cases where there is less hierarchy in spreading network routing information, with the result that many nodes have more information than they can use. In other words, it can result in IP routers and OXCs maintaining topological information that is not meaningful in its own network. For example, the IP routers would have to maintain optical domain-specific parameters such as link BER for which they would not have any use or understanding.

In the following we discuss how to set up a lightpath over an optical network where the integrated routing model is used. As an example, we consider an optical IP network that contains routers R1, R2, and OXCs O1, O2. Suppose that R1 is connected to O1 through ports (or wavelengths) P1, P2, and P3. Then the information that there are three possible links between R1 and O1 going over a single physical link must be distributed by the routing protocol. Let a lightpath be set up from R1 to another router R2 across the optical network, with port P1 used between R1 and O1. Then this information must be propagated into the link state routing protocol as a *forwarding adjacency* (FA) and must be advertised as a virtual link in the integrated routing protocol.[22] The FA must advertise that a link is now available between R1 and R2, and, in addition, that P1 is now available between R1 and O1. As such, this overlaid set of virtual links represented by the FA can replace the optical connectivity information distributed by the OXCs in the optical IP network.

Usually we can expect that the lightpath may contain more bandwidth than a single LSP can use, and in this case the availability of that bandwidth must be signaled in the routing protocol. The specific methods and messages for conveying such information will be covered in the following sections where the changes being studied for Internet routing protocols such as OSPF are discussed.

8.4.1.2 Aspects of the Domain-Specific Routing Model

As mentioned previously, the domain-specific routing model assumes that separate routing protocols are being run on the optical networks and the IP

---

22. An FA refers to a virtual link advertised into a link state routing protocol. It is formed when an optical lightpath is formed across an optical network between two IP routers. It essentially is a virtual link in the link state routing database. Refer to [25] for more details.

network. However, the two routing protocols are able to exchange information over the boundary between the two networks by using another routing or signaling protocol. The resulting topological and link state information of all the IP routers will then be sufficient to route through the optical networks. So the domain-specific routing model may be considered to be a mix of the integrated routing model and the overlay routing model.

An advantage of using the domain-specific model is that it may be easily extended to encompass the cases of NNI routing protocol exchanges between different optical subnetworks. This is based on the possibility of future optical network models where the optical networks themselves are multiple but under different administrative control.

There are two possible ways of supporting a domain-specific routing model as described above. One is to use BGP and the other is to use OSPF (or IS-IS) with extensions to support hierarchical routing.

Domain-specific routing using BGP. An obvious method of supporting domain-specific routing is to use BGP. As discussed in Chapter 3, BGP is the routing protocol developed in IP networks to enable the exchange of routing information between adjacent networks running different routing protocols or different instances of the same protocol.

The use of BGP would be achieved if each router on the edge of the optical domain advertises the IP addresses that it can reach into the optical domain by using BGP messages. The routers will also receive information on what IP addresses are reachable through the optical domain by receiving BGP messages. The border OXCs would exchange the reachability information by using BGP as well. The edge routers would be running the *external BGP* (EBGP) to communicate with the border OXCs, while the border OXCs would have to use *internal BGP* (IBGP).

It is necessary to identify the egress points in the optical network corresponding to externally reachable IP addresses. This mapping is necessary in the cases when an edge router sets up a path through the optical domain to a destination first and then sets up another path to another destination later. The first path will pass through the ingress border OXC, the optical domain, and finally the egress border OXC. The edge router must know whether or not the path to the second destination passes through the same egress border OXC as the first path. If so, another lightpath setup is not needed; otherwise, a new path is needed. By keeping track of the egress border OXC for an IP destination, this problem may be solved. This is similar to the problem that happens when implementing support for VPNs in BGP [34].

Domain-specific routing using OSPF/ IS-IS. Both OSPF and IS-IS support hierarchical link state routing. We consider the OSPF case only as the IS-IS case is very similar. In the case of OSPF it has a two-level hierarchy model, consisting of a backbone OSPF network and all other OSPF networks that are connected to the backbone. The routers on the edges of the backbone networks, called *area border routers* (ABRs) exchange *summary LSAs* that contain information on what IP addresses are available in one area to another (refer to Section 3.2.2).

In the case of the domain-specific routing model using OSPF, the client networks and the optical networks may operate as separate OSPF areas and exchange routing information by using a similar mechanism to that described above. The easiest way to see this is to assume that the backbone network of an OSPF network is changed into an optical network, with each of the connected optical switches participating in the OSPF protocol. Additionally, each of the IP routers on the edge of optical network also participates in the OSPF hierarchy as ABRs and floods summary LSAs into the "backbone" optical network. The optical network passes these summary LSAs into each of the connected client networks by way of the edge router to which the client network is connected. The information includes the address of the source ABR (for this summary LSA) and a metric describing distance to the source ABR.

8.4.1.3 Aspects of Overlay Routing Model

As mentioned previously, the overlay model assumes that the IP network is unaware of the routing topology of the optical network. There are two approaches to manage routing in this environment—one by using a *static method* and the other by using a *signaled model.*

The static method assumes that the optical paths between IP endpoints are set up over the optical NMS. The NMS system may change the optical paths over time but, overall, the time scale may be expected to be long, for example, more than 24 hours.

The signaled model assumes that there are methods by which the edge routers can register their addresses and also methods by which the addresses of other edge routers may be found. Based on the findings on the address of the opposite end point to which the connection must be made, the edge router may set up a connection. This basically assumes that it is easy to set up optical paths between optical network endpoints in a dynamic fashion.

As the overlay routing model is similar to the IPoA overlay model, similar architectural methods may be used to solve the routing problem. For example, there may be a server in the optical network that can be used by edge routers to register and query addresses. The edge router would query the server for the router address of the edge router with which it would like to connect. A successful response would come back with the exit OXC and port number. The signaling is defined over the UNI. The optical paths themselves must be set up or released dynamically by a separate signaling protocol that is specific to the optical network itself.

Both the ODSI and the OIF are defining optical UNIs. The ODSI's solution is based on the use of existing IP protocols. The end points are specified by using an <IP address, port number> pair. PPP is used for service discovery so that the endpoint can discover whether it can use ODSI signaling protocols.[23] The OIF's approach is to use the G-MPLS (or a variant) as the signaling protocol at the UNI. This approach would result in the use of a single signaling protocol over both the UNI and the internal optical networks in the near future, assuming that G-MPLS is adopted by optical network cores. In such a case, this path would be advantageous as only a single signaling protocol would be used over the whole network, which will later facilitate an evolution to an integrated routing model.

8.4.1.4 Partial Peer Routing Model

The complete solutions discussed above are complicated and may not be simple to implement. Even the domain-specific routing model, which forms the architectural basis of the UNI signaling based approaches (ODSI and OIF), may be hard to implement in the near future. Partial peer routing is an attempt to define a simpler model that can be implemented more easily in the near future [31].

Partial peer routing is similar to the PNNI PAR routing and also to BGP/MPLS VPN routing, but has a simplified form. It uses the concept of virtual private networks specific to optical networks, or *virtual private optical networks* (VPONs).

Each border router belonging to a VPON registers a set of <IP Address, VPON identifier> pairs with a border OXC by using UNI signaling. This information is propagated across the whole optical network, so that all border routers will know the mappings between the VPONs and IP addresses. Based on this information lightpaths may be set up between the border routers belonging to the same VPON. Appropriate routing protocols for each

---

23. The ODSI uses a new signaling protocol that is defined in [17] to request the establishment of optical paths.

VPON may be run on top of this first layer virtual network. Each of the optical links between the border routers of each VPON would be viewed as a virtual link in the OSPF database.

As separate routing protocols can be run for each VPON, the partial peer routing method may be viewed as a way of running multiple overlay routing models over a single optical network.

8.4.2 Extensions to Existing Routing Protocols

Currently there are two main link state protocols, OSPF and IS-IS. Both are being extended to support traffic engineering by using extensions. Note that the routing protocols can support optical networks by adding optical-specific information fields to the *traffic engineering* (TE) extensions. In this section we consider the OSPF TE extension case only as the IS-IS TE extension is expected to closely follow this [35, 36].

8.4.2.1 OSPF Extensions for TE

OSPF TE extensions essentially provide a method for describing and distributing parameters and topology information related to traffic engineering, such as bandwidth and administrative constraints.[24] This can be done by using opaque LSAs, which are special LSAs defined with a standard LSA header followed by 32-bit application-specific information field (see Section 3.2.2). The traffic engineering LSA is a type 10 opaque LSA with an LSA ID field containing "0x0100" followed by a 16-bit instance number. TE links are advertised as being additional properties of normal OSPF links. As such, an OSPF adjacency is first brought up and then TE properties of the link are advertised.

The payload of the traffic engineering LSA contains two *type length value* (TLV) attributes. The first TLV is the router address TLV identifying the IP address of the advertising router. The second TLV is the link TLV composed of a number of sub-TLVs. The sub-TLVs contain information that is useful in TE considerations, such as link type, link ID, traffic engineering metric, and others.[25] The link type differentiates between point-to-

24. Refer to [36] on the basic extensions to OSPF to support TE.

25. More specifically, the sub-TLVs contain link type, link ID, local interface IP address, remote address IP address, traffic engineering metric, maximum bandwidth, maximum reservable bandwidth, unreserved bandwidth, and resource class/color. Refer to [36] for more details.

point and multiaccess links. The link ID identifies the other end of the link, which can be either the router ID of the neighbor or the interface address of the designated router in multiple access networks.

8.4.2.2 OSPF Extensions for G-MPLS

The OSPF TE extensions described above assume that the nodes at each end of the link are normal routers that are capable of understanding IP packets and, furthermore, view each other as being OSPF adjacent nodes. However, the optical networks of the future, as envisioned under G-MPLS, would modify these constraints on the TE link in several ways. First, an LSP may be advertised as TE links, which means that the nodes at each end of the link will not be OSPF adjacent nodes even though they are the two ends of a TE link. Second, TE links may be comprised of more than one link between nodes as DWDM is introduced. Third, in many optical networks, though there is an LSP between two nodes, the nodes may not be able to understand packets as they may not be packet-switching capable LSRs. Consequently, G-MPLS TE links extend this model to include more general LSPs and TE characteristics made available in optical networks.[26]

OSPF Routing Enhancements

The basic approach is to further extend the TE extensions that were added to OSPF so that G-MPLS may be supported more easily. The outgoing interface identifier, incoming interface identifier, maximum LSP bandwidth, link protection type, link descriptor, and shared risk link group are added to the sub-TLVs available for inclusion in the link TLV to support G-MPLS.

The outgoing interface identifier and the incoming interface identifier identify the interfaces of the links at an LSR. The maximum LSP bandwidth corresponds to the maximum link bandwidth sub-TLV defined before. The link protection type sub-TLV indicates the protection capability of the link. The link descriptor sub-TLV contains various characteristics of the link, including the link type, priority, and bit transmission rate in terms of maximum-reservable and minimum-reservable rates. The shared risk link group sub-TLV contains the SRLG values of the link.

---

26. The G-MPLS-related features are closely related to the G-MPLS concepts introduced in Section 8.6. The reader is advised to read Section 8.6 for concepts that are not clear. Refer to [25] for details about the extensions of OSPF to support G-MPLS.

SRLGs and Routing

SRLGs are useful when attempting to route two paths between a source-destination pair such that the paths do not share any single link. In this way, the failure of any single link will cause only one of the paths to fail, leaving the other to be used. Links may belong to multiple SRLGs, so the SRLG sub-TLV must be able to indicate all the SRLGs to which the link belongs.

When an LSR attempts to route multiple diversely routed LSPs to another LSR, the path computation must attempt to route the paths so that they do not have any links in common—that is, so that the path SRLGs are disjoint.

8.5 Switching and Multiplexing

Switching and multiplexing are two of the basic functions of all network systems. The two functions are closely related in that the switching mechanisms are strongly governed by the multiplexing mechanisms. Depending on the format of data generated by multiplexers, the methodology of switching and the structure of switches are determined. Multiplexing itself originates from the sharing concept, such that the transmission channel or bandwidth are shared among multiple different users. As a consequence multiplexers generate a time or frequency shared format of data stream and transmit through transmission systems to switches where connection is established among the desired user pairs or groups.

In the case of IP over optics, or IPoW systems, multiplexing function is readily incorporated in the communication architectures and the relevant protocol formats. Revisiting the architecture to realize IPoW in Figure 8.7, we can recognize that, in this case, multiplexing, or bandwidth sharing, can take the form of ATM, SDH/SONET, or WDM. Accordingly, switching in the case of IPoW may take place in the level of ATM, SDH/SONET, or WDM/optics. Accordingly, to discuss switching for IPoW it suffices to consider switching in those three levels.

First, as to switching in the ATM level, there are various ATM switches readily available. ATM switching renders a fast switching by taking advantage of the well-established circuit-switching technologies. The small and fixed-sized ATM cells make this possible. Therefore ATM is definitely a strong candidate for switching if IPoW relies on IPoA or IP-over-ATM over SDH/SONET architecture. However, it is not applicable if other choices of architectures are made such as IP over SDH/SONET or direct IPoW. (Refer to Section 4.3 for further discussion of ATM switching.)

Second, switching in the SDH/SONET level is more for ADM or *digital cross-connecting* (DXC) in the VC/VT level of the STM-*n*/STS-*m* signals than switching in the low-level data carried in the VC/VT. In case the carried data is TDM data such as voice traffic, the ADM or DXC can help the traffic reshuffled and sent to the desired voice traffic switches. Likewise, if data traffic is carried, a similar arrangement is possible to send them to the desired packet switches. In fact, ADM and DCSs play the central role in the SDH/SONET networks, and are widely deployed in central offices. (Refer to Section 5.4 for more detailed discussions on ADM and DCS for SDH/SONET.)

Third, the switching in the WDM level may be similar to that in SDH/SONET if the carried data is in SDH/SONET format. That is, there may be more interest in add/drop multiplexing or optically cross-connecting the constituent high-level data, if they are in STM-*n*/STS-*m* signal formats. Unlike the SDH/SONET case, however, the incoming signals are supposed to be spread in different wavelengths, so OADM and OXC systems are needed for the add/drop and cross-connect operations, respectively, in the wavelength level or STM-1/STS-1 level. Lower-level add/drop and cross-connect functions may be supplemented by putting in additional ADM and DXC systems, respectively. There are OADM and OXC systems readily developed but are still evolving. (Refer to Section 6.4 for more details on OADM and OXC.)

When considering IPoW/optics, more interest lies in the case of direct multiplexing where IP packets are directly multiplexed into wavelengths. In this case, packets may be encapsulated in gigabit Ethernet frames or SDL frames, but not in ATM or SDH/SONET frames. This is the case where optical packet switching is very much in demand. Unfortunately, however, there are high technological barriers to overcome to make this happen, including the optical buffering or storage problem [37–41].

In this section we discuss two approaches to switching for direct IPoW among the multitude of proposals available in the literature—one using fixed-length packets and a synchronous mode of operation, and the other one using variable length packets and an asynchronous mode of operation. The former is the OTPN switching and the latter is the OBS [42–45].

8.5.1 OTPN

OTPN is an optical packet network that basically uses fixed-length packets in synchronous mode operation [42]. OTPN uses the packet format shown in Figure 8.17(a). As shown, the optical packet has the following characteristics: It uses a fixed time slot model where each packet takes up a fixed amount

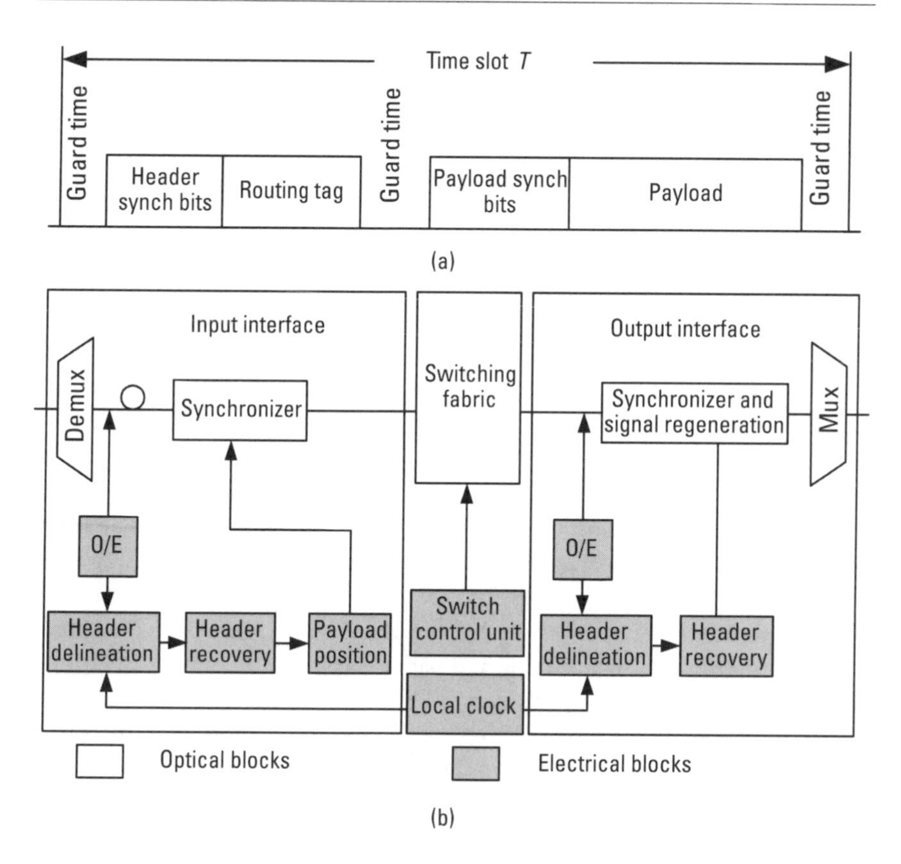

Figure 8.17 Switching in optical transport packet network (OTPN): (a) OTPN packet format, and (b) OTPN switch architecture [46].

of time. The packet header contains the routing tag and is encoded at a fixed bit rate. The packet payload has a fixed duration but is encoded at a variable bit rate. A distinctive characteristic of the structure is that guard times are specifically inserted into the structure. This is to handle the random jitter and the non-ideal nature of the synchronizers in the input/output interfaces.

Figure 8.17(b) shows the OTPN switch architecture. The switch is composed of three main blocks—*input interface, switch fabric,* and *output interface.*

The input interface uses a demultiplexer to split the incoming header from the payload. The payload is delineated and aligned with the packets coming in from all the other input ports by using an optical synchronizer. The optical synchronizer is used in order to keep payload transparency and to minimize O/E conversions. The header part is converted into an electronic signal, which is then used to control the switch fabric for the optical switching.

The optical switch fabric routes the packet payload toward the correct output port. It retains the ability to solve the contention resolution problems that may arise when packets are simultaneously routed to the same output port. The broadcast-and-select type of switch is used often for the optical switch fabric (refer to Section 6.3.1).

The output interface resynchronizes the output packets as they might have taken different paths with varying delays while passing through the switching fabric. Additionally the output interface rewrites the optical packet header and regenerates the optical signal.

As can be inferred from the above descriptions, synchronizers are critical in the operation of OTPN systems. All-optical synchronizers are the key to guaranteeing payload transparency. Synchronizers are important in the input interface as they must compensate for the jitter among the signals arriving over the serial input. The jitter may be caused by temperature variations, wavelength speed variations, and other reasons. The output interface synchronizes the jitter caused by buffering and path delays within the switch fabric.

There are two major problems to OTPN. One is that it must use constant packet size. As mentioned above, the packet size is constant with respect to time and uses variable-rate encoding to support different rates or packet sizes. This increases the complexity of the end systems. The other problem is that it must use optical packet synchronizers. Fabrication of all-optical packet synchronizers is a difficult task as they still take the form of optical delay lines.

8.5.2 OBS

OBS is another form of optical packet switching that contrasts to OTPN; as opposed to the fixed-length and synchronous mode of operation of OTPN, OBS takes variable-length and asynchronous mode of operation. "Burst" in the name OBS refers to the variable-length packets. It also uses a dual-wavelength approach for decoupling the control packet (or packet header) from the packet payload.

In OBS data are switched in the unit of bursts. Before each burst arrives at the OBS, connections of the OBS are set up by a control packet that is sent ahead of the optical burst. This control packet is usually sent in a separate channel. Figure 8.18 illustrates the operation of the OBS.

The front part of an OBS includes a demultiplexer that separates out the control header from the data burst. The control packet is converted into

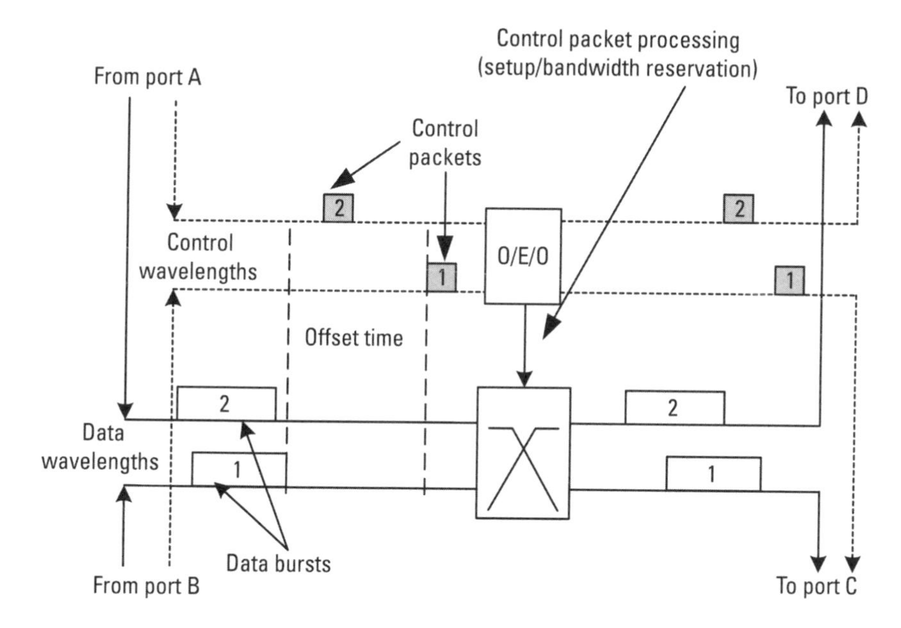

Figure 8.18 Operation of OBS [43].

an electronic signal and processed electronically. It is reinserted into the front part of its related data burst at the output port. The control packet contains various pieces of information regarding the data burst, including the routing information. Based on the offset time and other information in the control packet, the controller sets up resources and connections in the switch fabric to appropriately route or delay the data burst.

QoS in an OBS depends on how successful the controller is in scheduling the various data bursts that arrive at the OBS. A critical factor is the contention resolution ability of the switch fabric, for example, how much buffering is available. Another is the offset time between the control packet and the data burst. Depending on the length of time between the control header and the data burst, the controller may have more time to schedule resources for critical time-dependent packets.

The basic OBS structure can easily realize IP routing-like functionality by using IP addresses in the control headers and IP address lookup mechanisms along with IP routing protocols in the control plane. The controller would just use the IP address lookup tables to decide where to route an incoming data burst. However, another more efficient method of integrating OBS and IP networks can be found in utilizing the MPLS.

OBS can be easily integrated with MPLS to construct an integrated IP optical switch. Figure 8.19 shows the functions of the OXC that support the integrated operation of OBS and MPLS. As shown, the basic model of label swapping can be applied to the output port and wavelength scheduling function. Routing information in the control packet would just be in the form of MPLS labels, according to which the control plane of the OBS would select wavelengths, output ports, and new control packets.[27]

8.6 Signaling and Control

The signaling and control models heavily depend on the architecture assumed. As noted in Section 8.1, there are a number of organizations that are currently in the process of defining signaling and control protocols for

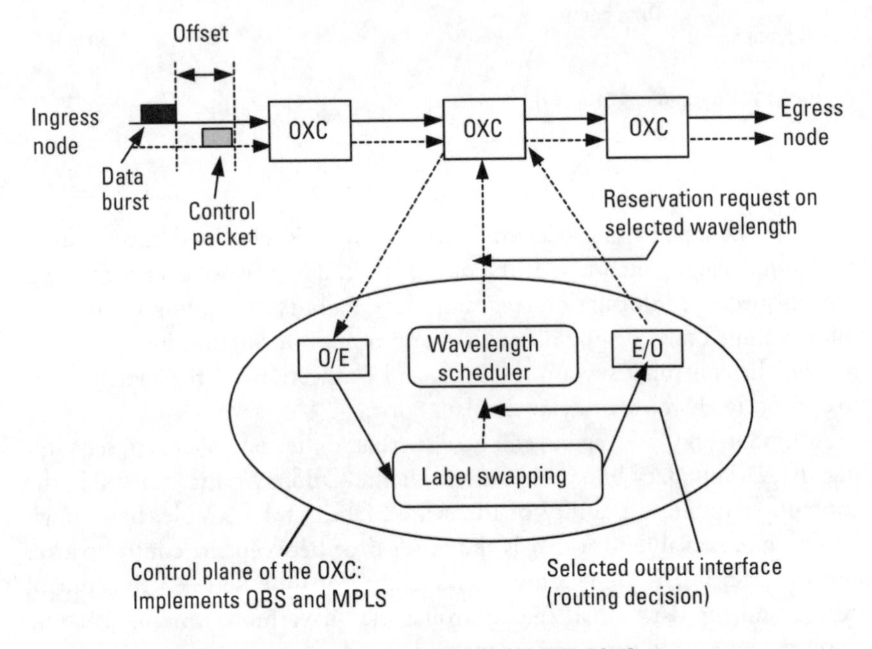

Figure 8.19 Functions at the OXC supporting OBS and MPLS [45].

---

27. Note that in the case of a non-MPLS OBS OXC, the routing information in the control packet would be in the form of routing addresses based on which the OBS controller would route the packets. In contrast, the MPLS OBS OXC would expect to find labels in the control packet.

the optical Internet. Among them we will consider the protocols being worked out by the IETF, the OIF, and the ODSI in this section.

First, the IETF's approach is to define an extended version of the MPLS protocol suite, called G-MPLS, to use as the signaling and control protocols in all future optical networks [22]. It is expected that by using a single control plane it should be possible to control both MPLS-based IP LSRs and G-MPLS-based OXCs as well. This IETF approach concentrates only on defining the G-MPLS and the relevant extensions needed for optical networks. It does not define any architecture for the optical networks.

The OIF's approach is also founded on the IETF's approach using G-MPLS. The OIF is working closely with the IETF to define the G-MPLS standards and at the same time it defines two architectures—the integrated model and the domain-specific model [20].

The ODSI only supports the domain-specific model and does not try to use the G-MPLS specification. Note this does not necessarily mean that the OXCs inside the optical network may not use the G-MPLS control plane. The ODSI is concentrating on defining a couple of new extensions to existing IP protocols, such as PPP, to enable a quick standardization of a UNI specification with the aim of facilitating the implementation of a domain-specific routing model–based optical Internet.

8.6.1 G-MPLS and IETF's Signaling and Control Protocols

As originally defined, MPLS can only be used for systems where the packets or cells carry labels based on which the switching or routing function is to be performed. These switches or routers are called LSRs reflecting the fact that label is the key element. These systems transmit data in delineable formats such as cells or packets, assuming that the LSRs are able to process the packet or cell headers that carry the labels.

In reality, there are many systems that are logically similar to MPLS networks and LSRs but that do not transport packets or cells and do not carry out processing at the LSRs. Specifically, there are many cases where the forwarding decision at a switching system is made, not based on information carried in the packet or cell, but based on the time slots in which the data arrived, the wavelengths over which the data was transported, or physical ports/fibers through which the data arrived.

From those examples we realize that the label-based forwarding and the LSP concept, which form the basis of MPLS, were essentially conceived as a way of introducing connection-oriented mechanisms into connectionless IP networks. As such, the root structure is similar to traditional connection-oriented networks that are basically circuit-switched networks. Time slot switching networks and WDM networks are some variants of them.

G-MPLS extends MPLS in such a way that it can be used in time division networks (e.g., SDH/SONET), in optical networks (e.g., WDM networks) and in spatial switching networks (e.g., incoming port or fiber to outgoing port or fiber). G-MPLS originates from the concept of *multiprotocol lambda switching* (MPλS), which was the first attempt to extend MPLS to other switching systems, primarily to optical networks containing fast dynamically configurable OXCs. Figure 8.20 shows an MPλS network structure. As shown, it consists of an optical lambda-labeling MPLS subnetwork that forms the core network and an exterior IP network at the edge, which is not necessarily MPLS-capable. There are multiple types of LSRs. Normal routers with MPLS LSR functionality can exist in the exterior IP network. Edge MPLS *optical LSRs* (O-LSRs) can exist on the edges of the core network, performing label merging and tunneling to and from larger bandwidth (or coarser lambda) optical LSPs. Core MPLS O-LSRs can exist in the core network, which would be pure optical switches (or OXCs) that perform lambda switching.

Based on the concepts presented above, we may classify switching systems into four categories: *Fiber-switch–capable* (FSC) *switches* based on the

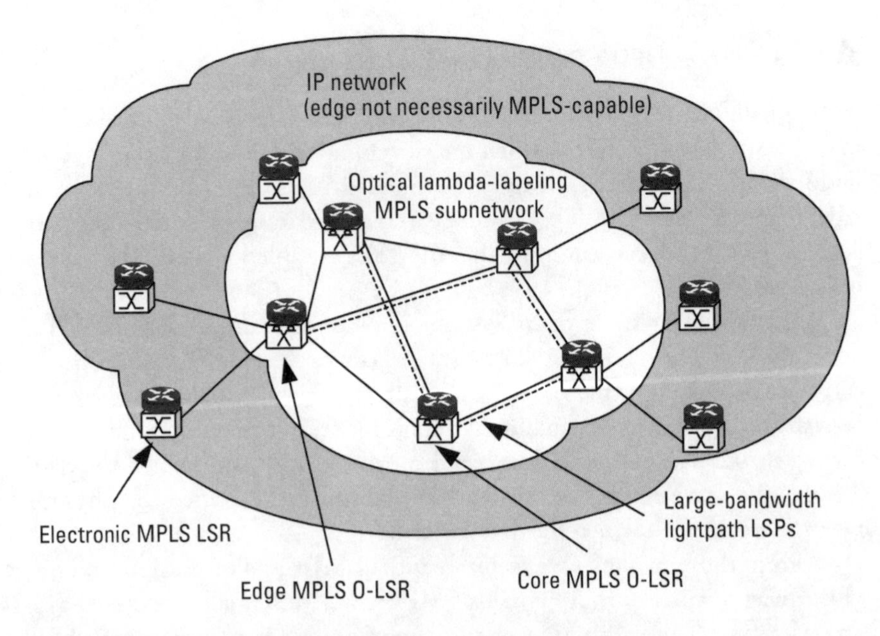

Figure 8.20 Illustration of MPλS and MPLS networks interaction [4].

physical port or fiber (e.g., OXC); *lambda-switch–capable* (LSC) *switches* based on the optical wavelength (e.g., WDM switches); TDM-*capable switches* based on the time slots (e.g., all TDM switches related to SDH/SONET); and *packet switch–capable* (PSC) *switches* based on the contents of the packet headers or cells (e.g., packet switches and routers). In terms of these four categories, the original MPLS dealt only with the establishment of LSPs over PSC switches. However, G-MPLS aims to extend the signaling and control functions to the other types of switches.

The conception of expanding to multiple types of LSPs naturally leads to the concept of expanding the forwarding hierarchy or LSP hierarchy, as different types of LSRs will generate different LSPs, which may in turn be nested to form a hierarchy of LSPs. For example, packet routers on the edge of an optical network may initiate and terminate LSPs. The packet routers may also be connected to SDH/SONET ADMs, which also initiate and terminate LSPs over part of the PSC defined LSP. The SDH/SONET LSP may in turn be made up partially of LSPs constructed between WDM nodes that form a network beneath the SDH/SONET ADMs [26].

The main changes to MPLS to realize G-MPLS can be grouped into four classes. First, there are a number of enhancements made on the signaling and basic control protocols which are carried out mainly by extending the existing RSVP and LDP protocols. Second, a new LMP is designed to solve various problems related to link management. Third, the IGP (i.e., routing protocols, such as OSPF and IS-IS) are extended to support optical networks in general. Fourth, various scalability enhancements are added to support optical networks and their consequences.

Figure 8.21 shows the resulting protocols and functions in an MPλS network, in which the constituent LSRs and MPLS-aware OXCs are interconnected by optical links. There are multiple links (i.e., link bundles) connecting the OXCs and the LSRs in Figure 8.21, but between each OXC and LSR pair there is a single control channel (the dotted line). The signaling protocols and other control protocols including the routing protocols are carried over this control channel. In Figure 8.21, the signaling protocol used between the OXCs and LSRs is either RSVP or LDP, while the routing protocol used is either OSPF or IS-IS. The signaling packets are sent end-to-end while the routing protocol packets are broadcast and thereby both types of packets have significance beyond the immediate neighbors. In contrast, the LMP protocol packets have significance only between two neighboring nodes. This is because the LMP aims to monitor only the immediate low-level links between two adjacent nodes and not to monitor any higher-level paths.

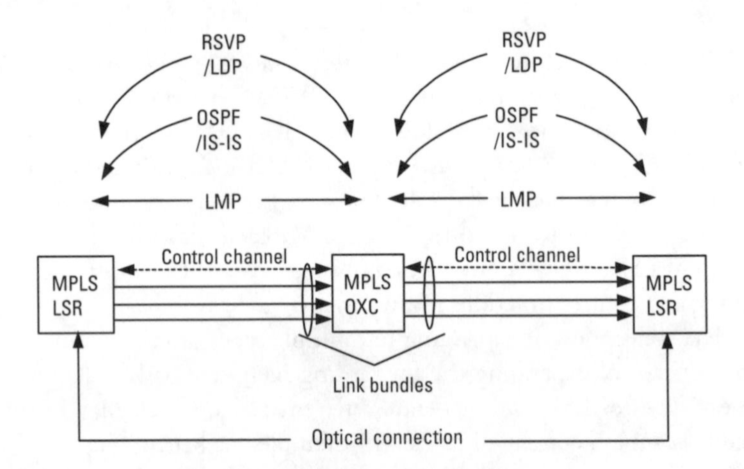

Figure 8.21 Protocols and functions in MPλS networks [47].

8.6.1.1 Signaling-Related Enhancements

Basically G-MPLS is obtained by newly defining some extensions to the basic MPLS messages. Some fields are added and some new methods are to be defined. All changes are made to both RSVP and CR-LDP so that either one of the signaling protocols may be used in G-MPLS networks. The changes are related to how labels are requested, how the requests are communicated, the unidirectional nature of LSPs, how errors are propagated, and so on [21, 23, 24].

In G-MPLS all LSPs must start and end on similar LSRs. This is similar in logic and reasoning to the requirement that all LSPs that carry IP traffic must start and end on some routers. That is, such LSPs may not start or end on ATM-LSRs having no IP modules. This implies that the forwarding planes of MPLS LSRs need not have the ability to recognize IP packet boundaries.

G-MPLS extends MPLS in that the LSP may be defined to include not only heterogeneous label encoding links (for example, links between ATM LSRs may be followed by links between traditional routers) but also the cases where the links are label-switched based on wavelength or time slots. However, G-MPLS LSP has the restriction of being able to support only a limited number of bandwidths. Consequently only a limited number of labels can be supported by a G-MPLS LSR.

G-MPLS extends the signaling ability of MPLS so that the upstream node may suggest the label used by the downstream node. This can be a valuable feature in the cases where the LSRs are limited in the rate at which LSPs can be set up. In addition, G-MPLS also enables upstream nodes to limit the range of labels that can be selected by downstream nodes. This may be applied to the whole or part of an LSP path. This feature would be useful in optical networks where only a small number of wavelengths are available.

G-MPLS supports the establishment of bidirectional LSPs. This is frequently needed in the case of non-PSC applications. By introducing the ability to set up bidirectional LSPs, setup latency may be reduced and setup messages may become shorter as well. Also G-MPLS aims to support rapid failure notification by adding some messages.

Among the changes the G-MPLS adds to MPLS, the most important are the changes on the labels and the way the labels are set up and released. The information in the new generalized labels must be sufficient for the generalized LSRs to set up its switches or cross-connects correctly between the related input and output "ports," whether the "ports" are fiber ports, time slots, or traditional MPLS LSPs.

However, the basic operation does not change; the ingress node generates a generalized label request containing information pertaining to the type and characteristics of the LSP needed. This is sent downstream in the normal path (in the RSVP case) or REQUEST (in the CR-LDP case) message. Each receiving node checks whether the parameters are supportable by the incoming and outgoing interfaces. If there is any problem, a path error/notification message is generated. Otherwise, the egress node generates a reservation/mapping message and sends it upstream toward the ingress node.

8.6.1.2 Routing-Related Enhancements

One of the main enhancements to MPLS introduced by G-MPLS is related to the concept of LSP hierarchy. This is based on the concept that an LSP may be contained in another LSP. This concept leads to the use of LSPs as links in the OSPF/IS-IS link state database [22, 25].

Figure 8.22 shows the organization of LSPs in relation to the switching categories discussed above. The inner FSC cloud offers links that can be used by the higher LSC cloud. As such, the links are called *forwarding adjacency LSCs* (FA-LSCs). The LSC cloud in turn offers a TDM link to the TDM cloud that can be considered as an FA-TDM. The TDM cloud in turn offers an FA-PSC to the PSC cloud. We observe that each LSP in the lower layers can bundle or carry multiple LSPs to form the higher layer. For example, multiple PSC LSPs can be tunneled in a single TDM LSP. In a similar manner, multiple TDM LSPs can be carried in a single LSC LSP.

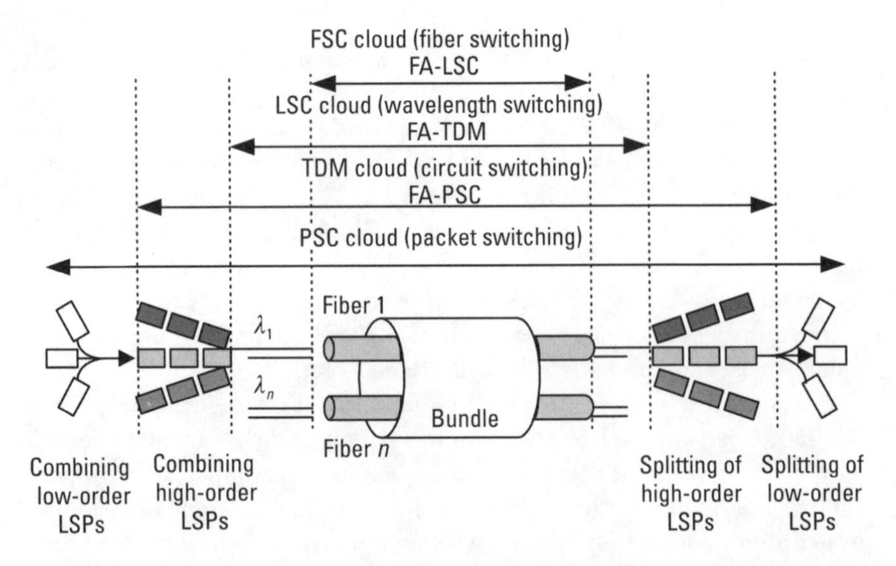

Figure 8.22 Organization of LSPs in relation to switching categories.

The hierarchical arrangement of LSPs helps to resolve the mismatch between the number of MPLS labels and the limited number of labels available in G-MPLS, as multiple MPLS LSPs can be aggregated into a single optical LSP at an optical node.

The hierarchical approach also helps to resolve the bandwidth mismatch problem. As mentioned before, the optical LSPs may only support discrete sets of bandwidth, whereas MPLS LSPs will support a much wider granularity. By grouping an appropriate number of MPLS LSPs it is possible to ensure that as a group the MPLS LSPs fully utilize the bandwidth made available by the optical LSP.

The LSPs are kept in the OSPF/IS-IS link state database of each router or OXC. New link types are defined for OSPF/IS-IS and included in the flooding messages. For computing new paths based on this information, the fact that there are some constraints in using links to route paths must be incorporated in the computing algorithm. For example, the order of LSP tunneling must be kept [25, 48].

8.6.1.3 Management-Related Enhancements

The LMP provides four basic services—namely control channel management function, link connectivity verification, link property correlation, and fault isolation [49].

Control channel management is built on a hello type protocol and is used to establish and maintain connectivity between two adjacent nodes. LMP allows the control channel to be decoupled from the data channel. Link connectivity verification verifies the physical connectivity between two nodes. By using LMP link summary message, it is possible for adjacent nodes to keep track of the correlation functions of links. This is always done on start-up and any time thereafter. LMP enables a mechanism to isolate link and channel failures.

LMP basically allows two adjacent nodes to automatically define and maintain component link IDs that may be used as labels for physical resources such as the ports of switch, wavelengths, and time slots. This solves the major management problem that can arise with respect to the addressing of each individual end point.

8.6.1.4 Scalability-Related Enhancements

There are two basic problems for using MPLS when optical networks are widely used, namely, *link bundling* and *unnumbered link* problems. Both are due to the possibility of many fibers being used to simultaneously connect to switches or routers.

In the situation where hundreds of fibers with each containing hundreds of wavelengths exist in the network, the link state database that each node has to maintain has to increase to a huge size. A solution to this situation may be found in aggregating the link attributes of several parallel links of similar characteristics and adding them to the network's link state database as a single "bundled" link [50].

Another problem in this heavily congested fiber/wavelength situation is that it is not possible to assign IP addresses to each link; this is called the unnumbered link problem. A solution to this problem may be found by assigning each node and link a unique identifier in the following two steps: First, identify each node in the network uniquely by using the IP router ID.[28] Second, identify the links from a particular node. This is done by giving each unnumbered link on an LSR a unique 32-bit identifier and defining extensions to RSVP-TE/CR-LDP to carry information on unnumbered links.[29]

---

28. Note that all IP router IDs are basically IP addresses that are globally unique.
29. Specifically, the explicit route object and record route object are extended to carry this 32-bit identifier. In addition, the routing protocols must also be extended to carry information on these unnumbered links. These are defined in the OSPF/IS-IS TE extensions. Refer to [50–52].

8.6.2 OIF's UNI Signaling Protocols

As mentioned earlier, the OIF is one of the organizations that is driving standardization efforts in support of optical networks based on IP protocols. The OIF is concentrating first on the domain service model described in Section 8.2. Accordingly, it concentrates on defining signaling and routing interfaces between the client and the optical networks through the definition of a UNI signaling protocol.

The OIF is defining the UNI protocol based on the G-MPLS efforts of the IETF. Among the standards documents it has generated and submitted to the IETF, there is a base document on signaling requirements regarding the *optical UNI* (O-UNI) [15]. This document is based on the UNI service model defined by the OIF. It aims to make interoperation with the G-MPLS-based framework currently being defined by the IETF relatively easy. There are two main items of interest in this UNI document: First, the document aims to ensure that the new signaling abilities being developed in the G-MPLS group are in harmony with the requirements of the optical UNI defined by the OIF. Second, the document aims to harmonize the lightpath parameters used in the signaling messages so that they will be the same as in the OIF UNI and the IETF G-MPLS.

As discussed in Section 8.2, realizing the domain service model involves defining a set of optical network services at the optical UNI that the client can access to set up LSPs across the optical network. The services include lightpath creation, lightpath deletion, lightpath modification, and lightpath status enquiry.

In addition, an address resolution procedure may also be needed so that the client may find out the address of the nodes that can be reached across the optical network. Address resolution requires a number of messages. First, all clients must use a client registration message to register the address. Second, a client deregistration message must be available. Third, to actually resolve an address, an address query message is needed.

End system discovery and service discovery procedures may also be used. End system discovery allows clients and network nodes to bootstrap themselves across the UNI. Service discovery allows for the discovery of what services are available across the UNI. Both of them may be considered to be part of the problems being solved in the G-MPLS architecture by developing the LMP protocol. Note that these procedures themselves are separate from the signaling protocols across the UNI.

One important fact to note is that the OIF protocols aim to be a subset of the overall G-MPLS standards, and thereby enable operators with a smooth transition path to future systems that implement the full G-MPLS standards. This is one of the main reasons why the OIF is concentrating on the domain service model first.

8.6.2.1 Direct Interface and Indirect Interface

The O-UNI defined by the OIF has flexibility in the type of interface: It allows for direct interface as well as indirect interfaces. A direct interface implies that there is a direct connection between the client and the network OXC over which all control messages are transmitted. This connection may use in-band or out-of-band signaling, but the connection must be directly between the both sides of the O-UNI. In contrast, the indirect interface allows the client to communicate with a third party with regard to the data connection between the client and the network node. This would be the case when optical lightpaths are set up through communication with a management station in the optical network.

8.6.2.2 IP Control Channel

The OIF requires that an IP control channel be available across the O-UNI to carry the signaling and control messages. This control channel is supposed to carry the signaling and control messages in the form of IP packets between the client and the network. Accordingly, the link should be securely managed by putting some authentication means across the UNI. In addition, it should be arranged such that any failure in the link be quickly detected and alerted to the opposite side.

8.6.2.3 O-UNI Signaling

The signaling messages needed for the domain service model include lightpath create request/response, lightpath delete request/response, lightpath modify request/response, lightpath status request, and notification messages. Note that these are abstract messages, so the actual messages that are defined in LDP or RSVP will be different in name and format. It is important that changes to CR-LDP and RSVP must be made in such a way that the end result is functionally equivalent to the above [15, 19].

The UNI signaling messages must also contain a number of parameters. These include information on identification, services, and other related needed items. The identification parameters include lightpath ID, source/destination point of attachment, user group ID, and UNI client IDs. The service-related parameters include the directionality of the lightpath, the framing type, the type of overhead termination, available lightpath bandwidth, propagation delay, and service level. Other signaling parameters include diversity, result codes, and status.

8.6.2.4 LDP-Based UNI Signaling Protocol

By using the LDP for the optical UNI signaling a number of advantages are gained, mainly in that it is possible to reuse LDP messages and formats, management and control procedures [53]. In support of this, only a couple of changes are needed: a few new TLVs to support the attributes needed by the O-UNI and a couple of new LDP messages for the exchange of lightpath status information.

The LDPs can be extended for use as the O-UNI signaling protocol. As noted, most of the abstract messages may be defined easily by using existing LDP messages. For example, a lightpath create request/response can be realized by using the LDP label request message and the LDP label mapping message. The lightpath delete request/response can be realized by using the LDP label release request message and the LDP label withdraw message, while the lightpath modify request/response can be realized by modifying LSP characteristics with CR-LDP. The notification message can be realized by using the LDP notification message, but the abstract signaling message and lightpath status request/response cannot be realized by any pre-existing method in LDP. Consequently, a new lightpath status enquiry and status response message or equivalent must be defined [23, 53].

8.6.3 ODSI's Signaling and Control Protocols

The ODSI develops signaling and control protocols that aim at quickly facilitating a bandwidth-on-demand paradigm at the boundaries of the electrical and optical network. Essentially this would be where electrical network nodes such as IP routers would interconnect with optical switched networks. Using the ODSI protocols the network elements at the edge of the new intelligent optical networks should be able to establish new lightpaths or modify existing lightpaths in real time. This maps most naturally to a domain service model from the service model point of view, and an overlay routing model from the routing architecture point of view. Functionally, it is similar to the basic form of the OIF initiative.

The ODSI basically assumes that in the optical network there exist intelligent optical switching nodes or OXCs that are able to support dynamic reconfiguration and setup of lightpaths. It also assumes a highly software-centric architecture in contrast to the existing optical transmission equipment such as SDH/SONET ADMs. In fact, this software is probably the most important factor in the new optical switches as they enable the new signaling and control architectures being planned to support circuit provisioning, protection, and restoration functionalities. In addition, the new OXCs also differ from traditional OXCs in that they support not only sophisticated software control interfaces but also a much larger number of ports.

The ODSI initiative aims to use IP protocols and addresses. This is based on the assumption that the control network used in the optical network is also IP based. Consequently, by using the same protocols and addressing over the UNI, interoperation between the two will become much simpler.

In addition, instead of defining new IP protocols, the ODSI aims to use pre-existing IP based protocols as much as possible and only add modifications or extensions as needed. This highlights a difference between the ODSI initiative and the OIF and IETF initiatives. The latter two aim to define and use a new protocol suite based on G-MPLS. Though the protocols developed by the ODSI such as the extensions to PPP may be viewed as being new as well, it should be noted that G-MPLS is based on a completely new protocol MPLS, which in itself is still in the process of development. Notice that the ODSI initiative does not rule out the possibility of using G-MPLS as the control plane architecture internal to the optical network.

8.6.3.1 Functional Description

Figure 8.23(a) shows the basic framework of the ODSI [18]. As mentioned before, it assumes a domain service model. The main network components are routers, the *optical network controller* (ONC), and the optical network. The optical network is used to set up point-to-point links between user devices on the edges of the optical network. The point-to-point link that is set up as a result is similar to a leased line.

An ONC exists in the network to participate in the ODSI signaling on behalf of the optical network. This may be a centralized management station or a distributed function that exists across all optical switches. Figure 8.23(a) indicates this by locating the ONC outside the optical network and connecting it to all the optical switches by dashed lines. This entity understands the signaling messages and can set up, release, and modify connections as requested.

Figure 8.23(b) shows an example of on-demand establishment of high-capacity leased lines that the ODSI protocols aims to support. Router *A*

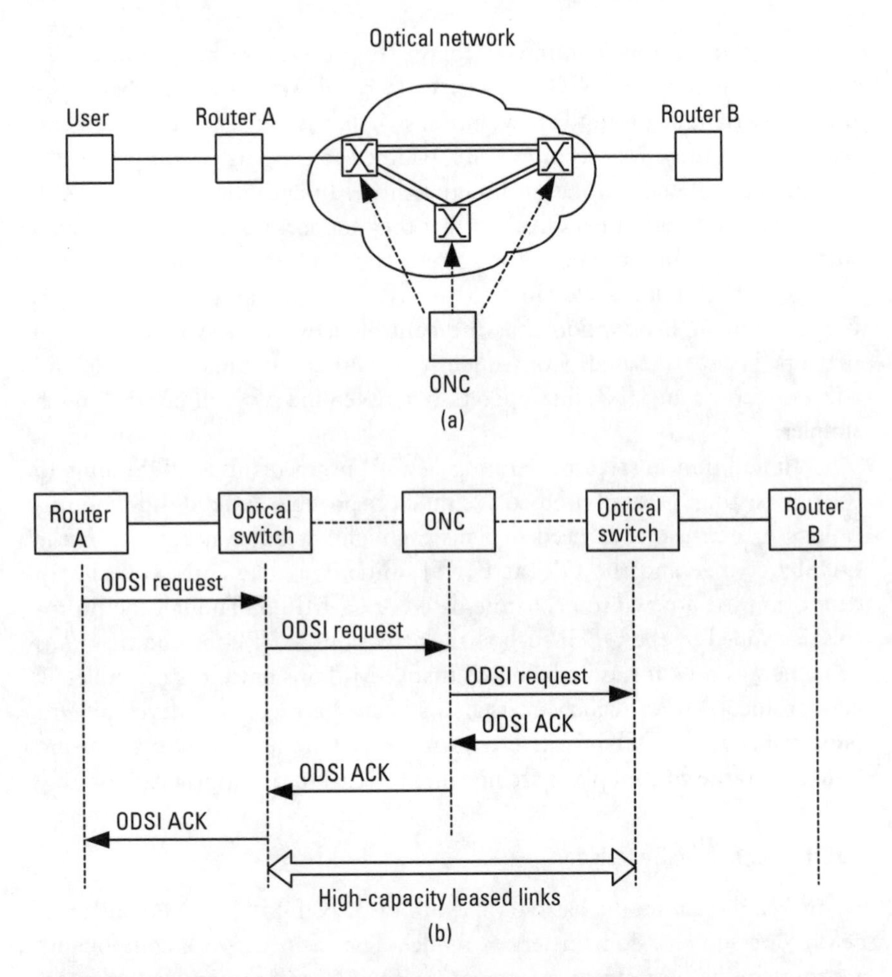

Figure 8.23 Basic framework of the ODSI: (a) network configuration, and (b) establishing leased-links by ODSI protocols.

sends an ODSI request message to the optical switching node that is nearest to it, and the optical switching node sets up an on-demand optical circuit to the optical switch nearest to the destination router *B* by going through the ONC. Before setting up the optical connection, the ONC first validates the connection request and determines whether a path exists for the connection that meets the user's requirements. The ONC also informs the two endpoints of the request, and only upon getting acceptance from the two endpoints does it set up the connection.[30]

In a similar manner to the OIF, there are a number of parameters used for describing the characteristics of these channels. They include the physical layer type, size, priority, protection, propagation delay, jitter, and error rate. The physical layer type indicates what type of encoding is to be used including SDH/SONET and Ethernet. The size indicates the bandwidth size, and the protection indicates the level and type of protection the circuit needs.

The use of the protocols defined by the ODSI is limited to the boundaries between the optical network and the electrical network. The optical network itself would use an internal control architecture separately from that of the electrical network.

As the optical network only provides a lightpath, it is basically circuit-like in behavior and does not offer any functions that are packet-based in nature. All packet-based processing must be carried out at the user devices on the edge of the optical network. Channelization of the link is also done by the user level protocols. An example of this would be the *data communication channel* (DCC) in the SOH of SDH/SONET.

As mentioned before, the ODSI architecture basically uses IP protocols and addresses so as to simplify interactions with the control plane of the optical network. This implies that the IP addresses assigned to the user devices serve two purposes: one as the IP address for ODSI control packets and the other as the endpoint identifiers for circuit setup. For the endpoint identifier, port index, channel, and user group are also used. User group is a qualifier for an IP address that limits which ports can be attached to it. This is used to divide user groups in a manner similar to VPN.

8.6.3.2 UNI Functions and Protocols

The ODSI protocols are basically grouped into a number of functions, such as *service discovery, address registration, signaling, access control* and *accounting*. Additionally MIBs are defined so that the network may be managed. Service discovery and address registration rely on the use of an in-band control channel between the user device and the connected optical network node. In contrast, signaling can be run through either the in-band or out-band channels through the IP-based control network. For example, an Ethernet connection may exist between the user device and the control network of the optical network.

30. This case would be an example of a third-party traditional UNI interface. Other methods can also be supported, including the support for legacy devices that do not understand the ODSI interface.

Service discovery protocol enables a user device to find out if it is connected to an optical network. It also helps to know if the user device can use ODSI protocols and certain parameters that it needs to know to use those services. The parameters passed over include the IP address and port identifiers of the user device and the network device, and the user group of the user device's port. Additionally the network device provides the user device with the address of the optical network controller to which signaling requests must be sent. The service discovery protocol is based on PPP running over the in-band channel.

Address registration enables a user device to assign IP addresses to a port. The IP addresses are used as the endpoint identifiers when setting up connections over the optical network. As for service discovery, address registration protocol also uses PPP running over the in-band channel. It is possible for the user device to advertise one IP address for a number of ports or multiple IP addresses for a single port.

The signaling function enables the user device to set up connections to other user devices connected to the optical network. For the signaling function, the ODSI interface offers, similar to the UNI defined by the OIF, a set of interface actions that the client may request. These are *create*, *destroy*, *query*, *modify*, and *directory lookup*. The create request is used to setup a point-to-point link between any two user devices, while the destroy request is used to release such a connection. The query request is used to query the status of the connection and the modify request is used to modify the current parameters of the connection such as the bandwidth or priority. The directory lookup request is used to obtain a list of end points to which the user device may set up a connection. Notice that these functions are functionally equivalent to the abstract signaling messages (i.e., *lightpath creation*, *lightpath deletion*, *lightpath modification*, and *lightpath status enquiry*) defined in Section 8.2.

The ODSI signaling has a number of characteristics: First, the ODSI signaling protocol is a TCP-based application, so the data elements carried include the endpoint identifiers and bandwidth characteristics of the connection. Secondly, ODSI signaling is intrinsically third party signaling in that neither of the connection endpoints need to be the original requesting end points.

The ODSI signaling messages may follow a different path from the user data connection. This is easily observed in cases where an out-of-band control channel is used between the user device and the optical network. While the user devices and network devices are presumed to be completely reachable and routable in the control network, the same assumptions do not apply to the optical paths that are set up over the optical network.[31]

8.6.4 Using G-MPLS as the Control Plane for SDH/SONET

Another research item that deserves special attention is the use of G-MPLS as the control function for SDH/SONET networks. This refers to the use of G-MPLS as a control and signaling protocol to set up and control SDH/SONET paths. As such, it deals purely with the control function of SDH/SONET networks. This can be viewed as one method of setting up and maintaining the SDH/SONET paths that need to be set up in the IP over SDH/SONET over WDM network case.[32]

As explained before, the use of G-MPLS for SDH/SONET networks is complicated when SDH/SONET labels are considered. SDH/SONET labels have the property that they would have to support a method for identifying timeslots in the synchronous multiplexing structure of SDH/SONET. As explained in Section 5.2, this multiplexing structure is very complex to build up basic DS-0 channels up to STM-16/OC-48 and beyond.

Additionally, if signal concatenation or inverse-multiplexing is used in the SDH/SONET link, the generalized label may contain a list of labels. For concatenation the list would be the explicit list of all signals in the concatenation. For inverse-multiplexing, the list of all signals that are part of the inverse-multiplexing must be included in the generalized label as well. Inverse-multiplexing requires that multiple LSPs must be open in parallel simultaneously [6].

8.7 Traffic Management and QoS

As noted in Chapter 2, traffic management and QoS guarantees in networks are a combined result of data plane functions and control plane functions.

The data plane functions consist of various peer packet functions and actions that occur at the data path of each node on small time scale, usually acting on a packet level. Consequently, unless optical processing and

31. Admission control and accounting are optional features. If implemented, they are to be based on the common open policy service (COPS) protocol being defined by the IETF [54].

32. As noted in Section 8.3.3, for the IPoW case that only uses SDH/SONET encapsulation, there may be cases where sophisticated SDH/SONET ADMs or DXCs are not needed.

buffering become possible, there will be very little traffic management or QoS strategies implemented inside the optical network.[33] The control plane functions consist of various signaling and control actions that are needed to collect and distribute information on the status and availability of resources in the network, and the distribution of policy and various methods for setting up and allocating resources for specific needs. Obviously this entails a very large amount of interaction among various entities of the network, including the optical layer, higher layer, and various management and control functions.

Figure 8.24 illustrates this interaction between the various entities of the optical layer and high layer (or IP layer). The two layers basically follow the client-server model of interaction. The higher layer routing and traffic engineering protocols request the setup or modification of lightpaths to the optical layer. The optical layer would be running various WDM RWA

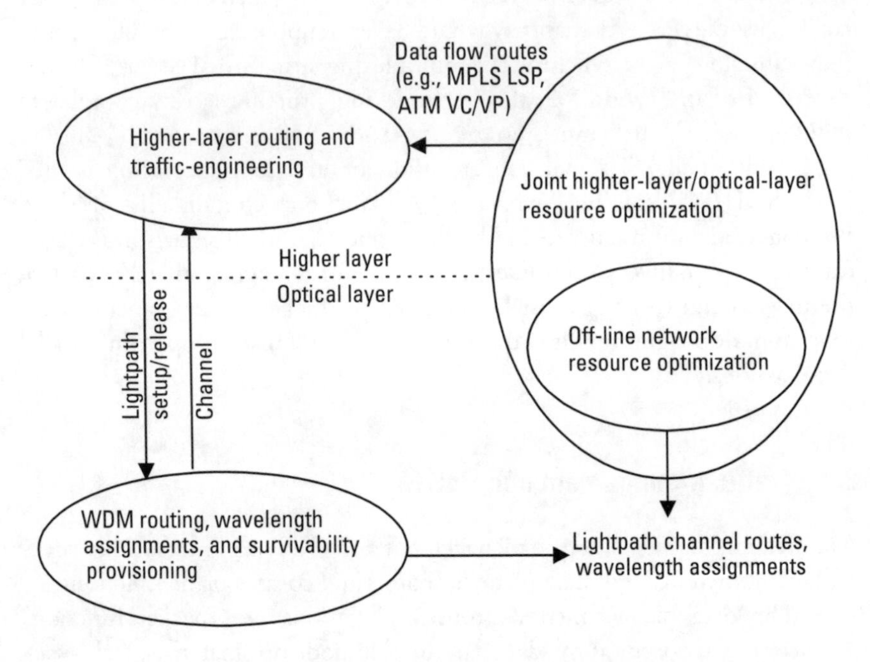

Figure 8.24 Illustration of interactions between optical layer and higher layer [4].

33. Note that the OTPN and OBS network architectures examined in Section 8.5 offer some possibility of implementing such complex functionality in the electronic part of their data plane. However, it should be noted that these essentially enable complex control algorithms, but without solving the more complex problem of buffering the optical packets [55].

algorithms and survivability provisioning algorithms. These would help in allocating specific routes and resources for specific lightpaths in an optimal fashion. At the same time an offline resource optimization algorithm would be running continuously, calculating the best resource allocation for the network by taking into account the current network status, current and predicted loads, and other data. Both the off-line and the higher layer planes would receive real-time updates from the optical layer on the current status of the network. Then the higher layer would react in a more timely but less optimal manner, while the off-line planes would react more slowly but more optimal way.

As can be seen, the problem of traffic engineering is a big one. Basically it involves three stages. The first stage involves the distribution of information relevant to TE. The second stage involves the analysis of the current network state based on this information. The third stage involves the network state modification based on the previous analysis. The second stage is a process that each network operator will do separately and as such is not subject to standardization. We show in the following how by adapting MPLS TE methods the first and third stages of the basic TE problems can be solved in optical IP networks that use a G-MPLS control plane. In addition, we show in a similar vein how optical IP networks that use the OTN model and a G-MPLS control plane can achieve QoS along the DiffServ QoS model.

8.7.1 Traffic Engineering

It is one of the main objectives of the MPLS to enable operators to implement means for traffic engineering in IP networks. The control plane of MPLS traffic engineering is composed of several components including *resource discovery*, *path selection*, *path management*, and others [28].

Resource discovery refers to the dissemination of static information to distribute information regarding the current state of the network, such as resource availability. This is achieved by using TE extensions to routing protocols as described in Section 8.4. Path selection is used to select a path through the network that is constrained by the imposed constraints. This is achieved by using constraint-based routing algorithms. Path management, which includes label distribution, path placement, maintenance, and revocation, is used to establish, maintain, and tear down LSPs as needed. This can be done by using RSVP TE or CR-LDP TE extensions [56].

The concept of a *traffic trunk* refers to the aggregation of traffic belonging to the same class that is forwarded through a common path. For example, LSPs form traffic trunks in MPLS networks and optical channels form traffic trunks in OTNs. Traffic trunks support various attributes such as traffic parameters indicating the bandwidth requirements of the trunk, adaptivity attributes specifying the sensitivity of the trunk to changes in the state of the network, and priority attributes indicating the priority of the trunk relative to other trunks.

In considering the problem of traffic engineering in IP-over-optical networks, we can try to leverage as much of this work on TE in MPLS based IP networks as possible. This can be done fairly easily if, as we have shown so far, we assume that the optical IP network uses G-MPLS control plane. In such a network, the different optical channels or lightpaths would correspond to LSPs and the wavelengths would correspond to labels in an MPLS network.

If this is the case, we arrange them such that the MPLS TE control plane is used as the single control plane for both LSRs and OXCs. While it is possible to put in separate control planes for the LSRs and OXCs, it is beneficial if a single control plane is used for both of them. This brings forth several advantages. First, it helps to eliminate the administrative complexity of managing a hybrid system. Second, the problem of addressing for OXCs is resolved as the devices are now IP addressable devices. Third, the distribution of topology state information may be done using existing IP routing protocols with extensions. Fourth, the OXCs may be managed by IP-centric methods such as SNMP.

The transportation and distribution of the control traffic may be accomplished by using a separate out-of-band IP communication network. This could be an entirely separate network from the optical network itself, or it could be a dedicated optical channel based on wavelengths or timeslots.

An OTN controlled through an MPLS control plane, as described above, could be compared with an IP network by the analogy that the optical fibers between OXCs are analogous to the links between routers. The fact that multiple fibers can exist between two OXCs may be handled by employing a process called *bundling*, which groups the fibers logically into a single link in the OTN topology (refer to Section 8.6).

The OTN topology and other information would be distributed by using the TE extensions being developed for OSPF and IS-IS, with possible optical network specific extensions. Based on this information the border routers or clients would be able to calculate explicit paths for the optical channels. For setting these paths, signaling protocols based on MPLS, such as RSVP or CR-LDP, would be used. As the label information in these cases would have to handle the case where the label is a more general feature of the path, such as wavelength, as G-MPLS would have to be used.

However there are some important differences between MPLS LSRs and OXCs with regard to the functions described above. Whereas LSRs maintain a *forwarding information base* (FIB), OXCs would maintain *wavelength FIB* (WFIB) per interface. The interface would be a fiber port. The WFIB would be used to map incoming wavelengths and fiber ports to outgoing wavelengths and fiber ports. This is analogous to the *label FIB* (LFIB) that an LSR based on ATM switch maintains.

As for the OXC, it cannot carry out any operation that is analogous to label merging and cannot carry out label pop/push either. Further, it is limited in the granularity of the number of labels that it supports. For example, MPLS LSRs use a 20-bit label space with potentially 2^{20} labels, but an OXC is limited in the number of wavelengths that it can support. Currently, the number is in the range of hundreds at most. On the other hand, while LSRs would be able to support arbitrary bandwidth per LSP, an OXC would be constrained in the types of OCh it can support based on the bandwidth. For example, STM-1/OC-3, STM-4/OC-12, STM-16/OC-48, and STM-64/OC-192 would be typical.

This leads to problems when only one LSP is mapped to a single lightpath. It is very unlikely that the single LSP will be able to fully utilize the bandwidth available. By using the concept of nested LSPs, where multiple low bandwidth LSPs are mapped into a single OCh, this problem can be resolved. However, this solution is based on using the label pop/push functionality, so nested LSPs can be started and terminated only on traditional IP routers/LSRs.

In this manner, the basic methods of MPLS TE can be extended to solve the TE problems of optical IP networks. However, the solutions are limited only to the problems of information dissemination, resource discovery, path selection, and path management. The actual algorithms for deciding what resources to use or not to use must be developed by the network operators and used in the manner shown in Figure 8.24.

8.7.2 Differentiated Optical Service Model

Various issues of scalable end-to-end QoS in metropolitan DWDM optical networks that are used as transit networks for IP access networks have been discussed. In particular, a QoS service model for the optical domain, called *differentiated optical services* (DOS), which is based on a set of optical parameters that reflect various aspects of the lightpath [57, 58], has been introduced. Figure 8.25 shows how this DOS domain fits in between two DiffServ-aware access networks. The main objective of the DOS concept is to ensure an easy mapping between DiffServ QoS models that are used by the IP layer and the services being offered by the optical layer.

In a DWDM optical network, there may be a number of DWDM lightpaths between a source and a destination. Each lightpath may be different in the values and characteristics of its optical parameters such as BER, delay, and jitter, and also different in the services that are offered, such as the types of protection, monitoring ability and security. DOS proposes to divide the lightpaths along the lines of these characteristics and map the DiffServ QoS to these lightpaths. This can be accomplished by introducing a new QoS-aware layer, which maps DiffServ QoS model functionality to the optical services, between the upper layers (such as IP, ATM, and SDH/SONET) and the lower layer OCh links. Figure 8.26 shows how this QoS-aware layer can be introduced into the protocol stack. Notice that there is an end-to-end DOS layer that is based on the OCh lightpath.

The DOS model is basically concentrated at the edge nodes of the optical network. So the core of an all-optical network should not be aware of the DOS model. Consequently the DOS interface layer implements QoS-aware functions that aggregate and map DiffServ flows originating from the access networks onto the equivalent optical flows with the QoS parameters enforced by the optical domain. These enforceable optical QoS parameters and characteristics may be defined by a set of parameters that characterize the quality of the optical signal carried over the lightpath.

The parameters and functional capabilities that DOS defines as being important fall roughly into five classes—*lightpath characteristics, lightpath protection, lightpath monitoring, lightpath security,* and *lightpath transparency.*

Lightpath characteristics reflect the various impairments that the signal experiences while being transmitted, including jitter, wander, crosstalk, loss

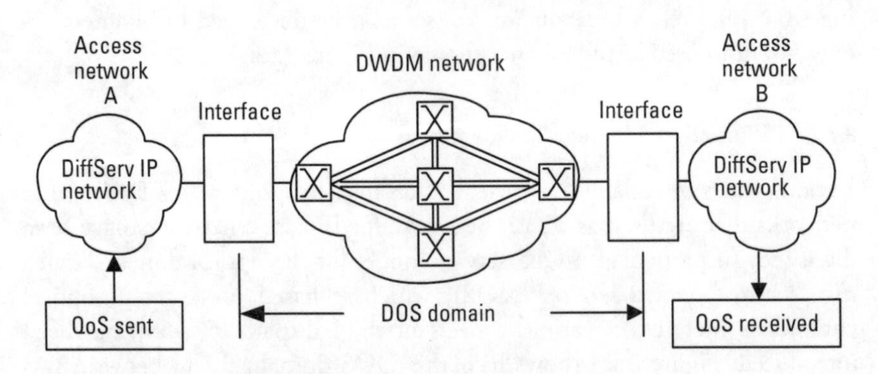

Figure 8.25 DOS domain model [57].

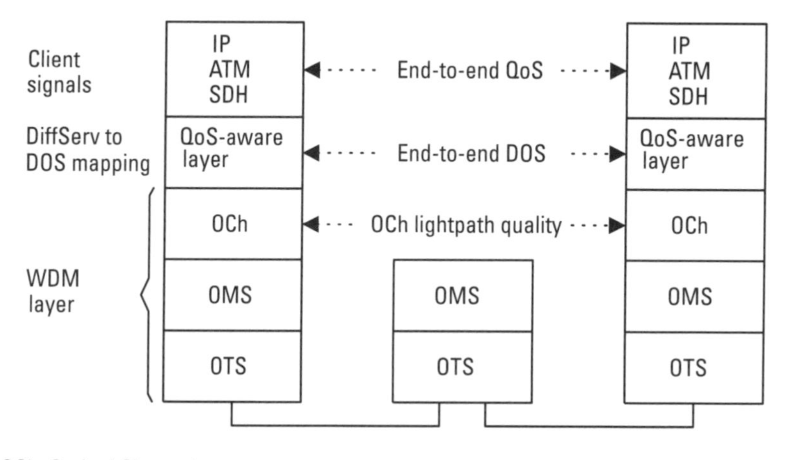

Figure 8.26 OTN architecture including the QoS-aware layer [57].

and *amplified spontaneous emission* level. These are usually reflected in the BER of the signal at the receiving node.

Lightpath protection reflects the protection or restoration characteristics of the lightpath. This is more of a service than any other physical characteristics. Depending on the functionality offered by the OXCs and edge routers, the lightpaths may offer various types of protection such as 1:1, 1:*n*, or 1+1.

Lightpath monitoring can be defined as being the ability to monitor trails for validity, integrity, and quality. Without such ability OTN control and network management would be impossible. A number of possible methods exist, including intrusive monitoring (i.e., breaking in the trail), inherent monitoring (i.e., relying on upper layers), and nonintrusive monitoring (i.e., passively monitoring).

Lightpath security is related to the various aspects of security in transmitting data of optical channels. This includes ensuring confidentiality of the transmitted data. Additionally it covers preventing denial of service attacks caused by disrupting the optical signal and/or QoS degradation attacks caused by interfering with the physical characteristics of the optical signal.

Lightpath transparency reflects the degree of transparency that the lightpath offers with respect to the signal format and bit rate. This is heavily influenced by the type of signal regeneration technique employed.

As noted above, the main goal of the DOS is to provide a mechanism for offering a spectrum of services in the optical domain through a classification of lightpaths according to the end-to-end quality of transmission. To achieve this goal the following arrangements are needed at the interface: The incoming DiffServ flows must be aggregated into a small number of flows that are compatible to the rates corresponding to the lightpaths, for example, STM-4/OC-12, STM-16/OC-48, and so on. The lightpaths themselves must be grouped into classes that reflect their QoS characteristics according to the classification methods and parameters defined above. The incoming aggregated DiffServ classes must then be mapped to the corresponding optical classes and channels. In addition, an aggregated admission control is needed to ensure that the OTN does not accept more DiffServ flows than it can handle.

Figure 8.27 shows how the above arrangements can be done. As can be seen, the different service classes are grouped into different optical channels based on the DiffServ class and also the characteristics of the optical channel themselves.

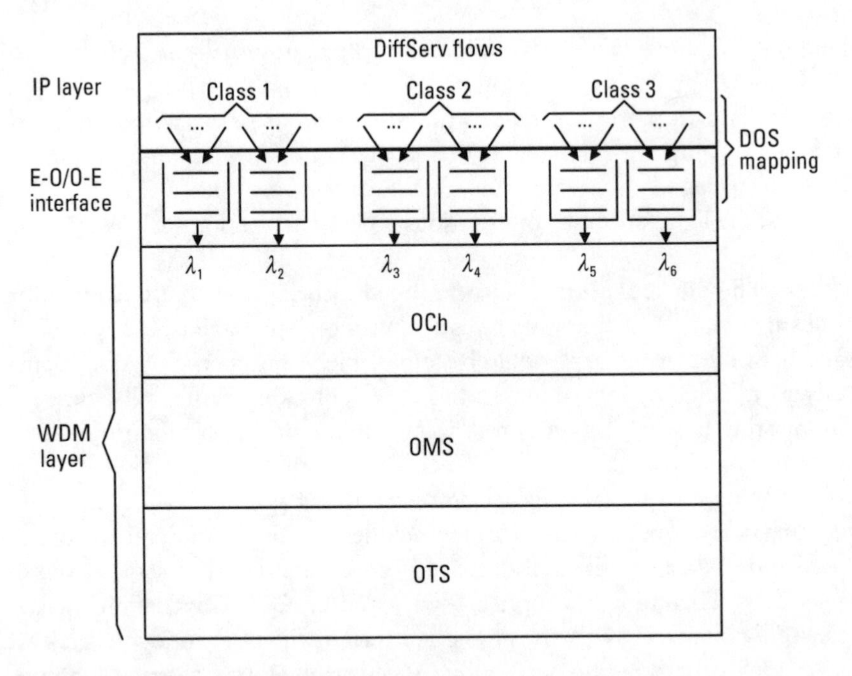

Figure 8.27 IP DiffServ mapping for DOS provisioning [57].

References

[1] Anderson, J., et al., "Protocols and Architectures for IP Optical Networking," *Bell Labs Technical Journal,* Vol. 4, No. 1, January–March 1999.

[2] Doshi, B., et al., "Optical Network Design and Restoration," *Bell Labs Technical Journal,* Vol. 4, No. 1, January–March 1999.

[3] Bonenfant, P., and A. Rodriguez-Moral, "Optical Data Networking," *IEEE Communications Magazine,* Vol. 38, No. 3, March 2000, pp. 63–70.

[4] Ghani, N., S. Dixit, and T. Wang, "On IP-Over-WDM Integration," *IEEE Communications Magazine,* Vol. 38, No. 3, March 2000, pp. 72–84.

[5] ITU-T Rec. G.872 (02/99) "Architecture of Optical Transport Networks."

[6] Rajagopalan, B., et al., "IP Over Optical Networks: A Framework," draft-many-ip-optical-framework-02.txt, work in progress, December 2000.

[7] Rajgopalan, B., et al., "IP Over Optical: Architectural Aspects," *IEEE Communications Magazine,* Vol. 38, No. 9, September 2000, pp. 94–102.

[8] ITU-T Rec. G.709 (02/2001) "Interface for the Optical Transport Networks (OTN)."

[9] Malis, A., and W. Simpson, "PPP Over SDH/SONET," *RFC2615,* June 1999.

[10] Simpson, W., "The Point-to-Point Protocol (PPP)," *RFC 1661,* July 1994.

[11] Simpson, W., "PPP Over SDH/SONET," *RFC1619,* May 1994.

[12] Doshi, B., et al., "A Simple Data Link (SDL) Protocol for Next Generation Packet Networks," *IEEE Journal on Selected Areas in Communications,* Vol. 18, No. 10, October 2000, pp. 1825–1837.

[13] IEEE P802.3: Part 3, "IEEE Standard: Carrier Sense Multiple Access with Collision Detection (CSMA/CD) Access Method and Physical Layer Specifications," *IEEE Std. 803.3,* 1998 edition.

[14] Unapproved draft P802.3ae, D3.1, "Draft Supplement to Carrier Sense Multiple Access with Collision Detection (CSMA/CD) Access Method and Physical Layer Specifications—Media Access Control (MAC) Parameters, Physical Layer, and Management Parameters for 10 Gb/s Operation."

[15] Abul-Magd, O., et al., "Signaling Requirements at the Optical UNI," draft-bala-mpls-optical-uni-signaling-01.txt, Internet draft, work in progress, November 2000.

[16] Arvind, K., et al, "Optical Domain Services Interconnect (ODSI) Signaling Control Specification, Version 1.4.5," ODSI Coalition, http://www.odsi-coalition.com, March 2000.

[17] Bernstein, R., et al., "Optical Domain Service Interconnect (ODSI) Functional Specification," ODSI Coalition, March 2000.

[18] Copley, A., "Optical Domain Service Interconnect (ODSI): Defining Mechanisms for Enabling On-Demand High-Speed from the Optical Domain," *IEEE Communications Magazine,* Vol. 38, No. 10, October 2000, pp. 168–174.

[19] Gray, E., et al., "RSVP Extensions in Support of OIF Optical UNI Signaling," draft-gray-mpls-rsvp-oif-uni-ext-00.txt, Internet draft, work in progress, October 2000.

[20] Optical Internetworking Forum Web site, http://www.oiforum.org.

[21] Ashwood-Smith, P., et al, "Generalized MPLS—Signaling Functional Description," draft-ietf-mpls-generalized-signaling-02.txt, Internet draft, work in progress, November 2000.

[22] Banerjee, A., et al., "Generalized Multiprotocol Label Switching: An Overview of Routing Management Enhancements," *IEEE Communications Magazine,* Vol. 39, No. 1, January 2001, pp. 144–150.

[23] Ashwood-Smith, P., et al., "Generalized MPLS—CR-LDP Signaling Functional Description," draft-ietf-mpls-generalized-cr-ldp-00.txt, Internet draft, work in progress, November 2000.

[24] Ashwood-Smith, P., et al., "Generalized MPLS—RSVP-TE Signaling Functional Description," draft-ietf-mpls-generalized-rsvp-te-00.txt, Internet draft, work in progress, November 2000.

[25] Kompella, K., et al., "OSPF Extensions in Support of Generalized MPLS," draft-kompella-ospf-gmpls-extensions-01.txt, work in progress, November 2000.

[26] Kompella, K., and Y. Rekhter, "LSP Hierarchy with MPLS TE," draft-ietf-mpls-lsp-hierarchy-01.txt, work in progress, November 2000.

[27] Mack-Crane, B., et al., "Enhancements to GMPLS Signaling for Optical Technologies," draft-mack-crane-gmpls-signaling-enchancements-00.txt, Internet draft, work in progress, November 2000.

[28] Awduche, D., et al., "Requirements for Traffic Engineering over MPLS," *RFC 2702,* September 1999.

[29] Awduche, D., and Y. Rekhter, "Multiprotocol Lambda Switching: Combining MPLS Traffic Engineering Control with Optical Crossconnects," *IEEE Communications Magazine,* Vol. 39, No. 3, March 2001, pp. 111–116.

[30] Awduche, D., et al., "Multiprotocol Lambda Switching: Combining MPLS Traffic Engineering Control with Optical Crossconnects," draft-awduche-mpls-te-optical-02.txt, work in progress, July 2000.

[31] Pendarakis, D., B. Rajagopalan, and D. Saha, "Routing Information Exchange in Optical Networks," draft-prs-optical-routing-01.ps, Internet draft, work in progress, November 2000.

[32] Lee, B. G., M. Kang, and J. Lee, *Broadband Telecommunications Technology,* Norwood, MA: Artech House, 1996.

[33] Lee, B. G., and S. C. Kim, *Scrambling Techniques for Digital Transmission,* Berlin: Springer Verlag, 1994.

[34] Rosen, E., and Y. Rekhter, "BGP/MPLS VPNs," *RFC 2547,* March 1999.

[35] Smit, H., and T. Li, "IS-IS Extensions for Traffic Engineering," draft-ietf-isis-traffic-02.txt, work in progress.

[36] Smit, H., and T. Li, "OSPF Extensions for Traffic Engineering," draft-ietf-ospf-traffic-02.txt, work in progress.

[37] Hu, L., et al., "Techniques for Optical Packet Switching and Optical Burst Switching," *IEEE Communications Magazine,* Vol. 39, No. 1, January 2001, pp. 136–142.

[38] Hunter, D. K., et al., "WASPNET: A Wavelength Switched Packet Network," *IEEE Communications Magazine,* Vol. 37, No. 3, March 1999, pp. 120–129.

[39] Hunter, D. K., and I. Andonovic, "Approaches to Optical Internet Packet Switching," *IEEE Communications Magazine,* Vol. 38, No. 9, September 2000, pp. 116–122.

[40] Modiano, E., "WDM-Based Packet Networks," *IEEE Communications Magazine,* Vol. 37, No. 3, March 1999, pp. 130–135.

[41] Yao, S., et al., "Advances in Photonic Packet Switching: An Overview," *IEEE Communications Magazine,* Vol. 38, No. 2, February 2000, pp. 84–94.

[42] Guillemot, C., et al., "Transparent Optical Packet Switching: The European ACTS KEOPS Project Approach," *Journal of Lightwave Technology,* Vol. 16, No. 12, December 1998, pp. 2117–2133.

[43] Qiao, C., "Labeled Optical Burst Switching for IP-Over-WDM Integration," *IEEE Communications Magazine,* Vol. 38, No. 9, September 2000, pp. 104–114.

[44] Qiao, C., and M. Yoo, "Optical Burst Switching (OBS)—A New Paradigm for an Optical Internet," *Journal of High Speed Networks,* Vol. 8, 1999, pp. 69–84.

[45] Verma, S., et al., "Optical Burst Switching: A Viable Solution for Terabit IP Backbone," *IEEE Network Magazine,* Vol. 14, No. 6, November/December 2000, pp. 48–53.

[46] Listanti, M., et al., "Architectural and Technological Issues for Future Optical Internet Networks," *IEEE Communications Magazine,* Vol. 38, No. 9, September 2000, pp. 82–92.

[47] Metz, C., "IP Over Optical: From Packets to Photons," *IEEE Internet Computing Magazine,* Vol. 4, No. 6, November/December 2000, pp. 76–82.

[48] Kompella, K., et al., "IS-IS Extensions in Support of Generalized MPLS," draft-ietf-isis-gmpls-extensions-02.txt, work in progress.

[49] Lang, J. P., et al., "Link Management Protocol," draft-ietf-mpls-lmp-01.txt, Internet draft, work in progress, November 2000.

[50] Rajagopalan, B., and D. Saha, "Link Bundling in Optical Networks," draft-rs-optical-bundling-01.txt, Internet draft, work in progress, October 2000.

[51] Kompella, K., Y. Rekhter, and A. Kullberg, "Signaling Unnumbered Links in CR-LDP," draft-ietf-mpls-crldp-unnum-01.txt, work in progress.

[52] Kompella, K., and Y. Rekhter, "Signaling Unnumbered Links in RSVP-TE," draft-ietf-mpls-rsvp-unnum-01.txt, work in progress.

[53] Abul-Magd, O., et al., "LDP Extensions for Optical UNI Signaling," draft-ietf-mpls-ldp-optical-uni-00.txt, Internet Draft, work in progress, October 2000.

[54] Boyle, J., A. Sastry, et al., "The COPS (Common Open Policy Service) Protocol," work in progress, November 1999.

[55] Yoo, M., C. Qiao, and S. Dixit "Optical Burst Switching for Service Differentiation in the Next Generation Optical Internet," *IEEE Communications Magazine,* Vol. 39, No. 2, February 2001, pp. 98–104.

[56] Awduche, D., et al., "RSVP-TE: Extensions to RSVP for LSP Tunnels," draft-ietf-mpls-rsvp-lsp-tunnel-07.txt, Internet draft, work in progress, October 2000.

[57] Golmie, N., et al., "A Differentiated Optical Services Model for WDM Networks," *IEEE Communications Magazine,* Vol. 38, No. 2, February 2000, pp. 68–72.

[58] Ndousee, T., and N. Golmie, "Differentiated Optical Services: A Quality of Optical Service Model for WDM Networks," *Proc. SPID All-Optical Net. Architecture,* Cont. And Mgmt Issues, Boston, MA, Vol. 3843, September 1999, pp. 79–87.

Selected Bibliography

Doverspike, R., and J. Yates, "Challenges for MPLS in Optical Network Restoration," *IEEE Communications Magazine,* Vol. 39, No.2, February 2001, pp. 89–96.

Ghani, N., "A Framework for IP-Over-WDM Using MPLS," *Opt. Networks Mag.,* Vol. 1, No. 2, April 2000, pp. 39–52.

ITU-T Rec. G.650 (10/2000) "Definition and Test Methods for the Relevant Parameters of Single-Mode Fiber."

ITU-T Rec. G.651 (02/98) "Characteristics of a 50/125μm Multimode Graded Index Optical Fiber Cable."

ITU-T Rec. G.652 (10/2000) "Characteristics of a Single-Mode Optical Fiber Cable."

ITU-T Rec. G.653 (10/2000) "Characteristics of a Dispersion-Shifted Single-Mode Optical Fiber Cable."

ITU-T Rec. G.654 (10/2000) "Characteristics of a Cut-Off Shifted Single-Mode Optical Fiber Cable."

ITU-T Rec. G.655 (10/2000) "Characteristics of a Non-Zero Dispersion Shifted Single-Mode Optical Fiber Cable."

ITU-T Rec. G.661 (10/98) "Definition and Test Methods for the Relevant Generic Parameters of Optical Amplifier Devices and Subsystems."

ITU-T Rec. G.662 (10/98) "Generic Characteristics of Optical Fiber Amplifier Devices and Subsystems."

ITU-T Rec. G.663 (04/2000) "Application Related Aspects of Optical Amplifier Devices and Subsystems."

ITU-T Rec. G.664 (06/99) "Optical Safety Procedures and Requirements for Optical Transport Systems."

ITU-T Rec. G.671 (02/2001) "Transmission Characteristics of Optical Components and Subsystems."

ITU-T Rec. G.681 (10/96) "Functional Characteristics of Interoffice and Long-Haul Line Systems Using Optical Amplifiers, Including Optical Multiplexing."

ITU-T Rec. G. 691 (10/2000) "Optical Interfaces for Single-Channel STM-64, STM-256, and Other SDH Systems with Optical Amplifiers."

ITU-T Rec. G.692 (10/98) "Optical Interfaces for Multichannel Systems with Optical Amplifiers."

ITU-T Rec. G.692 Corrigendum 1 (01/2000) "Optical Interfaces for Multichannel Systems with Optical Amplifiers."

ITU-T Rec. G.871 (10/2000) "Framework of Optical Transport Network Recommendations."

ITU-T Rec. G.955 (11/96) "Digital Line Systems Based on the 1544-Kbps and the 2048-Kbps Hierarchy on Optical Fiber Cables."

ITU-T Rec. G.958 (11/94) "Digital Line Systems Based on the Synchronous Digital Hierarchy for Use on Optical Fiber Cables."

ITU-T Rec. G.959.1 (02/2001) "Optical Transport Network Physical Layer Interfaces."

ITU-T Rec. G.982 (11/96) "Optical Access Networks to Support Services Up to the ISDN Primary Rate or Equivalent Bit Rates."

ITU-T Rec. G.983.1 (06/99) "Broadband Optical Access Systems Based on Passive Optical Networks (PON)."

ITU-T Rec. G.983.1 Corrigendum 1 (07/99) "Broadband Optical Access Systems Based on Passive Optical Networks (PON)."

ITU-T Rec. G.983.2 (04/2000) "ONT Management and Control Interface Specification for ATM PON."

ITU-T Rec. G.983.4 (03/2001) "A Broadband Optical Access System with Increased Service Capability by Wavelength Allocation."

Manchester, J., et al., "IP Over SONET," *IEEE Communications Magazine,* Vol. 36, No. 5, May 1998, pp. 136–142.

Optical Domain Service Interconnection, http://www.odsi.org.

Strand, J., A. Chiu, and R. Tkach, "Issues for Routing in the Optical Layer," *IEEE Communications Magazine,* Vol. 39, No. 2, February 2001, pp. 81–87.

List of Acronyms

AAL	ATM adaptation layer
AAL-IDU	AAL interface data unit
AB	area border
ABR	area border router
ABR	available bit rate
ABT	ATM block transfer
ABT/DT	ABT with delayed transmission
ABT/IT	ABT with immediate transmission
ACK	acknowledgment
ACR	allowed cell rate
ADM	add/drop multiplexer (multiplexing)
ADTM	asynchronous time division multiplexing

AESA	ATM end system address
AF	assured forwarding
AF	address filter
AFI	authorization and format identifier
ALG	application-level gateway
AIS	alarm indication signal
ANSI	American Natural Standard Institute
AOTF	acousto-optic tunable filter
APD	avalanche photodiode
API	access point identifier
APPN	advanced peer-to-peer networking
APS	automatic protection switching
ALG	application-level gateway
AR	antireflection
ARP	address resolution protocol
ARPANET	Advanced Research Agency network
AS	autonomous system
ASB	autonomous system boundary
ASE	amplified spontaneous emission
ASIC	application-specific IC

ASP	ATM service provider
A-SW	ATM switch
ATDM	asynchronous time division multiplex
ATM	asynchronous transfer mode
ATMARP	ATM address resolution protocol
AU	administrative unit
AUG	administrative unit group
AWG	arranged waveguide grating
BA	behavior aggregate
BAsize	buffer allocation size
BBN	broadcast Banyan network
BCS	behavior class selector
BDCS	broadband digital cross-connect system
BDI	backward defect indication
BECN	backward explicit congestion notification
BE	best effort
BEI	backward error indication
BER	bit error rate
BGP	border gateway protocol
BIM	byte-interleaved multiplexing

BIP	bit-interleave parity
B-ICI	broadband intercarrier interface
B-ISDN	broadband integrated services digital network
B-ISUP	broadband ISDN user part
B-LLI	broadband low-layer information
BLSR	bidirectional line-switched ring
B-NT	B-ISDN network terminal
BOM	beginning of message
BSHR	bidirectional self-healing ring
BSI	British Standard Institute
BSN	broadcast-and-select network
BSRF	basic synchronization reference frequency
BT	burst tolerance
BUS	broadcast and unknown server
C	container
CAC	call admission control
CAM	content-addressable memory
CAS	channel-associated signaling
CBR	constant bit rate
CCITT	International Telegraph and Telephone Consultative Committee

CCS	common channel signaling
CDMA	code division multiple access
CDV	cell delay variation
CERN	European Organization for Nuclear Research
CI	congestion indication
CID	channel identifier
CIDR	classless Internet domain routing
CL	connectionless
CLNAP	connectionless network access protocol
CLNIP	connectionless network interface protocol
CLNP	connectionless network protocol
CLP	cell loss priority
CLR	cell loss rate
CLSF	connectionless service function
CMOS	complementary metal-oxide-semiconductor
CO	connection-oriented
COM	continuation of message
COS	class of service
COPS	common open policy service
CP	call processor

CPE	customer premises equipment
CPS	common part sublayer
CPCS	common part convergence sublayer
CRC	cyclic redundancy check
CR-LDP	constrained-route LDP
CR-SPF	constrained-route SPF
CRM	cell rate margin
CS	convergence sublayer
CSI	convergence sublayer indication
CSMA/CD	carrier sense multiple access with collision detection
C-SW	circuit switch
CTD	cell transmission delay
CU	currently unused
CWDM	coarse wavelength division multiplexing
cwnd	congestion window
CX	coaxial
D	decrement
DA	destination address
DAE	dummy adder encoder
DAPI	destination access point identifier

DARPA	Defense Advanced Research Projects Agency
DBR	deterministic bit rate
DBR	distributed Bragg reflector
DCC	data communication channel
DCC	domain country code
DCN	data communication network
DCS	digital cross-connect system
DEMUX	demultiplexer
DFB	distributed feedback
DH	double heterostructure
DHCP	dynamic host configuration protocol
DiffServ	differentiated services
DOS	differentiated optical services
DNHR	dynamic nonhierarchical routing
DNS	domain name service
DSCP	DiffServ code point
DSL	digital subscriber line
DS-*n*	digital signal level-*n*
DSP	domain-specific part
DSS	distributed sample scrambling

DSX	digital system cross-connect
DTDM	dynamic time division multiplexing
DTL	designated transit list
DWDM	dense wavelength division multiplexing
DVMRP	distance vector multicast routing protocol
DXC	digital cross-connecting
EBCN	explicit backward congestion notification
EBGP	external BGP
ECC	embedded control channel
EDFA	erbium-doped fiber amplifier
EF	expedited forwarding
EFCI	explicit forward congestion indication
EFCN	explicit forward congestion notification
EGP	exterior gateway protocol
EIGRP	enhanced interior gateway routing protocol
ELAN	emulated LAN
E-LSP	exp bit-inferred PSC LSP
E/O	electronic/optical
EOF	end of frame
EOM	end of message

EPD	early packet discard
ER	explicit rate
ESI	end system identifier
EWMA	exponentially weighted moving average
EXP	experimental
FA	forwarding adjacency
FA-LSC	forwarding adjacency LSC
FB	fair buffer
FBG	fiber-Bragg grating
FC	fiber channel
FCFS	first-come first-serve
FCS	frame check sequence
FDDI	fiber-distributed data interface
FDM	frequency division multiplexing
FEC	forward error correction
FECN	forward explicit congestion notification
FF	fixed filter
F-GCRA	frame-based GCRA
FFQ	frame-based fair queuing
FIB	forwarding information base

FIFO	first-in first-out
FOH	fixed overhead
FPGA	field-programmable gate array
FSC	fiber-switch capable
FSS	frame synchronous scrambler
ftp	file transfer protocol
FWC	fixed wavelength converter
FWM	four-wave mixing
GCC	general communication channel
GCRA	generic cell rate algorithm
GFC	generic flow control
GFR	guaranteed frame rate
G-MPLS	generalized MPLS
GPS	generalized processor sharing
GSMP	general switch management protocol
GUI	graphic user interface
HDLC	high-level data link control
HEC	header error control
HL	header length
HO-DSP	higher-order domain-specific part

HOL	head of line
HO-POH	higher-order path overhead
HTML	hypertext markup language
HTTP	hypertext transport protocol
I	increment
IA	internal area
IAE	incoming alignment error
IASG	internetwork address subgroup
IaDI	intradomain interface
IBGP	internal BGP
ICC	ITU carrier code
ICMP	Internet control message protocol
ICD	international code designator
IDP	initial domain part
IDI	initial domain identifier
IDU	interface data unit
IE	information element
IETF	Internet Engineering Task Force
IFMP	Ipsilon flow management protocol
IG	information group

IGMP	Internet group management protocol
IGP	interior gateway protocol
IISP	interim interswitch signaling protocol
ILMI	interim local management interface
InATMARP	inverse ATMARP
IntServ	integrated services
IP	Internet protocol
I-PNNI	integrated PNNI
IPoA	IP over ATM
IPoS	IP over SDH/SONET
IPoW	IP over WDM
IPX	internetwork packet exchange
IrDI	interdomain interface
IRTF	Internet Research Task Force
ISDN	integrated services digital network
ISO	Internet Standard Organization
IS-IS	intermediate system to intermediate system
ISP	Internet service provider
ISUP	ISDN user part
ITU	International Telecommunication Union

ITU-T	ITU Telecommunication (sector)
IWU	interworking unit
JC	justification control
L2	layer 2
L3	layer 3
L4	layer 4
L7	layer 7
LAN	local area network
LANE	LAN emulation
LASER	light amplification by stimulated emission of radiation
LB	leaky bucket
LC-ATM	label switch–controlled ATM
LCN	local communication network
LDP	label distribution protocol
LE-ARP	LANE ARP
LEC	LANE client
LECS	LANE configuration server
LED	light-emitting diode
LER	label edge router
LES	LANE server

LFIB	label forwarding information base
LDP	label distribution protocol
LGN	logical group node
LI	length indicator
LI	line interface
LIJ	leaf initiated join
LIS	logical IP subnetwork
LLC	logical link control
LLN	linear lightwave network
L-LSP	label-inferred PSC LSP
LMP	link management protocol
LNNI	LANE NNI
LOF	loss of frame
LOP	loss of packet
LO-POH	lower-order path overhead
LOS	loss of signal
LSA	link state advertisement
LSC	lambda-switch capable
LSC	linear splitter and combiner
LSR	label switch router

LSP	label switching path
LTE	line terminal equipment
LUNI	LANE UNI
LX	long wavelength
MAC	medium access control
MARS	multicast address resolution server
MBS	maximum burst size
MCDV	maximum cell delay variation
MCF	message communication function
MCR	minimum cell rate
MCTD	maximum cell transfer delay
MDCR	minimum desired cell rate
MDF	main distribution frame
ME	mapping entity
MEMS	microelectro-mechanical system
MF	multifield
MFS	maximum frame size
MI	multilayer interference
MIB	management information base
MID	multiplexing identifier

MIN	multistage interconnection network
MLM	multiple longitudinal mode
MPC	MPOA client
MPLS	multiprotocol label switching
MPλS	multiprotocol lambda switching
MPOA	multiprotocol over ATM
MPS	MPOA server
MPU	microprocessor unit
MS	maintenance signal
MSB	most significant bit
MSOH	multiplexer section overhead
MSP	multiplexer section protection
MST	multiplexer section termination
MTPI	multiplexer timing physical interface
MTS	multiplexer timing source
MTU	maximum transmission unit
MUX	multiplexer
MZI	Mach-Zehnder interferometer
NBMA	nonbroadcasting multiple access
NDF	new data flag

NDP neighbor discovery protocol

NE network element

NHC next-hop client

NHRP next-hop resolution protocol

NHS next-hop server

NI nonincrease

N-ISDN narrowband ISDN

NJ negative justification

NJO negative justification opportunity

NMS network management system

NNI network node interface

NOMC network operator maintenance channel

NPC network parameter control

nrt non-real-time

nrt-VBR non-real time VBR

NSAP network SAP

NSF National Science Foundation

NTP network timing protocol

OADM optical add/drop multiplexing

OAM operations and management

OBS	optical burst switching
OCC	optical channel carrier
OCCo	optical channel carrier overhead
OCCp	optical channel carrier payload
OCDM	optical code division multiplexing
OCDMA	optical code division multiple access
OCG	optical carrier group
OCh	optical channel
OC-*m*	optical carrier level *m*
ODL	optical delay line
ODSI	optical domain service interconnect
ODU	optical-channel data unit
O/E	optical/electronic
OEIC	optoelectronic integrated circuit
OH	overhead
OIF	Optical Internetworking Forum
O-LSR	optical LSR
OLT	optical line termination
OMS	optical multiplex section
OMU-*n*	optical multiplex unit-*n*

ONC	optical network controller
ONNI	optical network-network interface
OOS	OTM overhead signal
OPWA	one path with advertising
OS	operating system
OSF	offset field
OSI	open system interconnection
OSPF	open shortest path first
OTDM	optical time division multiplexing
OTM	optical transport module
OTN	optical transport network
OTPN	optical transport packet network
OTS	optical transmission section
OTU	optical-channel transport unit
OUI	organizational unique identifier
O-UNI	optical UNI
OXC	optical cross-connect
P	parity
PAD	padding
PAR	PNNI-augmented routing

PCC	protection-communication control channel
PCR	peak cell rate
PCS	physical coding sublayer
PDH	plesiochronous digital hierarchy
PDU	protocol data unit
PG	peer group
PGL	peer group leader
PGPS	packet-by-packet generalized processor sharing
PHB	per-hop behavior
PHY	physical
PJ	positive justification
PJO	positive justification opportunity
PID	protocol ID
PIM	protocol-independent multicasting
PL	physical layer
PLI	PDU length indicator
PLL	phase-locked loop
PLOAM	physical layer OAM
PM	path monitor, path monitoring
PMD	physical medium–dependent

PMTU	path MTU
PNNI	private network-network interface
POH	path overhead
POS	packet over SDH/SONET
POTS	plain old telephone service
PPAR	proxy PAR
PPD	partial packet discard
PRM	protocol reference model
PPP	point-to-point protocol
PSC	packet-switch capable
PSI	payload structure identifier
PSTN	public switched telephone network
PT	payload type
PTE	path termination equipment
PTR	pointer
PTSP	protocol topology state packet
PTSE	protocol topology state element
PVC	permanent VC
PVPC	permanent VPC
PVCC	permanent VCC

P/Z/N	positive-zero-negative
QoS	quality of service
RA	running adder
RARP	reverse ARP
RDI	remote defect indication
RED	random early detection
REI	remote error indication
RES	reserved
RFI	remote failure indication
RFC	request for comments
RIO	RED IN/OUT
RIP	routing information protocol
RM	resource management
RS	Reed-Solomon
RSOH	regenerator section overhead
RST	regenerator section termination
RSVP	resource reservation protocol
RSVP-TE	RSVP tunneling extension
rt	real-time
RTNR	real-time network routing

RTP	real-time transport protocol
rt-VBR	real-time VBR
RWA	routing and wavelength assignment
SA	source address
SAAL	signaling AAL
SACK	selective ACK
SAP	service access point
SAPI	source access point identifier
SAR	segmentation and reassembly
SBR	statistical bit rate
SBS	stimulated Brillouin scattering
SC	sequence count
SCFQ	self-clocked fair queuing
SCMA	subcarrier multiple access
SCR	sustainable cell rate
SDH	synchronous digital hierarchy
SDL	simple data link
SDT	structured data transfer
SDU	service data unit
SE	shared explicit

SEMF	synchronous equipment management function
SFD	start frame delimiter
SGML	standard generalized markup language
SHR	self-healing ring
SLA	service-level agreement
SLIP	serial line IP
SLM	signal label mismatch
SM	section monitor
SMDS	switched megabit data service
SMN	SDH/SONET Management Network
SMTP	simple mail transfer protocol
S-MUX	synchronous multiplexer
SN	sequence number
SNA	system network architecture
SNAP	subnetwork attaching point
SNMP	simple network management protocol
SNP	sequence number protection
SOA	semiconductor optical amplification
SOH	section overhead
SONET	synchronous optical network

SP	signal processor
SPE	synchronous payload envelope
SPF	shortest path first
SPFQ	starting potential-based fair queuing
SPM	self-phase modulation
SRLG	shared-risk link group
SRTS	synchronous residual time stamp
SRS	stimulated Raman scattering
SS7	signaling system no. 7
SSCOP	service-specific communication-oriented protocol
SSCP	server synchronization cache protocol
SSCF	service-specific coordinate function
SSCS	service-specific convergence sublayer
SSM	short segment message
SSS	self-synchronous scrambler
ssthresh	slow-start threshold
ST	segment type
STF	start field
STM-*n*	synchronous transport module level *n*
STS-*m*	synchronous transport signal level *m*

STG	synchronous timing generation
SigVC	signaling VC
SVC	switched VC
SX	short wavelength
TAT	theoretical arrival time
TC	transmission convergence
TCA	traffic conditioning agreement
TCM	tandem connection monitoring
TCP	transmission control protocol
TDM	time division multiplexing
TE	transverse electric
TEI	terminal equipment identifier
TIM	trace identification mismatch
TLV	type length value
TM	transverse magnetic
TMN	telecommunication management network
TNT	trunk number translator
TOS	type of service
TSMR	trunk status map routing
TSR	tag switch router

TSW	time sliding window
TTI	trail trace identifier
TTL	time to live
TU	tributary unit
TUG	tributary unit group
TWC	tunable wavelength converter
T-WDMA	time-wavelength division multiple access
UBR	unspecified bit rate
UAPC	unique access point code
UDP	user datagram protocol
UNEQ	unequipped
UNI	user network interface
URL	universal resource locator
UPC	usage parameter control
UPSR	unidirectional path-switched ring
U-SDU	user SDU
USHR	unidirectional self-healing ring
UUI	user-to-user indication
UV	ultraviolet
VBR	variable bit rate

VC	virtual container
VC	virtual channel
VCC	virtual channel connection
VCI	virtual channel identifier
VD	virtual destination
VLAN	virtual LAN
VLSM	variable-length subnet mask
VoIP	voice over IP
VP	virtual path
VPC	virtual path connection
VPI	virtual path identifier
VPN	virtual private network
VPON	virtual private optical network
VS	virtual source
VT	virtual tributary
WAN	wide area network
WAP	wireless application protocol
WADM	wavelength add/drop multiplexing
WC	wavelength converter
WDM	wavelength division multiplexing

WF	wildcard filter
WFIB	wavelength forwarding information base
WFQ	weighted fair queuing
WF^2Q	worst-case fair WFQ
WRED	weighted RED
WRN	wavelength-routing networks
WRR	weighted round-robin
WWW	World Wide Web
WXC	wavelength cross-connect

About the Authors

Byeong Gi Lee received his B.S. and M.E. in electronics engineering in 1974 and 1978 from Seoul National University, Seoul, Korea, and Kyungpook National University, Taegu, Korea, respectively. He received his Ph.D. in electrical engineering in 1982 from the University of California, Los Angeles. He was with the Electronics Engineering Department of the ROK Naval Academy as an instructor and naval officer in the active service from 1974 to 1979. He worked for Granger Associates, Santa Clara, California, from 1982 to 1984 as a senior engineer responsible for applications of digital signal processing to digital transmission and for AT&T Bell Laboratories, North Andover, Massachusetts, from 1984 to 1986 as a member of the technical staff responsible for optical transmission system development along with related standards works. In 1986, he joined the faculty of Seoul University, where he is a professor and vice chancellor for research affairs.

Dr. Lee is the editor-in-chief of the *Journal of Communications and Networks* (JCN), a past editor of *the IEEE Global Communications Newsletter*, and a past associate editor of the *IEEE Transactions on Circuits and Systems for Video Technology*. He is the director of the magazine, the past director for membership programs development, a past director of the Asia-Pacific Region, and a member-at-large of the IEEE Communications Society (ComSoc). He served as the chair of the Asia Pacific Conference on Communications (APCC) steering committee and the chair of the Accreditation Board for Engineering Education of Korea (ABEEK). His current fields of interest include broadband networks, communications systems, and signal processing. Dr. Lee coauthored *Broadband Telecommunication Technology,* [Artech House, 1993 (1st ed.) and 1996 (2nd ed.)], *Scrambling Techniques for Digital Transmission* (Springer Verlag, 1994), and *Scrambling Techniques for CDMA Communications* (Kluwer, 2001). He holds seven U.S. patents and has three more pending.

Dr. Lee is a Fellow of the IEEE, a member of the board of governors of IEEE ComSoc, a member of Sigma Xi, and a member of the National Academy of Engineering of Korea. He holds the 1984 Myril B. Reed Best Paper Award from the Midwest Symposium on Circuits and Systems, Exceptional Contribution awards from AT&T Bell Laboratories, a Distinguished Achievement Award from the Korean Institute of Communication Sciences (KICS), and the 46th National Academy of Sciences of Korea Award (2001).

Woojune Kim received his B.S., M.S., and Ph.D. in 1991, 1993, and 1998, respectively, from Seoul National University—all in electrical engineering—and received a Distinguished Dissertation Award for his Ph.D. dissertation. He worked for Lucent Technologies, Murray Hill, New Jersey, from 1998 to 1999 as a PMTS responsible for various aspects of ATM PON systems. From 1999 to 2001, he worked for Samsung Electronics, Seoul, Korea, as a senior engineer, leading a team of engineers in the design and implementation of the first cdma2000 packet data system. Since 2001, he has been with Airvana, Boston, Massachusetts. Dr. Kim has four U.S. patents pending.

Index

Recent Titles in the Artech House Telecommunications Library

Vinton G. Cerf, Senior Series Editor

Access Networks: Technology and V5 Interfacing, Alex Gillespie

Achieving Global Information Networking, Eve L. Varma, et al.

Advanced High-Frequency Radio Communications, Eric E. Johnson, et al.

ATM Switches, Edwin R. Coover

ATM Switching Systems, Thomas M. Chen and Stephen S. Liu

Broadband Access Technology, Interfaces, and Management, Alex Gillespie

Broadband Networking: ATM, SDH, and SONET, Mike Sexton and Andy Reid

Broadband Telecommunications Technology, Second Edition, Byeong Lee, Minho Kang, and Jonghee Lee

The Business Case for Web-Based Training, Tammy Whalen and David Wright

Communication and Computing for Distributed Multimedia Systems, Guojun Lu

Communications Technology Guide for Business, Richard Downey, Seán Boland, and Phillip Walsh

Community Networks: Lessons from Blacksburg, Virginia, Second Edition, Andrew M. Cohill and Andrea Kavanaugh, editors

Component-Based Network System Engineering, Mark Norris, Rob Davis, and Alan Pengelly

Computer Telephony Integration, Second Edition, Rob Walters

Desktop Encyclopedia of the Internet, Nathan J. Muller

Digital Modulation Techniques, Fuqin Xiong

E-Commerce Systems Architecture and Applications, Wasim E. Rajput

Error-Control Block Codes for Communications Engineers, L. H. Charles Lee

FAX: Facsimile Technology and Systems, Third Edition, Kenneth R. McConnell, Dennis Bodson, and Stephen Urban

Fundamentals of Network Security, John E. Canavan

Guide to ATM Systems and Technology, Mohammad A. Rahman

A Guide to the TCP/IP Protocol Suite, Floyd Wilder

Information Superhighways Revisited: The Economics of Multimedia, Bruce Egan

Integrated Broadband Networks: TCP/IP, ATM, SDH/SONET, and WDM/Optics, Byeong Gi Lee and Woojune Kim

Internet E-mail: Protocols, Standards, and Implementation, Lawrence Hughes

Introduction to Telecommunications Network Engineering, Tarmo Anttalainen

Introduction to Telephones and Telephone Systems, Third Edition, A. Michael Noll

An Introduction to U.S. Telecommunications Law, Second Edition Charles H. Kennedy

IP Convergence: The Next Revolution in Telecommunications, Nathan J. Muller

The Law and Regulation of Telecommunications Carriers, Henk Brands and Evan T. Leo

Managing Internet-Driven Change in International Telecommunications, Rob Frieden

Marketing Telecommunications Services: New Approaches for a Changing Environment, Karen G. Strouse

Multimedia Communications Networks: Technologies and Services, Mallikarjun Tatipamula and Bhumip Khashnabish, editors

Performance Evaluation of Communication Networks, Gary N. Higginbottom

Performance of TCP/IP over ATM Networks, Mahbub Hassan and Mohammed Atiquzzaman

Practical Guide for Implementing Secure Intranets and Extranets, Kaustubh M. Phaltankar

Practical Internet Law for Business, Kurt M. Saunders

Practical Multiservice LANs: ATM and RF Broadband, Ernest O. Tunmann

Principles of Modern Communications Technology, A. Michael Noll

Protocol Management in Computer Networking, Philippe Byrnes

Pulse Code Modulation Systems Design, William N. Waggener

Security, Rights, and Liabilities in E-Commerce, Jeffrey H. Matsuura

Service Level Management for Enterprise Networks, Lundy Lewis

SIP: Understanding the Session Initiation Protocol, Alan B. Johnston

Smart Card Security and Applications, Second Edition, Mike Hendry

SNMP-Based ATM Network Management, Heng Pan

Strategic Management in Telecommunications, James K. Shaw

Strategies for Success in the New Telecommunications Marketplace, Karen G. Strouse

Successful Business Strategies Using Telecommunications Services, Martin F. Bartholomew

Telecommunications Cost Management, S. C. Strother

Telecommunications Department Management, Robert A. Gable

Telecommunications Deregulation and the Information Economy, Second Edition, James K. Shaw

Telemetry Systems Engineering, Frank Carden, Russell Jedlicka, and Robert Henry

Telephone Switching Systems, Richard A. Thompson

Understanding Modern Telecommunications and the Information Superhighway, John G. Nellist, and Elliott M. Gilbert

Understanding Networking Technology: Concepts, Terms, and Trends, Second Edition, Mark Norris

Videoconferencing and Videotelephony: Technology and Standards, Second Edition, Richard Schaphorst

Visual Telephony, Edward A. Daly and Kathleen J. Hansell

Wide-Area Data Network Performance Engineering, Robert G. Cole and Ravi Ramaswamy

Winning Telco Customers Using Marketing Databases, Rob Mattison

World-Class Telecommunications Service Development, Ellen P. Ward

For further information on these and other Artech House titles, including previously considered out-of-print books now available through our In-Print-Forever® (IPF®) program, contact:

Artech House
685 Canton Street
Norwood, MA 02062
Phone: 781-769-9750
Fax: 781-769-6334
e-mail: artech@artechhouse.com

Artech House
46 Gillingham Street
London SW1V 1AH UK
Phone: +44 (0)20 7596-8750
Fax: +44 (0)20 7630-0166
e-mail: artech-uk@artechhouse.com

Find us on the World Wide Web at:
www.artechhouse.com

WITHDRAWN FROM
OHIO NORTHERN
UNIVERSITY LIBRARY